# Computational Aerodynamics and Aeroacoustics

Tapan K. Sengupta · Yogesh G. Bhumkar

# Computational Aerodynamics and Aeroacoustics

 Springer

Tapan K. Sengupta
Department of Aerospace Engineering
Indian Institute of Technology Kanpur
Kanpur, Uttar Pradesh, India

Yogesh G. Bhumkar
School of Mechanical Sciences
Indian Institute of Technology Bhubaneswar
Bhubaneswar, Odisha, India

ISBN 978-981-15-4286-2     ISBN 978-981-15-4284-8    (eBook)
https://doi.org/10.1007/978-981-15-4284-8

This Springer imprint is published by the registered company Springer Nature Singapore Pte Ltd.
The registered company address is: 152 Beach Road, #21-01/04 Gateway East, Singapore 189721, Singapore

# Preface

There are many books and monographs written on aerodynamics and aeroacoustics separately. These have not been dealt with together due to some inherent issues with the approaches adopted in these subjects. It is easy to understand that in aerodynamics, one deals with pressure variations, which is of the order of one atmosphere (about $10^5$ Pascals). In contrast, perceptible acoustic signals involve variation of pressure that has amplitude around 20 μPa. Thus, to treat these, two physical events together would involve handling pressure variations which span across ten orders of magnitudes. Thus, it is uncommon to see the same being treated together computationally resolving such a large dynamic range of pressure. Disparity in the hydrodynamic and acoustic scales demands highly accurate numerical methods.

Additionally, there are other complicating factors in treating these two subjects and are related to the formulations used. Aerodynamicists relied on theoretical (potential flow analysis including panel methods) and experimental studies till 1980s in designing aerodynamic surfaces. With the availability of high-speed computers, one can now perform aerodynamic simulations to obtain important flow details as well as various aerodynamic coefficients for designing aircrafts. For solving engineering problems, researchers still rely on approximate turbulence modeling approach for solving Reynolds averaged Navier–Stokes equations. With improved methods and faster computers, large eddy simulation (LES) of Navier–Stokes equation appears feasible now for solving problems of aerodynamics.

Simulations of acoustic problems pose a different set of challenges, with equations of motion written as inhomogeneous wave equation, using Lighthill's analogy. The aerodynamic input is the source of inhomogeneity applicable for only simple geometries. The linearized Euler equation is used in computational aeroacoustics (CAA) as the improvement of Lighthill's analogy, posing it as a perturbation problem. Usually, full Euler equation has been solved or hybrid methods have been used to predict jet noise problem. Similarly, stochastic noise generation and radiation (SNGR) models are also used in a two-step approach.

The presence of low frequency requires computations to be performed for a large time interval so that low-frequency variations can be captured accurately. On the other hand, computing high-frequency waves demand resolving small scales by using refined grid. In consequence, one is forced to compute with a smaller time step to avoid numerical instabilities. Hydrodynamic fluctuations decay quickly as compared to acoustic disturbances. Usually, one chooses a computational domain with a size of 40–50 times characteristic dimension of the body for flow simulations as vortices decay quickly in the wake region. Acoustic disturbances travel much longer without significant attenuation and need computation to be performed on a larger domain to capture acoustic details accurately.

Despite such disparities of scales in computing, there are such unifying physical features also. While the aerodynamic phenomenon can be dispersive and dissipative physically, the acoustic phenomenon is non-dispersive and non-dissipative (to a large extent). It is for these reasons and for the reasons of accuracy that one must adopt numerical methods which are truly dispersion relation preserving (DRP)—an attribute required minimally! Also for the preservation of accuracy in scientific investigations, one must know all the sources of error. For this reason, there is a need to study error dynamics for different physical processes and their numerical solutions. This is one of the motivating factors, for presenting the book here with a major emphasis on high accuracy computing of aerodynamics and acoustics. As both the phenomena are dominated by wave propagation of the field, one must seek methods of solution which preclude reflections from finite domain over which the solution is sought. Additionally, we just mention in passing that sound propagates via longitudinal waves, and hence a neglected physical aspect of physical modeling is the bulk viscosity. This is highlighted only in the book, with the hope that this will be pursued further to explain how sound signal attenuates.

Appropriate boundary conditions must be prescribed at the domain inflow/outflow, as well as from the wall surface to obtain acoustic field correctly. Diffraction, reflection, and absorption of acoustic waves from the surface depend on the frequency/wavelength of disturbances. For a wide-band signal, prescription of such frequency-dependent conditions at the wall is challenging. On the other hand, avoiding spurious reflections of acoustic waves from the domain inflow and outflow boundaries is equally challenging. The present book talks about all these important aspects and provides a concise information about the correct way of computing aerodynamics and acoustics problems accurately. Data and solution methodology given for various test cases and examples can be used for calibrating new accurate numerical methods as well as for learning scientific computing altogether from the first principle.

The present book can be used for a course work integrating the parts: from Chaps. 1 to 5 for aerodynamics and from Chaps. 6 to 8 for CAA. As these parts are independent of each other, the materials could also be decoupled. However, for CAA one can also use the discussions given in the first chapter, Sects. 1.1–1.4 as introductory material. The main features of this book are in its emphasis on fundamental theoretical developments for the governing equation, computational methods and their analysis, and high accuracy computing aspects. Whether we have been successful or not will be judged by the users of this book.

We like to acknowledge all our present and past students, whose works are featured in the book. We are also very indebted to Trust for Aerodynamic Activities in India (TAAI) for supporting us in the preparation of the manuscript. The constant encouragement provided by Dr. Vidyadhar Mudkavi is gratefully acknowledged in completing the manuscript. Lastly and not the least, both the authors have benefited from the tireless efforts of Mrs. Baby Gaur to not only type the manuscript, but also helping prepare many figures and other secretarial assistance. Our family members have also supported us with patience, while we have been writing this manuscript.

Kanpur, India                                                                 Tapan K. Sengupta
Bhubaneswar, India                                                        Yogesh G. Bhumkar

# Contents

# About the Authors

**Tapan K. Sengupta** is a Professor in the department of Aerospace Engineering, Indian Institute of Technology (IIT) Kanpur, where he heads the High performance Computing Laboratory. He has received degrees from IIT Kharagpur, IISc Bangalore and Georgia Tech., Atlanta, USA and has worked in various research organizations and educational institutes including NAL Bangalore, India, University of Cambridge, U.K. and National University of Singapore, Singapore. His research interest spans across fields of scientific and high performance computing, fundamental fluids mechanics and aerodynamics, and transition and turbulence. His research teams have refined areas of scientific and high performance computing, receptivity/instability, transition and turbulence. He has authored over 200 scientific papers in internationally acclaimed journals and international conferences in addition to writing 6 books. He has also co-authored and edited two monographs.

**Yogesh G. Bhumkar** is an Assistant Professor in the School of Mechanical Sciences at IIT Bhubaneswar. After receiving his Ph.D. in Aerospace Engineering from IIT Kanpur, he worked as a post-doctoral researcher at the National Taiwan University, Taipei. He has 26 published papers in internationally acclaimed journals and conferences in areas of computational aeroacoustics and aerodynamics.

# List of Figures

# List of Tables

# Chapter 1
# Elements of Continuum Mechanics for Fluid Flow and General Stress–Strain System

## 1.1 Introduction

The fundamental physical laws in fluid flows are obtained from the conservation of mass, translational momentum, and energy, with the latter being nothing but the first law of thermodynamics applied to a control volume system. While for solids and liquids, the properties vary continuously in their macroscopic states, it is not so apparent for gases, which is characterized by mobility at the molecular level. One describes the flow of gases from the average behavior of the molecular ensemble in the control volume. This approach presupposes the existence of very many large number of molecules in the control volume, so that the statistical description is feasible. This is the essence of the continuum assumption that despite the presence of discrete molecules constituting the system, one can approximate the assembly by continuous variation of macroscopic properties. Continuous variation of properties arises due to momentum and energy exchanges by incessant collisions of molecules taking place. One of the most relevant non-dimensional numbers, determining whether a gaseous flow can be considered continuum or one must use statistical approaches, is the Knudsen number.

Knudsen number is the ratio of mean free in the gas divided by the characteristic physical length scale of the problem. If these two length scales are comparable or the mean free path is more than the characteristic length, then the number of molecular collisions will be fewer and one cannot expect a statistical homogeneity of the macroscopic properties. In contrast, if the Knudsen number is small, then the molecules will suffer collisions more often and via exchange of momentum and energy of constituent molecules the macroscopic properties will display continuous variation, making a good case for applying continuum approach. It can be shown [19] that the Knudsen number is also proportional to the ratio of Mach and Reynolds numbers, and this aspect highlights the importance of using molecular flow approach for hypersonic regime and also for microfluidics.

For air at normal temperature and pressure at sea level contains $2.5 \times 10^{25}$ molecules in a volume of 1 m$^3$, while the mean free path is $6.6 \times 10^{-8}$ m. With

© Springer Nature Singapore Pte Ltd. 2020
T. K. Sengupta and Y. G. Bhumkar, *Computational Aerodynamics and Aeroacoustics*,
https://doi.org/10.1007/978-981-15-4284-8_1

increase in altitude, the number density comes down, so that at an altitude of 130 km, the unit volume contains $1.67 \times 10^{15}$ molecules, while mean path increases to 10.2 m. However, all the present-day aircrafts fly in troposphere and lower reaches of stratosphere, where the mean free path is significantly lower, which makes continuum assumption very viable. The macroscopic description of air is adequate via properties like density $(\rho)$, pressure $(p)$ and temperature $(T)$, which do not develop discontinuity due to enormous number of collisions suffered by the constituent molecules, even when we consider the medium to be air. Random molecular motion and collisions tend to homogenize the molecular motion, thereby producing stable pressure and temperature, which are statistical manifestations of molecular motion. Thus, the continuum approach will satisfy our requirement, if we exclude motion at hypersonic speeds and motion of air at very high altitudes during re-entry. For the flow of gases, one needs to use equation of state as an additional constitutive relation among thermodynamic properties,

$$p = \rho R T$$

where $R$ is the universal gas constant. It is often stated for studying aerodynamic flows past a vehicle traveling at a speed of less than about $100 m/sec$ (for which Mach number less than equal to 0.3) one can assume the density to remain constant and we say that the flow is incompressible. This is despite the fact that the modulus of elasticity of air is about 20000 times smaller than that of water.

### 1.1.1 Compressibility of Flows

Compressibility states the ability of fluid to change its volume $(V)$ when an external force is applied and is given by the modulus of elasticity $(E)$, relating the applied pressure $(\Delta p)$ with relative change of volume $(\frac{\Delta V}{V_0})$ by

$$\Delta p = -E \frac{\Delta V}{V_0} \tag{1.1}$$

The compressibility is negligible for liquid, e.g., an increase in pressure of 1 atmosphere causes relative change of volume by about 0.005%. Hence, liquid is considered as incompressible for engineering applications. For flow of gases, modulus of elasticity is obtained following the process under consideration. If the process is isothermal, following Boyle's law,

$$(p_0 + \Delta p)(V_0 + \Delta V) = p_0 V_0$$

For isothermal flow of gases the modulus of elasticity is thus given by $E = p_0$. At sea level, under natural temperature and pressure (14.7 psi) condition, isothermal flow of air is about 20 000 times more compressible than water (for which $E = 280000$

psi). Similarly, for adiabatic flow the modulus of elasticity is $E = \gamma p_0$ and with $a$ as the speed of sound. One alternately gets $E = a^2 \rho$. Using continuity equation,

$$\frac{\Delta V}{V_0} = -\frac{\Delta \rho}{\rho_0}$$

Thus, Eq. (1.1) is written as

$$\Delta p = E \frac{\Delta \rho}{\rho_0} \tag{1.2}$$

The flow of gases is considered incompressible, when $\frac{\Delta \rho}{\rho_0} \ll 1$. This happens for a gas flowing with speed, $V$ and dynamic pressure, $q = \frac{1}{2}\rho V^2$, and then the maximum change in pressure is equal to the dynamic head and the above is written as

$$\frac{\Delta \rho}{\rho} \equiv \frac{q}{E} = \frac{1}{2}\left(\frac{V}{a}\right)^2 \tag{1.3}$$

One defines the Mach number by, $M = \frac{V}{a}$, the condition of incompressibility is equivalent to

$$\frac{1}{2}M^2 \ll 1$$

For $M \leq 0.3$, it is noted that $\frac{\Delta \rho}{\rho}$ is less than or equal to 0.05 or 5%. This is often used in aerodynamics to treat the flow as incompressible if $M \leq 0.3$.

## 1.2 Conservation of Momentum and General Stress System In Deformable Fluid

Consider an infinitesimal rectangular parallelepiped of volume $dx\,dy\,dz$, with the origin at the lower left-hand corner O with coordinate $(x, y, z)$. Two faces with area $(dy\,dz)$, perpendicular to $x$-axis, experience resultant stresses shown in Fig. 1.1 as $p_x$ and $p_x + \frac{\partial p_x}{\partial x}\,dx$.

Here, $p_x$, $p_y$, and $p_z$ are vectors which can be resolved into components along the coordinate directions acting on each face, i.e., into normal stress denoted by $\sigma$ with subscript indicating the direction, and into components parallel to each face, i.e., into shear stresses denoted by $\tau$ with two subscripts. In the shear stress, the first subscript indicates the axis which is perpendicular to the face and the second subscript indicates the direction along which the shearing stress is parallel. With this notation we have

$$\begin{aligned}
p_x &= \hat{i}\,\tau_{xx} + \hat{j}\,\tau_{xy} + \hat{k}\,\tau_{xz} \\
p_y &= \hat{i}\,\tau_{yx} + \hat{j}\,\tau_{yy} + \hat{k}\,\tau_{yz} \\
p_z &= \hat{i}\,\tau_{zx} + \hat{j}\,\tau_{zy} + \hat{k}\,\tau_{zz}
\end{aligned} \tag{1.4}$$

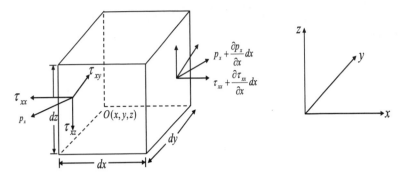

**Fig. 1.1** Control volume analysis for momentum of three-dimensional viscous flows

Similar terms can be obtained for the faces $(dx \, dz)$ and $(dx \, dy)$ which are perpendicular to $y$- and $z$-axes, respectively, in Fig. 1.1. Net components of the surface force are obtained as

$$\text{On surface with direction } \hat{i} : \quad \frac{\partial p_x}{\partial x} \, dx \, dy \, dz$$

$$\text{On surface with direction } \hat{j} : \quad \frac{\partial p_y}{\partial y} \, dx \, dy \, dz$$

$$\text{On surface with direction } \hat{k} : \quad \frac{\partial p_z}{\partial z} \, dx \, dy \, dz$$

The resultant surface force per unit volume can be obtained using the above along with Eq. (1.4).

These nine quantities in Eq. (1.4) constitute the stress tensor. One can show that the stress tensor is symmetric under special condition, by noting angular motions about the coordinate directions. For the angular motion about $y$-axis with a rate $\dot{\omega}_y$, this is given by

$$\dot{\omega}_y \, dI_y = (\tau_{xz} \, dy \, dz)dx - (\tau_{zx} \, dx \, dy) \, dz$$

with $dI_y$ indicating the moment of inertia of the fluid element about the $y$-axis. This can be rewritten as

$$\tau_{xz} - \tau_{zx} = \dot{\omega}_y \frac{dI_y}{dx \, dy \, dz}$$

If $dl$ is the typical length scale of the elementary volume, then the moment of inertia about the $y$-axis $(dI_y)$ is proportional to $(dl)^5$.

Thus,

$$\tau_{xz} - \tau_{zx} \simeq \dot{\omega}_y \, (dl)^2$$

Thus, with the elementary volume shrinking to a point, the difference in shear stress tensor tends to zero, provided $\dot{\omega}_y$ is not infinite. We emphasize that the elementary volume will never shrink to a point to make physical sense of stress tensor. Thus, the symmetry of stress tensor is a valid concept, only in the absence of unsteady variation of vorticity vector. For a general flow, therefore it is not right to expect the stress tensor to be symmetric. This is an under-emphasized aspect of continuum mechanics. Also note that unsteadiness can naturally arise due to distributed couple by applied torque.

Finally, the surface force per unit volume is

$$\vec{f}_s = \hat{i}\left(\frac{\partial \tau_{xx}}{\partial x} + \frac{\partial \tau_{xy}}{\partial y} + \frac{\partial \tau_{xz}}{\partial z}\right) + \hat{j}\left(\frac{\partial \tau_{xy}}{\partial x} + \frac{\partial \tau_{yy}}{\partial y} + \frac{\partial \tau_{yz}}{\partial z}\right)$$
$$+ \hat{k}\left(\frac{\partial \tau_{xz}}{\partial x} + \frac{\partial \tau_{yz}}{\partial y} + \frac{\partial \tau_{zz}}{\partial z}\right) \tag{1.5}$$

If the components of the body force $\vec{f}_b$ are $X, Y,$ and $Z$, then the equations of motion are

$$\rho\frac{Du}{Dt} = X + \frac{\partial \tau_{xx}}{\partial x} + \frac{\partial \tau_{xy}}{\partial y} + \frac{\partial \tau_{xz}}{\partial z}$$
$$\rho\frac{Dv}{Dt} = Y + \frac{\partial \tau_{xy}}{\partial y} + \frac{\partial \tau_{yy}}{\partial y} + \frac{\partial \tau_{yz}}{\partial z} \tag{1.6}$$
$$\rho\frac{Dw}{Dt} = Z + \frac{\partial \tau_{zx}}{\partial x} + \frac{\partial \tau_{yz}}{\partial y} + \frac{\partial \tau_{zz}}{\partial z}$$

These are the Cauchy equations. For a frictionless fluid, $\tau_{ij} = 0$ for $i \neq j$ and $\tau_{ii} = -p$, where $p$ is the arithmetic mean of normal stresses.

$$\text{i.e.} \quad p = \frac{1}{3}(\tau_{xx} + \tau_{xy} + \tau_{xz})$$

In general, Cauchy equations cannot be solved unless one has closure relations for the stress tensor. Next, even when we assume the stress tensor to be symmetric, then we need to relate the independent six stress components with the associated rate of strain field, as one of the ways this is achieved for a special class of fluids.

## 1.3  Strain Rate of Fluid Element in Flows

The relative motion of $B$ in Fig. 1.2, with respect to A caused by applied stresses, is described by the following matrix of nine partial derivatives of the local velocity field

$$\begin{bmatrix} u_x & u_y & u_z \\ v_x & v_y & v_z \\ w_x & w_y & w_z \end{bmatrix} \tag{1.7}$$

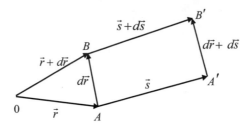

**Fig. 1.2** Relative motion of particles at $A$ and $B$ due to applied stresses [Reproduced with permission from Theoretical and Computational Aerodynamics. First Edition Tapan K. Sengupta Copyright (2015) John Wiley & Sons. Ltd. UK]

The relative velocity components $du, dv,$ and $dw$ between $A$ and $B$ can be alternately presented as

$$du = (\dot{\varepsilon}_x \, dx + \dot{\varepsilon}_{xy} \, dy + \dot{\varepsilon}_{xz} \, dz) + (\eta \, dz - \zeta \, dy)/2$$

$$dv = (\dot{\varepsilon}_{xy} \, dx + \dot{\varepsilon}_y \, dy + \dot{\varepsilon}_{yz} \, dz) + (\zeta \, dx - \xi \, dy)/2 \qquad (1.8)$$

$$dw = (\dot{\varepsilon}_{zx} \, dx + \dot{\varepsilon}_{zy} \, dy + \dot{\varepsilon}_z \, dz) + (\xi \, dy - \eta \, dx)/2$$

where $\dot{\varepsilon}_{ij} = \frac{1}{2}\left(\frac{\partial u_i}{\partial x_j} + \frac{\partial u_j}{\partial x_i}\right)$ for $i, j = 1$ to 3 represent the symmetric part of the strain tensor constructed from the partial derivatives of velocity components. These $\dot{\varepsilon}_{ij}$'s are symmetric with respect to the subscripts.

The antisymmetric part of the strain tensor is obtained in terms of the vorticity vector defined by

$$\vec{\omega} = \nabla \times \vec{V} = \xi \, \hat{i} + \eta \, \hat{j} + \zeta \, \hat{k}$$

### 1.3.1 Kinematic Interpretation of Strain Tensor

The terms $\dot{\varepsilon}_{ii}$ represent the rate of elongation or contraction in the $i$-direction suffered by an element due to the applied strain in the same direction. While this preserves shape, it contracts or dilates the fluid element. Applied general strains lead to alteration of shape of the fluid element.

The change in volume imparted to a fluid element by simultaneous action of all three diagonal elements of the strain tensor results in the element expanding (considering all partial derivatives of velocity components with corresponding coordinates to be positive, without loss of generality) in all the three directions, and the change in the length of its three sides produces a change in volume at a relative rate

$$\dot{e} = \frac{\{dx + \frac{\partial u}{\partial x}dx\ dt\}\ \{dy + \frac{\partial v}{\partial y}dy\ dt\}\ \{dz + \frac{\partial w}{\partial z}dz\ dt\} - dx\ dy\ dz}{dx\ dy\ dz\ dt}$$

$$= \frac{\partial u}{\partial x} + \frac{\partial v}{\partial y} + \frac{\partial w}{\partial z} = \mathrm{div}\ \vec{V} \qquad (1.9)$$

in the limit of $dt \to 0$. During this change, the shape of the element described by the angles at its vortices remains unchanged. Thus, $\dot{e}$ describes the local instantaneous volumetric dilatation of a fluid element.

The relative velocity field is different when one of the off-diagonal terms in the matrix of Eq. (1.7) is nonzero. Consider, for example, a case for which $u_y > 0$. The corresponding field is one of the pure shear strains due to which a rectangular element of fluid distorts into a parallelogram, as shown in Fig. 1.3(a). The original right angle at $A$ in this figure changes at a rate measured by the angle

$$d\gamma_{xy} = \frac{\frac{\partial u}{\partial y} dy\ dt}{dy}$$

Thus, the included angle at $A$ changes at a rate of $\frac{\partial u}{\partial y}$, i.e., the angular deformation rate due to pure strain is given by, $\dot{\gamma}_{xy} = \frac{\partial u}{\partial y}$

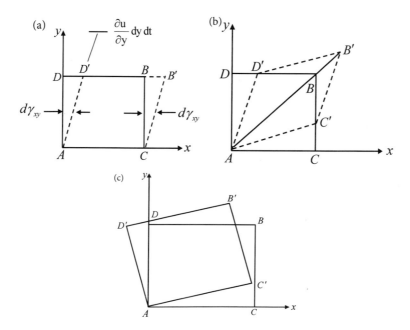

**Fig. 1.3   a** Case of pure strain applied to the fluid. **b** Case of irrotational flow. Note that the diagonal of the element does not rotate due to applied strain. **c** Case of rigid body rotation for which $u_y = v_x$

Consider next a case, when both $\frac{\partial u}{\partial y}$ and $\frac{\partial v}{\partial x} > 0$. The right angle at $A$ will then distort owing to the superposition of these two motions, as shown in Fig. 1.3(b).

Figure 1.3(b) is drawn such that

$$\frac{\partial u}{\partial y} = \frac{\partial v}{\partial x}$$

This is a case when

$$\dot{\varepsilon}_{xy} = \dot{\varepsilon}_{yx} = \frac{1}{2}\left\{\left(\frac{\partial u}{\partial y}\right) + \left(\frac{\partial v}{\partial x}\right)\right\} > 0,$$

with all other terms being zero and this gives rise to distortion in shape.

For the imposed strain field, one notes the $z$-component of the vorticity field to be zero, i.e., $\zeta = \frac{1}{2}(-u_y + v_x) = 0$.

For the case of Fig. 1.3(b), the right angle at $A$ now distorts at the rate

$$\dot{\varepsilon}_{xy} = \dot{\varepsilon}_{yx} = \frac{1}{2}\left(u_y + v_x\right)$$

such that this distortion is volume preserving and affects only the shape. With $B'$ now on the extension of $AB$, i.e., the diagonal does not rotate. This is the case of irrotational flows, with vorticity identically zero.

Next, consider when $u_y = -v_x$, so that $\dot{\varepsilon}_{xy} = \dot{\varepsilon}_{yx} = 0$, i.e., the right angles at $A$ remain a right angle. This is shown in Fig. 1.3(c), with fluid element rotating about $A$. This takes place without distortion and described as rigid body rotation, by the rotation rate of the diagonal $AB$ and is $v_x$ or $-u_y$. Such rigid body rotation is also represented by the vorticity $\frac{1}{2}(\nabla \times \vec{V})$ of the velocity field.

When all the elements of the matrix given by Eq. (1.7) are nonzero, the overall relative motion is a combination of all cases discussed above, mathematically represented by Eq. (1.8) by the relative motion of fluid particles.

## 1.4  Relation Between Stress and Rate of Strain Tensors in Fluid Flow

To simplify the Cauchy equation, we need to relate stress with rate of strain tensors, as was done by Navier and Stokes independently. When the fluid is at rest, it develops a uniform field of hydrostatic stress $(-p)$, identical to the thermodynamic pressure. Here, the connotation of thermodynamics is related to equilibrium thermodynamics and hence no losses. The losses suffered due to non-equilibrium nature of the flow in motion by normal stress will be associated with the diagonal elements of stress tensor matrix, indicated here by $\tau_{ii}$ in the following. Thus, when the fluid is in motion, the equation of state still determines a pressure at every point and one defines conveniently the deviatoric normal stresses by

$$\tau'_{xx} = \tau_{xx} + p, \quad \tau'_{yy} = \tau_{yy} + p \text{ and } \tau'_{zz} = \tau_{zz} + p \tag{1.10}$$

From Eq. (1.8), the velocity gradient tensor can be written as

$$\frac{\partial u_k}{\partial x_l} = \dot{\varepsilon}_{kl} - \frac{1}{2}\varepsilon_{klm}\omega_m \tag{1.11}$$

where the symmetric part is defined by $\dot{\varepsilon}_{ij} = \frac{1}{2}\left(\frac{\partial u_i}{\partial x_j} + \frac{\partial u_j}{\partial x_i}\right)$ and the second term on the right-hand side represent the antisymmetrical part in tensor notation. We have already defined the vorticity vector by the components: $\omega_1 = \xi$; $\omega_2 = \eta$, and $\omega_3 = \zeta$ following Eq. (1.8). We have also noted that the vorticity vector is related to the angular rotation rate or rigid body rotation as the former is twice the value of the latter.

However, one notes that pure translation and rigid body rotation of an element of fluid produces no surface forces on it. It is for this reason that the deviatoric stress components will be related to velocity gradients containing the symmetric part only in Eq. (1.11). The relations between the stress and strain rate are considered as linear for Newtonian fluids, which remain unchanged by rotation of the coordinate system to ensure isotropy. Isotropy also requires that the principle axes of strain rate and stress are identical; otherwise, a preferred direction is introduced. This is formally written as, following the developments in theory of elasticity [16] for the deviatoric stress,

$$\tau_{ij} = C_{ijkl}\dot{\varepsilon}_{kl} \tag{1.12}$$

Here, a tensor of the second rank on the left-hand side is written as equal to the contraction of the tetradic ($C$) with the symmetric part of rate of strain tensor, which is also of rank two. This tetradic or tensor of rank four, ($C_{ijkl}$) has $3^4 = 81$ components. These components depend upon the molecular structure of the fluids and become easier to handle for Newtonian flows. As the medium is expected to be statistically isotropic, so will be $C_{ijkl}$. From energy considerations, Sokolnikoff [16] has established a reflective symmetry along any arbitrary axis, so that

$$C_{ijkl} = C_{klij} = C_{jikl} = C_{ijlk}$$

in which case, one has 21 independent components. If the deviatoric stress has zero average contribution, then one further reduces the unknown components to 9, out which only 3 are independent. It is noted in Batchelor [2] that the Kronecker delta as the basic isotropic tensor can be used to represent Cartesian tensor of even order from the following identity as

$$C_{ijkl} = \mu\delta_{ik}\delta_{jl} + \mu'\delta_{il}\delta_{jk} + \lambda\delta_{ij}\delta_{kl} \tag{1.13}$$

Due to symmetry of $C_{ijlk}$, one requires that $\mu = \mu'$, which furthermore reduces the independent component to two unknowns, $\mu$ and $\lambda$, relating stress and rate of strain. Thus, the deviatoric stress tensor can be written as

$$\tau'_{ij} = 2\mu\dot{\varepsilon}_{kl}\delta_{jl} + \lambda\partial_k v_k\delta_{ij} \tag{1.14}$$

Thus, isotropy can be secured only, if each one of the three normal stresses $\bar{\tau}'_{xx}$, $\bar{\tau}'_{yy}$, and $\bar{\tau}'_{zz}$ is made to depend on the same component of rate of strain and on the trace of the strain tensor, each with a different factor of proportionality. Thus, we write one of the principal stresses as

$$\bar{\tau}'_{xx} = \lambda\left(\frac{\partial\bar{u}}{\partial\bar{x}} + \frac{\partial\bar{v}}{\partial\bar{y}} + \frac{\partial\bar{w}}{\partial\bar{z}}\right) + 2\mu\frac{\partial\bar{u}}{\partial\bar{x}} \tag{1.15}$$

Here $\bar{x}$, $\bar{y}$, and $\bar{z}$ are the principal axes directions and $\bar{u}$, $\bar{v}$, and $\bar{w}$ are the velocity components along these directions. This is a relation with two properties of fluids, $\lambda$ and $\mu$, instead of the original 81 components.

Relations given in Eq. (1.15) can be extended to any arbitrary coordinate systems obtained by a general rotation with appropriate linear transformation to show the stress system as

$$\tau'_{xx} = \lambda \operatorname{div} \overrightarrow{V} + 2\mu\frac{\partial u}{\partial x}$$

$$\tau'_{yy} = \lambda \operatorname{div} \overrightarrow{V} + 2\mu\frac{\partial v}{\partial y}$$

$$\tau'_{zz} = \lambda \operatorname{div} \overrightarrow{V} + 2\mu\frac{\partial w}{\partial z} \tag{1.16}$$

$$\tau_{xy} = \tau_{yx} = \mu\left(\frac{\partial v}{\partial x} + \frac{\partial u}{\partial y}\right)$$

$$\tau_{yz} = \tau_{zy} = \mu\left(\frac{\partial w}{\partial y} + \frac{\partial v}{\partial z}\right)$$

$$\tau_{zx} = \tau_{xz} = \mu\left(\frac{\partial u}{\partial z} + \frac{\partial w}{\partial x}\right)$$

The above relations can be written in general by $\tau_{ij} = -p\,\delta_{ij} + \lambda\,\partial_k v_k\,\delta_{ij} + 2\mu\,\partial_i\,(v_j)$, where $-p$ is the thermodynamic pressure as defined in equilibrium thermodynamics. In contrast, the mechanical pressure is given by $p_m = \tau_{ii}/3$ and should account for losses, a part of which will be due to $\tau_{ii}$.

Therefore, the difference between thermodynamic and mechanical pressure is given by

$$(-p) - p_m = -\left(\lambda + \frac{2}{3}\mu\right)\partial_k v_k = \left(\lambda + \frac{2}{3}\mu\right)\frac{1}{\rho}\frac{D\rho}{Dt}$$

**Stokes' hypothesis:** Considering only reversible work for the fluid dynamic system in this hypothesis, thermodynamic and mechanical pressure is postulated to be identical. Thus, according to this hypothesis,

$$3\lambda + 2\mu = 0 \tag{1.17}$$

This provides the constitutive relation for an isotropic, Newtonian fluid as

$$\tau_{xx} = -p - \frac{2}{3}\mu \operatorname{div} \overrightarrow{V} + 2\mu\frac{\partial u}{\partial x}$$

$$\tau_{yy} = -p - \frac{2}{3}\mu \operatorname{div} \overrightarrow{V} + 2\mu\frac{\partial v}{\partial y}$$

$$\tau_{zz} = -p - \frac{2}{3}\mu \operatorname{div} \overrightarrow{V} + 2\mu\frac{\partial w}{\partial z} \tag{1.18}$$

$$\tau_{xy} = \tau_{yx} = \mu\left(\frac{\partial v}{\partial x} + \frac{\partial u}{\partial y}\right)$$

$$\tau_{yz} = \tau_{zy} = \mu\left(\frac{\partial w}{\partial y} + \frac{\partial v}{\partial z}\right)$$

$$\tau_{xz} = \tau_{zx} = \mu\left(\frac{\partial u}{\partial z} + \frac{\partial w}{\partial x}\right)$$

Using the above in Cauchy's equation, one gets

$$\rho\frac{Du}{Dt} = X - \frac{\partial p}{\partial x} + \frac{\partial}{\partial x}\left[\mu\left(2\frac{\partial u}{\partial x} - \frac{2}{3}\operatorname{div}\overrightarrow{V}\right)\right] + \frac{\partial}{\partial y}\left[\mu\left(\frac{\partial u}{\partial y} + \frac{\partial v}{\partial x}\right)\right]$$
$$+ \frac{\partial}{\partial z}\left[\mu\left(\frac{\partial w}{\partial x} + \frac{\partial u}{\partial z}\right)\right]$$

$$\rho\frac{Dv}{Dt} = Y - \frac{\partial p}{\partial y} + \frac{\partial}{\partial y}\left[\mu\left(2\frac{\partial v}{\partial y} - \frac{2}{3}\operatorname{div}\overrightarrow{V}\right)\right] + \frac{\partial}{\partial z}\left[\mu\left(\frac{\partial v}{\partial z} + \frac{\partial w}{\partial y}\right)\right]$$

$$+ \frac{\partial}{\partial x}\left[\mu\left(\frac{\partial u}{\partial y} + \frac{\partial v}{\partial x}\right)\right] \tag{1.19}$$

$$\rho \frac{Dw}{Dt} = Z - \frac{\partial p}{\partial z} + \frac{\partial}{\partial z}\left[\mu\left(2\frac{\partial w}{\partial z} - \frac{2}{3}\ \text{div}\ \vec{V}\right)\right] + \frac{\partial}{\partial x}\left[\mu\left(\frac{\partial w}{\partial x} + \frac{\partial u}{\partial z}\right)\right]$$
$$+ \frac{\partial}{\partial y}\left[\mu\left(\frac{\partial v}{\partial z} + \frac{\partial w}{\partial y}\right)\right]$$

These are the well-known Navier–Stokes equation for compressible flows written down using Stokes' hypothesis.

### 1.4.1   Bulk Viscosity and Non-Equilibrium Thermodynamics

In equating thermodynamic and mechanical pressure in Stokes' hypothesis, an implicit assumption was made that even for viscous flow, compression and dilatation do not cause loss. However, any transport process like fluid flow or heat transfer is due to finite gradient, which by its very definition would be in the realm of non-equilibrium thermodynamics. Viewed in this perspective, we will term $p$ instead as the static pressure and its difference from the mechanical pressure is given as

$$(-p) - p_m = -\left(\lambda + \frac{2}{3}\mu\right)\partial_k v_k$$

In the literature, the combination of $\mu$ and $\lambda$ in the above is termed as the bulk viscosity,

$$\bar{\kappa} = \lambda + \frac{2}{3}\mu$$

And the divergence of velocity is $D = \partial_k v_k$. It is interesting to note that for incompressible flows, the equation of continuity simplifies to $D \equiv 0$. For such an approximation, the difference between static pressure and mechanical pressure vanishes. Thus, Stokes' hypothesis relates second coefficient of viscosity ($\lambda$) with the dynamic viscosity ($\mu$) by $\bar{\kappa}$, the bulk viscosity of the fluid, and set it to zero for all compressible and incompressible flow. It is to be highlighted that the acoustic signal propagation is via longitudinal waves, which physically involves compression and dilatation of the fluid parcel in the direction of propagation about its mean position. In acoustics, however small the mean convection speed be, propagation of sound is associated with $D$ being not equal to zero and there Stokes' hypothesis inhibits correct modeling of the phenomenon. To integrate the subjects of fluid flow and acoustics on a common platform, we need to remove Stokes' hypothesis from Navier–Stokes equation. It has been stated [15] in reporting the non-equilibrium aspect of Rayleigh–Taylor instability that the shortcoming of Stokes' hypothesis is observed experimentally for sound absorption. Another good compilation of estimates of bulk viscosity is given in [5], where bulk viscosity of ideal gases has been reported. In this kinetic theory of gases approach, the author considers non-equilibrium by invoking rotational and vibrational degrees of freedom via the relaxation times associated with these unit processes in achieving equilibrium via collisions among the polyatomic molecules. It is shown that unlike $\mu$, $\bar{\kappa}$ display wide variations of numerical value (mainly with temperature), even in ideal gas limit. *At sufficiently low temperatures, the vibrational mode is deactivated and bulk viscosity is due to the relaxation of the rotational mode typically yielding a bulk viscosity which is $O(\mu)$ and is gradually increasing with temperature. At sufficiently high temperatures, the vibrational mode is activated* [5]. Compilation by Liebermann [7], Karim and Rosenhead [6], and Rosenhead [11] noted that $\lambda$ is independent of $\mu$ and are orders of magnitude higher in amplitude, while the sign is reversed.

The Navier–Stokes equation without Stokes' hypothesis can be written down with the help of bulk viscosity as

$$\rho\frac{Du}{Dt} = X - \frac{\partial p}{\partial x} + \frac{\partial}{\partial x}\left[\mu\left(2\frac{\partial u}{\partial x} - \frac{2\bar{\kappa}}{3\mu}\,\partial_k v_k\right)\right] + \frac{\partial}{\partial y}\left[\mu\left(\frac{\partial u}{\partial y} + \frac{\partial v}{\partial x}\right)\right]$$
$$+ \frac{\partial}{\partial z}\left[\mu\left(\frac{\partial w}{\partial x} + \frac{\partial u}{\partial z}\right)\right]$$

$$\rho\frac{Dv}{Dt} = Y - \frac{\partial p}{\partial y} + \frac{\partial}{\partial y}\left[\mu\left(2\frac{\partial v}{\partial y} - \frac{2\bar{\kappa}}{3\mu}\,\partial_k v_k\right)\right] + \frac{\partial}{\partial z}\left[\mu\left(\frac{\partial v}{\partial z} + \frac{\partial w}{\partial y}\right)\right]$$
$$+ \frac{\partial}{\partial x}\left[\mu\left(\frac{\partial u}{\partial y} + \frac{\partial v}{\partial x}\right)\right] \tag{1.20}$$

$$\rho\frac{Dw}{Dt} = Z - \frac{\partial p}{\partial z} + \frac{\partial}{\partial z}\left[\mu\left(2\frac{\partial w}{\partial z} - \frac{2\bar{\kappa}}{3\mu}\,\partial_k v_k\right)\right] + \frac{\partial}{\partial x}\left[\mu\left(\frac{\partial w}{\partial x} + \frac{\partial u}{\partial z}\right)\right]$$
$$+ \frac{\partial}{\partial y}\left[\mu\left(\frac{\partial v}{\partial z} + \frac{\partial w}{\partial y}\right)\right]$$

With the second coefficient of viscosity, it is given as $\lambda = (-\frac{2}{3} + \frac{\bar{\kappa}}{\mu})\mu$, where $\bar{\kappa} = -7.383 \times 10^{-4} + 3.381 \times 10^{-4}\,T$ (with $T$ in K), which is obtained by a regression analysis of experimental data in [1], as shown in Fig. 1.4. The experimental $\lambda$ variation with temperature is shown for dry and saturated air in Fig. 1.4, as discrete points and the curved line are obtained by using regression analysis of these data. We specifically focus on the role of Stokes' hypothesis on non-equilibrium thermodynamics related to RTI and its computation in [15]. The reported results are indicative of the ability to perform calculations very accurately using proposed formulation and numerical methods to capture thermodynamic process, which is far from equilibrium.

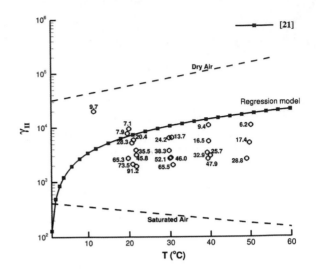

**Fig. 1.4** Variation of non-dimensional λ with $T$ and relative humidity with experimental data from [1], along with the variation obtained by regression analysis of these data

## 1.4.2  Non-Equilibrium Effects in Rayleigh–Taylor Instability: Role of Bulk Viscosity

For Rayleigh–Taylor instability, one considers a heavy fluid resting over a lighter fluid, such that the gravitational and external imposed acceleration directed toward the lighter fluid. This gives rise to baroclinic vorticity at the interface of the fluids, which is given by the baroclinic term: $-\frac{1}{\rho^2}\nabla\rho \times \nabla p$. This imposed acceleration distorts the interface, giving rise to instability showing growth of mixing, with the onset displaying distinct patterns. The Rayleigh–Taylor instability (RTI) has been variously studied [4, 9, 15, 17], with density difference between the fluids in non-dimensional parameter form of the Atwood number, $A = \left(\frac{\rho_1 - \rho_2}{\rho_2 + \rho_1}\right)$, playing a major role. Here, the subscripts 1 and 2 are used for the upper and lower fluid, as shown in the schematic of the domain in Fig. 1.5.

The present study in an isolated box undergoes RTI for the non-periodic boundary conditions used. The inhomogeneity of the system for the two air masses is due to the temperature difference ($\Delta T = T_2 - T_1$) of $70K$, which is purposely used to invalidate the Boussinesq assumption that is often used. Initially, the fluids are separated by an interface, which does not allow any transport of mass, momentum, and energy at $t = 0$. If we consider the air as calorically perfect, then also the Boussinesq approximation will be erroneous. The cold fluid (at $T_1$) resting on top fluid (at $T_2$) is an unstable configuration. With $\Delta T$ considered makes the Boussinesq approximation unacceptable [4]. Experimental investigation in Read [10] studied the onset of RTI, without creating any disturbance at the interface. This aspect is relevant to such problems studied numerically. Many computing efforts impose an external perturbation at the interface.

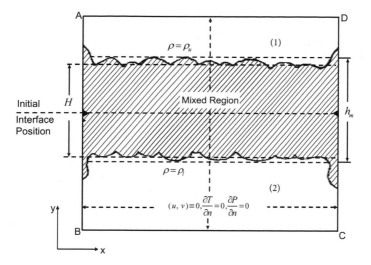

**Fig. 1.5**  Representation of Rayleigh–Taylor instability in a box is shown schematically. The mixing region in the middle, with boundary conditions, and the initial condition are shown

For the numerical solution of Navier–Stokes equation, Stokes' hypothesis has been questioned by fluid dynamicists and researchers in heat-transfer areas [5–8, 11]. Results have been presented in [15] without using Stokes' hypothesis, by using the second coefficient of viscosity $\lambda$ modeled using a regression analysis, presented for the dispersion and absorption of acoustic data by measuring acoustic attenuation in Ash *et al.* [1]. This naturally requires solving compressible NS equations to trace the growth of mixing layer at the interface. The total entropy of this isolated system is computed in a rectangular domain to present results, with and without Stokes' hypothesis. The pressure waves associated with the compressible flow formulation and non-periodic boundary condition at the walls are signatures of total entropy change in the isolated system.

Presented investigation for non-equilibrium thermodynamic process in [15] used air as the fluid media, with the non-equilibrium feature due to large thermal gradient, while the flow speeds are negligibly small. Such a change in entropy of the system at any point $(x_m, y_n)$ in the system is calculated using an equivalent instantaneous equilibrium state. Thus, at the advanced time the change in entropy is given by

$$\Delta \bar{s}_{mn}(t + \Delta t) = c_p \ln \frac{T(x_m, y_n, t + \Delta t)}{T(x_m, y_n, t)} - R \ln \frac{p(x_m, y_n, t + \Delta t)}{p(x_m, y_n, t)} \quad (1.21)$$

The total entropy variation with time of the isolated system is shown in Fig. 1.6. This figure compares the entropy change using with and without the Stokes' hypothesis, as computed with same grid and time step ($\Delta t = 3 \times 10^{-5}$). The computational results show consistently the entropy to be higher without Stokes' hypothesis, due to higher normal stress. One notices the time variation of entropy to be same for

**Fig. 1.6** Time variations of entropy during RTI inside an isolated box are shown for computations with and without Stokes' hypothesis using the model of Ash *et al.* with the time step used for the computation is $\Delta t = 3 \times 10^{-5}$ [Reproduced with permission from AIP from [15], *Physics of Fluids*, **28**, 094102 (2016)]

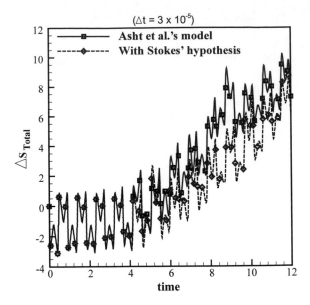

with and without Stokes' hypothesis, up to $t \approx 4$, a time till the convection of flow is negligible. The actual reason for higher entropy for without Stokes' hypothesis case is explained in Buresti [3] as .... *if we renounce to any form of Stokes' hypothesis, it is easily seen that a further term equal to* $\bar{\kappa}(div V)^2$ *would appear in the internal energy balance.* The attributed internal energy is the unavailable energy, whose increase causes the entropy to increase. For the discretely stratified fluids far from the state of equilibrium, when are allowed to mix one notices contribution from bulk viscosity term as non-negligible. Interestingly in Fig. 1.6 fluctuating entropy (with M-shape like structures) is noted from the onset of RTI, while in the mean, the entropy of the isolated system increases monotonically.

The authors in [15] have noted that the *early time interval is characterized by pressure terms dominating over the temperature-dependent terms. This is the time interval when convection is virtually negligible and hence the contribution by* $\bar{\kappa} \nabla \cdot \vec{V}$ *on normal viscous stresses are negligible. However, when convection ensues due to creation of discrete vortices via the formation of spikes and bubbles,* $\nabla \cdot \vec{V}$ *is not negligible anymore. Then one starts noticing the difference between the results obtained with and without Stokes' hypothesis that includes the effects on entropy variation with time. Due to higher diffusive action for the case without Stokes' hypothesis, one notices smoother time variation of entropy, while the entropy is always higher due to higher internal energy by the contribution of* $\bar{\kappa}(\nabla \cdot \vec{V})^2$.

## 1.5  Potential Flow

So far, we have tried to discuss flow fields in terms of conservation principles and tried to distinguish between equilibrium and non-equilibrium thermodynamic aspects of the flow field. In the process, we have tried to incorporate all sources of losses, including those which can arise due to bulk viscosity. For an aircraft in cruise configuration, the angle of attack is usually low, such that the flow over lifting surfaces remain attached, with viscous effects confined within the boundary layer. Due to the thinness of the boundary layer at such high Reynolds numbers, one would even neglect the presence of the thin shear layers in hydrodynamic studies. This is the basis of studying inviscid flow, and furthermore it is also assumed that the flow is irrotational.

With the continuity equation, $\nabla \cdot \vec{V} = 0$, one can define the vector potential ($\vec{\Psi}$), such that $\vec{V} = \nabla \times \vec{\Psi}$. Using the kinematic definition of vorticity ($\vec{\omega} = \nabla \times \vec{V}$), and enforcing divergence-free condition or solenoidality of the vector potential, then one gets the following Poisson equation:

$$\vec{\omega} = -\nabla^2 \vec{\Psi} \tag{1.22}$$

These are three equations for the three components of vorticity vector, written in terms of the three components of vector potential for 3D flow field. The complementary vorticity transport equations are obtained by taking curl of the Navier–Stokes equation. Such formulations for incompressible flows are often used and called derived variable formulation [12]. While these are related to viscous flows, the other extreme simplification is noted for 2D flows which are considered inviscid and irrotational.

For 2D flows, the vector potential has a single component perpendicular to the plane of the flow, and this is the stream function $\psi$, whose governing equation is obtained for irrotational flow from Eq. (1.22) as

$$\nabla^2 \psi = 0 \tag{1.23}$$

Also, from the requirement of zero vorticity for the inviscid, irrotational flow, one postulates the existence of velocity potential $\phi$, whose governing equation also is the Laplace equation:

$$\nabla^2 \phi = 0 \tag{1.24}$$

In the following part of this chapter, we only discuss 2D inviscid, irrotational flow using $\psi$ and $\phi$, and introduce the complex potential as an analytic function given by

$$W = \phi(x, y) + i\psi(x, y)$$

where $i$ is the square root of (-1), defined as the iota.

## 1.6   Preliminaries of Complex Analysis for 2D Irrotational Flows: Cauchy–Riemann Relations

Any complex variable can be represented either as $z = x + iy$ or $z = re^{i\theta}$. The complex conjugate of the same variable can be written either as $\bar{z} = x - iy$ or $re^{-i\theta}$, so that $r = (z\bar{z})^{1/2}$. The angle, $\theta$, is the argument, and one takes the principal value to lie within $\pm\pi$.

In complex analysis, one comes across analytic functions of $z$. Let $f(z)$ be one such function defined inside a closed contour C, and this must have the following properties: (i) $f(z)$ is finite and unique within C and (ii) for any $z$ within C, it has a single-valued finite derivative defined by

$$\frac{df'}{dz}(z) = \lim_{z' \to z} \frac{f(z') - f(z)}{z' - z}$$

The real and imaginary parts of an analytic function are called *conjugate functions*.

As $\phi$ and $\psi$ are continuous functions of their argument, so is $W$ of its argument $z$. Thus, $\phi$ and $\psi$ are the conjugate functions for $W$. One obtains the following partial derivatives of $W$:

$$\frac{\partial\phi}{\partial x} + i\frac{\partial\psi}{\partial x} = \frac{\partial W}{\partial x} = \frac{dW}{dz}\frac{\partial z}{\partial x} = W'(z) \tag{1.25}$$

and

$$\frac{\partial\phi}{\partial y} + i\frac{\partial\psi}{\partial y} = \frac{\partial W}{\partial y} = \frac{dW}{dz}\frac{\partial z}{\partial y} = iW'(z) \tag{1.26}$$

From the above two equations, one obtains the following, known as the Cauchy–Riemann relations::

$$\frac{\partial\phi}{\partial x} = \frac{\partial\psi}{\partial y} \quad and \quad \frac{\partial\phi}{\partial y} = -\frac{\partial\psi}{\partial x} \tag{1.27}$$

A corollary of Cauchy–Riemann relations is that the $\phi = constant$ and $\psi = constant$ lines meet at right angles to each other.

**Cauchy's Integral Theorem:** states that for an analytic function defined inside a closed curve C, $\oint f(z)\, dz = 0$.

In contour integrals considered, the counter-clockwise direction is considered as positive. The proof of this theorem also follows from Cauchy–Riemann relations for if $f(z) = \phi + i\psi$, then one notes that

$$\oint f(z)\, dz = \oint (\phi + i\psi)(dx + idy) = \oint (\phi\, dx - \psi\, dy) + i\oint (\psi\, dx + \phi\, dy)$$

Using Stokes' theorem: $\oint (P\, dy - Q\, dx) = \iint \left( \frac{\partial P}{\partial x} + \frac{\partial Q}{\partial y} \right) dx\, dy$, one obtains by Cauchy–Riemann relation

$$\oint (\phi \, dx - \psi \, dy) = -\iint \left( \frac{\partial \psi}{\partial x} + \frac{\partial \phi}{\partial y} \right) dx \, dy = 0$$

$$\oint (\psi \, dx + \phi \, dy) = \iint \left( \frac{\partial \phi}{\partial x} - \frac{\partial \psi}{\partial y} \right) dx \, dy = 0$$

**Singularities**:

Singularities are the points in the z-plane, where the complex function, $f(z)$, ceases to be analytic. In the neighborhood of such a point, indicated here by $z^*$, the function can be expressed in terms of positive and negative powers of $(z - z^*)$ by Laurent series expansion

$$f(z) = .... + a_3(z - z^*)^3 + a_2(z - z^*)^2 + a_1(z - z^*) + a_0$$
$$+ b_1(z - z^*)^{-1} + b_2(z - z^*)^{-2} + .... \qquad (1.28)$$

If the number of terms with negative exponent in the above Laurent series expansion is finite in number, then $z = z^*$ is called the pole. Furthermore, if all $b_j$'s are zero except $b_1$, then the point $z = z^*$ is called the simple pole. This coefficient $b_1$ is called the residue of the function at $z^*$. The residue is very important, as it can only produce nonzero contribution for a contour integral performed for the function given in Eq. (1.28) over the closed curve that includes the pole. To show this, consider the contour integral of the powers of $(z - z^*)$ on a circle of radius $R$ centered at $z^*$ by

$$I = \oint (z - z^*)^n dz = \int_0^{2\pi} R^{n+1} e^{i(n+1)\theta} \, i \, d\theta$$

as one can express $(z - z^*) = Re^{i\theta}$ along the contour of the circle. The above integral can be further simplified to

$$I = \frac{R^{n+1}}{(n+1)} \left[ e^{(n+1) \, i\theta} \right]_0^{2\pi} = 0 \text{ if } n \neq -1$$

However, for $n = -1$

$$I = \oint \frac{dz}{(z - z^*)} = \int_0^{2\pi} i \, d\theta = 2\pi i$$

Thus the contour integral of the function given in Eq. (1.28) is simply

$$\oint f(z) dz = 2\pi i b_1 \qquad (1.29)$$

### 1.6.1  Cauchy's Residue Theorem

Let C be a closed contour inside upon which $f(z)$ is analytic, except at finite number of singular points within C, for which if the residues are given as $r_1, r_2, r_3, \ldots r_n$, then

$$\oint f(z)dz = 2\pi i(r_1 + r_2 + r_3 + \ldots + r_n) \tag{1.30}$$

### 1.6.2  Complex Potential and Complex Velocity

As complex potential is an analytic function everywhere except at the singularities, $\phi$ and $\psi$ satisfy the Cauchy–Riemann relation. Conversely, any analytical function can be viewed to represent a fluid flow, whose real part is a valid velocity potential and imaginary part is a legitimate stream function. Now from the Cauchy–Riemann relation given in Eq. (1.27), we get

$$\frac{dW}{dz} = \frac{\partial \phi}{\partial x} + i\frac{\partial \psi}{\partial x} = u - iv \tag{1.31}$$

Thus, the derivative $\frac{dW}{dz}$ is called the complex velocity. If we represent $\overline{W}$ as the complex conjugate of $W$, then it is easy to see

$$\frac{dW}{dz}\frac{d\overline{W}}{dz} = (u - iv)(u + iv) = u^2 + v^2 \tag{1.32}$$

with the right-hand side representing the speed squared.

For uniform flow with velocity $U_\infty$ in the $x$-direction, the complex potential can be obtained as

$$W_\infty = U_\infty z$$

## 1.7   Elementary Singularities in Fluid Flows

In fluid mechanics, primary singularities are the source, sink, vortices, and their derivatives. For 2D flows, one considers line vortex. There are two ways of analyzing potential flow problems: either by posing these in the physical plane and obtaining the solutions heuristically or from analytic expression of $W$, obtaining the streamlines to infer possible flow fields. Obtained $W(z)$ due to uniform wind implies a source and sink pair positioned at infinity, with respect to field of view. Let us obtain $\phi$ and $\psi$ for source/sink and line vortex heuristically, assuming isotropy and symmetry.

**Flow field due to source/sink:**
A source at the origin of a Cartesian system will emit fluid in all directions isotropically for which the flow field is strictly radial and symmetric. The strength of the source is equal to the volume flow rate of the flow emanating from the source, which has the strength $m$. Using polar coordinates, the radial velocity component can be written in terms of $\phi$ as $v_r = \frac{\partial \phi}{\partial r}$. A circular control volume at a distance $r$ from the source the velocity will have zero azimuthal component and same radial component, $v_r$. Hence, the mass flow through the control surface is, $2\pi r v_r$ and is the strength of the source, i.e.,

$$\frac{\partial \phi}{\partial r} = \frac{m}{2\pi r} \tag{1.33}$$

which upon integration yields

$$\phi = \frac{m}{2\pi} \, \text{Ln} \, \frac{r}{r_0} \tag{1.34}$$

where $r = r_0$ is the radius of the equi-potential line, $\phi = 0$. The radial velocity can also be written down in terms of the stream function by

$$v_r = \frac{1}{r} \frac{\partial \psi}{\partial \theta}$$

This along with Eq. (1.33) provides the stream function as

$$\psi = \frac{m}{2\pi} (\theta - \theta_0) \tag{1.35}$$

where $\theta_0$ indicates the datum at which $\psi = 0$. The $\psi = constant$ lines are radial, and the equi-potential lines are azimuthal. With the help of $\phi$ and $\psi$, one constructs $W$ as

$$W = \frac{m}{2\pi} \left( \text{Ln} \frac{r}{r_0} + i(\theta - \theta_0) \right) = \frac{m}{2\pi} \text{Ln} \frac{z}{z_0} \tag{1.36}$$

where $z_0 = r_0 e^{i\theta_0}$ represents the datum for $W$. For a sink the complex potential will be of same magnitude, but of opposite sign. As $v_r \to \infty$ as $r \to 0$ (the location of the source/sink), the velocity at the singularity is indefinite. Mathematically, the source/sink is a distribution given by Dirac delta function, such that $\underset{r \to 0}{\text{Lim}} \, 2\pi r v_r = m$.

**Flow field due to a line vortex:**
A potential line vortex as a singularity is characterized by associated constant circulation, which is defined as $\Gamma = \oint \vec{V} \cdot d\vec{r}$. This circulation is positive, when the path in the contour integral is traversed in counter-clockwise direction and the line integral is positive. Although the streamlines describe closed circular paths, vorticity in the flow is zero everywhere excluding the singular line vortex element. Thus, the induced velocity is strictly azimuthal, without radial component. Considering a

cylindrical control volume of radius $r$, shown by any circle along which one can perform a contour integral, one gets

$$v_\theta = \frac{1}{r}\frac{\partial\phi}{\partial\theta} = \frac{\Gamma}{2\pi r} \tag{1.37}$$

which upon integration yields

$$\phi = \frac{\Gamma}{2\pi}(\theta - \theta_0) \tag{1.38}$$

As $v_\theta = -\frac{\partial\psi}{\partial r}$, one can obtain, using Eq. (1.37), $\psi$ as

$$\psi = -\frac{\Gamma}{2\pi}\text{Ln}\frac{r}{r_0} \tag{1.39}$$

Using $\phi$ and $\psi$, one obtains $W$ as

$$W = \phi + i\psi = \frac{\Gamma}{2\pi}\left(\theta - \theta_0 - i\text{Ln}\frac{r}{r_0}\right) = -\frac{i\Gamma}{2\pi}\left[i(\theta - \theta_0) + \text{Ln}\frac{r}{r_0}\right] = -\frac{i\Gamma}{2\pi}\text{Ln}\frac{z}{z_0} \tag{1.40}$$

Here also $z_0$ defines the datum for $W$. One notes from Eq. (1.39) that $v_\theta \to \infty$ as $r \to 0$ (the location of the vortex). Mathematically, the vortex also is a Dirac delta function, such that $\underset{r\to 0}{\text{Lim}}\, 2\pi r v_\theta = \Gamma$.

### 1.7.1   Superposing Solutions of Irrotational Flows

The governing equation for incompressible, irrotational flows is given by the Laplace's equation for $\phi$ and $\psi$, the conjugate functions of $W$. As these equations are linear, one can superpose solutions of elementary singularities to construct other flow fields, as performed in the following.

**(a) Flow field due to a pair of source and sink:**
Here, a pair of source and sink of identical strength $m$ is kept at a distance of $2s$ from each other, as shown in Fig. 1.7. The sink is placed to the left of the source. Consider an arbitrary point P with coordinate given by $(x, y)$. The lines joining P with the source and sink subtend angles $\theta_1$ and $\theta_2$, respectively, with the $x$-axis. The included angle between these two lines is $\theta$. The stream function at the point P is obtained by superposing the solution due to source and sink as

$$\psi = \frac{m}{2\pi}(\theta_1 - \theta_2) = \frac{m\theta}{2\pi}$$

In terms of Cartesian coordinates, this is

$$\psi = \frac{m}{2\pi}\left(\tan^{-1}\frac{y}{x-s} - \tan^{-1}\frac{y}{x+s}\right)$$

using the trigonometric identity

$$(\theta_1 - \theta_2) = \tan^{-1}\frac{\tan\theta_1 - \tan\theta_2}{1 + \tan\theta_1 \tan\theta_2}$$

we simplify $\psi$ as

$$\psi = \frac{m}{2\pi}\tan^{-1}\frac{2sy}{x^2 + y^2 - s^2} \tag{1.41}$$

The $\psi = $ constant lines can be obtained as

$$x^2 + \left(y - s\cot\frac{2\pi\psi}{m}\right)^2 = s^2\mathrm{cosec}^2\frac{2\pi\psi}{m}$$

which represents circles for different values of $\psi$ with center on the $y$-axis at $s\cot\frac{2\pi\psi}{m}$ and radius as $s\,\mathrm{cosec}\frac{2\pi\psi}{m}$.

One can obtain $\phi$ as

$$\phi = \frac{m}{2\pi}\mathrm{Ln}\frac{r_1}{r_2} = \frac{m}{4\pi}\mathrm{Ln}\frac{x^2 + y^2 + s^2 - 2xs}{x^2 + y^2 + s^2 + 2xs} \tag{1.42}$$

**(b) Flow field due to a doublet**:
A doublet is a special case with the distance between the source and sink vanishing, i.e., $2s \to 0$. Thus, from Fig. 1.7, one notes that $\theta \to 0$, so that $\tan\theta \simeq \theta$. Thus, from Eq. (1.41) one obtains

$$\psi_{\text{doublet}} = \mathop{\mathrm{Lim}}_{2s\to 0}\frac{m}{2\pi}\frac{2sy}{x^2 + y^2 - s^2} \tag{1.43}$$

We define the strength of the doublet as $\mu = \mathop{\mathrm{Lim}}_{2s\to 0} 2ms$, so that the denominator of Eq. (1.43) becomes $r^2$ and in the numerator we substitute $y = r\sin\theta$, to obtain

$$\psi_{\text{doublet}} = \frac{\mu y}{2\pi r^2} = \frac{\mu\sin\theta}{2\pi r} \tag{1.44}$$

The $\phi$ is obtained from the definition of the radial velocity, $\frac{1}{r}\frac{\partial\psi}{\partial\theta} = \frac{\partial\phi}{\partial r}$ as

$$\phi_{\text{doublet}} = -\frac{\mu\cos\theta}{2\pi r} \tag{1.45}$$

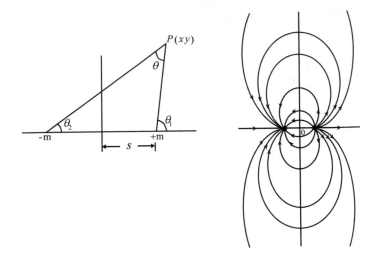

**Fig. 1.7** A source placed in the upstream of a sink, both having the same strength and separated by a distance $2s$ [Reproduced with permission from Theoretical and Computational Aerodynamics. First Edition Tapan K. Sengupta Copyright (2015) John Wiley & Sons. Ltd. UK]

**Fig. 1.8** Streamlines created by the doublet. The axis is pointed from the source to the sink, as indicated by the thick line with the arrowhead [Reproduced with permission from Theoretical and Computational Aerodynamics. First Edition Tapan K. Sengupta Copyright (2015) John Wiley & Sons. Ltd. UK]

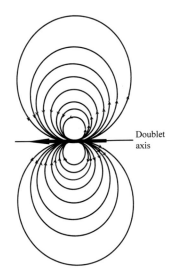

The complex potential is obtained as

$$W = -\frac{\mu \cos \theta}{2\pi r} + \frac{i\mu}{2\pi r} \sin \theta = -\frac{\mu}{2\pi r}(\cos \theta - i \sin \theta) = -\frac{\mu}{2\pi r e^{i\theta}} = -\frac{\mu}{2\pi z} \tag{1.46}$$

The streamlines due to the doublet are shown in Fig. 1.8, with the specific direction from the source to sink. Mathematically, a doublet is the derivative of a delta function and is also an elementary singularity.

**(d) Flow field due to a doublet in uniform horizontal stream**:
Next, we synthesize the flow field by combining $W$ due to uniform horizontal velocity with a doublet. The complex potential is $-Uz$, for the uniform flow going right to left. From Eq. (1.46), $W$ for the doublet is $-\mu/(2\pi z)$. Thus, $W$ for the combination is given by

$$W = -Uz - \frac{\mu}{2\pi\,z} = -U\left(z + \frac{a^2}{z}\right) \tag{1.47}$$

where $a = \sqrt{\frac{\mu}{2\pi U}}$. From Eq. (1.47), one can also write the stream function as

$$\psi = \frac{\mu}{2\pi}\frac{y}{x^2 + y^2} - Uy$$

Thus $\psi = 0$ is represented by

$$y\left(\frac{\mu}{2\pi(x^2 + y^2)} - U\right) = 0$$

which implies that

$$\psi = 0 \text{ is for } y = 0 \text{ and } x^2 + y^2 = \frac{\mu}{2\pi U}$$

The same condition of $\psi = 0$ in polar coordinate provides

$$\left(\frac{\mu}{2\pi r} - Ur\right)\sin\theta = 0$$

Thus, $\psi = 0$ represents the following possibilities:

$$\theta = 0 \text{ and } \pm\pi \text{ (the } x - \text{axis)}$$

The quantity in first bracket defines a circle of radius $a$ obtained from

$$a = \sqrt{\frac{\mu}{2\pi U}}$$

Therefore, the stream function $\psi = 0$ is given by the circle of the radius $a$ and the $x$-axis. Thus, this represents a flow past a circular cylinder of radius $a$ with the points of intersections with the $x$-axis as the stagnation points. The $\psi$ is also given in polar form as

$$\psi = U\sin\theta\left(-r + \frac{a^2}{r}\right) \tag{1.48}$$

**Fig. 1.9** Flow past a
stationary cylinder modeled
as a doublet in uniform
horizontal stream from right
to left. Note the front and
rear stagnation points
marked on the circle as A
and B [Reproduced with
permission from Theoretical
and Computational
Aerodynamics. First Edition
Tapan K. Sengupta
Copyright (2015) John Wiley
& Sons. Ltd. UK]

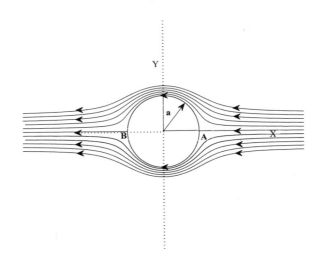

The streamlines of the flow field are depicted in Fig. 1.9, with the stagnation points marked as A and B. Such a flow field is not seen experimentally due to massive separation the flow suffers in the regions of adverse pressure gradient. The presence of massive separation is characteristics of flow past bluff bodies. Flow past a circular cylinder is a canonical example. The solution given by Eq. (1.48) is used as initial condition for solving impulsively started flow for solution of Navier–Stokes equation.

The velocity components are obtained in the flow as

$$v_r = \frac{1}{r}\frac{\partial \psi}{\partial \theta} = U\left(-1 + \frac{a^2}{r^2}\right)\cos\theta \qquad (1.49)$$

and

$$v_\theta = -\frac{\partial \psi}{\partial r} = U\left(1 + \frac{a^2}{r^2}\right)\sin\theta \qquad (1.50)$$

so that on the surface of the cylinder ($r = a$): $v_r = 0$ and $v_\theta = 2U\sin\theta$

This can be used to obtain the pressure ($p$) distribution on the surface of the cylinder by using Bernoulli's equation. The speed of the inviscid flow ($q$) on the surface of the cylinder is given by $v_\theta$ at $r = a$. Far away from the cylinder, the effect of it will disappear and the free-stream speed is therefore $U$. If the free-stream pressure is represented as $p_\infty$, then an application of Bernoulli's equation for steady irrotational flow provides

$$p_\infty + \frac{1}{2}\rho U^2 = p + \frac{1}{2}\rho q^2.$$

The coefficient of pressure is defined as

$$C_p = \frac{p - p_\infty}{\frac{1}{2}\rho U^2} = 1 - 4\sin^2\theta \qquad (1.51)$$

The force acting on the cylinder by the irrotational flow can be obtained by integrating the pressure distribution given in Eq. (1.51). However, an elegant theorem due to Blasius helps to obtain the force components and pitching moment directly, once $W$ of the flow past any arbitrary 2D body is known.

## 1.8  Blasius' Theorem: Forces and Moment for Potential Flows

Let a fixed body be placed in an irrotational steady flow and let $X$, $Y$, and $M$ denote the Cartesian components of force and moment about the origin due to pressure distribution, and then neglecting external forces one obtains

$$X - iY = \frac{i\rho}{2} \oint \left(\frac{dW}{dz}\right)^2 dz \qquad (1.52)$$

$$M = \text{Real}\left[-\frac{\rho}{2} \oint z \left(\frac{dW}{dz}\right)^2 dz\right] \qquad (1.53)$$

**Proof of Eqs.** (1.52) **and** (1.53):
Consider an arbitrary body shown in Fig. 1.10 (in this case an aerofoil is shown for reference), kept in an irrotational flow with a small element shown on which the differential force components are shown to act due to the pressure distribution.

The force components on the element $ds$ (with projections $dx$ and $dy$ in $x$- and $y$-directions, respectively) along the coordinate axes are given by

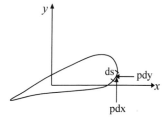

**Fig. 1.10** An arbitrary body shown with differential force acting on a small element for the application of Blasius' theorem [Reproduced with permission from Theoretical and Computational Aerodynamics. First Edition Tapan K. Sengupta Copyright (2015) John Wiley & Sons. Ltd. UK]

$$dX = -p \, dy \quad \text{and} \quad dY = p \, dx$$

Similarly, the moment of these force components about the origin is given by

$$dM = p(x \, dx + y \, dy)$$

where the counter-clockwise direction is considered as positive. Using the complex conjugate $\bar{z}$, one can write the force components together as

$$d(X - iY) = -ip \, d\bar{z} \tag{1.54}$$

Similarly, the pitching moment about the origin can also be written as

$$dM = \text{Real} \, [pz \, d\bar{z}] \tag{1.55}$$

The pressure using Bernoulli's equation is

$$p = C_1 - \frac{1}{2} \rho \, q^2$$

where $C_1$ is the Bernoulli head and the speed on any point on the surface of the body is obtained from

$$q^2 = (u^2 + v^2) = (u + iv)(u - iv) = \frac{dW}{dz} \frac{d\overline{W}}{d\bar{z}}$$

Using these in terms of $W$ and its derivative, one can further write compactly the complex force and moment in Eqs. (1.54) and (1.55) as

$$(X - iY) = \frac{i\rho}{2} \oint \frac{dW}{dz} d\overline{W} - i \oint C_1 d\bar{z}$$

The last contour integral of the above is identically zero, as a constant pressure cannot contribute to force and moment integrals. Also

$$M = \text{Real} \left[ \oint -\frac{\rho z}{2} \frac{dW}{dz} d\overline{W} \right]$$

As the surface of the body represents $\psi = $ constant on it, one notes $d\psi = 0$ and as a consequence $dW = d\overline{W}$ on the surface of the body in Fig. 1.10. Thus

$$d\overline{W} = \frac{dW}{dz} dz,$$

and one can further simplify Eqs. (1.54) and (1.55) as

$$(X - iY) = \frac{i\rho}{2} \oint \left(\frac{dW}{dz}\right)^2 dz$$

and

$$M = \text{Real}\left[-\oint \frac{\rho z}{2}\left(\frac{dW}{dz}\right)^2 dz\right]$$

### 1.8.1 Force Acting on a Vortex in a Uniform Flow

The Blasius' theorem is used to calculate the force acting on a clockwise vortex of strength $\Gamma$ placed in a horizontal stream flowing from left to right. The complex potential of this combination is given by

$$W = Uz + \frac{i\Gamma}{2\pi}\text{Ln}z \tag{1.56}$$

The derivative of this complex potential is given by

$$\frac{dW}{dz} = U + \frac{i\Gamma}{2\pi z} \tag{1.57}$$

Hence

$$\left(\frac{dW}{dz}\right)^2 = U^2\left[1 + \frac{i\Gamma}{\pi Uz} - \frac{\Gamma^2}{4\pi^2 U^2 z^2}\right] \tag{1.58}$$

.

For the function $W$, the singularity is at the origin of the coordinate system. Hence, to apply Blasius' theorem, we consider a closed contour given by a unit circle $z = e^{i\theta}$. Therefore, from Eqs. (1.52) and (1.58), we write

$$X - iY = \int_0^{2\pi} \frac{i\rho}{2}\left(\frac{dW}{dz}\right)^2 dz$$

$$= \frac{i\rho U^2}{2}\int_0^{2\pi}\left[1 + \frac{i\Gamma}{\pi U}e^{-i\theta} - \frac{\Gamma^2}{4\pi^2 U^2}e^{-2i\theta}\right]ie^{i\theta}d\theta$$

$$= -\frac{\rho U^2}{2}\left[-ie^{i\theta}\Big|_0^{2\pi} + \frac{i\Gamma}{\pi U}\theta\Big|_0^{2\pi} - \frac{\Gamma^2}{4\pi^2 U^2}ie^{-i\theta}\Big|_0^{2\pi}\right]$$

The first and third terms do not contribute to the integral, with the middle term contributing to

$$X - iY = -i\rho U\Gamma \tag{1.59}$$

Thus, $X = 0$ and $Y = \rho U \Gamma$. The same result is obtained if one uses Cauchy's residue theorem using Eq. (1.58) in Eq. (1.52) for the pole at $z = 0$. In Eq. (1.58), we note that the second term constitutes a simple pole and the third term represents a pole of second order. Equation (1.59) is also known as the **Kutta–Jukowski theorem**, which states that there is a transverse force (in $y$-direction) for a uniform horizontal flow (in $x$-direction) on $\Gamma$, in the $z$-direction. Thus, the force acting ($\overrightarrow{F}$) can be written for a general flow ($\overrightarrow{V}$) on the vortex ($\overrightarrow{\Gamma} = \Gamma \hat{k}$) in vectorial notation by

$$\overrightarrow{F} = \rho \overrightarrow{V} \times \overrightarrow{\Gamma} \tag{1.60}$$

Physically, we cannot conceive of a force acting without a physical body, and the above derivation is abstract. However, this can be attributed to a physical body like a cylinder and $\Gamma$ lumped at the center caused by spinning the cylinder.

## 1.8.2  Flow Past a Translating and Rotating Cylinder: Lift Generation Mechanism

Consider a cylinder of radius $a$ translating with speed $U$ and spinning about its axis to create a lumped vortex of circulation $\Gamma$ at its center. The flow is from left to right, and the circulation is in clockwise direction. One constructs the flow past translating cylinder from the following complex potential:

$$W_1 = U\left(z + \frac{a^2}{z}\right) \tag{1.61}$$

Additional $W_2$ due to spinning the cylinder to create a circulation $\Gamma$ is given by

$$W_2 = \frac{i\Gamma}{2\pi}\text{Ln}\frac{z}{a} \tag{1.62}$$

Thus, the cylinder ($z = ae^{i\theta}$) represents the streamline $\psi = 0$ in this case, and the singularities are doublet for the translating cylinder and the vortex due to spinning. Thus, one applies Blasius' theorem by considering a contour along the surface of the cylinder. The complex potential for the translating and spinning cylinder is obtained by superposition of Eqs. (1.61) and (1.62) as

$$W = U\left(z + \frac{a^2}{z}\right) + \frac{i\Gamma}{2\pi}\text{Ln}\frac{z}{a} \tag{1.63}$$

To obtain general form of the streamlines, one first obtains the stagnation points obtained from the complex velocity being equal to zero, i.e.,

$$\frac{dW}{dz} = 0 = U\left(1 - \frac{a^2}{z^2}\right) + \frac{i\Gamma}{2\pi z} \tag{1.64}$$

or

$$\frac{z^2}{a^2} + \frac{z}{a}\frac{i\Gamma}{2\pi Ua} - 1 = 0$$

Whence

$$\frac{z_{1,2}}{a} = -\frac{i\Gamma}{4\pi Ua} \pm \sqrt{1 - \frac{\Gamma^2}{16\pi^2 a^2 U^2}} \tag{1.65}$$

Depending upon the sign of the quantity under the radical sign, few sub-cases are considered next.

**(a) The circulation magnitude,** $\Gamma < 4\pi Ua$: Here, we introduce an auxiliary variable $\beta$, such that

$$\frac{\Gamma}{4\pi Ua} = \sin\beta$$

This is used to simplify Eq. (1.65), to obtain

$$\frac{z_{1,2}}{a} = -i\sin\beta \pm \cos\beta = e^{-i\beta} \text{ and } e^{i(\beta+\pi)} \tag{1.66}$$

The stagnation points and few representative streamlines are shown in Fig. 1.11 for this sub-case. Here, $\beta$ represents an angle and the figure shows two stagnation points symmetrically located below the $x$-axis, depressed by the angle $\beta$ with respect to flow direction, and marked as $A$ and $B$ in the figure. One notes the existence of a limiting circulation for which $\beta = \pi/2$ when the two stagnation points coalesce. We consider the value of $\Gamma$ for which $\beta < \pi/2$ as belonging to subcritical cases. Criticality would correspond to when $A$ and $B$ will merge to create maximum lift in this inviscid irrotational model of the flow.

For a stationary cylinder, $\beta = 0$ and the stagnation points would be at $(-a, \ 0)$ and $(a, \ 0)$. Circulation forces these stagnation points to move downward. For steady flow, one notes the fluid particles along and above $APB$ to traverse a much longer distances as compared to particles which remain along and below $AQB$. This implies immediately that the flow velocity will be higher for the streamlines along $APB$, as compared to $U$. In the same way, the flow velocity will be lower for the streamline along $AQB$. From Bernoulli's equation, this state of affair would indicate pressure to be lower above $APB$ (suction) and higher below $AQB$, with respect to free-stream pressure, $p_\infty$. This creates a lift force in the positive $y$-direction. Also, the drawn streamlines indicate a perfect fore-aft symmetry with respect to the $y$-axis, implying zero drag experienced. For irrotational flow, this is a consequence of D'Alembert's paradox (see [20]).

The lift experienced by the cylinder can be obtained from Kutta–Jukowski theorem as

$$L = \rho U\Gamma = 4\pi\rho U^2 a \sin\beta$$

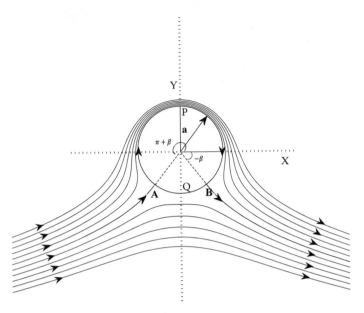

**Fig. 1.11** Streamlines past the rotating and translating cylinder, with the subcritical circulation [Reproduced with permission from Theoretical and Computational Aerodynamics. First Edition Tapan K. Sengupta Copyright (2015) John Wiley & Sons. Ltd. UK]

Choosing the diameter as the length scale, one can obtain the lift coefficient as

$$C_l = \frac{L}{\frac{1}{2}\rho U^2 2a} = 4\pi \sin \beta \qquad (1.67)$$

We will indicate the 2D lift coefficient with the subscript $l$ and the corresponding 3D value denoted by $C_L$. One can try to relate the rotation rate ($\Omega^*$) with circulation by calculating the latter from the basic definition as $\Gamma = 2\pi a^2 \Omega^*$. If the corresponding surface speed is $U_s = a\Omega^*$, then $\Gamma = 2\pi a U_s$.

**(b) For the circulation magnitude,** $\Gamma = 4\pi U a$: This is a critical case, when $\beta = \pi/2$ and the stagnation points $A$ and $B$ coincide. Equation (1.67) would indicate a maximum lift, i.e., $C_{l,max} = 4\pi$. This is known as the Prandtl's limit. This appears far in excess of that is used in aeronautical devices. However, in the above derivation, the circulation is overestimated by the inviscid, irrotational flow model. It is understood that the actual circulation is a consequence of conservation laws and cannot be achieved by spinning at a fixed rate and increased indefinitely. It must be sought from the solution of Navier–Stokes equation, and attention is drawn to the discussion about it in [20]. The flow field is depicted in Fig. 1.12 for a qualitative understanding. It is noted that the extent over which suction is now acting covers the full cylinder, with the stagnation points coincident for this critical case. Calling this as $\Gamma_{cr}$ and

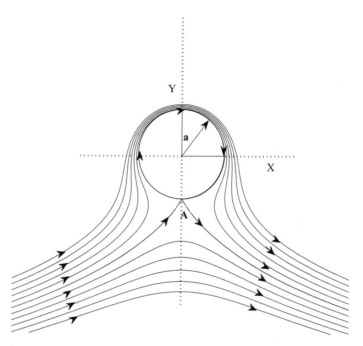

**Fig. 1.12** Streamlines past the rotating and translating cylinder, with the critical circulation ($\Omega = 2$) [Reproduced with permission from Theoretical and Computational Aerodynamics. First Edition Tapan K. Sengupta Copyright (2015) John Wiley & Sons. Ltd. UK]

corresponding oncoming flow as $U_{cr}$, then one can equate this circulation with $\Gamma = 2\pi a U_s$ to get $U_{cr} = 2U_s$ and the non-dimensional rotational rate, $\Omega = \Omega^* a/U = 2$.

We note the classification of singular points and their effects. The definition of streamline precludes intersection of these and the slope variation must be smooth. In the previous two sub-cases, the body streamlines do not have continuous slopes/ derivatives. Viscous actions smooth out such discontinuities, yet such streamlines are noted for viscous flows at the stagnation, separation, and reattachment points. These points on the surface of a physical body are referred to as half-saddle points. One of the features of these stagnation points is that at these points stream functions are continuous and its derivatives are zero. For the case (b), these two half-saddle points coincide. The following case is discussed corresponding to supercritical circulation.

**(c) The circulation magnitude, $\Gamma > 4\pi Ua$:** Here, we define

$$\frac{\Gamma}{4\pi Ua} = \cosh\beta$$

with $\beta$ just is a parameter. From Eq. (1.65) one gets

$$\frac{z_{1,2}}{a} = i(-\cosh\beta \pm \sinh\beta) = -ie^{\beta} \text{ and } -ie^{-\beta} \tag{1.68}$$

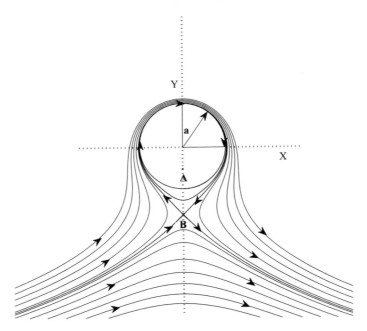

**Fig. 1.13** Streamlines past the rotating and translating cylinder, with the circulation supercritical ($\Omega > 2$) [Reproduced with permission from Theoretical and Computational Aerodynamics. First Edition Tapan K. Sengupta Copyright (2015) John Wiley & Sons. Ltd. UK]

This case also can be classified as $U > 2U_s$ or by the non-dimensional rotation rate given by $\Omega > 2$. Now for this case, both the stagnation points are on the $y$-axis. As $z_1 z_2 = a^2$, this implies that if one of the points is outside the cylinder $|z| > a$, then the other point must necessarily be inside the cylinder, so that the product remains always equal to $a^2$. Thus, one of the stagnation points now descends along the $y$-axis outside the cylinder, and this is the full saddle point. The streamlines are shown for this case in Fig. 1.13, for the supercritical circulation, with the saddle point (B) clearly outside the cylinder.

### 1.8.3   Prandtl's Limit on Maximum Circulation and Its Violation

For circulation exceeding the critical case creates a closed recirculating streamline, as shown in Fig. 1.13. Prandtl reasoned that as there is no flow across the streamline, so there cannot be any momentum transfer also across this limiting streamline, and this will limit the lift experienced by the cylinder beyond $U = 2U_s$. Generating lift by a rotating cylinder in a uniform flow is known as Magnus–Robins effect (see [13] and [20] for more details). It is possible to calculate actual lift experienced by a rotating cylinder by solving Navier–Stokes equation, and it has been noted that for subcritical case of rotation, the actual value is lower than that of the inviscid, irrotational flow model. Research continues on the topic of supercritical rotation rate and it has been shown both experimentally ([18]) and computationally [14], which demonstrate that for rotation rates significantly higher than the critical rate, Prandtl's limit is violated! It has been explained in [14] and [13] that this is due to viscous diffusion across the limiting streamline and the process of continuous generation of vorticity at the wall, due to no-slip condition.

### 1.8.4 Pressure Distribution on Spinning and Translating Cylinder

The stream function for this flow is obtained from $W$ as

$$\psi = Ur\left(\frac{a^2}{r^2} - 1\right)\sin\theta - \frac{\Gamma}{2\pi}\text{Ln}\frac{r}{a} \qquad (1.69)$$

from which the radial and azimuthal components of velocity are obtained as

$$u_\theta = -\frac{\partial\psi}{\partial r} = U\left(\frac{a^2}{r^2} + 1\right)\sin\theta - \frac{\Gamma}{2\pi r} \qquad (1.70)$$

$$u_r = \frac{1}{r}\frac{\partial\psi}{\partial\theta} = U\left(\frac{a^2}{r^2} - 1\right)\cos\theta \qquad (1.71)$$

Thus, on the surface of the cylinder $(r = a)$

$$u_r = 0 \text{ and } u_\theta = 2U\sin\theta - \frac{\Gamma}{2\pi a}$$

These can be used in Bernoulli's equation to relate local pressure $(p)$ with free-stream pressure $(p_\infty)$ as

$$p_\infty + \frac{1}{2}\rho U^2 = p + \frac{\rho}{2}\left(2U\sin\theta - \frac{\Gamma}{2\pi a}\right)^2$$

The pressure coefficient is defined as

$$C_p = \frac{p - p_\infty}{\frac{1}{2}\rho U^2} = 1 - \left(2\sin\theta - \frac{\Gamma}{2\pi aU}\right)^2 \qquad (1.72)$$

(a) For the case of $\Gamma < 4\pi aU$: As noted before

$$\frac{\Gamma}{2\pi a U} = 2\sin\beta$$

and the pressure coefficient is given by

$$C_p = 1 - 4(\sin\theta - \sin\beta)^2$$

(b) For the critical rotation rate $\Gamma = 4\pi a U$: The stagnation points move such that $\beta = \pi/2$ and then
$$C_p = 1 - 4(\sin\theta - 1)^2$$

(c) Supercritical rotation rate $\Gamma > 4\pi a U$: we defined

$$\frac{\Gamma}{4\pi a U} = \cosh\beta,$$

and then
$$C_p = 1 - 4(\sin\theta - \cosh\beta)^2$$

These pressure distributions can be integrated from $\theta = 0$ to $2\pi$ to obtain the lift generated and establish that $L = \rho U\Gamma$ in all the cases.

## 1.9   Method of Images

Superposition of elementary flows to obtain flows of practical interest can be further extended by using method of images. This is possible due to linear governing Laplace's equation for the irrotational flows. The method of images utilizes the presence of walls or bodies by using the principle that at the solid wall, the wall-normal component of velocity must be equal to zero, due to elementary singularities and the image system. Its implementation in fluid mechanics is due to Kelvin, Helmholtz, and Stokes, among many others.

The working principle of method of images is shown next with the help of elementary singularities placed in front of a solid wall. As there cannot be any flow

**Fig. 1.14** Method of images used for a source in front of a plane wall

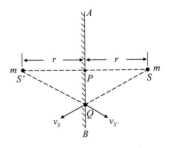

across the wall, this wall must by itself be a streamline. This is used as the principle for locating the image singularity for the flow field to be modeled. In Fig. 1.14 we show a source of strength $m$ in front of a plane wall $AB$. Now, the wall $AB$ being a streamline, the source placed at a normal distance $r$ from the wall will induce a velocity at the foot of this normal at $P$. To make the wall a streamline, one must cancel this induced normal velocity component. This can be achieved by placing another source of the same strength $(m)$, at the same distance from the wall $(r)$ at $S'$; on the other side of the wall, an image of the original source placed in front of the wall. This image source at $S'$ will induce a wall-normal velocity component, which is the same in magnitude to which was created by $S$ at $P$, but of opposite sign. Superposition of these two velocities will ensure zero wall-normal component of velocity everywhere along $AB$. For a general point $Q$, the velocity due to the source and its image is drawn as $v_S$ and $v_{S'}$ in Fig. 1.14. Resolving these two velocities into components, along and perpendicular to the wall, one notices that the wall-normal components of $v_S$ and $v_{S'}$ are equal and opposite, while the component along the wall adds up. This latter component does not violate the definition of the streamline. However, this violates no-slip condition on the tangential component. For inviscid flow, this is of no concern.

## Problems

**P1.1**: On a foggy day, will the speed of sound be lower or higher, as compared to speed of sound in dry air?

**P1.2**: To observe the unsteady separated flow in a diverging channel, bubbles are injected from point A at 10 ms interval, as shown in Fig. 1.15. These bubbles serve as tracer particles. AB line at any time instant represents which of the following?

(A) Streamline, streakline and pathline
(B) Streamline and pathline
(C) Only pathline
(D) Only streakline

**Fig. 1.15** Flow diagram for P1.2

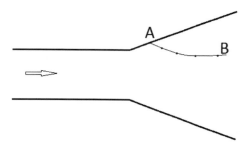

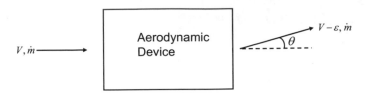

**Fig. 1.16** Block diagram for problem

**Fig. 1.17** Flow diagram for
problem

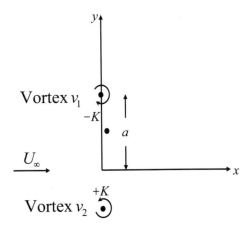

**P1.3**: One can measure the free-stream speed in a low-speed wind tunnel using a

(1) Pitot tube alone, aligned with the flow direction.
(2) Pitot tube aligned with the flow direction, along with the static pressure mea-
    surement at some appropriate position on the tunnel wall.
(3) Pitot tube aligned with the flow direction, along with the barometer pressure
    reading of the outside ambient condition.
(4) Pitot static tube alone, aligned with the flow direction.

**P1.4**: With reference to the following block diagrams, answer the questions that
follow (Fig. 1.16).

(a) Calculate lift and drag in terms of $\varepsilon$ and $\theta$.
(b) What is the physical implication of $\dot{m}$. How is it related to the efficiency of the
    device?
(c) What is the role of $\theta$ here?

**P1.5**: The figure below shows two counter-rotating pair of line vortices of equal
strength $K$, separated by a distance $2a$, subjected to a uniform normal flow. Find
the combined stream function and draw its contours for various values of $K/(U_\infty a)$
(Fig. 1.17).

**Fig. 1.18** Flow diagram for problem

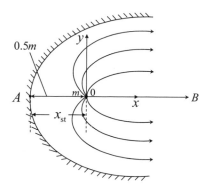

**P1.6**: A small fish pond is approximated by a half-body shape, as shown in Fig. 1.18. A point O, which is 0.5 m from the left edge of the pond, is a source of additional water delivering about 0.35 m³/s per meter of depth into the paper. Find the point $B$ along the axis where the water velocity is approximately 25 cm/s.

# References

1. R.L. Ash, A.J. Zuckerwar, Z. Zheng, *Second Coefficient of Viscosity in Air*. NASA-CR-187783 (1991)
2. G.K. Batchelor, *An Introduction to Fluid Dynamics* (Cambridge University Press, UK, 1988)
3. G. Buresti, A note on Stokes hypothesis. Acta Mechanica **226**(10), 3555–3559 (2015)
4. S. Chandrasekhar, *Hydrodynamics and Hydrodynamic stability* (Clarendon, Oxford, 1961)
5. M.S. Cramer, Numerical estimates for the bulk viscosity of ideal gases. Phys. Fluid **24**, 066102 (2012)
6. S.M. Karim, L. Rosenhead, The second coefficient of viscosity of liquids and gases. Rev. Modern Phys. **24**, 108–16 (1952)
7. L. Liebermann, The second viscosity of liquids. Phys. Rev. **75**, 1415 (1949)
8. K.R. Rajagopal, A new development and interpretation of the Navier-Stokes fluid which reveals why the "Stokes' assumption" is inapt. Int. J. Non-linear Mech. **50**, 141–151 (2013)
9. L. Rayleigh, Investigation of the character of the equilibrium of an incompressible heavy fluid of variable density. Proc. Lond. Math. Soc. **14**(1), 8 (1883)
10. K.I. Read, Experimental investigation of turbulent mixing by Rayleigh-Taylor instability. Physica D: Nonlinear Phenomena **12**(1), 45–58 (1984)
11. L. Rosenhead, Introduction—the second coefficient of viscosity: a brief review of fundamentals. Proc. R. Soc. Lond. A **226**, 1–6 (1954)
12. T.K. Sengupta, *High Accuracy Computing Methods: Fluid Flows and Wave Phenomenon* (Cambridge University Press, USA, 2013)
13. T.K. Sengupta, S.B. Talla, Robins-Magnus effect: a continuing saga. Curr. Sci. **86**(7), 1033–1036 (2004)
14. T.K. Sengupta, A. Kasliwal, S. De, M. Nair, Temporal flow instability for Magnus-Robins effect at high rotation rates. J. Fluids and Struct. **17**, 941–953 (2003)
15. T.K. Sengupta, A. Sengupta, N. Sharma, S. Sengupta, A. Bhole, K.S. Shruti, Roles of bulk viscosity on Rayleigh-Taylor instability: Non-equilibrium thermodynamics due to spatio-temporal pressure fronts. Phys. Fluids **28**, 094102 (2016)
16. I.S. Sokolnikoff, *Mathematical Theory of Elasticity* (McGraw-Hill, USA, 1946)

17. G. Taylor, The instability of liquid surfaces when accelerated in a direction perpendicular to their planes. I. Proc. R. Soc. Lond. A: Math. Phys. Eng. Sci. **201**(1065), 192–196 (1950)
18. P.T. Tokumaru, P.E. Dimotakis, The lift of a cylinder rotary motion in a uniform flow. J. Fluid Mech. **255**, 1–10 (1993)
19. W.G. Vincenti, C.H. Kruger, *Introduction to Physical Gas Dynamics* (Wiley, USA, 1965)
20. F.M. White, *Fluid Mech.*, 6th edn. (McGraw-Hill, New York, 2008)

# Chapter 2
# Elementary Aerodynamics

In Chap. 1, we have laid down the governing equations at different hierarchy levels, starting from the Navier–Stokes equation without the limiting Stokes' hypothesis down to the inviscid, irrotational flow modeled by Laplace's equation for velocity potential $(\phi)$, and stream function $(\psi)$ for 2D flows. We introduced complex variables to ease the synthesis of these two variables into a single complex potential $(W)$, which can be represented as analytic or holomorphic function of $z = x + iy$. In the present chapter, we show how this analysis is used originally to analyze and design of aerofoil, the quintessential element of aerodynamics.

## 2.1 Conformal Mapping: Use of Cauchy–Riemann Relation

Conformal mapping is the method by which the flow past a complicated body can be analyzed by studying flow past a simpler body, while tying the two flows using convenient transformation. We have seen in the previous chapter how to analyze flow past a circular cylinder, supporting finite amount of circulation or lift. Thus, it would be useful if one can link the flow past circular cylinder with the flow around a lifting aerofoil by transformation between these two flow domains, with appropriate auxiliary condition. To achieve this, we have to show that the governing equation in the physical $z$-plane (circle plane) retains the same form in the transformed $\zeta$-plane (aerofoil plane), so that $W$ described in the physical plane can be used in the transformed plane. Thus, we need a complex function

$$\zeta = f(z) \tag{2.1}$$

which transforms the geometric shape in the $z$-plane to another in the $\zeta$-plane, whose real and imaginary parts are given by $\xi$ and $\eta$, respectively. We assume the above transformation be analytic, so that a simple relationship between the physical and

© Springer Nature Singapore Pte Ltd. 2020
T. K. Sengupta and Y. G. Bhumkar, *Computational Aerodynamics and Aeroacoustics*,
https://doi.org/10.1007/978-981-15-4284-8_2

transformed plane flow properties exists. For any point $z_0$ in $z$-plane, we term $\zeta_0 = f(z_0)$ as the image of $z_0$ with the help of the map, $f(z)$. This implies that lines and planes are transformed uniquely by this mapping function. This is the central role of conformal mapping, by which we relate specific shapes in the $z$-plane to unique shapes in the $\zeta$-plane.

Given the transform in Eq. (2.1), we note that

$$d\zeta = f'(z)\, dz$$

If $f'(z) = re^{i\theta}$, then $|d\zeta| = r|dz|$, i.e., the transformed element equals the original element rotated through $\theta$ and multiplied by $r$. It is to be noted that a conformal mapping is one, which preserves angles between any curves both in magnitude and orientation.

Using chain rule of partial differentiation, one can obtain the Laplacian operator as

$$\nabla^2 = [\xi_{xx} + \xi_{yy}]\frac{\partial}{\partial \xi} + [\eta_{xx} + \eta_{yy}]\frac{\partial}{\partial \eta} + [\xi_x^2 + \xi_y^2]\frac{\partial^2}{\partial \xi^2}$$

$$+ [\eta_x^2 + \eta_y^2]\frac{\partial^2}{\partial \eta^2} + 2[\xi_x\eta_x + \xi_y\eta_y]\frac{\partial^2}{\partial \xi \partial \eta} \qquad (2.2)$$

As we are considering $\zeta = \xi + i\eta$ an analytic function of $z$, therefore $\xi$ and $\eta$ satisfy Cauchy–Riemann relations given by

$$\frac{\partial \xi}{\partial x} = \frac{\partial \eta}{\partial y} \quad \text{and} \quad \frac{\partial \xi}{\partial y} = -\frac{\partial \eta}{\partial x} \qquad (2.3)$$

One consequence of Eq. (2.3) is that $\xi_x\eta_x + \xi_y\eta_y = 0$. As shown in [15], this is a condition of orthogonal map. Another set of results follow from Eq. (2.3) that using the Cauchy–Riemann relations we can show

$$\frac{\partial^2 \xi}{\partial x^2} + \frac{\partial^2 \xi}{\partial y^2} = 0 \quad \text{and} \quad \frac{\partial^2 \eta}{\partial x^2} + \frac{\partial^2 \eta}{\partial y^2} = 0$$

And the equality of scale factors implicit in conformal mapping produces

$$J = [\xi_x^2 + \xi_y^2] = [\eta_x^2 + \eta_y^2] \qquad (2.4)$$

Hence, the Laplacian operator in the transformed $\zeta$-plane is simply given by

$$\nabla^2 = J\left(\frac{\partial^2}{\partial \xi^2} + \frac{\partial^2}{\partial \eta^2}\right)$$

$$\nabla^2 = J\left(\frac{\partial^2}{\partial \xi^2} + \frac{\partial^2}{\partial \eta^2}\right) \tag{2.5}$$

so that the governing equation for incompressible, irrotational flow in terms of $\phi$ is given by

$$\nabla^2\phi = J\left(\frac{\partial^2\phi}{\partial \xi^2} + \frac{\partial^2\phi}{\partial \eta^2}\right) = 0 \tag{2.6}$$

which is in the same form, as in the $z$-plane. Same observation holds for the governing equation involving $\psi$.

Singular or critical points of conformal mapping for $f(z)$ occur, where $f'(z)$ is zero or infinity. Take, for example, the Jukowski transformation is given by

$$\zeta = z + \frac{b^2}{z} \tag{2.7}$$

For this transformation

$$\frac{d\zeta}{dz} = 1 - \frac{b^2}{z^2}$$

Thus, the derivative is zero at $z = \pm b$. The concept of singularity is noted when one relates velocities in the two planes, next.

### 2.1.1 Relation Between Complex Velocity in Two Planes

The complex velocity in the $\zeta$-plane is given using Eq. (1.31) by

$$(u - iv)_\zeta = \frac{dW}{d\zeta} = \frac{dW}{dz}\frac{dz}{d\zeta} = (u - iv)_z \frac{dz}{d\zeta} \tag{2.8}$$

The complex velocities in the two planes are thus related by $\frac{dz}{d\zeta}$. Velocity in $z$-plane is magnified by $|\frac{dz}{d\zeta}|$ and rotated by $\arg(\frac{dz}{d\zeta})$. In general, $|\frac{dz}{d\zeta}|$ has finite value differing from zero, except at singular or critical points. One of the central ideas for conformal mapping is that the two flow fields must be the same, far away from the respective bodies in the two planes. This is equivalent to the condition

$$\lim_{\zeta,z\to\infty} \frac{dz}{d\zeta} \to 1$$

This is readily seen for the Jukowski transformation given in Eq. (2.7) as

$$\lim_{\zeta,z\to\infty} \left|\frac{dz}{d\zeta}\right| = \lim_{z\to\infty} \frac{1}{1 - \frac{b^2}{z^2}} = 1$$

## 2.1.2  Application of Conformal Transformation

We study Jukowski transformation, in particular, as it is capable of producing family of airfoils. The Jukowski transformation in polar representation is given for Eq. (2.7) with $b = c$ as

$$\zeta = \xi + i\eta = re^{i\theta} + \frac{c^2}{re^{i\theta}} \tag{2.9}$$

Thus,

$$\xi = \left(r + \frac{c^2}{r}\right)\cos\theta \quad \text{and} \quad \eta = \left(r - \frac{c^2}{r}\right)\sin\theta \tag{2.10}$$

### (a) Mapping a circular cylinder to a flat plate:

We note that the point of singularity brings in a slope discontinuity, even when it is located on the surface of a body. However, a singularity on a closed smooth curve can transform to another closed smooth curve without discontinuity. This shows that a body with sharp discontinuity can be transformed to a smooth geometry. If we map such that a point of singularity falls exactly on a smooth curve in the $z$-plane, then we obtain a transformed shape with sharp corner in the $\zeta$-plane.

Using the Jukowski transformation of the flat plate to a circular cylinder, we must use $c = a$ in Eq. (2.7), where $a$ is the radius of the circle in the $z$-plane with its center at the origin. Thus, both the singular points $z = \pm c$ fall on the surface of the cylinder. From Eqs. (2.9) and (2.10), the circle $r = a$ transforms to

$$\xi = 2a\cos\theta \quad \text{and} \quad \eta = 0 \tag{2.11}$$

The chord of the flat plate is $4a$ as $-2a \leq \xi \leq +2a$. The singularities at $z = \pm a$ produce the sharp edges at $\xi = \pm 2a$, respectively. In Fig. 2.1, we note the mapping between the circle and flat plate planes.

### (b) Transforming a circular cylinder to a symmetric airfoil:

For this Jukowski transformation, the circle and the airfoil planes are as shown in Fig. 2.2, with one of the singular points of the transformation at $z = -c$, lying on the

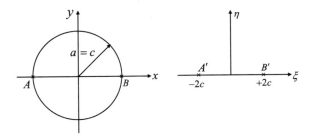

**Fig. 2.1**  Mapping a circle to a flat plate by Jukowski transform. Reproduced with permission from Theoretical and Computational Aerodynamics. First Edition Tapan K. Sengupta Copyright (2015) John Wiley & Sons. Ltd. UK

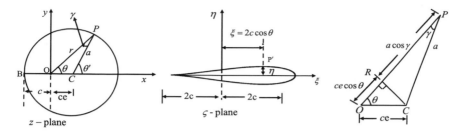

**Fig. 2.2** Conformal mapping between circle and symmetric airfoil planes by Jukowski transform. Reproduced with permission from Theoretical and Computational Aerodynamics. First Edition Tapan K. Sengupta Copyright (2015) John Wiley & Sons. Ltd. UK

circle of radius $a$. The center of the circle at $C$ is offset from the origin $O$, by $c * e$ in the $z$-plane. The offset being very small, we must have $e \ll 1$. A general point $P$, with polar coordinate $(r, \theta)$, joined with C; the line $CP$ subtends an angle $\theta'$ with the $x$-axis. On the right side of the figure, we show an enlarged view of the triangle $OCP$, where $\angle OPC$ is given by $\gamma$. A perpendicular from $C$ dropped on the line $OP$ meets the line at $R$.

From the geometry in Fig. 2.2, we note that

$$OR = ce \cos \theta \text{ and } PR = a \cos \gamma \simeq a$$

Therefore, $OP = r = a + ce \cos \theta$, so that the radius of the circle is approximated as $a = c(1 + e)$. The radial distance is written as

$$\frac{r}{c} = 1 + e + e \cos \theta, \tag{2.12}$$

which can be alternately written as

$$\frac{c}{r} = [1 + e + e \cos \theta]^{-1} = 1 - e(1 + \cos \theta) + \frac{e^2}{2}(1 + \cos \theta)^2$$

which upon linearization (for $e \ll 1$) simplify further as

$$\frac{c}{r} = 1 - e - e \cos \theta \tag{2.13}$$

As

$$\xi = c\left(\frac{r}{c} + \frac{c}{r}\right) \cos \theta \text{ and } \eta = b\left(\frac{r}{c} - \frac{c}{r}\right) \sin \theta \tag{2.14}$$

using Eqs. (2.12) and (2.13), one obtains

$$\xi = 2c \cos \theta \text{ and } \eta = 2ce(1 + \cos \theta) \sin \theta \tag{2.15}$$

**Fig. 2.3** Transforming a
circular cylinder to a circular
arc airfoil of zero thickness

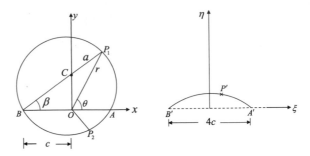

This represents a shape with chord $4c$ [with $-2c \leq \xi \leq +2c$], and it is symmetric as $\eta(\theta) = -\eta(-\theta)$. The maximum thickness is obtained from $\frac{d\eta}{d\theta} = 0$, which upon simplification yields $\cos^2 \theta - \sin^2 \theta + \cos \theta = 0$. This quadratic has the following solutions: $\cos \theta = \frac{1}{2}$ and $-1$. The value of $\cos \theta = \frac{1}{2}$ yields $\theta = \frac{\pi}{3}$, for which $\xi = c$, i.e., the quarter-chord point represents the maximum thickness location, whereas for $\cos \theta = -1$, the minimum occurs when $\theta = \pi$, which is the trailing edge, $\xi = -2c$. The maximum thickness is $2\eta_{max} = 4ce(1 + \cos \frac{\pi}{3}) \sin \frac{\pi}{3} = 3\sqrt{3}\ ce$. The thickness-to-chord ratio of the symmetric airfoil is given by $\frac{t}{c} = \frac{3\sqrt{3}ce}{4c} = 1.299e$. Thus, the parameter $e$ in the Jukowski transformation fixes the $t/c$ ratio of the airfoil. The slope at the leading edge is a right angle [as $\frac{d\eta}{d\xi}|_{\theta=0} = \frac{\pi}{2}$], while the slope at the trailing has a very special feature. The slope at the trailing edge is given by

$$\left.\frac{d\eta}{d\xi}\right|_{\xi=-2c} = \left.\frac{d\eta/d\theta}{d\xi/d\theta}\right|_{\theta=\pi} = \left.\frac{2ce[2\cos^2 \theta + \cos \theta - 1]}{2c \sin \theta}\right|_{\theta=\pi}$$

This can be evaluated by L'Hospital's rule as zero. This implies that the airfoil has a cusped trailing edge, a special feature of all Jukowski airfoils.

**(c) Transforming a circular cylinder to a circular arc:**

Here, Jukowski transformation is shown in Fig. 2.3, where we position the circle in such a way that the center is on the $y$-axis and the singular points $A$ and $B$ are on the circle in such a way that $OB = OA = c$. The radius of the circle $(CP)$ is given by $a$ as before, so that $CB = a = c \sec\beta$ and $OC = c \tan \beta$.

From the triangle $OCP$ in Fig. 2.3,

$$CP^2 = OC^2 + OP^2 - 2\ OC.\ OP\ \cos(\pi/2 - \theta)$$

so that

$$c^2\sec^2\beta = c^2 \tan^2 \beta + r^2 - 2cr \tan \beta\ \sin \theta$$

which upon simplification provides

$$\frac{r^2 - c^2}{r} = 2c \tan \beta\ \sin \theta \tag{2.16}$$

Now from Eq. (2.10), we have

$$\xi = (r + c^2/r) \cos\theta \quad \text{and} \quad \eta = (r - c^2/r) \sin\theta \qquad (2.17)$$

Using Eq. (2.16) in these, we get the simplified form of the ordinate as

$$\eta = 2c \tan\beta \, \sin^2\theta \qquad (2.18)$$

This shows that two points $P_1$ and $P_2$ symmetrically located about $AB$, indicated by $\pm\theta$, map to a single point $P'$ in the $\zeta$-plane. This establishes that the body in the $\zeta$-plane has zero thickness. We note that $A$ and $B$ map to $A'$ and $B'$, respectively, corresponding to $\theta = 0$ and $\pi$. Thus, $A'B' = 4c$ and the circle transforms to a cambered plate. Maximum value of $\eta$ occurs for $\theta = \pi/2$ given by

$$\eta_{max} = 2c \tan\beta$$

The camber ratio is defined as

$$\frac{\eta_{max}}{\text{chord}} = \frac{2c \tan\beta}{4c} = \frac{1}{2} \tan\beta \qquad (2.19)$$

From Eq. (2.17), one gets

$$\frac{\xi^2}{\cos^2\theta} - \frac{\eta^2}{\sin^2\theta} = \left(r^2 + \frac{c^4}{r^2} + 2c^2\right) - \left(r^2 + \frac{c^4}{r^2} - 2c^2\right) = 4c^2 \qquad (2.20)$$

Using $\sin^2\theta = \eta/(2c \, \tan\beta)$, from Eq. (2.18) in Eq. (2.20), one gets

$$\xi^2 - \eta^2 \frac{\left(1 - \frac{\eta}{2c \tan\beta}\right)}{\frac{\eta}{2c \tan\beta}} = 4c^2\left(1 - \frac{\eta}{2c \tan, \beta}\right) \qquad (2.21)$$

which can be further simplified as

$$\xi^2 + (\eta + 2c \cot 2\beta)^2 = 4c^2 \csc^2 2\beta \qquad (2.22)$$

This is a circular arc, with center at $(0, -2c \cot 2\beta)$, and radius $2c \csc 2\beta$. The value of the camber is decided by the vertical offset of the center of the cylinder from the origin.

**(d) Transforming circular cylinder to a cambered airfoil**:

It is noted that a singular point of the Jukowski transformation passing through the cylinder creates a transformed shape in $\zeta$-plane with a sharp edge at the corresponding point. The offset of the center in horizontal direction is related to thickness distribu-

**Fig. 2.4** Transforming a
circular cylinder to a general
cambered airfoil of finite
thickness

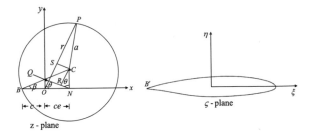

tion and offset in the vertical direction causes camber distribution; a combination of
these two offsets will produce a cambered airfoil with finite thickness.

The mapping is shown in Fig. 2.4, showing the circle in $z$-plane mapped to an
airfoil in the $\zeta$-planes. Here, the center is offset in horizontal and vertical directions,
with respect to the origin. In the circle plane, consider an arbitrary point $P$ with polar
coordinate $(r, \theta)$, and $CP = a$ and $OB = c$.

The horizontal shift $(ON)$ is given by $ce$, whereas the vertical shift of the
center $(CN)$ is given by $h$. The corresponding angular offset, $\angle CBO = \beta$, is
considered small and $\angle CPO = \gamma$, so that $\cos \beta \simeq 1 = \cos \gamma$. From the figure,
$CN = BC \sin \beta$, i.e., $h = a \sin \beta$. Also, $BN = BC \cos \beta$, which implies

$$c + ce = a \cos \beta \simeq a \tag{2.23}$$

Therefore

$$h = c(1 + e)\beta \tag{2.24}$$

From the circle plane shown in Fig. 2.4,

$$OP = ON \cos \theta + CN \sin \theta + CP \cos \gamma$$

Thus

$$r = ce \cos \theta + h \sin \theta + a \cos \gamma$$

$$= ce \cos \theta + c\beta (1 + e) \sin \theta + c(1 + e)$$

Since $c\beta e$ is a product of two small terms, it can be neglected in the linear analysis
and one obtains

$$\frac{r}{c} = 1 + e + e \cos \theta + \beta \sin \theta \tag{2.25}$$

$$\frac{c}{r} = 1 - (e + e \cos \theta + \beta \sin \theta) \tag{2.26}$$

From Jukowski transform in Eq. (2.14), one gets

$$\xi = 2c \, \cos\theta \quad \text{and} \quad \eta = 2c \, (e + e\cos\theta + \beta\sin\theta) \, \sin\theta \qquad (2.27)$$

The ordinate of the airfoil can also be written as

$$\eta = 2ce \, (1 + \cos\theta) \, \sin\theta + 2c\beta \, \sin^2\theta \qquad (2.28)$$

We note that $\beta$-term gives rise to asymmetry, which is always positive. The camber is given by $\beta$, as $\beta = 0$ reduces it to the symmetric airfoil case. Differentiating Eq. (2.28) with respect to $\theta$, one locates the maximum of $\eta$ to be at the quarter-chord, as shown below.

The thickness at any station is given by $t = \eta_1 - \eta_2$, where $\eta_1$ and $\eta_2$ are for the same $\xi$ coordinate, i.e., $\theta_2 = -\theta_1$. Therefore, the thickness is given by

$$t = 4ce \, (1 + \cos\theta_1) \, \sin\theta_1$$

As the chord of the airfoil is $4c$, the thickness-to-chord distribution is given by

$$\frac{t}{c} = e \, (1 + \cos\theta) \, \sin\theta \qquad (2.29)$$

Once again, $t/c$ is maximum at $\theta = \pi/3$. In general, camber of an airfoil is the maximum deviation of the mean camber line from the chord, which is given by $\frac{1}{2}(\eta_1 + \eta_2)$. Thus, the camber is defined as

$$\text{Camber} = \frac{(\eta_1 + \eta_2)_{\max}}{2 \text{ chord}} = \frac{(4c\beta \, \sin^2\theta)_{\max}}{8c} = \frac{\beta}{2} \qquad (2.30)$$

which occurs at $\theta = \pi/2$. These general cambered airfoils with thickness and camber are called the Jukowski airfoil. Thus, in constructing the airfoil, the camber line is drawn as a function of $x$, to which the thickness is added as a sheared distribution.

## 2.2  Lift Created by Jukowski Airfoil

The flow past Jukowski airfoil can be analytically related to flow past a circular cylinder for inviscid, irrotational flow. We obtain the flow field around the airfoil and calculate the lift force experienced by it. Lift force is generated here by angle of attack and fixed camber.

The oncoming flow to this airfoil creates steady circulation originating from viscous action of no-slip but can be estimated using experimental observations and heuristics. Application of transformation technique rests on the fact that far away from the body, in both the $z$- and $\zeta$-planes, the flow field remains identical, as shown

**Fig. 2.5** Flow past a
Jukowski airfoil without
circulation. Also shown is
the corresponding circle
plane flow with some
representative streamlines.
Reproduced with permission
from Theoretical and
Computational
Aerodynamics. First Edition
Tapan K. Sengupta
Copyright (2015) John Wiley
& Sons. Ltd. UK

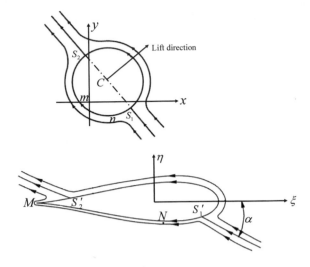

in Sect. 2.1.1, with respect to the Jukowski transform given in Eq. (2.7) applied to an airfoil. Few representative streamlines are obtained for the flow approaching the airfoil at an angle of attack and shown in Fig. 2.5.

The flow field shown in Fig. 2.5 is ideal, which does not create any circulation or lift on the airfoil, as shown with respect to the streamlines drawn in this figure. The flow field in both the planes can be inter-related and understood simultaneously. In the $z$-plane, we display a circle with center $C$, which is offset from the origin of $z$-plane in the horizontal and the vertical directions. The angle of attack ($\alpha$) of the flow in the airfoil plane is indicated in the figure, by the front and rear stagnation points being located in the circle plane at $S_1$ and $S_2$, at an angle $\alpha$ with respect to the $x$-axis. Corresponding stagnation points on the airfoil are marked at $S_1'$ and $S_2'$, respectively. Such a flow field will not support any circulation for the airfoil. However, for real flows, vorticity will be continually generated at the no-slip wall, which will be diffused and convected, governed by the Navier–Stokes equation. A steady state will be reached after an impulsive start and the steady state will account for the circulation in the airfoil plane.

Corresponding circulation in the circle plane will be as given by $W$, and one can show that $\oint_{C_z} (u - iv)_z \, dz = \oint_{C_\zeta} (u - iv)_\zeta \, d\zeta$, which shows that the singularity strength in one plane remains the same in the other plane. Flow past a circular cylinder with bound circulation has been discussed in Section 1.8.2, which showed that the presence of circulation will cause the stagnation points for flow past a stationary cylinder (as shown in Fig. 2.5 as $S_1$ and $S_2$) will be symmetrically depressed. For subcritical circulation, these two stagnation points will be distinct and depressed from the no-circulation case of Fig. 2.5 to that shown in Fig. 2.6 The pertinent quantity of circulation at the steady state can be estimated by invoking physical nature of the flow.

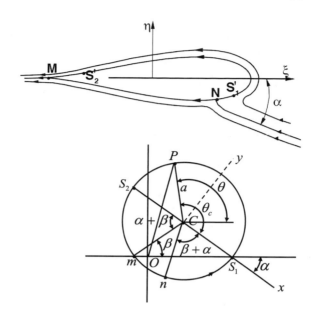

**Fig. 2.6** Flow past a Jukowski airfoil with circulation studied in the corresponding circle plane. Reproduced with permission from Theoretical and Computational Aerodynamics. First Edition Tapan K. Sengupta Copyright (2015) John Wiley & Sons. Ltd. UK

### 2.2.1  Kutta Condition and Circulation Generation

When the slope of the streamlines changes abruptly, the flow is strained enormously, as the rate of strain tensor develops singularity. This is the case for the flow field shown in Fig. 2.5, as the body streamline is forced to turn abruptly around the trailing edge. The curvature of streamlines is indicative of strain rates suffered by the fluid flow. One observes that such highly turning flows are possible in the steady state, if the energy supplied by the flow is sufficiently large. The energy dissipation rate is the product of stress and rates of strain and rapidly turning streamline imply very large strain rate. Thus, for finite energy fluid flows, abrupt turning of streamline is not feasible. This can be circumvented, if the rear stagnation point $S_2'$ is moved backward to the trailing edge, so that the flow will leave the trailing edge smoothly without suffering large strain rate. This physical consideration can be used to fix the circulation created by the flow field in negotiating the airfoil. We note that such a movement of rear stagnation point from $S_2'$ to $M$ in Fig. 2.6 would depress the rear stagnation point in the circle plane from $S_2$ to $m$, as indicated. We have already noted that the movement of rear stagnation to $m$ in the circle plane will also symmetrically depress the forward stagnation point from $S_1$ to $n$. This will also cause the forward stagnation point to move from $S_1'$ to $N$ in the airfoil plane, as a consequence of the movement of the rear stagnation point from $S_2'$ to $M$. The altered flow field via the movement of stagnation points will correspondingly fix the circulation experienced by the airfoil. This generated circulation ($\Gamma$) in a uniform fluid flow ($U_\infty$) will create a lift force given by the Kutta–Jukowski theorem by $L = \rho U_\infty \Gamma$, as described in Sect. 1.8.1. The direction of the lift force is perpendicular to $S_1 C S_2$ in the circle plane.

In the airfoil plane also, lift is perpendicular to the free-stream direction. Note that the free stream (measured by streamlines far away from the body) remains identically inclined as the complex velocities in the two planes are in the same direction, as $\frac{d\zeta}{dz} \to 1$, as $z, \zeta \to \infty$.

We have noted that the Jukowski transform creates an airfoil with a cusped trailing edge, i.e., the slope of the airfoil viewed from the top and bottom surfaces is zero. Thus, fixing of the rear stagnation point at the trailing edge causes the flow to emerge smoothly, which will cause the pressure to be also identical, i.e., the load at the trailing edge is zero. These observations of flow emerging smoothly from the trailing edge and zero loading at the trailing edge was proposed by Kutta to calculate the generated lift on the Jukowski airfoil and is known as the Kutta condition at the trailing edge. This heuristic argument may appear logical, but it should be borne in mind that this is essentially a model, whose strict validity remains in doubt, although the value of time-averaged lift may correlate well with experimental average value at small angles of attack. In an actual case, the flow past the airfoil need not be steady, especially when the flow is turbulent at higher Reynolds numbers. As a consequence in actual experiments, one can expect to notice that the rear stagnation points would move about the trailing edge. Many of our accepted theories are based on steady-state aerodynamics, and these are quite useful for high Reynolds number application for steady-state devices. For conventional aircraft at high Reynolds number, and cruise flight condition, such models yield realistic engineering results. For unsteady flows, or flows at low Reynolds numbers, as in case of micro air vehicle (MAV), one may require the solution of Navier–Stokes equation computationally.

### 2.2.2  Lift on Jukowski Airfoil

The analysis of flow past an airfoil and calculating the lift acting on it is performed with the help of Fig. 2.6, showing the various important points along the body streamline for the circulation fixed by Kutta condition.

We define an auxiliary angle ($\theta_c$) in Fig. 2.6 to locate an arbitrary point, P, such that $\theta_c = \theta + \alpha$. The point $P$ with coordinate ($r$, $\theta$) on the circle of radius $a$, i.e., $OP = r$, and $\theta$ is equal to $\angle POS_1$. Note that the line along the diameter $S_1 C S_2$ is set at the angle of attack of the airfoil ($\alpha$). The Jukowski transform relates the airfoil to the circle plane and has one of the singular points at $m$, so that the airfoil camber is defined by the angle $\beta$, which the line $Cm$ subtends with the $x$-axis. This angle is considered to be very small. From the geometry shown in Fig. 2.6, we note that $\angle S_2 Cm = \angle S_1 Cn = \alpha + \beta$. In the circle plane, the complex potential is given by

$$W = U_\infty \left( z + \frac{a^2}{z} \right) + \frac{i\Gamma}{2\pi} \operatorname{Ln} \frac{z}{a} \tag{2.31}$$

If we now define the circle by $z = a\, e^{i\theta_c}$, then the complex velocity on the circle is given by

$$\frac{dW}{dz} = U_\infty\left(1 - e^{-2i\theta_c}\right) + \frac{i\Gamma}{2\pi a}\, e^{-i\theta_c} = e^{-i\theta_c}\left[2iU_\infty \sin\theta_c + \frac{i\Gamma}{2\pi a}\right] \quad (2.32)$$

If the speed of the flow at $P$ is $q$, then

$$q^2 = \left(2U_\infty \sin\theta_c + \frac{\Gamma}{2\pi a}\right)^2$$

In this expression, the circulation $\Gamma$ is undetermined and is fixed analytically by the Kutta condition. Applying Kutta condition is equivalent to forcing the stagnation points to be located at $m$ and $n$, respectively, in Fig. 2.6. For the point $n$, $\theta_c = -(\alpha + \beta)$ and $q = 0$. Using this one fixes the circulation as

$$\Gamma = 4\pi U_\infty\, a\, \sin(\alpha + \beta) \quad (2.33)$$

Using this $\Gamma$, the surface speed distribution in the circle plane is fixed as

$$q = 2U_\infty[\sin\theta_c + \sin(\alpha + \beta)] \quad (2.34)$$

Also with the value of $\Gamma$ fixed, one can calculate the lift experienced by the Jukowski airfoil from Kutta–Jukowski theorem as

$$L = 4\pi\rho U_\infty^2 a\, \sin(\alpha + \beta) \quad (2.35)$$

As the chord of the airfoil is $4c$, where $a = c(1 + e)$, one obtains the lift coefficient from $L = \frac{1}{2}\rho U_\infty^2 c C_l$ as

$$C_l = 2\pi(1 + e)\, \sin(\alpha + \beta) \quad (2.36)$$

Neglecting $e$ in comparison, the maximum lift coefficient is given by $C_{l,max} = 2\pi$, when $\alpha + \beta = \pi/2$. It is a very unrealistic value, as the developed theory is with linearized assumption for which $\beta$ is small from the conditions set above and $\alpha$ should also be similarly restricted to negligibly small values to avoid flow separation to reduce lift to smaller value. Despite this, the lift curve slope is obtained, which is quite good and is a distinct contribution of this theory. The lift curve slope is given by

$$C_{l\alpha} = \frac{\partial C_l}{\partial\alpha} \simeq 2\pi \quad (2.37)$$

This is often quoted also as equivalent to producing a lift coefficient of 0.1 per degree, approximately. Having obtained the total lift generated, one is also interested in the velocity and pressure distribution on the Jukowski aerofoil, for detailed chordwise load distribution.

### 2.2.3   Velocity and Pressure Distribution on Jukowski Airfoil

Speed distribution in the circle plane is given by Eq. (2.34). From the Jukowski transformation

$$\frac{d\zeta}{dz} = 1 - \frac{c^2}{z^2} = 1 - \frac{c^2}{r^2}(\cos 2\theta - i \sin 2\theta)$$

$$\left|\frac{d\zeta}{dz}\right| = \left[1 - \frac{2c^2}{r^2}\cos 2\theta + \frac{c^4}{r^4}\right]^{1/2}$$

Therefore, the speed distribution in the airfoil plane is given by

$$v_a = 2U_\infty \frac{[\sin\theta_c + \sin(\alpha + \beta)]}{\left[1 - \frac{2c^2}{r^2}\cos 2\theta + \frac{c^4}{r^4}\right]^{1/2}} \tag{2.38}$$

For the point $P$ in Fig. 2.6,

$$y = r \sin\theta = a \sin(\theta_c - \alpha) + a \sin\beta$$

Therefore

$$\theta_c = \alpha + \sin^{-1}\left[\frac{r}{a}\sin\theta - \beta\right] \tag{2.39}$$

Also as

$$r = c + ce + ce \cos\theta + b\beta \sin\theta$$

One can obtain $r/c$ from this relation and use it in Eq. (2.38) to obtain the speed distribution on the surface of the airfoil.

The pressure coefficient over the airfoil is thereafter obtained from

$$C_p = 1 - \left(\frac{v_a}{U_\infty}\right)^2$$

## 2.3   Thin Airfoil Theory

So far we have obtained section characteristics of Jukowski airfoil by conformal mapping. For thin airfoils, a theory was propounded by Munk using regular perturbation theory, extending the above results. The essential element of the proposed theory assumes the size and shape of the airfoil to be thin, represented by its camber line, which in turn is slightly different from the chord line. In this theory, line vortices of infinitesimally variable strengths ($\bar{\gamma}$) is distributed along the camber line, so that the total circulation about the chord is the sum of the vortex elements' strength. Thus,

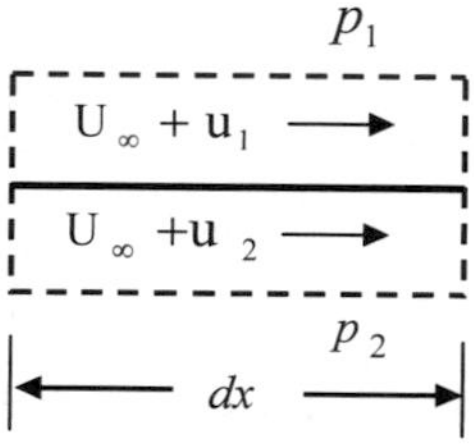

**Fig. 2.7** An element of vorticity distributed along the camber line and the flow variables surrounding the element. Reproduced with permission from Theoretical and Computational Aerodynamics. First Edition Tapan K. Sengupta Copyright (2015) John Wiley & Sons. Ltd. UK

if we measure the length along the camber line by $s$, then the circulation carried by the airfoil is given by

$$\Gamma = \int \bar{\gamma} \, ds \tag{2.40}$$

As we assume the chord and the camber line to be same, Eq. (2.40) is rewritten as

$$\Gamma = \int_0^c \bar{\gamma} \, dx \tag{2.41}$$

and using Kutta–Jukowski theorem, the lift experienced by the airfoil is

$$L = \rho \, U_\infty \int_0^c \bar{\gamma} \, dx$$

The pitching moment experienced by the airfoil about its leading edge is obtained as

$$M_{\mathrm{LE}} = -\rho \, U_\infty \int_0^c \bar{\gamma} \, x \, dx \tag{2.42}$$

with nose-up pitching moment considered as positive. As one estimates lift and pitching moment by considering inviscid, irrotational flow, the vortex element strength can be obtained by applying Bernoulli's equation on an infinitesimal element of length $dx$ shown in Fig. 2.7. The camber line represents a line of discontinuity with pressure above and below indicated by $p_1$ and $p_2$ and local perturbation velocity indicated by $u_1$ and $u_2$, respectively.

The streamlines below and above the camber line have identical reservoir condition and one applies Bernoulli's equation above and below the elementary vortex ($\bar{\gamma} \, dx$) to obtain

$$p_2 - p_1 = \frac{1}{2} \rho \, U_\infty^2 \left[ 2 \left( \frac{u_1}{U_\infty} - \frac{u_2}{U_\infty} \right) + \left( \frac{u_1}{U_\infty} \right)^2 - \left( \frac{u_2}{U_\infty} \right)^2 \right]$$

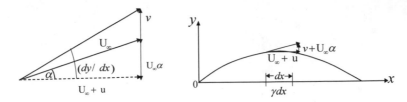

**Fig. 2.8** A thin airfoil shown represented by the camber line placed at an angle of attack, $\alpha$ and the corresponding velocity diagram for a point at a location, $x = x_1$. Reproduced with permission from Theoretical and Computational Aerodynamics. First Edition Tapan K. Sengupta Copyright (2015) John Wiley & Sons. Ltd. UK

For the general thin airfoil, the perturbation velocities are small and one can linearize to obtain the loading as

$$\Delta p = p_2 - p_1 \simeq \rho\, U_\infty(u_1 - u_2) \tag{2.43}$$

Hence

$$C_p = \frac{2}{U_\infty}(u_1 - u_2)$$

Performing a contour integral over the dotted contour in Fig. 2.7, one obtains the circulation on the element as

$$\bar{\gamma}\, dx = (U_\infty + u_1)dx - (U_\infty + u_2)dx$$

Thus, the loading distribution over the camber line is obtained easily. Next, the shape of the camber line is related to the loading distribution from the boundary condition applied for the airfoil placed at an angle of attack $\alpha$ in Fig. 2.8.

From Fig. 2.8, the camber line must be a streamline. Thus, the slope of the camber line must match the flow direction given by

$$\left(U_\infty + u\right)\frac{dy}{dx} = v + U_\infty\, \alpha$$

The term $u\, \frac{dy}{dx}$ is of lower order and can be neglected upon linearization to obtain

$$v \cong U_\infty\left(\frac{dy}{dx} - \alpha\right) \tag{2.44}$$

Now, the left-hand side of Eq. (2.44) can be obtained as the induced velocity at the point $x_1$, created by a line vortex element of strength $\bar{\gamma}\, dx$ by

$$v(x_1) = \frac{1}{2\pi}\int_0^c \frac{\bar{\gamma}\, dx}{x - x_1} \tag{2.45}$$

Equating Eqs. (2.44) and (2.45) on the camber line, one gets

$$U_\infty \left( \frac{dy}{dx} - \alpha \right)_{x=x_1} = \frac{1}{2\pi} \int \frac{\bar{\gamma}\, dx}{x - x_1} \tag{2.46}$$

Equation (2.46) is the key to thin airfoil theory, whereby the airfoil camber shape denoted by $\left( \frac{dy}{dx} \right)$ is related to the vorticity distribution. This is an integral equation, which forms the basis for studying various types of airfoil, as described next.

## 2.3.1  Thin Symmetric Flat Plate Airfoil

In this case, the camber line is along the $x$-axis, i.e., $\frac{dy}{dx} = 0$, and this Eq. (2.46) yields

$$U_\infty \alpha = \frac{1}{2\pi} \int_0^c \frac{\bar{\gamma}\, dx}{x_1 - x}$$

We simplify this by a change of variable

$$x = \frac{c}{2}(1 - \cos\theta)$$

so that
$$dx = \frac{c}{2} \sin\theta\, d\theta$$

Using these in the above equation, one gets

$$U_\infty \alpha = \frac{1}{2\pi} \int_0^\pi \frac{\bar{\gamma} \sin\theta\, d\theta}{(\cos\theta - \cos\theta_1)} \tag{2.47}$$

This is called the Glauert's integral. We demonstrate the solution of the integral equation by using the results of the conformal mapping introduced.

**Direct method of finding the loading distribution, $\bar{\gamma}$:**
With reference to Fig. 2.9, one obtains the velocity in the circle plane from Eq. (2.32). Note that $\theta_c$ is measured with respect to the diameter $S_1 S_2$ as $\theta_c = (\theta + \alpha)$. Thus,

$$v_c = 2U_\infty \sin(\theta + \alpha) + \frac{\Gamma}{2\pi a},$$

where $\Gamma$ is the total circulation in the airfoil plane lumped at the center of the circle, fixed by the Kutta condition of fixing the stagnation point at $m$, for which $\theta = \pi$.
Thus, $v_c = 0$ at $m$ produces

$$\Gamma = 4\pi a\, U_\infty \sin\alpha$$

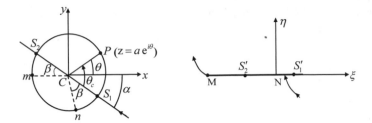

**Fig. 2.9** The circle plane and its map in the flat plate plane, indicating the front and rear stagnation points. Reproduced with permission from Theoretical and Computational Aerodynamics. First Edition Tapan K. Sengupta Copyright (2015) John Wiley & Sons. Ltd. UK

Therefore, the velocity in the circle plane is given by

$$v_c = 2\,U_\infty\,[\sin(\theta + \alpha) + \sin\alpha]$$

In the airfoil plane, the velocity is given by

$$v_a = \frac{v_c}{\left|\frac{d\zeta}{dz}\right|},$$

where the denominator evaluated on the surface of the cylinder is given by

$$\left|\frac{d\zeta}{dz}\right| = \left|1 - \frac{a^2}{z^2}\right| = \left|1 - e^{-i\,2\theta}\right| = 2\sin\theta$$

Thus, the velocity distribution in the airfoil plane simplifies to

$$\frac{v_a}{U_\infty} = \cos\alpha + \frac{\sin\alpha}{\sin\theta}(1 + \cos\theta) \tag{2.48}$$

For small angle of attack, $\cos\alpha \simeq 1$ and $\sin\alpha \simeq \alpha$, so that one obtains the non-dimensional velocity in the airfoil plane as

$$\frac{v_a}{U_\infty} = 1 + \alpha \cot\theta/2 \tag{2.49}$$

Having obtained the velocity distribution, one can obtain the loading for the flow from $\bar{\gamma} = v_{a_1} - v_{a_2}$, where the velocities are as shown in Fig. 2.7, for which $\theta_1 = -\theta_2 = \theta$ Thus,

$$\bar{\gamma} = 2\,U_\infty\,\alpha\,\frac{1 + \cos\theta}{\sin\theta} \tag{2.50}$$

One can integrate the loading distribution to obtain the lift acting on the section as

$$\text{Lift} = \rho \, U_\infty \int_0^\pi 2 \, U_\infty \alpha \, \frac{1 + \cos\theta}{\sin\theta} \, \frac{c}{2} \sin\theta \, d\theta$$

Thus, the lift acting on the flat plate is given by

$$\text{Lift} = \rho \, U_\infty^2 c \, \alpha \int_0^\pi (1 + \cos\theta) \, d\theta = \pi \alpha \rho \, U_\infty^2 c \qquad (2.51)$$

And the lift coefficient is given by

$$C_l = 2\pi \alpha \qquad (2.52)$$

Using the loading distribution given by Eq. (2.50) in Eq. (2.42), one gets the pitching moment about the leading edge as

$$M_{\text{LE}} = -\rho \, U_\infty \int 2 \, U_\infty \, \alpha \, \frac{1 + \cos\theta}{\sin\theta} \, \frac{c}{2} (1 - \cos\theta) \, \frac{c}{2} \sin\theta \, d\theta$$

$$= -\frac{1}{2}\rho \, U_\infty^2 \, \alpha \, c^2 \int_0^\pi (1 - \cos^2\theta) \, d\theta = C_{M_{\text{LE}}} \frac{1}{2}\rho \, U_\infty^2 c^2$$

Thus, the pitching moment coefficient about the leading edge can be written as

$$C_{M_{\text{LE}}} = -\frac{\pi}{2}\alpha = -\frac{C_l}{4} \qquad (2.53)$$

Before proceeding further, we will discuss two important aerodynamic parameters, which help us understand the section properties better and relate aerodynamic results with flight dynamics.

## 2.3.2  Aerodynamic Center and Center of Pressure

Consider an airfoil at an angle of attack $\alpha$, as shown in Fig. 2.10, with the sectional force components and pitching moment referred to a datum point, which is at a distance $p$ from the leading edge of the airfoil. Material in this subsection applies to any general case, for which the force and moment coefficients are considered obtained by theoretical, computational, or experimental means. The lift and drag force components are indicated as $L$ and $D$, respectively. The pitching moment about this datum point is indicated as $M_p$, with clockwise moment (nose-up moment) treated as positive.

The pitching moment about the leading edge ($M_{\text{LE}}$) can be written in terms of $M_p$ and the pitching moment about any other arbitrary point located at a distance $x$ from the leading edge by

**Fig. 2.10**  Representation of
section properties of lift,
drag, and pitching moment
referred to different points

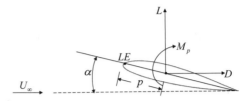

$$M_{LE} = M_p - L\, p\, \cos\alpha - D\, p\, \sin\alpha = M_x - L\, x\, \cos\alpha - D\, x\, \sin\alpha \quad (2.54)$$

Therefore,

$$M_x = M_p - (L\cos\alpha + D\sin\alpha)(p - x)$$

Defining the pitching moment coefficient as

$$C_M = \frac{M}{\frac{1}{2}\rho U_\infty^2 c^2}$$

one can write

$$C_{M_x} = C_{M_p} - (C_l \cos\alpha + C_d \sin\alpha)\left(\frac{p}{c} - \frac{x}{c}\right)$$

If $p$ is at the leading edge

$$C_{M_x} = C_{M_{LE}} + (C_l \cos\alpha + C_d \sin\alpha)\frac{x}{c} \quad (2.55)$$

**Aerodynamic center**:

Along the chord, there is one point for which $C_{M_x}$ is independent of $C_l$ and this point
is called the aerodynamic center. Let $x_{ac}$ be the position of the aerodynamic center
behind the leading edge, then

$$C_{M_p} = C_{M_{ac}} - (C_l \cos\alpha + C_d \sin\alpha)\left(\frac{x_{ac}}{c} - \frac{p}{c}\right)$$

For small angles of attack, $\cos\alpha \simeq 1$ and $\sin\alpha \simeq \alpha$ and for all streamlined body,
$C_l \gg C_d$. Therefore

$$C_{M_p} = C_{M_{ac}} - C_l\left(\frac{x_{ac}}{c} - \frac{p}{c}\right)$$

It is readily apparent that for $C_l = 0 : C_{M_p} = C_{M_{ac}}$ and for this reason, one often
indicates $C_{M_{ac}} = C_{M_0}$. According to the definition of aerodynamic center $C_{M_p}$ is
independent of $C_l$, i.e.,

$$\frac{dC_{M_p}}{dC_l} = \frac{dC_{M_{ac}}}{dC_l} - \left(\frac{x_{ac}}{c} - \frac{p}{c}\right)$$

Therefore the aerodynamic center can be located from the above equation as

$$\frac{x_{ac}}{c} = \frac{p}{c} - \frac{dC_{M_p}}{dC_l} \tag{2.56}$$

Thus, the aerodynamic center can be found by measuring $C_l$ and $C_{M_p}$ about the point $p$ and use Eq. (2.56) to obtain the aerodynamics center. For many high Reynolds number applications, if the angle of attack of an airfoil is low enough to avoid flow separation, then the aerodynamic center is seen experimentally to be located in the range $0.23 \leq \frac{x_{ac}}{c} \leq 0.25$. Thus, it is customary for incompressible flow past airfoil to take the aerodynamic center at the quarter-chord point. This is specifically true for flat or curved plate of negligible thickness. It is noted that thickness and viscous effects push the aerodynamic center forward. It is noted as reference that for supersonic flow past thin airfoils the aerodynamic center is at the mid-chord.

For the case of thin symmetric flat plate airfoil, from Eq. (2.53) one notes that the aerodynamic center can be obtained from Eq. (2.56) by using $p = 0$ as

$$\frac{x_{ac}}{c} = -\frac{dC_{M_{LE}}}{dC_l} = \frac{1}{4} \tag{2.57}$$

Thus, the aerodynamic center is located at the quarter-chord point for this case.

**Center of pressure**:

For every $C_l$, there is a point along the chord line about which the pitching moment is zero. This is referred to as the center of pressure. It is important to note that it is not necessary for the center of pressure to lie inside the airfoil.

From this definition of center of pressure and from Eq. (2.54), if we indicate the location of the center of pressure to be given by $x_{cp}$, then

$$M_{LE} = M_{ac} - (L \cos\alpha + D \sin\alpha) x_{ac} = -(L \cos\alpha + D \sin\alpha) x_{cp} \tag{2.58}$$

Therefore, from this equation,

$$\frac{x_{cp}}{c} = \frac{x_{ac}}{c} - \frac{C_{M_{ac}}}{C_l \cos\alpha + C_d \sin\alpha} \tag{2.59}$$

The thin airfoil theory is for small perturbation, and thus, we consider only small angle of attack, for which $\cos\alpha = 1$; $\sin\alpha = \alpha$ and $C_d \sin\alpha$ is negligibly small, so that the denominator in Eq. (2.59) can be approximated by $C_l$ only. Based on these observations and Eq. (2.59), one can also locate the center of pressure from

$$\frac{x_{cp}}{c} = \frac{x_{ac}}{c} - \frac{C_{M_{ac}}}{C_l} = -\frac{C_{M_{LE}}}{C_l} \tag{2.60}$$

For traditional airfoils, $C_{M_{ac}} < 0$ and so $x_{cp} > x_{ac}$, i.e., the center of pressure is behind the aerodynamic center. Specifically, one notes that for zero lift condition the center of pressure goes to $-\infty$.

For the symmetric flat plate using the results of Eq. (2.53) in Eq. (2.60), one notes

$$\frac{x_{cp}}{c} = -\frac{C_{M_{LE}}}{C_l} = \frac{1}{4} \tag{2.61}$$

Hence, the aerodynamic center and center of pressure are at identical location of quarter-chord for the symmetric flat plate airfoil.

### 2.3.3 The Circular Arc Airfoil

In Sect. 2.1.2, a circular arc is mapped to a circular cylinder by Jukowski transform, and these two planes have been shown in Fig. 2.3. The Jukowski transformation has been applied on the circle, with its center offset along the $y$-axis. Consider the transformation of a circle of radius $a$ with center at the point $y = h = a \sin \beta$ in the circle plane to the airfoil plane by the transformation $\zeta = z + \frac{c^2}{z}$, where $c = a \cos \beta \simeq a$, with the circle transforming to a circular arc of camber $\frac{1}{2} \tan \beta$.

The flow field is analyzed by looking at the circle plane shown in Fig. 2.11. Velocity on the circle plane is given by

$$v_c = 2\, U_\infty \left[ \sin \theta_c + \sin(\alpha + \beta) \right] \tag{2.62}$$

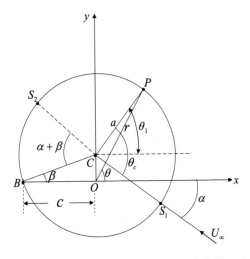

**Fig. 2.11** Flow in the circle plane for flow past a circular arc airfoil. Note the rear stagnation point is now located at $B$. Also, note that the front stagnation point is not as shown at $S_1$, which will also be correspondingly depressed. Reproduced with permission from Theoretical and Computational Aerodynamics. First Edition Tapan K. Sengupta Copyright (2015) John Wiley & Sons. Ltd. UK

where the circulation is obtained by applying the Kutta condition at $m$ in Eq. (2.32), i.e., at $\theta_c = \pi + (\alpha + \beta)$.

From Fig. 2.11, $\theta_c = \theta_1 + \alpha$. Thus using, $\sin \theta_c = \sin \theta_1 + \alpha \cos \theta_1$, in Eq. (2.62) one obtains

$$v_c = 2\, U_\infty (\sin \theta_1 + \alpha \cos \theta_1 + \alpha + \beta)$$

The point $P$ in Fig. 2.11 has the ordinate, $y = a \sin \theta_1 + c\beta = r \sin \theta$. Thus $\sin \theta_1 = \frac{r}{a} \sin \theta - \frac{c}{a} \beta \simeq \frac{r}{c} \sin \theta - \beta$ and abscissa, $x = a \cos \theta_1 = r \cos \theta$. Therefore, $\cos \theta_1 = \frac{r}{a} \cos \theta = \frac{r}{c} \cos \theta$.

From Fig. 2.11,

$$a^2 = a^2 \beta^2 + r^2 - 2a\beta r \sin \theta \tag{2.63}$$

$$a^2 = a^2 \beta^2 + c^2 \tag{2.64}$$

From Eqs. (2.63) and (2.64), one gets

$$c^2 = r^2 - 2ra\beta \sin \theta.$$

This equation for $r$ yields

$$r = a\beta \sin \theta \pm \sqrt{c^2 + a^2 \beta^2 \sin^2 \theta}$$

As $r$ is always positive, we consider

$$r = a\beta \sin \theta + c\sqrt{1 + \left(\frac{a\beta}{c}\right)^2 \sin^2 \theta}$$

$$r \simeq a\beta \sin \theta + c \tag{2.65}$$

Thus, $\quad \dfrac{r}{c} \simeq 1 + \dfrac{a}{c} \beta \, \sin \theta \simeq 1 + \beta \sin \theta$

This results in the circle plane velocity as

$$v_c = 2\, U_\infty \left[\sin \theta \, (1 + \beta \sin \theta) + \alpha(1 + \cos \theta)\right] \tag{2.66}$$

For the transformation to a circular arc airfoil, we have

$$\left|\frac{d\zeta}{dz}\right| = \left|1 - \frac{c^2}{z^2}\right| = \left|1 - \frac{c^2}{r^2} e^{-i2\theta}\right|$$

$$= \left[1 - \frac{2 \cos 2\theta}{(1 + \beta \sin \theta)^2} + \frac{1}{(1 + \beta \sin \theta)^4}\right]^{1/2}$$

As $\beta$ is small,

$$\left|\frac{d\zeta}{dz}\right| = \left[(1 + \beta \sin \theta)^2 - 2 \cos 2\theta + (1 + \beta \sin \theta)^{-2}\right]^{1/2} (1 + \beta \sin \theta)^{-1}$$

Using further simplifications, $(1 + \beta \sin \theta)^2 \simeq 1 + 2\beta \sin \theta$ and $(1 + \beta \sin \theta)^{-2} \simeq 1 - 2\beta \sin \theta$, one obtains the following:

$$\left|\frac{d\zeta}{dz}\right| = [2 - 2 \cos 2\theta]^{1/2}(1 + \beta \sin \theta)^{-1} = \frac{2 \sin \theta}{1 + \beta \sin \theta}$$

One finally obtains the speed on the surface of the airfoil as

$$v_a = \frac{v_c}{\left|\frac{d\zeta}{dz}\right|} = \frac{U_\infty(1 + \beta \sin \theta)}{\sin \theta} \left[\sin \theta(1 + \beta \sin \theta) + \alpha(1 + \cos \theta)\right]$$

$$\simeq U_\infty \left[(1 + 2\beta \sin \theta) + \alpha \frac{1 + \cos \theta}{\sin \theta}\right]$$

From the speed distribution on the surface of the airfoil, one obtains the circulation from $\bar{\gamma} = v_{a_1} - v_{a_2}$, where for the points on top and bottom of the camber line one has $\theta_1 = -\theta_2$, and hence

$$\bar{\gamma} = 2\, U_\infty \left[2\beta \sin \theta_1 + \alpha \frac{1 + \cos \theta_1}{\sin \theta_1}\right] \tag{2.67}$$

Comparing with the flat plate result, the camber increases $\bar{\gamma}$ by $4\, U_\infty\, \beta \sin \theta$. For the circular arc airfoil, using Eq. (2.67) one obtains the lift generated as

$$\text{Lift} = 2\rho\, U_\infty^2 \int_0^\pi \left[2\beta \sin \theta + \alpha \frac{1 + \cos \theta}{\sin \theta}\right] \frac{c}{2} \sin \theta \, d\theta$$

$$= \pi\rho\, U_\infty^2 c\, (\alpha + \beta)$$

Therefore, the lift coefficient is obtained as

$$C_l = 2\pi\, (\alpha + \beta) \tag{2.68}$$

and the lift curve slope is $2\pi$.

As before, one calculates the pitching moment about the leading edge as

$$M_{\text{LE}} = -2\rho\, U_\infty^2 \int_0^\pi \left[\alpha \frac{1 + \cos \theta}{\sin \theta} + 2\beta \sin \theta\right] \frac{c}{2}(1 - \cos \theta)\frac{c}{2} \sin \theta \, d\theta$$

Therefore,

$$C_{M_{LE}} = -\frac{\pi}{2}(\alpha + 2\beta) \tag{2.69}$$

and one obtains the center of pressure using

$$\frac{x_{cp}}{c} = -\frac{C_{M_{LE}}}{C_l} = \frac{1}{4} + \frac{\pi\beta}{2C_l} \tag{2.70}$$

using the lift coefficient value in Eqs. (2.69) and (2.70), one notices that

$$C_{M_{LE}} = -\frac{\pi}{2}(\alpha + 2\beta) = -\frac{C_l}{4} - \frac{\pi\beta}{2}$$

Thus,

$$\frac{dC_{M_{LE}}}{dC_l} = 1/4,$$

which also shows that the aerodynamic center is at the quarter-chord. At zero lift, the center of pressure is at an infinite distance behind the airfoil, which means that there will be a moment without lift.

In the limit of zero thickness, we noted that the load distribution for a circular arc airfoil is given by Eq. (2.67). The load distribution depends on camber and angle of attack. Thus, for any general airfoil with finite thickness, one superposes circulation distribution for the flat plate airfoil (indicated here by $\gamma_a$) and that due to the circular arc airfoil (indicated here by $\gamma_b$) in the linear theory to obtain

$$\bar{\gamma} = \gamma_a + \gamma_b$$

This can be further generalized for any general airfoil under linear theory assumption in the following.

## 2.4 General Thin Airfoil Theory

In this theory, $\gamma_a$ represents the load due to angle of given by

$$\gamma_a = 2\,U_\infty\,A_0\left(\frac{1 + \cos\theta}{\sin\theta}\right)$$

with $A_0$ as a constant to absorb differences between the 'equivalent' flat plate airfoil and the actual chord line, with the former distribution given by Eq. (2.50), as created by the angle of attack. The component $\gamma_b$ is due to camber effect, as noted in Eq. (2.67), and can be generalized by a Fourier series as

$$\gamma_b = 2\, U_\infty \sum_1^\infty A_n \sin n\theta$$

Therefore, the general load distribution is given by

$$\bar{\gamma} = 2\, U_\infty \left\{ A_0 \frac{1 + \cos\theta}{\sin\theta} + \sum_1^\infty A_n \sin n\theta \right\} \qquad (2.71)$$

The coefficients can be found by substituting Eq. (2.71) in Eq. (2.47) as

$$U_\infty \left[ \frac{dy}{dx} - \alpha \right] = \frac{1}{2\pi} \int_0^c \frac{\bar{\gamma}\, dx}{x - x_1}$$

As before with change of variable, $x = \frac{c}{2}(1 - \cos\theta)$ and $dx = \frac{c}{2} \sin\theta\, d\theta$ in the above to get the right-hand side as

$$= -\frac{1}{2\pi} \int_0^\pi \frac{\bar{\gamma} \sin\theta\, d\theta}{\cos\theta - \cos\theta_1}$$

Using Eq. (2.71) in this expression yields

$$= -\frac{U_\infty}{\pi} \int_0^\pi \left\{ \frac{A_0(1 + \cos\theta)}{\sin\theta} + \sum_1^\infty A_n \sin n\theta \right\} \frac{\sin\theta\, d\theta}{\cos\theta - \cos\theta_1}$$

Therefore,

$$\frac{dy}{dx} - \alpha = -\frac{A_0}{\pi} \int_0^\pi \frac{1 + \cos\theta}{\cos\theta - \cos\theta_1}\, d\theta - \frac{1}{\pi} \int_0^\pi \sum \frac{A_n \sin n\theta \sin\theta}{\cos\theta - \cos\theta_1}\, d\theta \qquad (2.72)$$

Using the identity

$$A_n \sin n\theta \sin\theta = \frac{A_n}{2}\, [\cos(n - 1)\theta - \cos(n + 1)\theta]$$

Equation (2.72) is represented symbolically as

$$\frac{dy}{dx} - \alpha = -\left\{ \frac{A_0}{\pi} G_0 + \frac{A_0}{\pi} G_1 + \sum \frac{A_n}{2\pi} G_{n-1} - \sum \frac{A_n}{2\pi} G_{n+1} \right\}$$

where $G_n$ is the Glauert integral given by

$$G_n = \int_0^\pi \frac{\cos n\theta}{\cos\theta - \cos\theta_1}\, d\theta = \frac{\pi \sin n\theta_1}{\sin\theta_1}$$

Proof of the above can be found in [16]. Using this in Eq. (2.72), one gets

$$\frac{dy}{dx} = -A_0 + \alpha + \sum A_n \cos n\theta \tag{2.73}$$

This integrated over the full range of $\theta$ yields

$$A_0 = \alpha - \frac{1}{\pi} \int_0^\pi \frac{dy}{dx} \, d\theta \tag{2.74}$$

Similarly by multiplying by $\cos m\theta$, where m is an integer, and integrating with respect to $\theta$, one gets

$$A_m = \frac{2}{\pi} \int_0^\pi \frac{dy}{dx} \cos m\theta \, d\theta \tag{2.75}$$

From the $\bar{\gamma}$-distribution given in Eq. (2.71), one can obtain the lift in terms of the coefficients as

$$\text{Lift} = \int_0^c \rho \, U_\infty \, \bar{\gamma} \, dx = 2\rho \, U_\infty^2 \, \frac{c}{2} \int_0^\pi \left[ A_0(1 + \cos \theta) + \sum_1^\infty A_n \sin n\theta \sin \theta \right] d\theta$$

$$= 2 \, \rho \, U_\infty^2 \, \frac{c}{2} \left[ \pi \, A_0 + \frac{\pi}{2} \, A_1 \right]$$

The lift coefficient is given by

$$C_l = 2\pi \left[ A_0 + \frac{A_1}{2} \right] \tag{2.76}$$

Thus, the lift coefficient is solely determined by $A_0$ and $A_1$.

**Zero lift angle**:

From Eqs. (2.74) and (2.75), one obtains

$$A_0 = \alpha - \frac{1}{\pi} \int \frac{dy}{dx} \, d\theta$$

and

$$A_1 = \frac{2}{\pi} \int_0^\pi \frac{dy}{dx} \cos \theta \, d\theta$$

Substituting these in Eq. (2.76) one gets

$$C_l = \frac{dC_l}{d\alpha}\left[\alpha - \frac{1}{\pi}\int \frac{dy}{dx}\,d\theta + \frac{1}{\pi}\int \frac{dy}{dx}\cos\theta\,d\theta\right]$$

where we have used $\frac{dC_l}{d\alpha} = 2\pi$ and therefore

$$C_l = \frac{dC_l}{d\alpha}(\alpha - \alpha_0),$$

with $\alpha_0$ as the zero lift angle given by

$$\alpha_0 = \frac{1}{\pi}\int \frac{dy}{dx}(1 - \cos\theta)\,d\theta \qquad (2.77)$$

Similarly from the load distribution given in Eq. (2.71), one obtains the pitching moment about the leading edge as

$$-M_{\mathrm{LE}} = -C_{M_{LE}}\frac{1}{2}\rho U_\infty^2\,c^2 = \rho\,U_\infty\int_0^c \bar{\gamma}\,x\,dx$$

Thus

$$C_{M_{LE}} = -\int_0^\pi \left[A_0\frac{1+\cos\theta}{\sin\theta} + \sum_1^\infty A_n\sin n\theta\right]\sin\theta(1 - \cos\theta)d\theta$$

Using trigonometric identities, one simplifies this to [16]

$$C_{M_{LE}} = -\frac{\pi}{2}\,A_0 - \frac{\pi}{2}\,A_1 + \frac{\pi}{4}\,A_2 \qquad (2.78)$$

Using Eq. (2.76) in Eq. (2.78), one can write

$$C_{M_{LE}} = -\frac{C_l}{4}\left[1 + \frac{A_1 - A_2}{C_l}\pi\right] \qquad (2.79)$$

Thus, the center of pressure is obtained from

$$\frac{x_{cp}}{c} = -\frac{C_{M_{LE}}}{C_l} = \frac{1}{4} + \frac{\pi}{4C_l}(A_1 - A_2) \qquad (2.80)$$

From Eq. (2.55),

$$C_{M_x} = C_{M_{LE}} + \frac{x}{c}(C_l\cos\alpha + C_d\sin\alpha)$$

For small angles of attack, $C_d \ll C_l$, and substituting $x = c/4$, one gets

$$C_{M_{c/4}} = C_{M_{LE}} + \frac{C_l}{4}$$

$$= -\frac{\pi}{2}A_0 - \frac{\pi}{2}A_1 + \frac{\pi}{4}A_2 + \frac{\pi}{2}(A_0 + A_{1/2})$$

$$= \frac{\pi}{4}(A_2 - A_1)$$

For symmetrical airfoil, it is easy to establish that all $A_m$'s are zero. Hence, $C_{M_{c/4}} = 0$.

One notes the location of the aerodynamic center from Eq. (2.79) as

$$\frac{x_{ac}}{c} = -\frac{dC_{M_{LE}}}{dC_l} = \frac{1}{4} \qquad (2.81)$$

## 2.5  Finite Wing Theory

This theory developed in [9, 14] is based on the idea to replace a finite wing by a system of vortices of particular types, which create a flow field similar to the actual one for the lift sustained by the wing. The vortex system representing the wing can be divided into (i) a bound vortex system for the sectional lift property; (ii) a starting vortex system necessary to satisfy a system of theorems for vortical field; and (iii) the trailing vortex to complement the other two into a single description of the load on a finite wing. The role of each element needed to develop the basics of elementary vortex motion is as following the laws given next.

### 2.5.1  Fundamental Laws of Vortex Motion

Vortex in a flow represents a singularity for ideal irrotational fluid. In real flows, this occupies a finite area and can be viewed as a distribution of vorticity. Localized vortex is an idealization, as there are flows which have distributed vorticity, yet do not have concentrated vortices. For example, in a boundary layer, one has primarily vorticity created to satisfy no-slip condition and does not necessarily display lumped vortex. On the other hand, wakes and free shear layer have regions of large concentrated vorticity distribution, which can be identified as vortices. Whether we have vortex or vorticity distribution, we can describe both by the same variable termed as the circulation. In general, the vortex is a curve in space and has a finite area normal to its axis given by, $S$. It may be convenient to consider the finite area vortex to be made up of several infinitesimal elements, as in theory of distribution, with the line vortex viewed as the Dirac delta function. Thus, the vortex in real flow may be thought to consist of a bundle of elemental line vortex or filaments, with the bundle often called a vortex tube. For inviscid flows, properties of vortex in motion are governed by the following theorems due to Helmholtz and Kelvin.

## 2.5.2  Kelvin–Helmholtz's Theorems of Vortex Motion

These are fundamental laws to help us enunciate properties of vortices for flows.
**Theorem 1**: It refers to a fluid particle in general motion possessing the kinematic
properties of linear velocity, vorticity, and distortion.
**Theorem 2**: It demonstrates constancy of strength of vortex along its length, which
at times is referred to as the equation of vortex continuity. It is demanded that the
strength of a vortex cannot grow or diminish along its axis or length.

We define circulation as the contour integral of the velocity vector, as $\Gamma = \oint \overrightarrow{V} \cdot$
$d\overrightarrow{r}$. If the velocity is expressed in terms of velocity potential as $\overrightarrow{V} = \nabla\phi$, then
one can represent the circulation as $\Gamma = \oint \nabla\phi \cdot d\overrightarrow{r} = \oint d\phi$. Thus, for a flow with
circulation, $\phi$ is not single-valued and is not defined.

**Corollary of Theorem 2**:

The strength of a vortex is the magnitude of the circulation around it, and this is also
equal to $\Gamma = \iint_S \omega \cdot dS$, with $\omega$ as the vorticity.

Since infinite vorticity is unacceptable, so $S \rightarrow 0$ is not a possibility. A vortex
therefore cannot abruptly end in a flow. In practice, any vortex must form a closed
loop and originate (or terminate) in a discontinuity in the fluid such as a solid body
or a surface of separation.

**Corollary of constancy of circulation**:

Since $\Gamma$ remains constant in an inviscid flow, then upon an impulsive start, the
circulation generated by a lifting body (known as the bound vortex) must be equally
balanced by a starting vortex leaving the body, so that the net circulation is zero.
This is shown in Fig. 2.12. As, at $t = 0^-$, there was no motion, and hence there was
no circulation, it must remain so thereafter, according to Theorem 2 above. This is
the foundation of finite wing theory, with no provision for creation of vorticity due
to real flow behavior.

**Theorems 3 and 4**:

These two theorems demonstrate, respectively, that a vortex tube consists of the same
particles of fluid, i.e., that there is no fluid interchange between the vortex tube and

**Fig. 2.12** The idealized
necessary vortex system for
an infinite aspect ratio wing,
which requires only a bound
and the starting vortex. Both
the vortices are of the same
strength and opposite in sign
and extend to infinity in the
spanwise direction

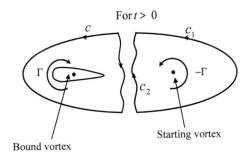

surrounding fluid and that the strength of a vortex remains the same as the vortex moves through fluid confines.

Having defined the fundamental vortex laws, we must also state the limitations of these theorems. First, in these theorems, there is no provision for the generation of circulation. For example, in Fig. 2.12, the appearance of $\Gamma_b$ is considered to be impulsive without any attribution to physical mechanism(s). In an actual flow, even when the flow is started impulsively, the physical mechanism is attributed to satisfaction of no-slip condition at the solid wall. One must, however, admit that there are no proofs for no-slip condition itself. Then, there is the central question of why should there be the requirement for the constancy of circulation, when the solid wall is the constant source of vorticity generation.

The more problematic aspects of Theorems 3 and 4 arise, where it is demanded that the vortex strength remains the same, even when the vortex tube turns, etc. After the following discussions on vortex elements, we will revisit this question again. The following discussion also pertains to steady flow consideration, even though it may have impulsive start as its origin. To facilitate our deliberations, let us fix our coordinate system, in which the $x$-axis is in the streamwise direction, the $y$-axis is in wall-normal direction, and the $z$-axis is in the spanwise direction over the wing.

## 2.6   The Bound Vortex Element

This is the circulation developing over a finite wing, responsible for lift on the wing. The corresponding vortex element is in the spanwise ($\omega_z$) direction. The finite aspect ratio of the wing means that the lift must decay to zero at the wing tip, as the absence of physical surface would necessitate that. Thus, there is a grading of spanwise loading, and the bound vortex system is to be appropriately constructed for this. This is an arrangement of vortices to represent the wing to support the lift distribution in every detail, except the thickness distribution. The equivalent bound vortex element explains sectional loading along the span, to reproduce the lift experienced by the wing.

For an aircraft in steady level flight, the spanwise load distribution has to be symmetric, such that the maximum load is experienced at the wing root, i.e., at the midspan. It will also represent the loading in a manner that explains the load variation with basic wing planform variations, i.e., the span, planform, and aerodynamic or geometric twist. This qualitatively explains the variation of load distribution with wing planform variations, while the sectional property can be obtained from the general thin airfoil theory. Thus, to represent this spanwise distribution, the bound vortex system should have spanwise vortex elements whose cumulative strength will be represented by a requisite number of such elements.

## 2.7  Starting Vortex Element

The second theorem, in showing constancy of circulation of a barotropic inviscid flow, necessitated that the sustained steady circulation following an impulsive start must also be associated with a starting vortex, so that the total of bound and starting vortices is zero. This is necessary, as before the impulsive start there is no circulation. To maintain the zero circulation, the starting vortex has equal strength to that of the bound vortex, but with the opposite sign—as pictorially shown in Fig. 2.12. For 2D flow, the bound vortex consists of a spanwise vortex extending from $-\infty$ to $+\infty$, for the infinite aspect ratio wing. To counterbalance this circulation, the 2D starting vortex is present at $x \to +\infty$, again infinitely long in the spanwise direction with opposite sign. Thus, these two vortex systems meet with each other at $\pm\infty$ and form a loop, without violating the second theorem.

The situation requires modification for a finite wing, which is more elaborate for two reasons. First, the loading is graded to fall off to zero at the wing tip. This tapering load toward tips cannot be replicated by taking vortex elements of differing finite spanwise lengths, as we cannot have vortices ending abruptly, according to the second theorem. This requires a third element of the vortex system.

## 2.8  Trailing Vortex Element

Consider an aircraft of finite aspect ratio wing flying level and steady, with symmetric loading. Such a wing will produce a graded lift for the bound vortex system. This must consist of bundle of line vortices of varying length, so that any closed contour at any spanwise station will account for the sectional load, as the sum total strength of the enclosed line vortices. As such line vortices cannot end abruptly, a plausible model proposed by Lanchester and Prandtl is depicted in Fig. 2.13.

We noted above that $\omega_z$ have finite length in $z$-direction to account for the falling lift, as one moves from wing root to tip section. To satisfy the second theorem, these finite spanwise line vortices are turned backward at each end symmetrically to form a pair of trailing vortex elements. This explains the formation of the trailing vortex system as shown in Fig. 2.13. This trailing vortex system simulates the wake seen behind the wing. Thus, the model of finite wing theory supports two physical attributes of a real wing in terms of spanwise lift variation and the wake behind the wing. Apparently, this turning of line vortices from spanwise to streamwise direction is the key feature of the finite wing theory. This is also the troublesome aspect of finite wing theory, which demands $|\omega_z| \equiv |\omega_x|$ during the turning, apparently to satisfy Kelvin's theorems! However, what is the physical basis of such assertion? Usually, one must use some conservation principle for some guideline. In general, in aerodynamics the governing equations are nothing but the conservation principles of mass, translational momentum, and energy. However, we do not satisfy conservation of angular momentum. As angular momentum is product of moment of inertia and

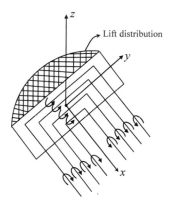

**Fig. 2.13**  Vortex system over a finite aspect ratio, as put forward by Lanchester and Prandtl. This vortex system has three elements as shown in the figure and discussed in the text. Reproduced with permission from Theoretical and Computational Aerodynamics. First Edition Tapan K. Sengupta Copyright (2015) John Wiley & Sons. Ltd. UK

angular rotation rate (the components of which are in turn proportional to vorticity components), it is apparent that the requirements of Kelvin's theorem are tantamount to the claim that the fluid packet affected by the aircraft in $x$- and $z$-directions have the same moment of inertia. This is a very speculative requirement and would require deeper understanding of the relevant issues. Thus, we note that the finite wing theory is merely a model and further refinement of which is a distinct possibility, especially in the context of computational aerodynamics now feasible with powerful computers and high accuracy methods.

## 2.8.1  Horseshoe Vortex

The vortex system consisting of the bound, trailing, and starting vortices form multiple loops, consistent with vortex theorems. It is noted that associated vortex tubes bend from bound spanwise vortex to streamwise trailing vortex and again to the spanwise starting vortex. The starting vortex is left behind, and the trailing pair stretches effectively to infinity as steady flight proceeds. This three-sided vortex has been called the horseshoe vortex. Having described qualitatively the vortex system in terms of the three elements, it is necessary to quantify the effects of such loading distribution. This is in essence the major contribution of finite wing theory. We note that the finite aspect ratio wing with bound vortex of finite length interacts with the semi-infinite trailing vortices on either side. The quantitative measure of such interactions between bound and trailing vortices is found by the Biot–Savart law.

## 2.8.2   *Biot–Savart Law for Simplified Cases*

In finite wing theory of unswept planform, the vortex system consists of elements which are straight. In further simplification, we can replace this vortex system consisting of countably infinite number of filaments to a single bound vortex turned backward to two trailing vortices extending to infinity.

We show a schematic diagram in Fig. 2.14 to calculate induced velocity at a field point $P$ by a line vortex of finite length and strength per unit length that is $\Gamma$. The field point $P$ is located with respect to an infinitesimal element $d\bar{s}$, given by the polar coordinate of $(R, \theta)$. Also the field point subtends angles $\theta_l$ and $\theta_r$ with the finite length vortex $AB$. From Biot–Savart law, one calculates the induced velocity as

$$dv = \frac{\Gamma}{4\pi R^2} \sin \theta \; d\bar{s} \tag{2.82}$$

This induced velocity is in the direction of the normal to the plane $APB$.

The total induced velocity at $P$ due to the complete line element $AB$ is the integration of effects of such successive elements obtained from Eq. (2.82), for the geometry shown in Fig. 2.14. The normal $PO$ drawn upon $AB$ from $P$ has a length $d$. Defining an auxiliary angle $\psi_s$ in the figure, we note that this varies from $\psi_A$ and $\psi_B$. The angle is given by $\psi_s = \frac{\pi}{2} - \theta$. Also,

$$\psi_A = -\left(\frac{\pi}{2} - \theta_l\right) \text{ and } \psi_B = +\left(\frac{\pi}{2} - \theta_r\right)$$

Note that $\psi_A$ is oriented in clockwise direction with respect to $PO$, and hence is negative and $\psi_B$ is in anti-clockwise direction with respect to $PO$ and is positive. Since

$$\sin \theta = \cos \psi_s, \quad R = d \sec \psi_s \text{ and } \bar{s} = d \tan \psi_s \tag{2.83}$$

Therefore

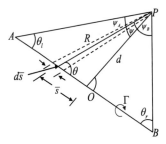

**Fig. 2.14**   Schematic to calculate induced flow field due to a straight-line vortex of constant strength. Reproduced with permission from Theoretical and Computational Aerodynamics. First Edition Tapan K. Sengupta Copyright (2015) John Wiley & Sons. Ltd. UK

$$d\bar{s} = d\,(d\tan\psi_s) = d\sec^2\psi_s\,d\psi_s$$

Using these in Eq. (2.82), one obtains

$$v = \int_{\psi_A}^{\psi_B} \frac{\Gamma}{4\pi d}\cos\psi_s\,d\psi_s = \frac{\Gamma}{4\pi d}\left[\sin(\pi/2 - \theta_r) + \sin(\pi/2 - \theta_l)\right]$$

Thus, the induced velocity at $P$ is obtained in terms $\theta_l$ and $\theta_r$ as

$$= \frac{\Gamma}{4\pi d}(\cos\theta_l + \cos\theta_r) \tag{2.84}$$

**Influence of a semi-infinite vortex:**

Consider that in Fig. 2.14, one end of the vortex stretches to $+\infty$ (say the end B). Then $\theta_r = 0$ and Eq. (2.84) reduces to

$$v = \frac{\Gamma}{4\pi d}(\cos\theta_l + 1) \tag{2.85}$$

Furthermore, if $\theta_l = \pi/2$, i.e., for semi-infinite vortex, then the induced velocity of this special sub-case is given by

$$v = \frac{\Gamma}{4\pi d} \tag{2.86}$$

**Influence of an infinite vortex:**

This is the case of 2D line vortex of infinite length. Hence, in this case $A$ and $B$ tend to infinite distance in the opposite directions, so that $\theta_l = \theta_r = 0$, and one gets

$$v = \frac{\Gamma}{2\pi d} \tag{2.87}$$

## 2.8.3 Relation Between Spanwise Loading and Trailing Vortices

Kelvin–Helmholtz's second theorem states that circulation round any section of the bundle of vortices is the sum of strengths of the vortex filaments cut by the sectional plane. Thus, from Fig. 2.13, one notices that the spanwise loading drops off toward the wing tip, as different elements have different spanwise lengths and turned backward. Now as the sectional plane is moved outward from the midspan, the bound vortex

**Fig. 2.15** Spanwise load
distribution for a straight
wing and shed vortices.
Reproduced with permission
from Theoretical and
Computational
Aerodynamics. First Edition
Tapan K. Sengupta
Copyright (2015) John Wiley
& Sons. Ltd. UK

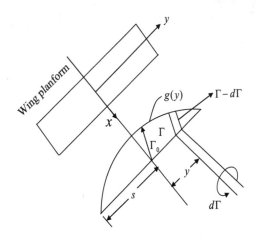

filaments and corresponding sum of the strength of the bundle decreases, as depicted in Fig. 2.15.

Finite aspect ratio of a wing causes the lift to fall off, and this is possible that the lost circulation between two successive spanwise stations should be shed in the wake of the wing in the form of trailing vortices. Thus, the strength of the vortex filaments shed in the wake must increase, as the rate of fall in circulation strength falls rapidly in the vicinity of the wing tip. Thus, the general rule of change in circulation from section to section is equal to the strength of the vortices shed between the sections, for example, for the spanwise load distribution shown in Fig. 2.15.

In the figure at section $y$, the sectional circulation is $\Gamma$, while at the adjoining section at $(y + dy)$, the sectional circulation reduced to $\Gamma - d\Gamma$. Therefore, between these two stations, trailing vortices of strength $d\Gamma$ are shed according to this steady-state finite wing theory. Remembering that this is just an artifact of a model, we proceed further. If the circulation varies as shown in Fig. 2.15 which is described as a function of $y$, given by $g(y)$, then the strength of circulation shed between the above-mentioned two sections $dy$ apart can be written as

$$d\Gamma = -g'(y)\,\mathrm{d}y$$

where the prime indicates derivative with respect to the argument.

As noted in Figs. 2.13 and 2.15, the velocity induced by $d\Gamma$ causes a downward component of velocity at an inboard station along the bound vortex. As the strength of trailing vortices increases toward the wing tip, the net contribution of induced effects by the trailing vortex sheet would introduce a downward velocity over the wing planform, represented by the bound vortex. This induced component is thus called the **downwash**. Consider the influence of $d\Gamma$ shed from the section at $y$, at another section at $y_1$ along the span in creating an induced velocity, is equal to

$$dw(y_1) = -\frac{g'(y)dy}{4\pi(y - y_1)} \tag{2.88}$$

The direction of this induced velocity is downward, if $y_1$ is inboard of $y$. The overall effect is given by

$$w(y_1) = -\frac{1}{4\pi} \int\limits_{-s}^{s} \frac{g'(y)dy}{(y - y_1)} \tag{2.89}$$

The spanwise load distribution over the complete wing is indicated by the range of integral.

## 2.9 Consequence of Downwash: Induced Drag

The downwash adds to the drag and is entirely due to the trailing vortices of the finite aspect ratio wing. As this is obtained by inviscid theory, this is often classified as inviscid and termed as induced drag.

At any spanwise station, the consequence is depicted in Fig. 2.16, showing the airfoil section at a geometric angle of attack of $\alpha$, which is the angle subtended by the chord line with the oncoming flow velocity ($U_\infty$) direction. The downwash component $w$ is vectorially added to $U_\infty$ to provide the resultant velocity, $U_R$. This resultant velocity vector makes a reduced angle of attack of $\alpha_\infty$ with chord line— an equivalent angle of attack made by an infinite aspect ratio wing section. The difference between these two angles of attack is indicated by $\varepsilon$, which is given for small downwash angles by

$$\varepsilon = \tan^{-1}\frac{w}{U_\infty} \simeq \frac{w}{U_\infty} \tag{2.90}$$

There is a consequence of this downwash angle in this inviscid theory. The lift acting on the wing section ($L_\infty$) will be perpendicular to $U_R$, and there is no drag for this potential flow. By standard convention, the actual lift should be referred to the actual free-stream direction, as indicated by the component $L$ of $L_\infty$. The other

**Fig. 2.16** The effect of induced downwash by trailing vortices and creation of induced drag

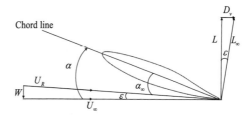

component of $L_\infty$, indicated by $D_v$ is against the direction of forward velocity and is therefore a drag, called the induced or trailing vortex drag.

There is an alternative viewpoint about the lift experienced by the finite aspect ratio wing. The induced velocity, $w$, in effect reduces the effective incidence of the wing section, so that to generate the same lift as created by an infinite aspect ratio wing with the same airfoil section at incidence $\alpha_\infty$, the corresponding angle of incidence for the finite aspect ratio wing at the same section must be set at an incidence of $\alpha = \alpha_\infty + \varepsilon$.

For small downwash, $L = L_\infty \cos \varepsilon \simeq L_\infty$, i.e., the lift remains virtually the same for the finite aspect ratio wing section. The other component, $D_v = L_\infty \sin \varepsilon \simeq L_\infty \varepsilon$ is normal to lift direction and is the induced drag.

If the wing experiences variable circulation distribution along the span, as shown in Fig. 2.15, then the total lift experienced by the wing of semi-span $s$ is

$$L_\infty = L = \int_{-s}^{s} \rho \, U_\infty \, \Gamma \, dy \tag{2.91}$$

and the induced drag is

$$D_v = \int_{-s}^{s} \rho \, U_\infty \left(\frac{w}{U_\infty}\right) \Gamma \, dy$$

which can also be written in terms of the downwash and circulation distribution as

$$D_v = \int_{-s}^{s} \rho \, w \, \Gamma \, dy \tag{2.92}$$

It is noted that for a long haul transport aircraft flying in turbulent flow condition, the induced drag of the aircraft is more than 30% of the total drag, implying this component as significant. Many commercial aircraft are fitted with a device called the winglet to reduce induced drag.

## 2.10  Simple Symmetric Loading: Elliptic Distribution

For a given load distribution described in Fig. 2.15, one can obtain the lift and induce drag for an aircraft flying level and steady with symmetric loading. For such a loading, the maximum circulation is at the root chord (let us say given by, $\Gamma_0$) and which progressively falls off to zero at the wing tips at $y = \pm s$. The simplest possible variation can be a quadratic variation given as

$$\left(\frac{\Gamma}{\Gamma_0}\right)^2 + \left(\frac{y}{s}\right)^2 = 1 \tag{2.93}$$

Note that this represents an equation for an ellipse and such loading is therefore called the elliptic loading, with only positive sign taken for the circulation, $\Gamma$. For such a loading, the lift experienced by the wing is given by

$$L = \int_{-s}^{s} \rho\, U_\infty\, \Gamma\, dy = \rho\, U_\infty\, \Gamma_0 \int_{-s}^{s} \sqrt{1 - \left(\frac{y}{s}\right)^2}\, dy$$

Using $y = -s \cos\phi_i$, from Eq. (2.93), one gets $\Gamma = \Gamma_0 \sin\phi_i$ with $0 \leq \phi_i \leq \pi$. Plugging these expressions one gets

$$L = \rho\, U_\infty \Gamma_0 \pi\, s/2 \tag{2.94}$$

Using the expression of lift coefficient for the finite wing ($C_L$) with planform area $S$ as

$$L = C_L \frac{1}{2} \rho U_\infty^2 S$$

gives the circulation at the midspan as

$$\Gamma_0 = \frac{C_L\, U_\infty S}{\pi s} \tag{2.95}$$

The lift coefficient for the finite wing is represented by $C_L$. For the circulation distribution in Eq. (2.93), one obtains the rate of spanwise variation of loading as

$$\frac{d\Gamma}{dy} = -\frac{\Gamma_0}{s} \frac{y}{\sqrt{s^2 - y^2}}$$

Using this expression, one gets the downwash as

$$w(y_1) = \frac{\Gamma_0}{4\pi s} \int_{-s}^{s} \frac{y\, dy}{\sqrt{s^2 - y^2}(y - y_1)}$$

This can be simplified further to

$$= \frac{\Gamma_0}{4\pi s} \left[ \int_{-s}^{s} \frac{dy}{\sqrt{s^2 - y^2}} + y_1 \int_{-s}^{s} \frac{dy}{(s^2 - y^2)^{1/2}(y - y_1)} \right]$$

The first integral on the right-hand side can be easily evaluated and representing the second integral on the right-hand side by $I$, we can write the downwash as

$$w(y_1) = \frac{\Gamma_0}{4\pi s} \left[ \pi + y_1 I \right]$$

For any symmetric flight, the loading will be symmetric on the wing and hence the induced downwash on two stations located symmetrically on either side of the midspan must be the same, i.e., $w(y_1) = w(-y_1)$.

This implies then

$$\frac{\Gamma_0}{4\pi s}\left[\pi + y_1\, I\right] = \frac{\Gamma_0}{4\pi s}\left[\pi - y_1\, I'\right]$$

We note that the two integrals are not the same, and therefore the above identity of symmetric downwash distribution on either side of the wing is satisfied only if $I \equiv 0 = I'$.

And thus

$$w(y_1) = \frac{\Gamma_0}{4s} \tag{2.96}$$

One of the key features of the elliptic load distribution in Eq. (2.93) is that the induced downwash is constant along the span. Such loading distribution is optimum in terms of induced drag, as shown next.

### 2.10.1  Induced Drag for Elliptic Loading

The induced drag for the elliptic loading is obtained as

$$D_v = \int_{-s}^{s} \rho\, w\, \Gamma\, dy = \rho\frac{\Gamma_0}{4s}\,\Gamma_0 \int_{-s}^{s}\sqrt{1 - \left(\frac{y}{s}\right)^2}\, dy$$

$$= \frac{\pi}{8}\rho\,\Gamma_0^2 = C_{D_v}\frac{1}{2}\,\rho\,U_\infty^2\,S$$

Thus, the induced drag coefficient, $C_{D_v}$, can be obtained in terms of the wing lift coefficient, by substituting the midspan circulation expression of $\Gamma_0 = \frac{C_L\,U_\infty S}{\pi s}$ in the above to get

$$C_{D_v} = \frac{C_L^2}{\pi\, \mathcal{R}} \tag{2.97}$$

where

$$\frac{4s^2}{S} = \frac{\text{(wing span)}^2}{\text{wing planform area}} = \mathcal{R}\ \text{(aspect ratio)}$$

The elliptic loading represented by Eq. (2.93) is also given by

$$\Gamma = \Gamma_0 \sin\phi_i \tag{2.98}$$

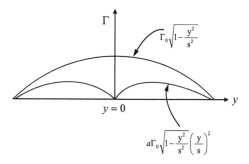

**Fig. 2.17** The individual harmonics of modified elliptic load distribution. Reproduced with permission from Theoretical and Computational Aerodynamics. First Edition Tapan K. Sengupta Copyright (2015) John Wiley & Sons. Ltd. UK

The circulation is represented by a periodic function described over half the wavelength. This lacks mathematical rigor, as the expression for the circulation implies the corresponding periodic extension of the same. It is noted that the elliptic loading distribution provides the least induced drag and is of interest. Such loading can be obtained in an actual wing by various means. For example, it is simply obtained by spanwise chord length distribution given by $c(y) = c_0 \sin \phi_i$, where $c_0$ is the wing root chord, with all airfoil sections identical. However, with different wing sections and geometric twist, a whole range of planform will provide elliptic loading.

### 2.10.2 Modified Elliptic Load Distribution

Symmetric loading, with the constraints of zero loading at the wing tip, can be obtained by all odd superharmonics of the fundamental given in Eq. (2.98). Next, we consider a modified load distribution given by

$$\Gamma = \Gamma_0 \sqrt{1 - \left(\frac{y}{s}\right)^2} \left[1 + a\left(\frac{y}{s}\right)^2\right] \tag{2.99}$$

Let us identify the modified elliptic loading by using $y = -s \cos \phi_i$ to rewrite Eq. (2.99) as (Fig. 2.17)

$$\Gamma = \Gamma_0 \sin \phi_i \left[1 + a(1 - \sin^2 \phi_i)\right]$$

The non-dimensional circulation is written as

$$\frac{\Gamma}{\Gamma_0} = (1 + a) \sin \phi - a \sin^3 \phi$$

Or

$$\frac{\Gamma}{\Gamma_0} = \left(1 + \frac{a}{4}\right) \sin \phi + \frac{a}{4} \sin 3\phi$$

Using $A_1 = (1 + \lambda)\Gamma_0$ and $A_3 = \lambda \Gamma_0$
the modified elliptic loading is given by

$$\Gamma = A_1 \sin \phi + A_3 \sin 3\phi \qquad (2.100)$$

These odd harmonics of sine function provide symmetric loading.


## 2.11  General Loading on a Wing

We have noted above that the loading represented by Eq. (2.100) provides symmetric loading. In contrast, even harmonics of sine function provide antisymmetric loading. This provides the way to represent any arbitrary loading over the wing by

$$\Gamma = 4 s \ U_\infty \sum_{n=1}^{N_1} A_n \sin n\theta \qquad (2.101)$$

where the spanwise coordinate is defined as $y = -s \cos \theta$, with $0 \le \theta \le \pi$. The starboard wing tip corresponds to $\theta = \pi$, and the port wing tip is given by $\theta = 0$. The circulation scale has been taken as $4s U_\infty$. Depending upon the prescribed loading pattern, one can appropriately choose the number of terms $(N_1)$ in the general loading function. The lift on a section of width $dy$ $(= s \sin \theta d\theta)$ at $y$ $(= -s \cos \theta)$ is obtained by application of Kutta–Jukowski theorem by

$$dL = 4\rho \ s^2 \ U_\infty^2 \sum_{n=1}^{N_1} A_n \sin n\theta \ \sin \theta \ d\theta \qquad (2.102)$$

Thus, the lift acting on the wing is given by

$$L = 4\rho \ s^2 \ U_\infty^2 \int_0^\pi \sum_{n=1}^{N_1} A_n \sin n\theta \ \sin \theta \ d\theta \qquad (2.103)$$

or

$$L = 2\rho \ s^2 \ U_\infty^2 \sum \left[ A_n \left( \frac{\sin(n-1)\theta}{n-1} - \frac{\sin(n+1)\theta}{n+1} \right) \right]_0^\pi$$

Both the terms on the right-hand side do not contribute for all $n$, except for $n = 1$, for which one gets

$$L = 2\rho \, s^2 \, U_\infty^2 \pi A_1 \tag{2.104}$$

Thus, for a general loading distribution, the lift is decided by the first odd harmonic of the loading distribution, with the lift coefficient obtained as

$$C_L = \pi A_1 \, \mathcal{R} \tag{2.105}$$

### 2.11.1  Downwash for General Loading

The general loading distribution given by Eq. (2.101) can be used in Eq. (2.89) to obtain the corresponding downwash, by using $\theta$ as the independent variable with $y = -s\cos\theta$ as

$$w(\theta_1) = \frac{1}{4\pi s} \int_0^\pi \frac{\frac{d\Gamma}{d\theta} d\theta}{\cos\theta - \cos\theta_1}$$

From Eq. (2.101), $\frac{d\Gamma}{d\theta} = 4s\, U_\infty \sum n\, A_n \cos n\theta$. Thus,

$$w(\theta_1) = \frac{4s\, U_\infty}{4\pi s} \int_0^\pi \frac{\sum n\, A_n \cos n\theta}{\cos\theta - \cos\theta_1} d\theta = \frac{U_\infty}{\pi} \sum n\, A_n\, G_n \tag{2.106}$$

with Glauert's integral already defined as

$$G_n = \frac{\pi \sin n\theta_1}{\sin\theta_1}$$

### 2.11.2  Induced Drag on a Finite Wing for General Loading

For the general loading of the wing given by $\Gamma$ and the downwash $w(\theta)$ acting on the finite wing, the induced drag is obtained from

$$D_v = \int_{-s}^{s} \rho\, w\, \Gamma\, dy$$

Using the expression of $\Gamma$ from Eq. (2.101) and $w(\theta)$ from Eq. (2.106) in the above, one gets the induced drag as

$$D_v = \int_0^\pi \rho \, \frac{U_\infty \sum n \, A_n \sin n\theta}{\sin \theta} \, 4s \, U_\infty \sum A_m \sin m\theta \, s \, \sin \theta \, d\theta$$

Or $D_v = 2\rho \, U_\infty^2 \, s^2 \sum n \, A_n^2 \, \pi$

Writing the induced drag in terms of a coefficient as

$$D_v = C_{D_v} \frac{1}{2} \rho \, U_\infty^2 \, S$$

one gets

$$C_{D_v} = \pi \, \mathcal{R} \sum n \, A_n^2 \tag{2.107}$$

From Eq. (2.105), $A_1^2 = \frac{C_L^2}{\pi^2 \mathcal{R}^2}$, and hence the induced drag coefficient can be expressed as

$$C_{D_v} = \frac{C_L^2}{\pi \, \mathcal{R}} \sum n \left( \frac{A_n}{A_1} \right)^2$$

Thus, the induced drag for general loading distribution can be obtained for symmetric loading from

$$C_{D_v} = \frac{C_L^2}{\pi \, \mathcal{R}} \left[ 1 + \left( \frac{3A_3^2}{A_1^2} + \frac{5A_5^2}{A_1^2} + \frac{7A_7^2}{A_1^2} + \cdots \right) \right] = \frac{C_L^2}{\pi \, \mathcal{R}} \left[ 1 + \delta \right] \tag{2.108}$$

### 2.11.3 Load Distribution for Minimum Drag

For the general loading distribution, the induced drag coefficient is

$$C_{D_v} = \pi \, \mathcal{R} \left[ A_1^2 + 2A_2^2 + 3A_3^2 + 4A_4^2 + \cdots \right] \tag{2.109}$$

This drag is a minimum, when $A_2 = A_3 = \cdots = A_n = \cdots = 0$ and $A_1 \neq 0$. For such a distribution,

$$\Gamma = 4s \, U_\infty \, A_1 \sin \theta \quad \text{with} \quad y = -s \cos \theta$$

which is nothing but the elliptic load distribution,

$$\Gamma = 4s \, U_\infty \, A_1 \sqrt{1 - \left( \frac{y}{s} \right)^2}$$

## 2.12  Asymmetric Loading: Rolling and Yawing Moment

Consider the asymmetric loading indicated in Fig. 2.18, for the flow model. We note that one gets an associated symmetric-induced drag, as sketched in the figure. Due to the asymmetry of loading, there will be a rolling moment, $L_R$ as shown in the figure. At the same time, due to asymmetric drag grading, there will be a yawing moment, $N$.

### 2.12.1  Rolling Moment ($L_R$)

Due to asymmetric loading, for a strip of width $dy$, at a distance $y$, the differential lift acting is

$$dL = \rho \, U_\infty \, \Gamma \, dy$$

which will contribute to $L_R$. With the convention that starboard wing going down as positive, $L_R$ is given by

$$dL_R = -dL \; y$$

The net $L_R$ acting on the wing is obtained as

$$L_R = - \int_{-s}^{s} \rho \, U_\infty \, \Gamma \, y \, dy \tag{2.110}$$

Using the loading distribution given in Eq. (2.101), this simplifies to

$$L_R = 4\rho s \, U_\infty^2 \int_0^\pi \sum s^2 \, A_n \sin n\theta \cos \theta \; \sin \theta \; d\theta$$

**Fig. 2.18** Asymmetric spanwise loading giving rise to rolling and yawing moments

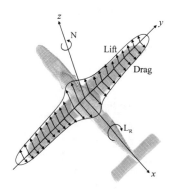

which upon further simplification yields

$$L_R = 2\rho s^3 \, U_\infty^2 \int_0^\pi \sum A_n \sin n\theta \, \sin 2\theta \, d\theta$$

$$= \rho \, U_\infty^2 \, s^3 \left[ \sum A_n \left( \frac{\sin(n-2)\theta}{n-2} - \frac{\sin(n+2)\theta}{n+2} \right) \right]_0^\pi$$

Writing the rolling moment in terms of the rolling moment coefficient $C_{lR}$ as

$$\rho s^3 \, U_\infty^2 \, A_2 \pi = C_{lR} \frac{1}{2} \, \rho \, U_\infty^2 \, S \, s$$

Thus,

$$C_{lR} = \frac{\pi}{2} \cancel{R} \, A_2 \qquad\qquad (2.111)$$

The rolling moment is given by the first even harmonics of the general loading distribution.

## 2.12.2  Yawing Moment (N)

As noted in Fig. 2.18, the asymmetric loading also causes asymmetric-induced drag, whose asymmetric grading will create a net yawing moment ($N$) which can be calculated from $D_v$ as

$$dN = dD_v \, y$$

where $dD_v = \rho \, w \, \Gamma \, dy$. This can be integrated over the whole wingspan to get

$$N = \int_{-s}^{s} \rho \, w \, \Gamma \, y \, dy$$

which can be simplified as

$$N = -4s^3 \rho \, U_\infty^2 \int_0^\pi \frac{\sum n \, A_n \sin n\theta}{\sin \theta} \sum A_m \sin m\theta \, \cos \theta \sin \theta \, d\theta$$

Equating this with the yawing moment in terms of moment coefficient, one gets

$$N = C_N \frac{1}{2} \rho \, U_\infty^2 \, Ss$$

Thus,

$$C_N = 2 \mathcal{R} \int_0^\pi \sum n \, A_n \sin n\theta \sum A_m \sin m\theta \cos \theta \, d\theta$$

The integral on the right-hand side has nonzero value only for $m = n + 1$, and thus

$$C_N = \frac{\pi}{2} \, \mathcal{R} \left[ 3A_1 \, A_2 + 5A_2 \, A_3 + 7A_3 \, A_4 + \cdots + (2n + 1) \, A_n \, A_{n+1} \right] \quad (2.112)$$

### 2.12.3   Effect of Aspect Ratio on Lift Curve Slope

Consider the $(C_L, \alpha)$-plot shown in Fig. 2.19, where the lift curve slope of the finite wing is defined by $a = \frac{dC_L}{d\alpha}$. Additional subscript $\infty$ refers to conditions for an infinite aspect ratio wing. The lift curve slope of a 3D wing will be lower compared to a 2D wing.

Thus, from Fig. 2.19, to generate the same $C_L$, we must have

$$C_L = a_\infty [\alpha_\infty - \alpha_0] = a[\alpha - \alpha_0]$$

where $\alpha_0$ is the zero lift angle. As the finite aspect ratio wing induces a downwash characterized by tilting the lift backward by an angle $\varepsilon$ so that $\alpha_\infty = \alpha - \varepsilon$ and

$$C_L = a_\infty \left[ (\alpha - \alpha_0) - \varepsilon \right] = a(\alpha - \alpha_0)$$

Thus,

$$a = a_\infty \left[ 1 - \frac{\varepsilon}{(\alpha - \alpha_0)} \right] \quad (2.113)$$

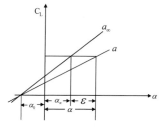

**Fig. 2.19** Lift curves for the finite $\mathcal{R}$ wing and the equivalent infinite aspect ratio wing. Reproduced with permission from Theoretical and Computational Aerodynamics. First Edition Tapan K. Sengupta Copyright (2015) John Wiley & Sons. Ltd. UK

Now for elliptic load distribution $\varepsilon = \frac{w}{U_\infty} = \frac{\Gamma_0}{4s\,U_\infty}$ with $\Gamma_0 = \frac{C_L\,U_\infty\,S}{\pi s}$ Therefore

$$\varepsilon = \frac{C_L S}{\pi 4s^2} = \frac{C_L}{\pi\,\mathcal{R}}$$

Using this in Eq. (2.113), we get

$$a = a_\infty\left[1 - \frac{C_L}{\pi\,\mathcal{R}(\alpha - \alpha_0)}\right] = a_\infty\left[1 - \frac{a}{\pi\,\mathcal{R}}\right] \tag{2.114}$$

Therefore, for elliptic loading,

$$a = \frac{a_\infty}{1 + a_\infty/\pi\,\mathcal{R}} \tag{2.115}$$

For a finite wing of general planform, the above equation is modified as

$$a = \frac{a_\infty}{1 + \frac{a_\infty}{\pi\,\mathcal{R}}(1 + \tau)} \tag{2.116}$$

The value of $\tau$ as a function of $A_n$ in the loading distribution has been calculated [3] and found to lie between 0.05 and 0.25.

## 2.13  Panel Method

If the flow is inviscid and irrotational, then the governing equation is

$$\nabla^2\phi = 0 \tag{2.117}$$

This Laplace equation was derived from the continuity equation. We also noted that the governing equation for irrotational flow can be written with a Laplace equation for the stream function. For 2D flows, $\phi$ and $\psi$ being analytic function of $(x, y)$, one defines a complex potential $(W)$ as a function of $z\ (= x + iy)$ and develop a thin airfoil theory, neglecting the thickness effects. This flow model was extended for finite wing, with the thickness of the geometry not included. These approaches perform poorly near the stagnation point. Also, these methods do not work for wing with multi-element airfoils. To eliminate these problems for full aircraft, using irrotational flow assumption one uses panel method.

To discuss panel method, we note the property and solution of Eq. (2.117). We have a boundary value problem, belonging to elliptic partial differential equation class. This class of problems has the highest derivative given by $\nabla^{2n}$, and such an equation requires $n$ boundary conditions. Thus, we must have one boundary condition prescribed on all the segments of the boundary.

In panel methods, sources, doublets, vortices, etc. are distributed either as basic units or in combination of these over the physical surface, represented by small panels and hence named as the panel method. The boundary condition on the wing/body and the Kutta condition at the wing trailing edge fixes the strength of such distributions. There is no unique distribution for such singularities, and the accuracy of the method rests on the adequate number of panels and relative placement. There are many methods in use, and we discuss only the method in [5]. But before doing this, we will discuss some issues related to singularity distribution. We also state without the proof that to represent a closed body, the sum of strengths of sources and sinks used to represent the body must be equal to zero. Also, the self-induced velocity caused by the singularities, at the beginning and at the end of each panel, has a logarithmic singularity. For these results and other details, look at Chap. 5 of [16].

## 2.14  Panel Method of Hess and Smith

The first useful exposition of panel method was provided by Hess and Smith by a distribution of sources and vortices on the surface of the geometry in [5]. In this method, the velocity potential is given by

$$\phi = \phi_\infty + \phi_s + \phi_v \tag{2.118}$$

where the total potential $\phi$ is composed of $\phi_\infty = U_\infty(x \cos \alpha + y \sin \alpha)$, due to the oncoming free-stream flow at an angle of attack, $\alpha$; $\phi_s$ is contributed by the sources and $\phi_v$ by the vorticity distribution, to model the lift generated. The last two terms are due to source and vortex strengths, $q(s)$ and $\bar{\gamma}(s)$ measured along the surface-conforming curve of each panel. Thus,

$$\phi_s = \int \frac{q(s)}{2\pi} \mathrm{Ln}\, r \,\mathrm{d}s \quad \text{and} \quad \phi_v = -\int \frac{\bar{\gamma}(s)}{2\pi} \theta \,\mathrm{d}s \tag{2.119}$$

Geometric terms appearing in the expressions are as defined in Fig. 2.20. Values of $q(s)$ and $\bar{\gamma}(s)$ are fixed from zero normal velocity boundary condition, along

**Fig. 2.20** Notations and schematic of panel method as applied for a lifting airfoil. Reproduced with permission from Theoretical and Computational Aerodynamics. First Edition Tapan K. Sengupta Copyright (2015) John Wiley & Sons. Ltd. UK

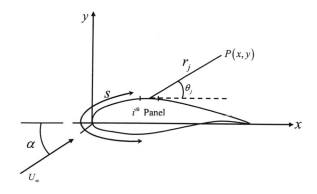

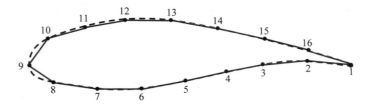

**Fig. 2.21** Straight panels approximate the curved airfoil surfaces, with the difference between the two can be minimized by taking more panels. Reproduced with permission from Theoretical and Computational Aerodynamics. First Edition Tapan K. Sengupta Copyright (2015) John Wiley & Sons. Ltd. UK

with the Kutta condition. In thin airfoil theory also, the Kutta condition fixed the net circulation. The same condition for different three-dimensional geometries is used, with the criterion of zero loading at the edges. In Hess and Smith's method, Kutta condition for the airfoil is specified by a constant strength vorticity over all the panels, while flow tangency condition is applied at that many points, as is the number of unknown source strength distribution. The source strength is a fixed constant over each panel, which varies from panel to panel. For the linear problem with discrete panels, one can superpose the solution as

$$\phi = \phi_\infty + \sum_{j=1}^{N_1} \left[ \frac{q_j(s)}{2\pi} \operatorname{Ln} r_j - \frac{\bar{\gamma}}{2\pi} \theta_j \right] ds \qquad (2.120)$$

The integrals of Eq. (2.119) will be difficult to evaluate for curved panels. Hence, to avoid this source of errors, the panels on which the sources and vortices are distributed are taken as straight lines, as shown in Fig. 2.21.

One selects $N_1$ number of points on the body, called the nodes. One connects the nodes by straight lines, forming the panels on which sources and vortices are distributed. It is proposed that the source strength is constant on each panel, and vorticity is constant over all panels. In Eq. (2.120), there are $(N_1 + 1)$ unknowns: $\bar{\gamma}$ and $(q_j, j = 1, 2...N_1)$, which are solved by satisfying flow tangency condition at $N_1$ control points, and additionally satisfying the Kutta condition.

As stated before, at $x \to 0$ or $c$, the induced velocity components by the constant strength source distribution, $u_s$ and $v_s$, display logarithmic singularities. Thus, the end of each panel (the nodes) cannot be treated as control points. Generally, the mid-point of each panel is used as control points. For Kutta condition, one equates the tangential velocities of the first and last panel at the respective mid-points.

The numerological sequence shown in Fig. 2.21 indicates that the $i$th panel is defined by the $i$th and (i+1)th nodes, as shown in Fig. 2.22. The panel is inclined to the $x$-axis (of the global axis system) by $\theta_i$. Therefore,

$$\sin \theta_i = \frac{y_{i+1} - y_i}{l_i} \quad \text{and} \quad \cos \theta_i = \frac{x_{i+1} - x_i}{l_i},$$

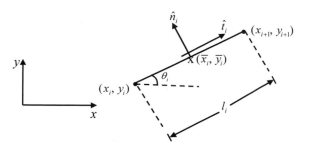

**Fig. 2.22** A straight panel showing the convention used in the global coordinate system. Reproduced with permission from Theoretical and Computational Aerodynamics. First Edition Tapan K. Sengupta Copyright (2015) John Wiley & Sons. Ltd. UK

with $l_i = [(x_{i+1} - x_i)^2 + (y_{i+1} - y_i)^2]^{1/2}$

The unit normal and unit tangent of the $i$th panel are, respectively, given by

$$\hat{n}_i = -\hat{i} \sin\theta_i + \hat{j} \cos\theta_i \qquad (2.121)$$

$$\hat{t}_i = \hat{i} \cos\theta_i + \hat{j} \sin\theta_i \qquad (2.122)$$

The unit normal is positive, if it is directed outward for the sequence of points adopted and the unit tangent is from the $i$th to the (i+1)th point, as indicated for the panel in Fig. 2.22 for the global coordinate system.

The mid-point of the ith panel is given by

$$\bar{x}_i = \frac{1}{2}(x_i + x_{i+1}) \text{ and } \bar{y}_i = \frac{1}{2}(y_i + y_{i+1})$$

One denotes the velocity ($\vec{V}$) components at the mid-point of $i$th panel as

$$u_i \equiv u(\bar{x}_i, \bar{y}_i) \text{ and } v_i \equiv v(\bar{x}_i, \bar{y}_i)$$

The flow tangency condition is applied at the panel mid-point as $\vec{V}_i \cdot \hat{n}_i \equiv 0$. Using Eq. (2.121), one obtains

$$0 = -u_i \sin\theta_i + v_i \cos\theta_i \qquad (2.123)$$

The Kutta condition at the trailing edge of the airfoil is given using Eq. (2.122) as

$$(\vec{V}_i \cdot \hat{t}_i)_{i=1} = -(\vec{V}_i \cdot \hat{t}_i)_{i=N_1}$$

Further simplification yields

$$u_1 \cos\theta_1 + v_1 \sin\theta_1 = -(u_N \cos\theta_{N_1} + v_N \sin\theta_{N_1}) \qquad (2.124)$$

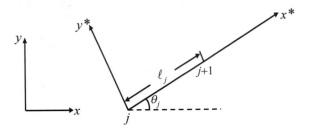

**Fig. 2.23** Local and global coordinate systems used to obtain influence coefficients. Reproduced with permission from Theoretical and Computational Aerodynamics. First Edition Tapan K. Sengupta Copyright (2015) John Wiley & Sons. Ltd. UK

Negative sign on the right-hand side of Eq. (2.124) is due to the tangent directions being opposite for the first and last panel ($N_1$th). As the induced velocities can be superposed, one can write the velocity components at the control point of the $i$th panel by summing individual effects of sources and vortices at all the $N_1$ panels by

$$u_i = U_\infty \cos\alpha + \sum_{j=1}^{N_1} q_j \, u_{ij}^{(s)} + \bar{\gamma} \sum_{j=1}^{N_1} u_{ij}^{(v)} \tag{2.125}$$

$$v_i = U_\infty \sin\alpha + \sum_{j=1}^{N_1} q_j \, v_{ij}^{(s)} + \bar{\gamma} \sum_{j=1}^{N_1} v_{ij}^{(v)} \tag{2.126}$$

where $u_{ij}^{(s)}$, $u_{ij}^{(v)}$, $v_{ij}^{(s)}$ and $v_{ij}^{(v)}$ are the influence coefficients. Physically, these represent respective velocity components due to unit source or vortex indicated by the superscript in parenthesis.

Note that $u_{ij}^{(s)}$ denotes $x$-component of velocity at the $i$th panel, caused by a unit source distributed on the $j$th panel. These coefficients are preferably obtained in the panel-fitted coordinate system and shown in Fig. 2.23. The quantities expressed in panel-fitted coordinate system are with asterisk, while quantities in the global coordinate system are without asterisks. As the $x^*$-axis of the $j$th panel subtends an angle $\theta_j$ with the $x$-axis of the global coordinate system, the global and the local velocities are thus related as

$$u = u^* \cos\theta_j - v^* \sin\theta_j \tag{2.127}$$

$$v = u^* \sin\theta_j + v^* \cos\theta_j \tag{2.128}$$

To use the above relation, one obtains the velocity components at $(x_i, \, y_i)$ created by a unit-strength source distribution on the $j$th panel as

$$u_{ij}^{*(s)} = \frac{1}{2\pi} \int_0^{lj} \frac{x_i^* - t}{(x_i^* - t)^2 + y_i^{*2}} dt = -\frac{1}{2\pi} \mathrm{Ln}\left[ (x_i^* - t)^2 + y_i^{*2} \right]^{1/2} \Big|_0^{lj}$$

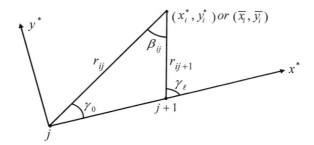

**Fig. 2.24** Local coordinate system indicating various distances and angles used in calculating influence coefficients. Reproduced with permission from Theoretical and Computational Aerodynamics. First Edition Tapan K. Sengupta Copyright (2015) John Wiley & Sons. Ltd. UK

$$v_{ij}^{*(s)} = \frac{1}{2\pi} \int_0^{lj} \frac{y_i^*}{(x_i^* - t)^2 + y_i^{*2}} dt = \frac{1}{2\pi} \tan^{-1} \frac{y_i^*}{x_i^* - t}\bigg|_0^{lj}$$

Here, $(x_i^*, y_i^*)$ is the local /panel-fitted coordinate corresponding to $(x_i, y_i)$ in the global coordinate system, as shown in Fig. 2.24.

The distance between $j$th and $i$th nodes is indicated by $r_{ij}$. Similarly, $r_{ij+1}$ is defined in Fig. 2.24. These two lines indicated by $r_{ij}$ and $r_{ij+1}$ make angles $\gamma_0$ and $\gamma_l$ with the $x^*$-axis. If the angle between $r_{ij}$ and $r_{ij+1}$ is $\beta_{ij}$, then the influence coefficients is given by

$$u_{ij}^{*(s)} = -\frac{1}{2\pi} \mathrm{Ln} \frac{r_{ij+1}}{r_{ij}} \tag{2.129}$$

$$v_{ij}^{*(s)} = \frac{1}{2\pi}(\gamma_l - \gamma_0) = \frac{\beta_{ij}}{2\pi} \tag{2.130}$$

These expressions are coordinate system independent. From Eqs. (2.129) and (2.130), one obtains self-induced velocity components due to a uniform source distribution as

$$u_{ii}^{*(s)} \equiv 0 \text{ and } v_{ii}^{*(s)} = \frac{1}{2}$$

if the point is approached from outside, for which $\beta_{ij} = \pi$. Instead if one approaches the panel from inside, then $\beta_{ij} = -\pi$, and then

$$v_{ii}^{*(s)} = -\frac{1}{2}$$

In aerodynamics, one is interested in the external flow problem.

## 2.14.1  Calculation of Influence Coefficients

Referring to Fig. 2.24, at the $i$th panel in the global coordinate system, influence of the $j$th panel is calculated by first noting the inclinations of $r_{ij}$ and $r_{ij+1}$ with respect to the $j$th panel as

$$\gamma_0 = \tan^{-1} \frac{\bar{y}_i - y_j}{\bar{x}_i - x_j} \quad \text{and} \quad \gamma_l = \tan^{-1} \frac{\bar{y}_i - y_{j+1}}{\bar{x}_i - x_{j+1}}$$

The included angle between $r_{ij}$ and $r_{ij+1}$ is

$$\beta_{ij} = \gamma_l - \gamma_0$$

$$\beta_{ij} = \tan^{-1} \frac{\bar{y}_i - y_{j+1}}{\bar{x}_i - x_{j+1}} - \tan^{-1} \frac{\bar{y}_i - y_j}{\bar{x}_i - x_j} = \tan^{-1} \frac{\tan \gamma_l - \tan \gamma_0}{1 + \tan \gamma_l \, \tan \gamma_0}$$

$$\beta_{ij} = \tan^{-1} \frac{(\bar{y}_i - y_{j+1})(\bar{x}_i - x_j) - (\bar{y}_i - y_j)(\bar{x}_i - x_{j+1})}{(\bar{x}_i - x_{j+1})(\bar{x}_i - x_j) + (\bar{y}_i - y_{j+1})(\bar{y}_i - y_j)} \tag{2.131}$$

Using Eq. (2.131) in Eqs. (2.130) and (2.129), the influence coefficients at the $i$th panel control point are obtained due to the unit source distribution on the $j$th panel.

After the influence coefficients due to source are obtained, one calculates the influence coefficients due to vortex distribution. Velocity component at the $(x_i^*,\ y_i^*)$ due to unit-strength vortex on the $j$th panel is given by

$$u_{ij}^{*(v)} = -\frac{1}{2\pi} \int_0^{l_j} \frac{y_i^*}{(x_i^* - t)^2 + y_i^{*2}} dt = \frac{\beta_{ij}}{2\pi} = v_{ij}^{*(s)} \tag{2.132}$$

$$v_{ij}^{*(v)} = -\frac{1}{2\pi} \int_0^{l_j} \frac{x_i^* - t}{(x_i^* - t)^2 + y_i^{*2}} dt = \frac{1}{2\pi} \operatorname{Ln} \frac{r_{ij+1}}{r_{ij}} = -u_{ij}^{*(s)} \tag{2.133}$$

These various influence coefficients are used in the flow tangency condition given by Eq. (2.123) as

$$-u_i \sin \theta_i + v_i \cos \theta_i = 0,$$

to get

$$-\sin \theta_i \left\{ U_\infty \cos \alpha + \sum_{j=1}^{N_1} q_j \, u_{ij}^{(s)} + \bar{\gamma} \sum_{j=1}^{N_1} u_{ij}^{(v)} \right\}$$

$$+ \cos \theta_i \left\{ U_\infty \sin \alpha + \sum_{j=1}^{N_1} q_j \, v_{ij}^{(s)} + \bar{\gamma} \sum_{j=1}^{N_1} v_{ij}^{(v)} \right\} = 0$$

Equations (2.127) and (2.128) are used to relate the velocity components in global and local coordinate system, to get

$$-\sin\theta_i\left\{\sum_{j=1}^{N_1}q_j\left(u_{ij}^{*(s)}\cos\theta_j-v_{ij}^{*(s)}\sin\theta_j\right)+\bar{\gamma}\sum_{j=1}^{N_1}\left(u_{ij}^{*(v)}\cos\theta_j-v_{ij}^{*(v)}\sin\theta_j\right)\right\}$$

$$+\cos\theta_i\left\{\sum_{j=1}^{N_1}q_j\left(u_{ij}^{*(s)}\sin\theta_j+v_{ij}^{*(s)}\cos\theta_j\right)+\bar{\gamma}\sum_{j=1}^{N_1}\left(u_{ij}^{*(v)}\sin\theta_j+v_{ij}^{*(v)}\cos\theta_j\right)\right\}$$

$$=U_\infty(\sin\theta_i\cos\alpha-\cos\theta_i\sin\alpha)$$

Reorganizing the terms with influence coefficients, one gets

$$\sum_{j=1}^{N_1}q_j\left\{u_{ij}^{*(s)}(-\sin\theta_i\cos\theta_j+\cos\theta_i\sin\theta_j)+v_{ij}^{*(s)}(\sin\theta_i\sin\theta_j+\cos\theta_i\cos\theta_j)\right\}$$

$$+\bar{\gamma}\sum_{j=1}^{N_1}\left\{u_{ij}^{*(v)}(-\cos\theta_j\sin\theta_i+\cos\theta_i\sin\theta_j)+v_{ij}^{*(v)}(\sin\theta_i\sin\theta_j+\cos\theta_i\cos\theta_j)\right\}$$

$$=U_\infty\sin(\theta_i-\alpha)$$

which upon further simplification yields

$$\sum_{j=1}^{N_1}q_j\left\{u_{ij}^{*(s)}\sin(\theta_j-\theta_i)+v_{ij}^{*(s)}\cos(\theta_i-\theta_j)\right\}$$

$$+\bar{\gamma}\sum_{j=1}^{N_1}\left\{u_{ij}^{*(v)}\sin(\theta_j-\theta_i)+v_{ij}^{*(v)}\cos(\theta_j-\theta_i)\right\}=U_\infty\sin(\theta_i-\alpha)$$

This is written as an algebraic equation given by

$$\sum_{j=1}^{N_1}q_j\,A_{ij}+\bar{\gamma}\,A_{iN_1+1}=b_i \tag{2.134}$$

where

$$A_{ij}=-u_{ij}^{*(s)}\sin(\theta_i-\theta_j)+v_{ij}^{*(s)}\cos(\theta_i-\theta_j)$$

$$A_{iN_1+1} = \sum_{j=1}^{N_1} \left[ -u_{ij}^{*(v)} \sin(\theta_i - \theta_j) + v_{ij}^{*(v)} \cos(\theta_i - \theta_j) \right]$$

$$\text{and} \quad b_i = U_\infty \sin(\theta_i - \alpha)$$

The above equation is supplemented by the Kutta condition to close the system of unknowns, where the Kutta condition is rewritten as

$$u_1 \cos\theta_1 + v_1 \sin\theta_1 + u_{N_1} \cos\theta_{N_1} + v_{N_1} \sin\theta_{N_1} = 0$$

which with the help of Eqs. (2.126) and (2.127) is obtained as

$$\cos\theta_1 \left[ U_\infty \cos\alpha + \sum_{j=1}^{N_1} (q_j\, u_{1j}^{(s)} + \bar{\gamma} u_{1j}^{(v)}) \right] + \sin\theta_1 \left[ U_\infty \sin\alpha + \sum_{j=1}^{N_1} (q_j\, v_{1j}^{(s)} + \bar{\gamma} v_{1j}^{(v)}) \right] +$$

$$\cos\theta_N \left[ U_\infty \cos\alpha + \sum_{j=1}^{N_1} (q_j\, u_{N_1 j}^{(s)} + \bar{\gamma} u_{N_1 j}^{(v)}) \right] + \sin\theta_{N_1} \left[ U_\infty \sin\alpha + \sum_{j=1}^{N_1} (q_j\, v_{N_1 j}^{(s)} + \bar{\gamma} v_{N_1 j}^{(v)}) \right] = 0$$

This is written in short form for the $j$th panel by summing over all the panels as

$$\sum_{k=1,N_1} \left\{ \cos\theta_k \sum_{j=1}^{N_1} q_j \left( u_{kj}^{*(s)} \cos\theta_j - v_{kj}^{*(s)} \sin\theta_j \right) + \sin\theta_k \sum_{j=1}^{N_1} q_j \left( u_{kj}^{*(s)} \sin\theta_j + v_{kj}^{*(s)} \cos\theta_j \right) \right.$$

$$\left. + \bar{\gamma} \cos\theta_k \sum_{j=1}^{N_1} \left( u_{kj}^{*(v)} \cos\theta_j - v_{kj}^{*(v)} \sin\theta_j \right) + \bar{\gamma} \sin\theta_k \sum_{j=1}^{N_1} \left( u_{kj}^{*(v)} \sin\theta_j + v_{kj}^{*(v)} \cos\theta_j \right) \right\} = R_1$$

where $R_1 = -U_\infty \left[ \cos(\theta_1 - \alpha) + \cos(\theta_{N_1} - \alpha) \right]$. This also can be written as

$$\sum_{k=1,N_1} \left[ \sum_{j=1}^{N_1} q_j \bar{A}_{kj} + \bar{\gamma} A_{kk} \right] = b_{N_1+1}$$

or

$$\sum_{j=1}^{N_1} A_{N+ij}\, q_j + \bar{\gamma} A_{N_1+1,N_1+1} = b_{N_1+1} \tag{2.135}$$

Upon collating all equations given by Eqs. (2.134) and (2.135) for all the $(N_1 + 1)$ unknowns: $q_j$ and $\bar{\gamma}$, once these singularity strengths are obtained, one evaluates the tangential velocity on the airfoil surface using Eq. (2.122) as

$$v_{t,i} = u_i \cos \theta_i + v_i \sin \theta_i$$

from which we obtain the pressure coefficient as

$$C_p(\bar{x}_i, \bar{y}_i) = 1 - \frac{v_{ti}^2}{U_\infty^2}$$

In the following, results obtained by Hess and Smith's method are shown for the AG24 aerofoil. The AG24 airfoil is designed for low Reynolds number applications, mostly for radio-controlled model aircraft and for micro air vehicle. At the low Reynolds number of operation, the flow over the section will be highly dominated by viscous action, and hence the panel method results would be of lesser relevance, yet it will help one understand qualitatively the variation of surface pressure with angles of attack.

One of the major advantages of panel methods is the automatic inclusion of thickness effects. Also, the poor accuracy of solution near the stagnation point is greatly reduced in panel methods by close-packing of panels in the near vicinity of the stagnation points, near the leading and trailing edges.

In Fig. 2.25, coefficient of pressure distribution is shown for the AG24 airfoil. This is a thinner airfoil, but more cambered, as compared to airfoils used for general aviation aircraft or even gliders. Due to thinner section, the zero lift angle is lower ($\alpha_0 = -2.4°$). But, existence of upward positive loading in the aft part of the airfoil, the pitching moment is nearly zero, making its better pitching moment property attractive, as compared to conventional airfoils. However, it must be remembered that this airfoil is used for low Reynolds numbers, and panel method may not be the best tool to use.

The information from the pressure distribution can be used to study the development of the boundary layer. Boundary layer solution is often used for predicting separation and transition, apart from calculating skin friction drag. For low Reynolds number, the shear layer will be thick and prone to separation, making the flow unsteady. For separation prediction, the criterion of vanishing shear stress at the wall cannot be used if one has unsteady separation. It is somewhat paradoxical that some practitioners use such boundary layer profiles for empirical transition prediction. A better transition prediction of flow over airfoil at such low Reynolds number would be by solving Navier–Stokes equation directly.

### 2.14.2  Panel Method Applied on a Full Aircraft

The panel method, known as the NPM [8], has been used for the Dornier DO-228 aircraft shown in Fig. 2.26. First, the clean baseline geometry is obtained in the form of CAD drawing shown in the figure. Next, the panels are considered in two types: one of which is non-lifting type (i.e., those which are assumed not to produce any lift), mostly on the fuselage and in its neighborhood. The main wing, horizontal and

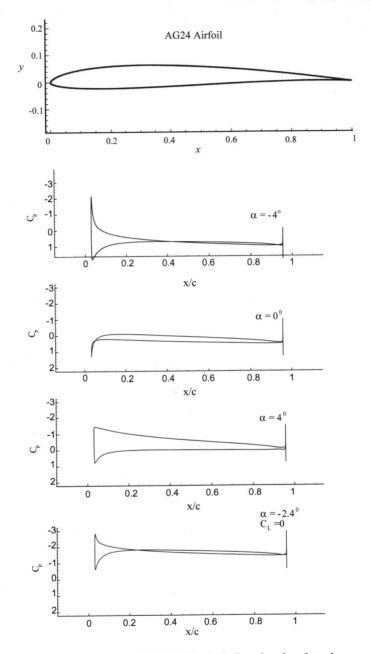

**Fig. 2.25** $C_p$ distributions for the AG24 airfoil for the indicated angles of attack

**Fig. 2.26** Clean baseline geometry of Dornier DO-228 aircraft, which has been used for panel method applied for the complete aircraft

vertical tailplanes, are considered as lifting surfaces, for which one includes vortex distribution to compute the lift generated by the full aircraft. It is to be noted that in NPM, the parts of horizontal and vertical tailplanes near the intersection with the fuselage are also treated as non-lifting panels. This is in contrast to past design practices where the main wing is considered to be joined together up to the wing planform centerline.

A typical paneling on this aircraft is shown in three views in Fig. 2.27. Roughly about 12,000 panels are taken for the full aircraft in the form of surface panels. Due to the complexities of the geometry, and due to the presence of multiple antennas for the aircraft on fuselage and wing tips, roughly about seven thousand panels are used for the fuselage and another three thousand two hundred panels are used for the wing surfaces. One of the versatile features of panel method is the lower number of unknowns handled by this method, as compared to other present-day methods like solving Reynolds-averaged Navier–Stokes (RANS) equation for the flow field. For the same geometry, about nine million cells have been used for the finite volume discretization to solve RANS equations.

In the NPM, the unknown singularity strength is obtained from which one calculates the tangential component of the velocity on the panels. These velocity components are used to obtain the pressure distribution over the complete aircraft. In Fig. 2.28, the top and bottom view perspective of pressure distribution over the entire aircraft in cruise configuration is shown.

Obtained inviscid pressure distribution over the full aircraft is integrated to calculate the lift, drag, and pitching moment. Such results are presented in coefficient form in Fig. 2.29, with wing planform area and free-stream speed used for non-dimensionalization. The used codes for RANS also require turbulence models and other artifacts. The formulations and/ or numerical methods used in these indicated approaches are also different. The experimental data are those supplied by original

**Fig. 2.27** Paneling of Dornier DO-228 aircraft used for panel method applied on the complete aircraft

OEM, for the baseline aircraft, which has been modified subsequently on multiple occasions.

What is of interest in the computed data in Fig. 2.29 by NPM is the ability of panel method to follow the qualitative trends of these force and moment coefficients, despite the absence of viscous effects in the panel method. We also note that panel methods can provide the induced drag effects, by incorporating the elements of lifting surface theory [16], which helps to calculate induced drag due to induced downwash on the lifting wing by modeled vortices in the wake. The trend provided by the drag coefficient is significantly similar to the experimental data, and those obtained by RANS. The same observation on qualitative match with other methods holds for the panel method in obtaining the pitching moment coefficient. We have already noted the flexibility of NPM in accounting for interference effects near wing–fuselage junction, by assigning non-lifting role to such proximal panels of the lifting surfaces.

## 2.15  Slender Wing and Body Theory

For low aspect ratio ($\mathcal{R}$), sweptback, and delta wings, analytical approaches and panel methods for high $\mathcal{R}$ wing are inappropriate, due to strong rotationality of the flow field, which originate at the leading edge. In the absence of rotationality, one

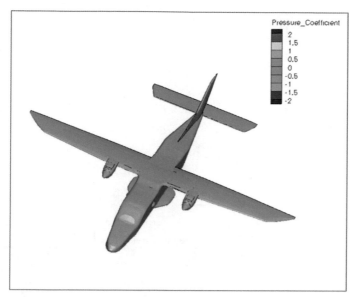

(a) Top view in perspective

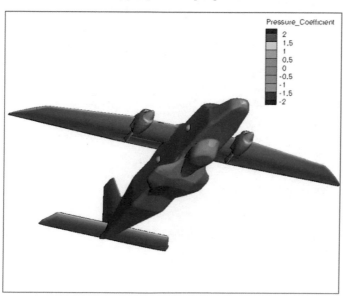

(b) Bottom view in perspective

**Fig. 2.28** Figure showing pressure distribution over the entire aircraft in top and bottom view in perspective for clean aircraft in cruise flight configuration

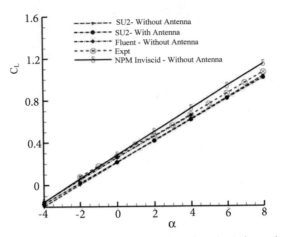

(a) $C_L$ versus $\alpha$ for cruise configuration using various codes.

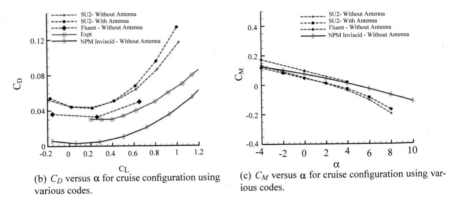

(b) $C_D$ versus $\alpha$ for cruise configuration using various codes.

(c) $C_M$ versus $\alpha$ for cruise configuration using various codes.

**Fig. 2.29** Computed lift, drag, and pitching moment coefficient by NPM are compared with other indicated methods using RANS equation, for the complete aircraft

can adopt lifting surface theory, simultaneously obtaining chord-wise and spanwise loading by the distribution of singularities on the lifting surface and its wake. One of such useful tools is due to Falkner [2], who proposed the vortex lattice method. In this method, wing surface is segmented into panels with a simple horseshoe vortex placed, as in finite wing theory. The bound element of the horseshoe vortex remains on the wing planform and the trailing parts extend to infinity in downstream direction. Thus, each panel carries one unknown singularity strength to be evaluated, by satisfying flow tangency condition. In the limit of infinitely many such horseshoe vortices, one obtains a vortex sheet. This is equivalent to a bound vortex element on the lifting surface, which extends in the streamwise direction representing the wake of the wing.

## 2.16  Slender Wing Theory

A simple, but elegant theory is described next that provides the aerodynamic characteristics of flat wing (with negligible thickness) of slender platform. General problems of wings and bodies of finite thickness distribution will be dealt with later in
the slender body theory.

The essential simplification arises for such flat plate wing at small angle of attack
$\alpha$ can be visualized by the flow pattern in any transverse plane approximated by
two-dimensional incompressible potential flow, with oncoming velocity $U_\infty \alpha$ for
the flow considered in Fig. 2.30. The planform is a delta wing with sharp leading
edge with negligible thickness. The potential flow at any streamwise section thus
represents a flow normal to a 2D flat plate, as shown in part (b) of the figure. A real
flow will readily separate from the sharp corners. But an assumption is made in the
slender wing theory, as the streamwise velocity component being much larger in
magnitude will subdue the tendency of the flow to separate. This component can be
approximated as $U_\infty$, and hence is significantly larger than the transverse component.
This is a very realistic approximation for very small angles of attacks. Although this
idea is extended often for relatively larger angles of attack, a delta wing aircraft
during landing and takeoff is used for angles of attack which are significantly higher
than that used for large aspect ratio wings. We emphasize the qualitatively different
aerodynamics of low aspect ratio wings, as compared to high aspect ratio wings.
We will note that actual operation of low aspect ratio wings at high angles of attack
displays a leading edge vortex noted at the leading edge of slender wings of delta or
similar planform.

The planform shown in Fig. 2.31 has the root chord denoted by $c_0$ and maximum
span as $b = 2S$, and the wing is at an angle of attack, $\alpha$, with the free-stream velocity in

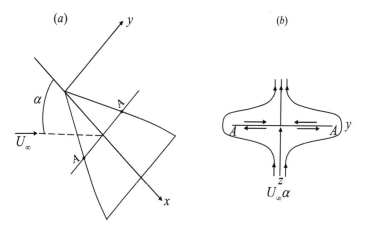

**Fig. 2.30**  Irrotational potential flow model for low aspect ratio plane wing. Reproduced with
permission from Theoretical and Computational Aerodynamics. First Edition Tapan K. Sengupta
Copyright (2015) John Wiley & Sons. Ltd. UK

**Fig. 2.31** The geometry of a
plane low aspect ratio wing
modeled by slender wing
theory. Reproduced with
permission from Theoretical
and Computational
Aerodynamics. First Edition
Tapan K. Sengupta
Copyright (2015) John Wiley
& Sons. Ltd. UK

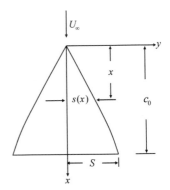

the plane of symmetry given by $U_\infty$. The $y$-axis is positive in the starboard direction, while $z$-axis is positive pointing downward. The trailing edge is straight and unswept. This theory leads to consideration that near the wing, the flow in the transverse section is that of a flow past flat plate of span $2s(x)$, in a uniform stream.

We note an interesting aspect of the present approach. Even though we are studying the three-dimensional flow field, we are going to use strip theory to render the problem two-dimensional and employ the tool of complex variable analysis which is strictly valid for two-dimensional flow field. The velocity normal to the plate on any chosen strip is $U_\infty\alpha$. The 2D problem is solved by the complex variable method of conformal mapping on the transverse flow plane ($Z = y + iz$) mapped to another $Z^*$ (= $y^*$ + $iz^*$) plane given by

$$Z^{*2} = Z^2 - s^2(x) \tag{2.136}$$

The flow field mapped from $Z$-plane to $Z^*$-plane is as indicated in Fig. 2.32. The complex potential of the flow in $Z^*$-plane is obtained in straightforward fashion as

$$W = -i\, U_\infty\alpha\, Z^* \tag{2.137}$$

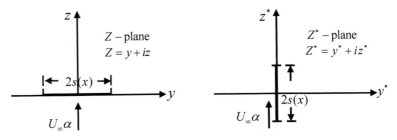

**Fig. 2.32** The flow in physical ($Z = y + iz$) and transformed plane ($Z^* = y^* + iz^*$) as modeled by slender wing theory. Reproduced with permission from Theoretical and Computational Aerodynamics. First Edition Tapan K. Sengupta Copyright (2015) John Wiley & Sons. Ltd. UK

$$= -i \ U_\infty \alpha \sqrt{Z^2 - s^2(x)}$$

The quantity under the radical sign is negative over the wing, for $z = 0$. The use of complex variable method enables one to define a complex potential for the incompressible, irrotational flow by

$$W = \phi + i\psi \tag{2.138}$$

Now using Eq. (2.137), one obtains, on the upper surface of the wing ($z = 0^+$), the velocity potential

$$\phi = U_\infty \alpha \left[ s^2(x) - y^2 \right]^{1/2} \tag{2.139}$$

Similarly, for the lower surface ($z = 0^-$), the velocity potential is obtained as

$$\phi = -U_\infty \alpha \left[ s^2(x) - y^2 \right]^{1/2} \tag{2.140}$$

The streamwise velocity component is obtained from the above expressions, which can be used next to calculate the pressure differential across the plate as follows:

$$\Delta C_p = C_{p_l} - C_{p_u} = \frac{2u_u}{U_\infty} - \frac{2u_l}{U_\infty}$$

It may appear as surprising that by inspecting the problem in $(y, z)$-plane, we are able to calculate $u_u$ and $u_l$, the $x$-component of velocity from velocity potential. This becomes feasible due to the implicit dependence of $\phi$ on $x$ through $s(x)$, the local semi-span. Thus, the pressure differential depends upon $\frac{ds}{dx} \neq 0$, such that

$$\Delta C_p = \frac{4\alpha \ s(x)}{[s^2 - y^2]^{1/2}} \frac{ds}{dx} \tag{2.141}$$

$$= \frac{4\alpha}{\sin \theta} \frac{ds}{dx}$$

where $y = s(x) \cos \theta$

As $\Delta C_p \to \infty$ when $y \to s(x)$, one notes the existence of infinite leading edge suction.

The lift acting over the strip of the wing can be obtained by integrating $\Delta C_p$, over the wingspan as

$$\delta L = \left[ \frac{1}{2} \rho \ U_\infty^2 \int\limits_{-s}^{s} \frac{4\alpha \ s(x)}{(s^2 - y^2)^{1/2}} \frac{ds}{dx} dy \right] \delta x$$

Or writing the non-dimensional form as

$$\frac{\delta L}{1/2 \, \rho \, U_\infty^2} = 4\alpha \, s(x) \frac{\mathrm{d}s(x)}{\mathrm{d}x} \delta x \int_{-s}^{s} \frac{\mathrm{d}y}{(s^2 - y^2)^{1/2}}$$

As $\mathrm{d}y = -s(x) \sin\theta \, \mathrm{d}\theta$, the above integral simplifies to

$$I = \int_{\pi}^{0} \frac{-s(x) \sin\theta \, \mathrm{d}\theta}{s(x) \sin\theta} = \pi$$

Therefore

$$\frac{\delta L}{1/2 \, \rho \, U_\infty^2} = 4\pi\alpha \, s(x) \frac{\mathrm{d}s(x)}{\mathrm{d}x} \delta x$$

If $S_w$ is the wing planform area, the lift coefficient is given by

$$C_L = \frac{\int^{c_0} \delta L}{1/2 \, \rho U_\infty^2 S_w} = \frac{4\pi\alpha}{S_w} \int_0^{c_0} s(x) \frac{\mathrm{d}s(x)}{\mathrm{d}x} \, \mathrm{d}x$$

$$= \frac{2\pi\alpha}{S_w} \int_0^{c_0} \frac{\mathrm{d}}{\mathrm{d}x}(s^2) \, \mathrm{d}x = \frac{2\pi\alpha}{S_w} s^2 = \frac{\pi\alpha}{2} \left( \frac{4S^2}{S_w} \right)$$

Hence

$$C_L = \frac{\pi\alpha}{2} \mathcal{R} \tag{2.142}$$

This provides the lift curve slope as

$$\frac{\mathrm{d}C_L}{\mathrm{d}\alpha} = \frac{\pi \mathcal{R}}{2} \tag{2.143}$$

This theoretical lift curve slope, $C_{L_\alpha}$, provides good match with experimental value for very small aspect ratio wing, $\mathcal{R} \to 0$ for very small angles of attack.

## 2.17  Spanwise Loading

For the low aspect ratio, flat wing with semi-span $S$ has pointed apex in Fig. 2.31 and has the straight trailing edge, as required in slender wing theory. To calculate the lift acting on a chord-wise strip of width $\delta y$ at a distance $y$ from the wing centerline, we note that the leading edge of the strip is given by $x_L(y)$. At the leading edge location, the local span is indicated in Fig. 2.33 as $s(x)$. The lift acting on this strip is obtained by using $\Delta C_p$ from Eq. (2.141) as

**Fig. 2.33**  Spanwise loading
of a low aspect ratio wing
obtained by slender wing
theory. Reproduced with
permission from Theoretical
and Computational
Aerodynamics. First Edition
Tapan K. Sengupta
Copyright (2015) John Wiley
& Sons. Ltd. UK

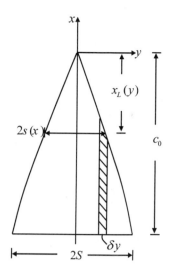

$$\delta L = \left[\frac{1}{2}\,\rho\,U_\infty^2\int_{x_L(y)}^{c_0}\Delta C_p\,\mathrm{d}x\right]\delta y = \left[\frac{1}{2}\,\rho\,U_\infty^2\int\frac{4\alpha\,s(x)}{[s^2-y^2]^{1/2}}\frac{\mathrm{d}s}{\mathrm{d}x}\mathrm{d}x\right]\delta y$$

$$\delta L == \frac{1}{2}\,\rho\,U_\infty^2 4\alpha\left[(S^2-y^2)^{1/2}-(s^2(x)-y^2)^{1/2}\right]\delta y \tag{2.144}$$

This spanwise loading is clearly noted to vary as an elliptic distribution.

The Kutta condition at the trailing edge is obtained for the unswept trailing edge, only when $\frac{\mathrm{d}s}{\mathrm{d}x}$ at $x = c_0$ is zero.

For the spanwise elliptic loading, we immediately obtain the induced drag of the slender wing to be given by

$$C_{D_v} = \frac{C_L^2}{\pi\,A\!\!R} = C_L\frac{\alpha}{2} \tag{2.145}$$

### 2.17.1  An Example of Delta or Triangular Wing Planform

For a delta or the triangular planform wing shown in Fig. 2.34, with the semi-apex angle $\Lambda$, the potential lift acting on is obtained from Eq. (2.141). From Fig. 2.34, one notes that at any streamwise station $x$,

$$s(x) = x\tan\Lambda$$

Thus,  $\dfrac{\mathrm{d}s}{\mathrm{d}x} = \tan\Lambda\ (= C_1)$

**Fig. 2.34** Delta wing or triangular wing characterized by the semi-apex angle $\Lambda$. Reproduced with permission from Theoretical and Computational Aerodynamics. First Edition Tapan K. Sengupta Copyright (2015) John Wiley & Sons. Ltd. UK

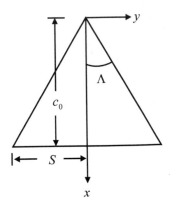

Therefore,

$$\Delta C_p = 4\alpha \, C_1 \frac{s(x)}{(s^2 - y^2)^{1/2}}$$

Using the transformation $y = s(x) \cos\theta$, one rewrites $\Delta C_p$ as

$$\Delta C_p = 4\alpha \, C_1 / \sin\theta$$

Thus, the loading is constant along rays, on which $\theta$ remains constant, starting from the apex. This constant loading along rays is known as the conical flow. Integrating Eq. (2.141) over the planform, one obtains the potential lift as

$$L = \frac{1}{2} \rho \, U_\infty^2 \int \frac{4\alpha \, s(x)}{\sqrt{s^2 - y^2}} \frac{ds}{dx} dy \, dx$$

This can also be directly obtained using the formula given in Eq. (2.142), for the planform in Fig. 2.34 to get

$$\mathcal{R} = \frac{4S^2}{S_w}, \quad \text{with } S_w = \frac{1}{2} \, 2S \, c_0$$

As $S = c_1 \tan\Lambda$, one obtains $\mathcal{R} = \frac{4S^2}{S \, c_0} = \frac{4S}{c_0} = 4\tan\Lambda$. Thus, the lift coefficient can be obtained for the delta wing as

$$C_L = 2\pi\alpha \tan\Lambda \tag{2.146}$$

The lift calculated by the pressure distribution, obtained from $\Delta C_p$, will be perpendicular to the planform, as indicated in Fig. 2.35. As shown in the figure, the other component of $C_{L_p}$ would have caused a drag, similar to induced drag. However, this is canceled by the leading edge suction created in the slender wing theory. At higher

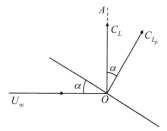

**Fig. 2.35** Potential flow lift acting on a low aspect ratio delta wing and its representation. Reproduced with permission from Theoretical and Computational Aerodynamics. First Edition Tapan K. Sengupta Copyright (2015) John Wiley & Sons. Ltd. UK

angles of attack, it is noted phenomenologically, an additional vortex at the leading edge. This actually creates additional lift, termed in the following as the vortex lift.

## 2.17.2 Low Aspect Ratio Wing Aerodynamics and Vortex Lift

The potential lift on the delta wing given in Eq. (2.146) is strictly by the pressure distribution acting along the normal to the surface of the wing. This resultant lift acting upon the wing at an angle of attack shown in the side view in Fig. 2.35, identified as $C_{L_p}$, can be decomposed to obtain the lift acting perpendicular to the oncoming flow direction as

$$C_{L_{pr}} \text{(along OA)} = C_{L_p} \cos \alpha$$

where from Eq. (2.146)

$$C_{L_p} = 2\pi \alpha \tan \Lambda$$

At significantly higher angles of attack, low aspect ratio delta wings in symmetric flight condition display a pair of vortices which remains firmly attached to the leading edge, as shown in Fig. 2.36. These attached primary vortices at the leading edge of the delta wing induce secondary vortices, also attached to the wing. However, these are essentially inboard of the primary vortex. As a consequence, these primary and secondary vortices induce additional flow in the streamwise direction. Looking at the surface flow pattern (more correctly the skin friction lines) indicates traces that originate from the core location of the vortices toward inboard and outboard directions. For even higher angles of attack, the secondary vortices can induce, even weaker vortex pair, inboard of the secondary vortices. These vortices form a very stable and robust vortex system creating higher lift without the flow displaying any tendencies for separation. The presence of these stable vortex systems over the delta

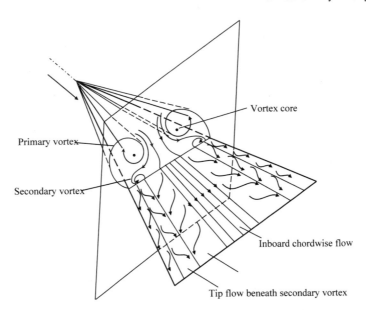

Vortex core

Primary vortex

Secondary vortex

Inboard chordwise flow

Tip flow beneath secondary vortex

**Fig. 2.36** Delta wing flow field in symmetric flight at high angle of incidence. Reproduced with permission from Theoretical and Computational Aerodynamics. First Edition Tapan K. Sengupta Copyright (2015) John Wiley & Sons. Ltd. UK

wing makes such wing planform very attractive for practical usage. However, such a benefit comes with the penalty of higher drag.

At low angles of attack, the correlation between potential lift with experimental value is very good, with the potential lift indicated above by $C_{L_p}$ that is obtained based on attached flow model with Kutta condition used. As angle of attack increases, formation of attached leading edge vortex system is noted, as explained above. This provides additional vortex lift, with major contribution from the primary vortex pair. The pair of vortex forms with its axis remaining aft of the leading edge, and at the trailing edge it merges with the tip vortices.

In [11, 13], a model is proposed for the noted flow features by hypothesizing the primary vortex pair to cause the stagnation line to move on the top surface, thereby rotating the suction force vector by 90 degrees to create the additional vortex lift. This turning of drag into lift led Pohlhamus to propose an empirical model, by which the drag caused in the cross-flow plane to be responsible for the additional lift.

For larger angles of attack, the potential flow lift in Eq. (2.146) is modified by replacing $\alpha$ by $\sin \alpha$, so that

$$C_{L_{pr}} = \underbrace{2\pi \tan \Lambda}_{K_P} \sin \alpha \cos \alpha \qquad (2.147)$$

The vortex lift is estimated by considering a spanwise strip on the delta wing of width $dx$ and length $2s(x)$, which experiences a cross-flow velocity $U_\infty \sin \alpha$. The drag experienced by this strip is represented as $C_{D_p}$, by which the cross-flow creates an equivalent *drag force* of magnitude

$$\frac{1}{2}\rho \, U_\infty^2 \sin^2 \alpha \, 2s(x) \, C_{D_p}$$

This equivalent flat plate drag coefficient is approximated as equal to 1.95, as suggested in [6, 16]. This *drag* acts perpendicular to the planform, and then the actual vortex lift is equal to this and given by

$$L_v = \frac{1}{2}\rho \, U_\infty^2 \sin^2 \alpha \, C_{D_p} \int_0^{c_0} 2s(x) \, \mathrm{d}x$$

As this lift also acts perpendicular to the plane wing and $s(x) = x \tan \Lambda$, then with respect to free-stream direction, the vortex lift coefficient is given by

$$C_{L_v} = C_{D_p} \sin^2 \alpha \cos \alpha \qquad (2.148)$$

Generalizing the above discussion, the total lift experienced by a delta wing is given by

$$C_L = K_p \, \sin \alpha \cos \alpha + K_v \sin^2 \alpha \cos \alpha \qquad (2.149)$$

In this generalized expression for total lift on the delta wing, $K_v$ is an unknown coefficient associated with the vortex lift, which needs to be parameterized based on experimental observations relating lift to wing planform shape and flow conditions. Similarly, $K_p$ is the modeled parameter for the component determined by pressure distribution obtained using the irrotational flow model. In the literature, this expression has been extended to other planform, such as arrow, delta, and diamond wings. A typical sketch for a composite geometry, which can represent these planform, is given in Fig. 2.37a. Corresponding lift coefficient functions $K_p$ and $K_v$ are shown in Figs. 2.37b and c. The variations of $K_p$ and $K_v$ with $R$ for arrow, delta, and diamond wings shown in these figures are reported in Pohlhamus [12]. However, this model of lift coefficient based on potential flow with leading edge suction analogy does not work well for moderate angles of attack.

For low aspect ratio wings, the delta planform is suitable for its low-speed and high-speed performances. During takeoff and landing of such aircrafts, the stall characteristics are vastly improved due to the presence of leading edge vortex systems shown in Fig. 2.36, which keeps the flow attached up to very high angles of attack without stall and abrupt loss of lift.

The lift curve slope of triangular wing is obtained from Eq. (2.149) as

$$\frac{dc_L}{d\alpha} = K_p \cos 2\alpha + K_v(2 \sin \alpha \cos^2 \alpha - \sin^3 \alpha)$$

$$= K_p \cos 2\alpha + K_v(2 \sin \alpha - 3 \sin^3 \alpha)$$

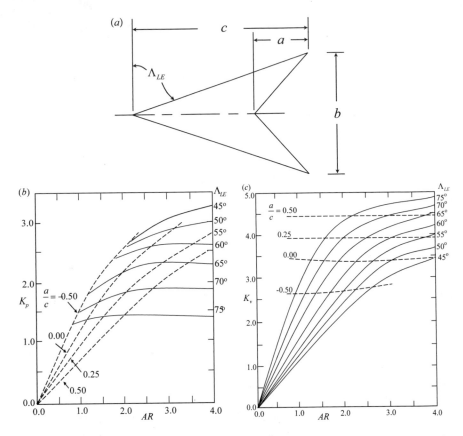

**Fig. 2.37** **a** Arrow, delta, and triangular wing characterized by the apex angle $\Lambda_{LE}$ and other dimensions given by $a$, $b$, and $c$. (**b**) Variation of $K_p$ with $\mathit{AR}$ for low aspect ratio triangular wing. The geometry is as defined in (**a**). (**c**) Variation of $K_v$ with $\mathit{AR}$ for low aspect ratio triangular wing geometry shown in (**a**). Reproduced with permission from Theoretical and Computational Aerodynamics. First Edition Tapan K. Sengupta Copyright (2015) John Wiley & Sons. Ltd. UK

In Fig. 2.38, the potential and vortex lift experienced by a 70-degree delta wing is shown for variation with angle of attack. The experimental results are due to Wentz and Kohlman [18]. Evidently, as angle of attack increases, the potential lift decreases due to breakdown of attached flow model used to evaluate $C_{L_p}$, while the vortex lift continues to increase, which is due to the leading edge primary vortex system. Actual total lift variation with angle of attack obtained experimentally is different from the correlated formula shown in Fig. 2.38. The experimental data points show saturation of total lift and subsequent gentle fall of the same, for angle of attack above $\alpha \geq 40°$.

The reason for the gentler drop of lift with high angles of attack of a delta wing is qualitatively different from that is noted for high aspect ratio wings, due to the presence of the leading edge vortex pairs. The primary vortex, which contributes most of the vortex lift, is extremely stable due to its axial and azimuthal components

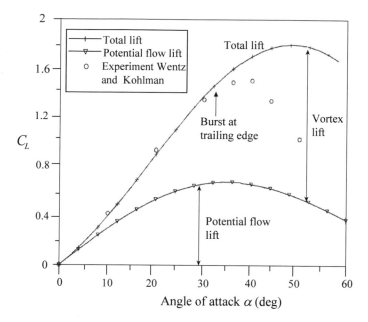

**Fig. 2.38** Comparison of modeled lift in Eq. (2.149) with experimental data in [4]. Total lift is split into potential and vortex lift contributions. Reproduced with permission from Theoretical and Computational Aerodynamics. First Edition Tapan K. Sengupta Copyright (2015) John Wiley & Sons. Ltd. UK

of the velocity distribution. This aspect of stability and instability of leading edge vortices is discussed in [16].

### 2.17.3  Slender Body Theory

This theory is a further generalization of slender wing theory described above to treat wing–body combinations. There are many applications including elongated bodies of revolution fitted with relatively smaller control surfaces. The flow field and loads are estimated for these three-dimensional bodies by considering slices of the wing–body analyzed in the cross-flow plane. These elementary contributions are summed up to provide quick estimate of the loads. A typical sketch of such a body is shown in Fig. 2.39. The body is considered at an angle of attack, $\alpha$, to the oncoming flow with speed, $U_\infty$. In the side view, only one control surface is visible, while in the cross section at the station $AA$, shown on the right in the $yz$-plane, the symmetrically placed pair of control surfaces is noted. As in slender wing theory, flow in the cross-flow plane is analyzed in isolation with the help of ideal potential flow model, using conformal mapping technique. The body of revolution is characterized by the radius, $R$, and wing by the semi-span, $s$.

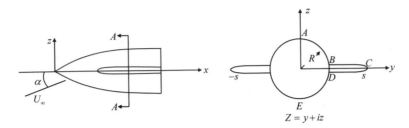

**Fig. 2.39** Flow past a wing–body combination analyzed by conformal mapping. Reproduced with permission from Theoretical and Computational Aerodynamics. First Edition Tapan K. Sengupta Copyright (2015) John Wiley & Sons. Ltd. UK

The complex potential for the flow field in the cross-flow plane is written using $Z = y + iz$, for the oncoming velocity $U_\infty \, \alpha$, as

$$W(Z) = \phi(y, z) + i\psi(y, z) \tag{2.150}$$

where $\phi$ is the velocity potential and $\psi$ is the stream function. We analyze the flow field by first making conformal mapping in two steps. The first step is by the Jukowski transformation, mapping the complex $Z$-plane to $Z_1$-plane by

$$Z_1 = Z + R^2/Z \tag{2.151}$$

The transformed body stretches the semi-span to $s_1 = s + R^2/s$ in the $Z_1$ plane. This makes the wing–body in $Z$-plane to a slit in the $Z_1$-plane, with the flow normal to this slit. Further simplification results in obtaining another complex potential, obtained by transforming the horizontal slit in $Z_1$-plane to a vertical slit in $Z_2$-plane, so that the flow is tangential to the vertical slit. This is performed by the second Jukowski transformation (Fig. 2.40):

$$Z_2 = (Z_1^2 - s_1^2)^{1/2} \tag{2.152}$$

It is easy to obtain the complex potential in $Z_2$-plane given by

$$W(Z_2) = -i \, U_\infty \, \alpha Z_2 \tag{2.153}$$

Using the transforms of Eqs. (2.151) and (2.152) in Eq. (2.153), one gets the requisite complex potential in the original complex $Z$-plane as

$$W(Z) = -i \, U_\infty \, \alpha \left[ (Z + R^2/Z)^2 - (s + R^2/s)^2 \right]^{1/2} \tag{2.154}$$

This complex potential can be further simplified by expanding Eq. (2.154) in inverse power of $Z$, as given in [1] by

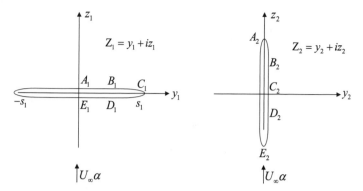

**Fig. 2.40** Jukowski transforms used to calculate potential lift for flow past a wing–body combination. Reproduced with permission from Theoretical and Computational Aerodynamics. First Edition Tapan K. Sengupta Copyright (2015) John Wiley & Sons. Ltd. UK

$$W = -i\alpha U_\infty \left\{ Z - \frac{1}{2Z}\left[ \left( s + \frac{R^2}{s} \right)^2 - 2R^2 \right] + \cdots \right\} \qquad (2.155)$$

This simplification helps us identify the simple pole and use Blasius' theorem to calculate the lift acting from the residue evaluated for the pole at $Z = 0$ [1],

$$L = \pi \rho_\infty \, U_\infty^2 \, \alpha \left( s^2 - R^2 + \frac{R^4}{s^2} \right)_{\text{(base)}} \qquad (2.156)$$

where the subscript on the right-hand side indicates the quantities to be evaluated at the base of the body. One evaluates the lift acting on the body alone by substituting $s_{\text{base}} = R_B$ as

$$L_B = \pi \rho_\infty \, U_\infty^2 \, \alpha R_B^2$$

Thus, the lift coefficient is obtained (based on the base area) as

$$C_{L_B} = 2\alpha$$

Similarly, the lift acting on the wing alone is obtained by substituting $R_B = 0$ in Eq. (2.156) as

$$L_W = \pi \rho_\infty \, U_\infty^2 \, \alpha S_B^2$$

where $S_B$ is the base span at the trailing edge.
Corresponding lift coefficient is obtained as

$$C_{L_W} = \frac{\pi}{2} \mathcal{R}\alpha$$

**Fig. 2.41** Flow diagram for
the problem

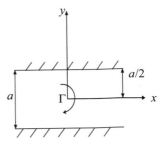

The body-alone result is due to Munk [10], and wing-alone result is due to Jones
[7]. These results are semi-analytic, and with the present-day methods of solving the
Navier–Stokes equation can provide superior results. The above estimates are to be
treated as ballpark quick estimates, to be augmented by proper experiments.

## Problems

**P2.1** A vortex of strength $\Gamma$ is located midway between two parallel walls separated
by distance $a$, as shown in Fig. 2.41. Calculate the resulting complex potential and
plot the streamline contours.

**P2.2** Find the slope at the leading edge of a symmetric Joukowski airfoil.

**P2.3** Consider an infinitely thin flat plate of chord c is held at an angle of attack $\alpha$
in a supersonic flow. The pressures on the upper and lower surfaces are different but
constant over each surface; that is, $p_u(s) = c_1$ and $p_l(s) = c_2$, where $c_1$ and $c_2$ are
constants and $c_2 > c_1$. Ignoring shear stress, calculate the location of the center of
pressure (Fig. 2.42).

**P2.4** Consider a 10% thick and 5% camber, Jukowski airfoil. Find out the Jukowski
transformation. Calculate the velocity on the upper and lower surface of the airfoil
40% behind the leading edge.

**Fig. 2.42** Flow diagram for
problem

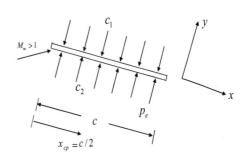

**Fig. 2.43** Circulation distribution over the finite wing

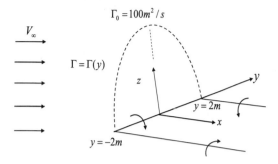

**P2.5** The NACA 4412 airfoil has a mean camber line given with chord as '$c$', by

$$\frac{y}{c} = \begin{cases} 0.25\left[0.8\frac{x}{c} - \left(\frac{x}{c}\right)^2\right] & \text{for}\quad 0.0 \le \frac{x}{c} \le 0.4 \\ 0.111\left[0.2 + 0.8\frac{x}{c} - \left(\frac{x}{c}\right)^2\right] & \text{for}\quad 0.4 \le \frac{x}{c} \le 1.0 \end{cases}$$

Using thin airfoil theory calculate,

(a) zero lift angle $\alpha_{L=0}$

(b) $C_l$ when $\alpha = 3^0$

**P2.6** For the airfoil given in problem P2.5, calculate $C_{m_{c/4}}$ and $x_{cp}/c$ when $\alpha = 2^0$.

**P2.7** Consider a potential flow over a finite wing with the following circulation distribution:

$$\Gamma(y) = 100\sqrt{1 - \left(\frac{2y}{4}\right)^2}\, m^2/s$$

If free stream velocity is 100 m/s, then what will be the downwash angle along the span (Fig. 2.43)?

**P2.8** Consider an untwisted wing of elliptical planform in inviscid, incompressible, irrotational flow at an angle of attack 4°. The wing has an aspect ratio of $R = 7$ and the zero lift angle of attack is -2°. What will be wing lift coefficient and induced drag coefficient?

**P2.9** Consider the case where the spanwise circulation distribution for a wing is parabolic,

$$\Gamma(y) = \Gamma_0\left(1 - \frac{y^2}{s^2}\right)$$

If the total lift generated by the wing with the parabolic circulation distribution is to be equal to the total lift generated by a wing with an elliptic circulation distribution, what is the relation between the $\Gamma_0$ values for the two distributions? What is the relation between the induced downwash velocities at the plane of symmetry for the two configurations?

**P2.10** A designer claims to fly with reduced drag by reducing the span of the aircraft wing that has a rectangular planform. For a given weight during the flight the span is reduced to 2/3 of its value. Estimate the ratio of drag before and after the change of span. Use a drag polar of the type: $C_D = C_{D_0} + kC_L^2$ and $L/D$ of the aircraft as 15 (a very conservative value!).

# References

1.  H. Ashley, M. Landahl, *Aerodynamics of Wings and Bodies* (Dover Publ. Inc., New York, USA, 1965)
2.  V.M. Falkner, The solution of lifting plane problem by vortex lattice method. *British ARC*, R & M **2591** (1948)
3.  H. Glauert, *The Elements of Aerofoil and Airscrew Theory* (Cambridge University Press, Cambridge, 1926)
4.  I. Heron, Vortex burst behaviour of a dynamically pitched delta wing under the influence of a von Karman vortex street and unsteady freestream. Ph.D. Thesis submitted to Department of Aerospace Engineering, Wichita State University, USA, 2007
5.  J.L. Hess, A.M.O. Smith, Calculation of potential flow about arbitrary bodies. Prog. Aerospace Sci. **8**, 1–138 (1967)
6.  E.L. Houghton, P.W. Carpenter, *Aerodynamics for Engineering Students*, 4th edn. (Edward Arnold Publishers Ltd., UK, 1993)
7.  R.T. Jones, Properties of low-aspect ratio pointed wings at speeds below and above the speed of sound. NACA Rep. **835** (1946)
8.  J.K. Puttam, K. Madhu Babu, V.K. Suman, V.Y. Mudkavi, T.K. Sengupta, Aerodynamic Analysis of Dornier 228 With and Without External Antennae. HPC Laboratory, Department of Aerospace Engineering, IIT Kanpur CTFD Division, CSIR-NAL, Bangalore (2014)
9.  F.W. Lanchester, *Aerodonetics* (Constable, UK, 1908)
10. M.M. Munk, The aerodynamic forces acting on airship hulls. NACA Rep. **184** (1924)
11. E.C. Pohlhamus, A concept of the vortex lift of sharp-edge delta wing based on a leading edge suction analogy. NASA TN D- **3767** (1966)
12. E.C. Pohlhamus, Charts for predicting the subsonic vortex-lift characteristic of arrow delta and diamond wings. NASA TN D- **6243** (1971)
13. E.C. Pohlhamus, Predictions of vortex-lift characteristics by a leading-edge suction analogy. J. Aircraft **8**(4), 193–199 (1971a)
14. L. Prandtl, Application of modern hydrodynamics to Aeronautics. NACA TR **116** (1921)
15. T.K. Sengupta, *High Accuracy Computing Methods: Fluid Flows and Wave Phenomenon* (Cambridge University Press, USA, 2013)
16. T.K. Sengupta, *Theoretical and Computational Aerodynamics*, Aerospace Series (Wiley, UK, 2015)
17. T.K. Sengupta, Y.G. Bhumkar, S. Sengupta, Dynamics and instability of a shielded vortex in close proximity of a wall. Comput. Fluids **70**, 166–175 (2012)
18. W.H. Wentz, D.L. Kohlman, Vortex breakdown on slender sharp-edged wings. J. Aircraft **8**(3), 156–161 (1971)
19. F.M. White, *Fluid Mechanics*, 6th edn. (McGraw-Hill, New York, USA, 2008)

# Chapter 3
# Governing Equations for Aerodynamics and Acoustics

## 3.1  Governing Equations in Fluid Dynamics

Fluid dynamical governing equations are given by the conservation principles of
mass, momentum, and energy as noted in Chap. 1. Although the Navier–Stokes equation is an application of Newton's second law for fluid flows, one also assumes the
relation between the stress and the rate of strain tensor. There are many versions of
Navier–Stokes equation (NSE), depending upon the constitutive relation of the fluid
medium. In the following, we will focus mainly on what is known as Newtonian
fluid, for which the stress and rate of strain have linear relation. Additionally, some
more assumptions are made, which requires rudimentary knowledge of equilibrium
thermodynamics, in the form of Stokes' hypothesis. For flows with heat-transfer and
compressibility effects, one also requires to solve the energy equation to determine
the temperature and density at each point. Sometimes heat-transfer effects are not too
strong for incompressible flows, enabling one to make simplifying assumption for
which the temperature gradient affects only the fluid density, and whose net effects are
via gravity. In incompressible NSE, this buoyancy effect by the temperature-induced
density variation is modeled by Boussinesq approximation [65]. For compressible
flows, Boussinesq assumption is not required, as the governing equations which are
the compressible NSE, with conservation of mass, momentum, and energy, are considered. Additionally, one requires to close the system of equations by the equation of
state relating temperature, density, and pressure. We have noted that the borderline
between compressible and incompressible flows in fluid mechanics is enunciated
empirically by the Mach number, to be below 0.3, then the relative density variation
is below 5%. However, when it comes to estimating the sound generation and its
propagation, incompressible assumption is unacceptable, as propagation of sound
wave is intimately related to bulk action of fluid flow to applied stresses. This has
two consequences: (i) an error of 5% in density variation is too high, such variations can be the consequence of sound and (ii) bulk viscosity should not be treated
as zero as it is done for classical Navier–Stokes equation with Stokes' hypothesis.
Thus, ideally speaking, one would be required to revisit Stokes' hypothesis, and

© Springer Nature Singapore Pte Ltd. 2020

T. K. Sengupta and Y. G. Bhumkar, *Computational Aerodynamics and Aeroacoustics*,
https://doi.org/10.1007/978-981-15-4284-8_3

retain bulk viscosity term. In the second part of the book, we will show some such calculations of acoustics problems starting from the first principle, without invoking any empiricism, which happens to be the state of the art in acoustics.

The conservation of mass, momentum, and energy for fluid flows is illustrated in the following.

### 3.1.1   Mass Conservation or Equation of Continuity

The conservation of mass for a *control volume* requires that the net rate of creation and destruction of mass inside it is given by the amount of mass flow rate through the *control surfaces*. For general unsteady compressible flows, this is stated as

$$\frac{D\rho}{Dt} + \rho \, \nabla \cdot \vec{V} = \frac{\partial \rho}{\partial t} + \nabla \cdot (\rho \, \vec{V}) = 0 \qquad (3.1)$$

where $D/Dt$ stands for the material derivative and $\vec{V}$ is the velocity vector, so that

$$\frac{D}{Dt} = \left( \frac{\partial}{\partial t} + \vec{V} \cdot \nabla \right)$$

Incompressible flows are characterized by constant density, i.e., $\rho = $ constant, and Eq. (3.1) simplifies to

$$\nabla \cdot (\vec{V}) = 0 \qquad (3.2)$$

Vectors satisfying similar condition given by Eq. (3.2) are called the *divergence-free* or *solenoidal* field. For problems of acoustics, the velocity field will not necessarily be required to be solenoidal.

### 3.1.2   Momentum Conservation Equation

The conservation of translational momentum is obtained by the Newton's second law applied to the control volume. Following its motion is equivalent to tracking the *control mass system* (for which mass is constant with time). Rate of change of momentum inside the *control volume* is obtained by summing net forces acting on the control volume, with the total rate of change of momentum which crosses (enters or leaves) the *control surface* elements. This equation is presented in Sects. 1.2–1.4.

In the context of non-equilibrium thermodynamics, one can define a *bulk viscosity* hypothesis in Eq. (1.7) by

$$\mu_b = \lambda + \frac{2}{3}\mu$$

The adoption of Stokes' hypothesis is equivalent to making the bulk viscosity to vanish. There is an alternate interpretation in that the action of bulk viscosity will be of no consequence, when $D\rho/Dt \equiv 0$, i.e., the effect of compressibility is negligible. This is the case for aerodynamics in the incompressibility limit of Mach number lower than the value of 0.3. However, even for very small values of Mach number, effect of compressibility cannot be neglected for problems of acoustics.

Unfortunately, even for transonic and supersonic flows, use of vanishing bulk viscosity is a very common practice, despite the fact that the flow compressibility is significant in these speed regimes. We may also like to note that in many aeroacoustic computations, one ends up solving Euler equation, which does not require using relation between stress and rate of strain. Thus, Stokes' hypothesis is not invoked at all in solving such acoustic problem. Thus, any serious attempt to use viscous flow equations for acoustic problems might require abandoning the use of Stokes' hypothesis to begin with.

Researchers have questioned the validity of Stokes' hypothesis by pointing out the serious shortcomings [78]. Stokes' hypothesis, which equates $\mu_b$ as zero for any flow, by equating thermodynamic and mechanical pressures has been identified as responsible for this inconsistency [11, 19, 20, 28, 49, 77]. Emanuel [19] and Buresti [11] have noted that Stokes' hypothesis may be weakly justified for monoatomic gases. However, Liebermann [40], Karim and Rosenhead [34] and Rosenhead [56] have noted that $\lambda$ is independent of $\mu$, while it can have not only higher amplitude, but also the sign reversed. Emanuel [19] and Gad-el-Hak [20] questioning Stokes' hypothesis, specifically noted its inapplicability for re-entry aerodynamics into planetary atmosphere. Cramer [15], in taking a kinetic theory of gas approach, noted that many diatomic gases can have bulk viscosity ($\mu_b$) thousand times larger than $\lambda$, which also prompted Rajagopal [49] to suggest that new approach be adopted for NSE, applicable for all fluids, without the use of Stokes' hypothesis.

As the first coefficient of viscosity depends on pressure and temperature, viewed from macroscopic and microscopic perspectives, the bulk viscosity will also depend on the same thermodynamic variables. Here the temperature dependence will be on absolute value. Action of bulk viscosity is simply not determined by state property of the fluid, it is also dependent on the flow process. For example, action of bulk viscosity will be a strong function of compression and dilatation involved in the process. For this reason, propagation of sound requires the presence of the divergence of the velocity field, however small the value may be.

According to Landau [38], for dilute gas in rapid compression or expansion, the process ceases to be in thermodynamic equilibrium, and internal processes start to restore local thermodynamic equilibrium. These rapidity of the processes cause the relaxation time to be short and the return to equilibrium makes the change in volume almost instantaneous, especially in the case of acoustic signal propagation. It may then imply that the process will be without significant loss of energy. However, if the relaxation time for return to equilibrium is long, then considerable dissipation of energy will occur for nonzero divergence of velocity field. This dissipation is determined by the second coefficient of viscosity ($\lambda$), and the bulk viscosity ($\mu_b$) is necessarily large.

According to Graves and Argrow [28], departure from thermodynamic equilibrium can occur from other routes also (apart from that is given above): *In a dense gas, bulk viscosity occurs from the viscous forces that arise when a volume of fluid is compressed or dilated without change of shape. Whereas a liquid bulk viscosity involves the structural relaxation associated with rearrangement of molecules during acoustic compression and rarefaction.* The authors [28] have even questioned whether all these sources of departure from equilibrium can be accommodated by a single transport coefficient.

Keeping in tune with the contents of this book, we will only focus on effects of bulk viscosity for dilute gases, for which kinetic theory of gases provides a framework [15]. The specific energy can be considered only as a function of absolute temperature, with the expression given as

$$e(T) = e_{tr} + \sum_{i=1}^{N} e_i$$

where $e_{tr}$ is the translation energy at the molecular level and the second term is contributed by internal energy modes associated with rotation and vibration. These latter quantities will not contribute for mono-atomic gases at up to moderate temperatures, or even when the temperature is raised to very high value, increased frequency of molecular collision returns the system to local thermodynamic equilibrium, whose time scale is of the order of collision time interval. In contrast, for polyatomic gases presence of rotational and vibrational modes is necessary due to the conceived atomic connections. As a consequence, the time scale for return to another equilibrium state will be significantly longer for polyatomic gas. It is not only that each mode will equilibrate with itself, but the rotational and vibrational mode will achieve equilibrium with the translational mode also. If $\tau_i$ is the relaxation time for the $i$th internal mode, then Tisza [89] reported a zero-frequency, near-equilibrium value for bulk viscosity as

$$\mu_b = (\gamma - 1)^2 \sum_{i=1}^{N} \frac{1}{R} \frac{de_i}{dT} p \tau_i$$

where the author reported that for $CO_2$ and $N_2O$, $\mu_b$ is roughly 2000 times $\mu$, while for air, $\mu_b = 0(\mu)$. In the context of aerodynamics and aeroacoustics, we are more concerned with air only. However, most of the times, presence of moisture changes the property of the humid air. In Fig. 4 of Cramer [15], temperature variation of $\mu_b/\mu$ is shown for water vapor, which shows a very typical behavior of this ratio increasing with reduction in temperature. In near-room temperature, it is noted that $\mu_b/\mu$ attains a value that lies between 7 and 8. This variation is contributed mostly by contributions from rotational and vibrational modes. In contrast, there are lighter gases like $H_2$, which at normal temperature range below $6000\,k$, and the non-equilibrium is sustained over a longer time due to lower weight and lower moment of inertia. For such reason, $\mu_b/\mu$ attains a value in the range of 36–55 [15]. In some gases and liquids, the vibrational mode contributes significantly to $\mu_b/\mu$, which are of the order of hundreds. The largest value of $\mu_b/\mu$ known for common gas is for $CO_2$, which is of the order of thousands, as obtained from relaxation time. However, such relaxation times are very sensitive to the presence of trace amount of water vapor. For example, at $300\,K$, the higher relaxation time yields a value of $\mu_b/\mu = 3849$.

The estimation of bulk viscosity is often measured from sound absorption experiments. Ash et al. [2] reported second coefficient of viscosity ($\lambda$) by similar measurement for air in a temperature range of 284–323 K, while the relative humidity varying between 6 and 91%. It has been noted that the *presence of the angular $H_2O$ molecule* due to humidity increases the relaxation time of vibrational mode for $N_2$ and $O_2$, which leads to the ratio of $\lambda/\mu$ to vary between 2000 and 20,000. Such a drastic variation of $\mu_b$ prompted Sengupta et al. [78] to study the effects of bulk viscosity on Rayleigh–Taylor instability, using the data provided in the experiment [2] using the regression analysis. This is believed to be the first exercise of its kind, where Stokes' hypothesis is dispensed with and a realistic macroscopic model for bulk viscosity has been used for solving compressible Navier–Stokes equation.

Landau and Lifshitz [38] report the Navier–Stokes equation with bulk viscosity (for constant $\mu$ and $\mu_b$) to be given by the following:

$$\rho\frac{D\vec{V}}{Dt} = -\nabla p + \mu\nabla\cdot[\nabla\vec{V} + (\nabla\vec{V})^T] + \left(\mu_b - \frac{2\mu}{3}\right)\nabla(\nabla\cdot\vec{V}) + \rho\vec{g}$$

As $\nabla\cdot(\nabla\vec{V}) = \nabla(\nabla\cdot\vec{V})$; $\nabla\cdot[(\nabla V)^T] = \nabla^2\vec{V}$ and $\nabla^2\vec{V} = \nabla(\nabla\cdot\vec{V}) - \nabla\times(\nabla\times\vec{V})$ One obtains the alternate form as

$$\rho\frac{DV}{Dt} = -\nabla p + \mu\nabla^2\vec{V} + (\frac{\mu}{3} + \mu_b)\nabla(\nabla\cdot\vec{V}) + \rho\vec{g}$$

It is written in rotational form as

$$\rho\frac{D\vec{V}}{Dt} = -\nabla p - \mu\nabla\times(\nabla\times\vec{V}) + \left(\frac{4\mu}{3} + \mu_b\right)\nabla(\nabla\cdot\vec{V}) + \vec{g}\rho$$

Using the constitutive relation in Eq. (1.16), one writes the final form of NSE for compressible flow as

$$\frac{\partial(\rho u)}{\partial t} + \frac{\partial(\rho u^2)}{\partial x} + \frac{\partial(\rho uv)}{\partial y} + \frac{\partial(\rho uw)}{\partial z} = -\frac{\partial p}{\partial x} + $$
$$\frac{\partial}{\partial x}\left(\lambda\nabla\cdot\vec{V} + 2\mu\frac{\partial u}{\partial x}\right) + \frac{\partial}{\partial y}\left(\mu\left[\frac{\partial v}{\partial x} + \frac{\partial u}{\partial y}\right]\right) + \frac{\partial}{\partial z}\left(\mu\left[\frac{\partial u}{\partial z} + \frac{\partial w}{\partial x}\right]\right) + \rho f_x \quad (3.3)$$

$$\frac{\partial(\rho v)}{\partial t} + \frac{\partial(\rho uv)}{\partial x} + \frac{\partial(\rho v^2)}{\partial y} + \frac{\partial(\rho vw)}{\partial z} = -\frac{\partial p}{\partial y} +$$

$$\frac{\partial}{\partial y}\left(\lambda \nabla \cdot \vec{V} + 2\mu \frac{\partial v}{\partial y}\right) + \frac{\partial}{\partial x}\left(\mu\left[\frac{\partial v}{\partial x} + \frac{\partial u}{\partial y}\right]\right) + \frac{\partial}{\partial z}\left(\mu\left[\frac{\partial v}{\partial z} + \frac{\partial w}{\partial y}\right]\right) + \rho f_y \quad (3.4)$$

$$\frac{\partial(\rho w)}{\partial t} + \frac{\partial(\rho uw)}{\partial x} + \frac{\partial(\rho wv)}{\partial y} + \frac{\partial(\rho w^2)}{\partial z} = -\frac{\partial p}{\partial z} +$$

$$\frac{\partial}{\partial z}\left(\lambda \nabla \cdot \vec{V} + 2\mu \frac{\partial w}{\partial z}\right) + \frac{\partial}{\partial x}\left(\mu\left[\frac{\partial w}{\partial x} + \frac{\partial u}{\partial u}\right]\right) + \frac{\partial}{\partial y}\left(\mu\left[\frac{\partial v}{\partial z} + \frac{\partial w}{\partial y}\right]\right) + \rho f_z \quad (3.5)$$

These equations are simplified for incompressible flows by noting $\nabla \cdot \vec{V} = 0$ that the terms multiplied by $\lambda$ drop out in Eqs. (3.3)–(3.5). Also for incompressible flows with little or no temperature variation, $\mu$ is treated as a constant. With these assumptions, the vector form of the NSE for incompressible flow is

$$\frac{\partial \vec{V}}{\partial t} + (\vec{V} \cdot \nabla)\vec{V} = -\frac{\nabla p}{\rho} + \nu \nabla^2 \vec{V} + \vec{F} \quad (3.6)$$

This is the equation using *primitive variables* and is noted as the *primitive variable formulation* of incompressible NSE.

### 3.1.3  Energy Conservation Equation

This is nothing but the first law of thermodynamics for a control volume system stated as: The rate of change of energy inside the *control volume* must be the sum of heat transfer across the *control surface* and the work done by body and surface forces, with the relevant terms in the energy equation as follows.

**Work done terms due to body and surface forces:**
The time rate of work done by body forces for the *control volume* of mass $\rho$ (dx dy dz) is given in mixed notation as

$$\left[-\frac{\partial}{\partial x_j}(pu_j) + \frac{\partial}{\partial x_j}(u_i \tau_{ji})\right] dx\ dy\ dz + \rho\ F_j u_j\ dx\ dy\ dz \quad (3.7)$$

**Heat-transfer terms:**
The net flux of heat is given by the volumetric heating (absorption or emission of radiation) and heat transfer through the control surface is due to conduction. Defining the rate of heat addition per unit mass by $\dot{q}$, the volumetric heating of the control volume is

$$= \rho\ \dot{q}\ dx\ dy\ dz \quad (3.8)$$

Fourier's law provides the directional conductive heat transfer through the control surfaces with the temperature gradient given by

$$\dot{q}_j = -\kappa \frac{\partial T}{\partial x_j}$$

where $\kappa$ is the thermal conductivity. The total heat interaction using Newton's law is obtained as

$$\left[ \rho \dot{q} + \frac{\partial}{\partial x_j} \left( \kappa \frac{\partial T}{\partial x_j} \right) \right] dx \, dy \, dz \tag{3.9}$$

**Substantive derivative of change of energy:**
For the fluid control volume, the total energy is written as $E = e + \frac{V^2}{2}$, where $E$, $e$, and $\frac{V^2}{2}$ represent specific total, internal, and kinetic energy (per unit mass). The substantive derivative of $E$ is given by

$$\rho \frac{D}{Dt} \left( e + \frac{V^2}{2} \right) dx \, dy \, dz \tag{3.10}$$

**The final energy equation:**
The final equation is obtained by collecting all the contributions listed above to obtain the *non-conservation* or *convective* form of the energy equation as

$$\rho \frac{D}{Dt} \left( e + \frac{V^2}{2} \right) = \left[ \rho \dot{q} + \frac{\partial}{\partial x_j} \left( \kappa \frac{\partial T}{\partial x_j} \right) \right] - \frac{\partial}{\partial x_j} (p u_j) + \frac{\partial}{\partial x_j} (u_i \tau_{ji}) + \rho \, F_j u_j \tag{3.11}$$

## 3.1.4 Other Forms of Energy Equation

The energy equation is often written in terms of $e$ as

$$\rho \frac{De}{Dt} = \rho \dot{q} + \frac{\partial}{\partial x_j} \left( \kappa \frac{\partial T}{\partial x_j} \right) - p \frac{\partial u_j}{\partial x_j} + \lambda \left( \frac{\partial u_j}{\partial x_j} \right)^2 + 2\mu \left[ \left( \frac{\partial u}{\partial x} \right)^2 + \left( \frac{\partial v}{\partial y} \right)^2 + \left( \frac{\partial w}{\partial z} \right)^2 \right]$$
$$+ \mu \left[ \left( \frac{\partial u}{\partial y} + \frac{\partial v}{\partial x} \right)^2 + \left( \frac{\partial u}{\partial z} + \frac{\partial w}{\partial x} \right)^2 + \left( \frac{\partial v}{\partial z} + \frac{\partial w}{\partial y} \right)^2 \right] \tag{3.12}$$

This can be written in conservation form from the above by noting

$$\rho \frac{De}{Dt} = \frac{\partial}{\partial t} (\rho e) + \nabla \cdot (\rho e \vec{V})$$

Terms involving $\mu$ constitute the dissipation term ($\Phi_0$) in Eq. (3.12).

## 3.1.5  Derived Variable Equation for Incompressible Flows

For fluid flows, vorticity $\vec{\omega}$, is the derived variable obtained as the curl of the velocity, i.e., $\vec{\omega} = \nabla \times \vec{V}$. The derived variable formulation avoids the problems of primitive variable equation, namely, the absence of Dirichlet condition for pressure, and this is often avoided by using derived variable formulation, which eliminates the pressure-dependent terms. For incompressible NSE, this is obtained by taking a curl of Eq. (3.6) and using the definition of vorticity to obtain the *non-conservative form* of vorticity transport equation (VTE) as

$$\frac{\partial \vec{\omega}}{\partial t} + (\vec{V} \cdot \nabla)\vec{\omega} = (\vec{\omega} \cdot \nabla)\vec{V} + \nu\nabla^2\vec{\omega} \tag{3.13}$$

The first term on the right-hand side of the above is due to vortex stretching, and is absent for 2D flows. One uses the vector identity $\nabla \times (\vec{A} \times \vec{B}) = \vec{A}(\nabla \cdot \vec{B}) - \vec{B}(\nabla \cdot \vec{A}) + (\vec{B} \cdot \nabla)\vec{A} - (\vec{A} \cdot \nabla)\vec{B}$ and solenoidality of velocity and vorticity $(D_v = \nabla \cdot \vec{V} = 0$ and $D_\omega = \nabla \cdot \vec{\omega} = 0)$, and obtains the so-called the *Laplacian form* of VTE as

$$\frac{\partial \vec{\omega}}{\partial t} + \nabla \times (\vec{\omega} \times \vec{V}) = \nu\nabla^2\vec{\omega} \tag{3.14}$$

The viscous term of Eq. (3.14) can also be modified by using the identity

$$\nabla^2\vec{\omega} = \nabla(\nabla \cdot \vec{\omega}) - \nabla \times (\nabla \times \vec{\omega})$$

so that,  $\nabla^2\vec{\omega} = -\nabla \times (\nabla \times \vec{\omega})$, as $D_\omega = 0$.

Further modifying the RHS of Eq. (3.14) using the above, we get the *rotational form* of VTE

$$\frac{\partial \vec{\omega}}{\partial t} + \nabla \times (\vec{\omega} \times \vec{V}) + \nu\nabla \times (\nabla \times \vec{\omega}) = 0 \tag{3.15a}$$

which is often written concisely by defining $\vec{H_\omega} = (\vec{\omega} \times \vec{V} + \nu\nabla \times \vec{\omega})$

The rotational form of VTE is expressed as

$$\frac{\partial \vec{\omega}}{\partial t} + \nabla \times \vec{H_\omega} = 0 \tag{3.15b}$$

Theoretically, the vorticity is divergence-free, which comes from the vector calculus relating velocity and vorticity. This is always true for 2D flows, as the vorticity vector is perpendicular to the plane of the flow. For 3D numerical computation using vorticity transport equation, this is not automatically guaranteed at all points in the flow domain. For wall-bounded flows, vorticity is created due to the no-slip condition at the wall for all time instants. If the wall vorticity is not made divergence-free, then wall vorticity itself will be the source of error at all times. Thus, the judicious choice of the form of VTE for 3D simulations is a vital requirement. Three different forms of VTE providing the governing equation for $D_\omega$ are given in Eqs. (3.13)–(3.15a), whose details can also be found in [7, 63].

The incompressible NSE is solved using *velocity–vorticity formulation*, and apart from solving the vorticity transport equations, one additionally solves Poisson equation for velocity components by taking a curl of vorticity to obtain

$$\nabla^2 \vec{V} = -\nabla \times \vec{\omega} \qquad (3.16)$$

Any form of VTE given in Eqs. (3.13)–(3.15a) along with the velocity Poisson equation Eq. (3.16) constitute the governing *velocity–vorticity* equations for the derived variable formulation. In this formulation of NSE, satisfaction of divergence-free condition of velocity ($D_v = 0$) depends on the accuracy with which Eq. (3.16) is solved [6, 7]. Some researchers did not use Eq. (3.16) due to accuracy issues in computing the velocity field. For example, authors in [24–26, 45] have instead solved continuity equation, along with the kinematic definition of vorticity, in addition to the VTE. Special measures are necessary in these approaches to avoid the issue of over-determined system of linear algebraic set of equations.

Another possibility is using the vector potential $\vec{\Psi}$, which is defined from the velocity which is the curl of $\vec{\Psi}$

$$\vec{V} = \nabla \times \vec{\Psi} \qquad (3.17)$$

This automatically ensures the velocity field $\vec{V}$, to satisfy the solenoidality condition for the velocity field, $D_v = 0$. Further, taking the curl of Eq. (3.17), and additionally imposing that $\vec{\Psi}$ is also divergence-free, ($\nabla \cdot \Psi = 0$), we get the vector equation for $\vec{\Psi}$ as

$$\nabla^2 \vec{\Psi} = -\vec{\omega} \qquad (3.18)$$

In 2D, $\vec{\Psi}$ has only one component, same as that for the vorticity, both of which are normal to the plane of the flow. These lead to the familiar *stream function vorticity* ($\psi, \omega$)-formulation for 2D flows. The ($\psi, \omega$)-formulation possess significant advantages over the primitive variable formulation via (i) only two unknowns, as compared to three unknowns for the ($p, \vec{V}$)-formulation; (ii) it also exactly satisfies mass conservation everywhere in the domain numerically; (iii) one obtains a primary quantity of interest, such as the vorticity, from the solution of VTE, and not as a post-processed quantity obtained by numerically differentiated form of velocity (this severely attenuates the vorticity magnitude corresponding to the high wavenumbers), and (iv) avoids the difficulty for the prescription of boundary condition for pressure.

However, it is not straightforward to specify boundary conditions for $\vec{\Psi}$ in solving 3D problems, for general cases. Mathematical formulation of boundary conditions for $\vec{\Psi}$ is discussed in Hirasaki and Hellums [30, 31]. Solution of 3D NSE using $\vec{\Psi}$ and $\vec{\omega}$) is given in [100] for 3D duct flows for $Re = 10$ and 100; for cubic-lid-driven cavity (CLDC) in [32, 98]; for computing nonlinear stability of rotating

Hagen–Poiseuille flow in [44]. The difficulty in prescribing the boundary condition for $\vec{\Psi}$, for multiply connected domains is addressed in [101], with an alternative proposed for boundary condition.

Equation (3.18) is derived with the solenoidality assumption: $\nabla \cdot \vec{\Psi} = 0$. As this is not guaranteed for the solution of $\vec{\Psi}$ obtained from Eq. (3.18), i.e., $\vec{\Psi}$ is divergence-free, and the associated vorticity field must also be solenoidal. If $\vec{\Psi}$ is not solenoidal, then one can use decomposition due to Helmholtz, to express it as

$$\vec{\Psi} = \nabla \Phi + \vec{\tilde{\psi}}$$

where $\vec{\tilde{\psi}}$ is solenoidal and at the domain boundary normal components of $\vec{\Psi}$ and $\vec{\tilde{\psi}}$ are identical. Because of the vector identity: $\nabla \times \nabla \Phi = 0$, one notes that $\vec{V} = \nabla \times \vec{\Psi} = \nabla \times \vec{\tilde{\psi}}$, and thus the computed vector potential field may not be divergence-free, still it will compute the associated velocity field. In this context, one may note a very high accurate solution of 3D Taylor–Green vortex periodic problem in [83] using velocity–vector potential formulation.

### 3.1.6  Computation of Derived Variable Formulation for 2D Incompressible NSE

For 2D incompressible flows, the non-dimensional governing NSE in $(\psi, \omega)$-formulation is given in coordinate-free vector form by

$$\frac{D\omega}{Dt} = \frac{1}{Re}\nabla^2\omega \tag{3.19}$$

$$\nabla^2\psi = -\omega \tag{3.20}$$

These equations are solved in the computational domain with appropriate grid clustering to resolve flow gradients. We recommend the use of orthogonal grid, so that in the computational $(\xi, \eta)$-plane, Eqs. (3.19) and (3.20) transform to the following form [59, 61]:

$$h_1 h_2 \frac{\partial \omega}{\partial t} + \frac{\partial \psi}{\partial \eta}\frac{\partial \omega}{\partial \xi} - \frac{\partial \psi}{\partial \xi}\frac{\partial \omega}{\partial \eta} = \frac{1}{Re}\left[\frac{\partial}{\partial \xi}\left(\frac{h_2}{h_1}\frac{\partial \omega}{\partial \xi}\right) + \frac{\partial}{\partial \eta}\left(\frac{h_1}{h_2}\frac{\partial \omega}{\partial \eta}\right)\right] \tag{3.21}$$

$$\frac{\partial}{\partial \xi}\left(\frac{h_2}{h_1}\frac{\partial \psi}{\partial \xi}\right) + \frac{\partial}{\partial \eta}\left(\frac{h_1}{h_2}\frac{\partial \psi}{\partial \eta}\right) = -h_1 h_2 \omega \tag{3.22}$$

In Eqs. (3.21) and (3.22), $h_1$ and $h_2$ are the scale factors of the orthogonal transformation and along $\xi$- and $\eta$-directions, these are given by

$$h_1 = \sqrt{\left(\frac{\partial x}{\partial \xi}\right)^2 + \left(\frac{\partial y}{\partial \xi}\right)^2}$$

$$h_2 = \sqrt{\left(\frac{\partial x}{\partial \eta}\right)^2 + \left(\frac{\partial y}{\partial \eta}\right)^2} \tag{3.23}$$

The contravariant components of the velocity vector are given by

$$u = \frac{1}{h_2}\frac{\partial \psi}{\partial \eta}$$

$$v = -\frac{1}{h_1}\frac{\partial \psi}{\partial \xi} \tag{3.24}$$

In Eqs. (3.19) and (3.21), we used the Reynolds number defined as

$$Re = \frac{U_{\text{ref}} L_{\text{ref}}}{\nu}$$

where $U_{\text{ref}}$ and $L_{\text{ref}}$ are reference velocity and length scales used for non-dimensionalization. The kinematic viscosity is denoted as $\nu$.

For the velocity–vorticity ($\vec{V}$, $\omega$)-formulation, VTE is written either in conservative or in non-conservative form. The often used non-conservative form of VTE in orthogonally transformed ($\xi$, $\eta$)-plane is obtained from Eq. (3.21) as

$$h_1 h_2 \frac{\partial \omega}{\partial t} + h_2 u \frac{\partial \omega}{\partial \xi} + h_1 v \frac{\partial \omega}{\partial \eta} = \frac{1}{Re}\left[\frac{\partial}{\partial \xi}\left(\frac{h_2}{h_1}\frac{\partial \omega}{\partial \xi}\right) + \frac{\partial}{\partial \eta}\left(\frac{h_1}{h_2}\frac{\partial \omega}{\partial \eta}\right)\right] \tag{3.25}$$

The conservative form of VTE is written as

$$h_1 h_2 \frac{\partial \omega}{\partial t} + \frac{\partial (h_2 u \omega)}{\partial \xi} + \frac{\partial (h_1 v \omega)}{\partial \eta} = \frac{1}{Re}\left[\frac{\partial}{\partial \xi}\left(\frac{h_2}{h_1}\frac{\partial \omega}{\partial \xi}\right) + \frac{\partial}{\partial \eta}\left(\frac{h_1}{h_2}\frac{\partial \omega}{\partial \eta}\right)\right] \tag{3.26}$$

The continuity equation ($\nabla \cdot \vec{V} = 0$) in the transformed plane is obtained from

$$\frac{1}{h_1 h_2}\left[\frac{\partial (h_2 u)}{\partial \xi} + \frac{\partial (h_1 v)}{\partial \eta}\right] = 0 \tag{3.27}$$

In the ($\vec{V}$, $\omega$)-formulation, the components of velocity are computed from Eq. (3.16) and in 2D, these equations for $u$ and $v$ are given as

$$\left(\frac{1}{h_1}\frac{\partial}{\partial\xi}\left[\frac{1}{h_1h_2}\frac{\partial(h_2u)}{\partial\xi}\right]+\frac{1}{h_2}\frac{\partial}{\partial\eta}\left[\frac{1}{h_1h_2}\frac{\partial(h_1u)}{\partial\eta}\right]\right)+$$

$$\left\{\frac{1}{h_1}\frac{\partial}{\partial\xi}\left[\frac{1}{h_1h_2}\frac{\partial(h_1v)}{\partial\eta}\right]-\frac{1}{h_2}\frac{\partial}{\partial\eta}\left[\frac{1}{h_1h_2}\frac{\partial(h_2v)}{\partial\xi}\right]\right\}=-\frac{1}{h_2}\frac{\partial\omega}{\partial\eta}$$

$$\left(\frac{1}{h_1}\frac{\partial}{\partial\xi}\left[\frac{1}{h_1h_2}\frac{\partial(h_2v)}{\partial\xi}\right]+\frac{1}{h_2}\frac{\partial}{\partial\eta}\left[\frac{1}{h_1h_2}\frac{\partial(h_1v)}{\partial\eta}\right]\right)+ \qquad (3.28)$$

$$\left\{\frac{1}{h_2}\frac{\partial}{\partial\eta}\left[\frac{1}{h_1h_2}\frac{\partial(h_2u)}{\partial\xi}\right]-\frac{1}{h_1}\frac{\partial}{\partial\xi}\left[\frac{1}{h_1h_2}\frac{\partial(h_1u)}{\partial\eta}\right]\right\}=\frac{1}{h_1}\frac{\partial\omega}{\partial\xi}$$

From Eq. (3.28), one notices that unlike Eq. (3.22), the governing Poisson equations in Eqs. (3.28) contain mixed derivatives. The discretized Eqs. (3.28) would have more entries, slowing the convergence of these Poisson equations, when an iterative method is used to solve the resultant linear algebraic equations. If $\xi$ and $\eta$ are functions of $x$- and $y$-directions, respectively, then one notes that $h_1 = h_1(\xi)$ and $h_2 = h_2(\eta)$, so that Eqs. (3.28) are simplified to

$$\frac{\partial}{\partial\xi}\left(\frac{h_2}{h_1}\frac{\partial u}{\partial\xi}\right)+\frac{\partial}{\partial\eta}\left(\frac{h_1}{h_2}\frac{\partial u}{\partial\eta}\right)=-h_1\frac{\partial\omega}{\partial\eta}$$

$$\frac{\partial}{\partial\xi}\left(\frac{h_1}{h_2}\frac{\partial v}{\partial\xi}\right)+\frac{\partial}{\partial\eta}\left(\frac{h_2}{h_1}\frac{\partial v}{\partial\eta}\right)=h_2\frac{\partial\omega}{\partial\xi} \qquad (3.29)$$

Equations (3.29) being in self-adjoint form, similar to Eq. (3.22) for stream function $\psi$.

## 3.2  Space and Time Scales in Transitional and Turbulent Flows

Unsteadiness is inherent in transitional and turbulent flows, with wide ranges of space and time scales excited. For turbulent flows, the existence of broadband spectrum is due to nonlinearity in the governing equations, with energy transferring from one wavenumber ($k$) (or frequency, $\omega_0$) component to neighboring wavenumbers (or frequencies). For 3D flows, vortex stretching term as the first term on the RHS of Eq. (3.13) is responsible for energy transfer from larger to smaller scales. These are intrinsic features of NSE, and one notices such transfers also in late stages of transition. For 2D flows also, there exists enstrophy cascade variously discussed in [18, 58, 79] which can also explain reverse transfer of energy/enstrophy from smaller to larger scales, which has been denoted as the inverse cascade.

It is customary to relate turbulent flow scales to the size of eddies in the flow field. The largest scale is given by the integral dimension of the flow, denoted as $l$, the flow is created by energy supplied at such low wavenumber. It has been shown by Kolmogorov that for *homogeneous and isotropic turbulent flows* [88] the smallest length scale is given by the *Kolmogorov scale* as

$$\eta_k = (\nu^3/\varepsilon)^{\frac{1}{4}} \tag{3.30}$$

Defining $u$ as the large-scale velocity (representing kinetic energy), one gets the Reynolds number as

$$Re = \frac{ul}{\nu}$$

Also, the dissipation, $\varepsilon$, is given by $\nu\|\nabla u\|_2^2$, allowing one to relate the largest and the smallest length scales in a turbulent flow by

$$\frac{l}{\eta_k} = (Re)^{\frac{3}{4}} \tag{3.31}$$

The most energetic structures in a turbulent flow is noted to be at lower wavenumbers, which fixes the peak of the energy spectrum there. The dissipation peak is located at a high wavenumber. In general, energy spectrum depends on $k$, $\varepsilon$, and $\nu$. This description explains the width of scales excited in transitional and primarily in turbulent flows. Thus, for computing high $Re$ flows by DNS, one resolves this wide range of scales. If the cutoff $k$ is denoted by $k_c$ (related to $\eta_k$), then Eq. (3.31) can be alternately written as

$$k_c l \approx Re^{\frac{3}{4}} \tag{3.32}$$

This tells one about the grid requirements for DNS, which for any 3D flow shows the grid requirement to scale as $(Re^{3/4})^3$. In deriving Kolmogorov's scaling theory, it is noted that there are length scales shorter than $l$, but larger than $\eta_k$, for which the energy spectrum is in a steady state, independent of how energy is extracted out by the viscous dissipation mechanism. At these intermediate scales—the *inertial subrange*—the structure of $E(k)$ is determined only by the nonlinear energy transfer by the cascade process, and the energy flux is shown to depend on

$$E(k, \varepsilon) = C_k k^{-\frac{5}{3}} \varepsilon^{\frac{2}{3}} \tag{3.33}$$

Existence of this *inertial subrange* suggests the universality of turbulent flow structure, for these length scales. This idea is used in large eddy simulation (LES), where the flow is computed resolving up to the *inertial subrange*, while the effects of smaller scales are modeled via sub-grid scale (SGS) stress model.

Inhomogeneous turbulent flows are identified with large-scale, low-frequency coherent structures, along with random high-frequency fluctuations. The organized structures in the flow carries about 20% of total turbulent kinetic energy (TKE). Hence, the role of the coherent structures to determine flow features is important. Researchers have identified the roles of coherent structures in determining various phenomenon, like noise radiation [87], low-frequency oscillation of the separated flow in shock-boundary layer interactions [90], mixing and entrainment in free shear layers and wakes [53], swirling jets [46], low-frequency buffeting of transonic flow over airfoils [17], among other phenomena. The accurate prediction of various features of coherent structures is important in obtaining information of the flow. At the same time, dynamics of coherent structures strongly depend upon random fluctuations too. Authors [53] have shown that one must accurately model the drain of energy (as in SGS stress models) from the coherent structures to random fluctuations to compute the flow.

For wall-bounded inhomogeneous flows, organized structures also show up in the time average, in the form of peaks and valleys in the near-wall region. In wall units, these have dimensions between 100 and 2000 units in the streamwise direction, while a spacing of about 50 units exists in the spanwise direction. The high energy events occur very close to the wall at about 20 to 50 units (in and around the buffer layer). These are statistical estimates, with some flow-to-flow variations. The near-wall events are punctuated by what is known as the bursting of coherent structures. Following which new intermediate scale motion starts again within the buffer layer, in the form of streamwise and hair-pin vortices. The time scale of the turbulent boundary layer is characterized by bursting frequency ($F_b$), even when the time-averaged flow indicates the shear layer to remain attached. For the zero pressure gradient boundary layer, this critical frequency is roughly between 0.20 and 1.00 times $F_b$, where $F_b = U_\infty/5\delta$, with $\delta$ as the shear layer thickness. For adverse pressure gradient flows, this critical frequency is between 0.06 and 0.28 times $F_b$.

**Fig. 3.1** Dynamic range of time scales noted in engineering flows for different speeds and the corresponding frequency ranges ($\Omega_f$) are shown. Indicated band of mean frequencies for the larger eddies in high $Re$ flows are indicated

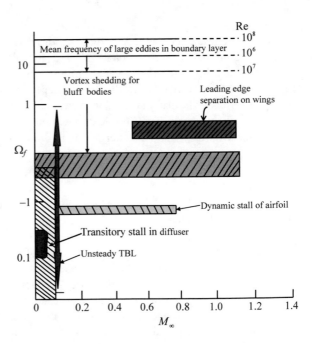

Different time scales noted in aeronautical applications are shown for different speed regimes in Fig. 3.1. For high $Re$ flows, the mean frequencies of larger eddies are only shown. Clearly, the ranges of non-dimensional frequencies span over three orders of magnitude, and computing methods have to resolve these scales for DNS/LES of high $Re$ flows.

## 3.3 Numerical Analysis for Methods Used in DNS/LES

The importance in resolving wideband space–time scales for computing transitional and turbulent flows has been described above. The numerical methods employed must preserve the physical relation of space–time scales. This aspect of numerical methods preserving the physical *dispersion relation* is the subject matter of many researches reported in [61, 70, 71].

### 3.3.1 Wave Mechanics—Understanding Propagation of Disturbance Field

We highlight the fact that disturbance evolution for turbulent flow can be explained in terms of wave propagation. Whether the flow is linear or nonlinear, one can explain disturbance as an ensemble over continuous Fourier–Laplace transforms in space and time, and the disturbances can be viewed as plane waves, which are either dispersive or non-dispersive, either governed by hyperbolic or non-hyperbolic partial differential equations. The hyperbolic waves can be represented by 1D advection equation given by

$$\frac{\partial u}{\partial t} + c \frac{\partial u}{\partial x} = 0 \tag{3.34}$$

Consider propagation of $u$ following the initial condition:

$$u(x, 0) = f(x) \quad \text{for} \quad -\infty < x < \infty \tag{3.35}$$

In Eq. (3.35), $f(x)$ can be piecewise continuous, whose space–time variation is given by

$$u(x, t) = f(x - ct)$$

The solution is thus the initial disturbance propagating to the right at the convection speed $c$, without attenuating or dispersing. This is the basic unit of any propagating disturbance as an aggregate of plane waves, which is defined by the wave parameters for a periodic function,

$$u(x, t) = a \sin\left[\frac{2\pi}{\lambda}(x - ct)\right] \tag{3.36}$$

One identifies this as the solution of Eq. (3.34). The square bracketed quantity is the phase, while $a$ is the amplitude of the wave. The quantity $\lambda$ is the wavelength, and one defines wavenumber $k$ ($= \frac{2\pi}{\lambda}$), as the number of waves in a length $2\pi$. Thus, Eq. (3.36) can also be given by

$$u(x, t) = a \sin\left[k(x - ct)\right] \tag{3.37}$$

Fixing attention at a single point, the least amount of time after which $u(x, t)$ attains the same value is the time period $T$, and thus $T = \frac{\lambda}{c}$. The number of oscillations at a point per unit time is the frequency given by $f_0 = \frac{1}{T}$. One defines the circular frequency $\bar{\omega}$ by noting

$$\bar{\omega} = kc \tag{3.38}$$

Thus, $c$ ($= \bar{\omega}/k$) has a dimension of speed and hence is called the phase speed, a rate at which the phase of the wave changes. Phase change is not necessarily physical. Equation (3.38) relates the time and space scales, indicated in the spectral plane by circular frequency and wavenumber. This is known as the physical dispersion relation.

One interesting aspect of dispersion relation is given by the group velocity $V_g$, whose implications were identified as the velocity with which energy of the disturbance travels for a system displaying a wideband spectrum. This is identified by researchers in many disciplines of science and engineering, and recorded in [3, 10, 33, 41, 61, 99]. Rayleigh [51] is credited to have introduced group velocity to be more important than the phase speed. However, it was discussed earlier in [29]. The carrier waves' phase variation is indicated by the phase speed, while the group velocity is associated with the *space–time varying amplitude*. Brillouin [10] described the group velocity is the velocity of energy propagation, while this was stated as the signal speed by Rayleigh [52]. For general wave systems, $V_g$ is defined as

$$V_g = \frac{d\bar{\omega}}{dk} \tag{3.39}$$

For the 1D advection equation, the dispersion relation gives,

$$\frac{d\bar{\omega}}{dk} = V_g = c \tag{3.40}$$

This shows that the phase speed and group velocity are indistinguishable for non-dispersive systems, as given in Eq. (3.34). For a general system, the phase speed and the group velocity are not identical. A more detailed discussion from physical and mathematical standpoints on group velocity can be found in [60, 61, 99].

## 3.3.2  Spatial Discretization and Its Resolution

NSE basically defines space–time evolution of primary variables as disturbances introduced via the placement of a body. In various forms, NSE contains primarily first and second spatial derivatives. The resolution of discretizing these derivatives would affect the accuracy of the overall solution of NSE. Thus, one introduces the resolution of spatial discretization while estimating first and second derivatives.

Let the prime in $u'_j = \left(\frac{\partial u}{\partial x}\right)_j$ represent a first derivative of the variable $u = u(x)$ at the $j$th-node of the independent variable, $x$. The nodes are spaced uniformly with $h$ giving the distance between adjacent nodes in Fig. 3.2. The most rudimentary second-order central difference ($CD_2$) scheme to evaluate $u'_j$ is given by

$$u'_j = \frac{u_{j+1} - u_{j-1}}{2h} \qquad (3.41)$$

with leading truncation error proportional to $h^2$. The fourth-order central difference ($CD_4$) scheme for spatial discretization of the first derivative is given as

$$u'_j = \frac{-u_{j+2} + 8u_{j+1} - 8u_{j-1} + u_{j-2}}{12h} \qquad (3.42)$$

Here, we will compare resolution of schemes in the spectral plane, by expressing the variable by its Fourier–Laplace transform [39, 59, 61],

$$u_j = \int U(k)e^{ikx_j}\,dk \qquad (3.43)$$

The limits of the integral ($\pm k_{max}$) are determined by $h$, with the help of the Nyquist limit given by $k_{max} = \pi/h$. From Eq. (3.43), one notes that ideally

$$(u'_j)_{exact} = \int ikU(k)e^{ikx_j}\,dk \qquad (3.44)$$

For the $CD_2$-scheme (Eq. (3.41)), the numerical derivative is written in spectral plane as

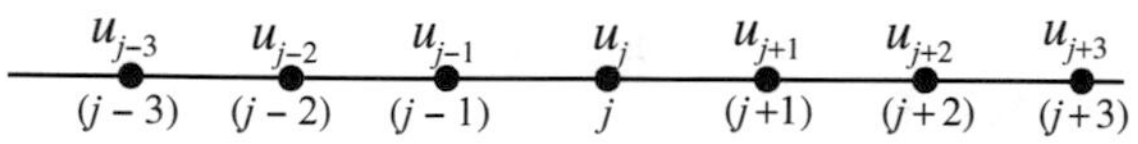

**Fig. 3.2** Schematic of a uniformly spaced grid with the nodes is at a distance $h$ apart between immediate neighboring points

$$(u'_j)_{CD_2} = \int \frac{e^{ikh} - e^{-ikh}}{2h} U(k) e^{ikx_j} dk \tag{3.45}$$

which yields

$$(u'_j)_{CD_2} = \int i \frac{\sin(kh)}{h} U(k) e^{ikx_j} dk = \int i k_{eq} U(k) e^{ikx_j} dk \tag{3.46}$$

where $k_{eq}$ is an equivalent wavenumber. The term $k_{eq}/k$ basically defines the resolution of the spatial discretization scheme. For $CD_2$- and $CD_4$-schemes, this is given in Eqs. (3.41) and (3.42) by

$$\left( \frac{k_{eq}}{k} \right)_{CD_2} = \frac{\sin(kh)}{kh} \tag{3.47}$$

$$\left( \frac{k_{eq}}{k} \right)_{CD_4} = \left[ \frac{(4 - \cos(kh))}{3} \right] \frac{\sin(kh)}{kh} \tag{3.48}$$

For both Eqs. (3.47) and (3.48), $(k_{eq}/k) \longrightarrow 1$ as, $kh \longrightarrow 0$, and is the *consistency condition* for any discretization. This must be true for any discretization scheme, since $kh \longrightarrow 0$ implies approaching the continuum limit from the discrete value. Furthermore, for the central $CD_2$ and $CD_4$ schemes, $(k_{eq}/k)$ must be real, as obtained for central schemes on uniform grid.

For non-central or upwinded spatial discretization, $(k_{eq}/k)$ turns out to be a complex quantity. For example, for a first-order upwind ($UD_1$) scheme

$$u'_j = \frac{u_{j+1} - u_j}{h} \tag{3.49}$$

so that $(k_{eq}/k)$ for $UD_1$ scheme is given by

$$\left( \frac{k_{eq}}{k} \right)_{UD_1} = \frac{\sin(kh)}{kh} + i \frac{(\cos(kh) - 1)}{kh} = \left( \frac{k_{eq}}{k} \right)_{real} + i \left( \frac{k_{eq}}{k} \right)_{img} \tag{3.50}$$

where $(k_{eq}/k)_{real}$ and $(k_{eq}/k)_{img}$ are the real and imaginary parts of $(k_{eq}/k)$.

The real and imaginary parts of $(k_{eq}/k)$ are plotted in Fig. 3.3, as a function of $kh$. As noted before, $(k_{eq}/k)_{real}$ indicate the phase of the equivalent spectral representation, and therefore indicates the resolution of the spatial scheme with respect to the Fourier spectral method of spatial discretization. The imaginary part, $(k_{eq}/k)_{img}$, indicates numerical diffusion that is added for the non-central schemes. For the 1D advection equation (Eq. (3.34))

$$\left( \frac{\partial u}{\partial x} \right)_j = \int i k_{eq} U(k, t) e^{ikx_j} dk \tag{3.51}$$

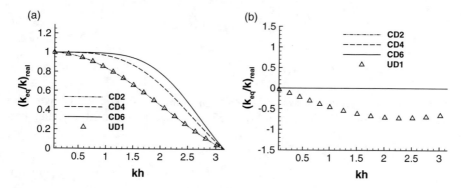

**Fig. 3.3** **a** $(k_{eq}/k)_{real}$ and **b** $(k_{eq}/k)_{img}$ plotted for $CD_2, CD_4, CD_6$, and $UD_1$ spatial discretization schemes as a function of $kh$ up to the Nyquist limit $\pi$

Using the spectral representation for $u(x, t)$ in terms of $U(k, t)$, Eq. (3.34) can be written as an ordinary differential equation given by

$$\left(\frac{dU}{dt} + ik_{eq}U(k, t)\right) = 0 \tag{3.52}$$

which can be solved exactly with the initial condition given by $U(k, 0)$ as

$$U(k, t) = U(k, 0) \, e^{-i(k_{eq}/k)_{real}kt} \, e^{(k_{eq}/k)_{img}kt} \tag{3.53}$$

This shows that $U(k, t)$ changes with time determined by $e^{(k_{eq}/k)_{img}kt}$. The numerical solution will grow if the exponent is positive and decay if it is negative. It has been established [59] that the presence of upwinding is equivalent to altering Eq. (3.34) to the convection–diffusion equation,

$$\frac{\partial u}{\partial t} + c\frac{\partial u}{\partial x} = \beta_1 \frac{\partial^2 u}{\partial x^2} \tag{3.54}$$

where $\beta_1 > 0$ indicates added diffusion, if $(k_{eq}/k)_{img} < 0$. In contrast, a positive $(k_{eq}/k)_{img}$ implies adding *anti-diffusion*, which will destabilize the convection dominated problem or high $Re$ flow solved using NSE.

Similarly, the second-order central schemes for the evaluation of second derivatives are given as

$$u''_j = \frac{u_{j+1} - 2u_j + u_{j-1}}{h^2} \tag{3.55}$$

where double prime indicates second derivative. As before, one can express the Fourier spectral and the numerical second derivative as

$$\left(\frac{\partial^2 u_j}{\partial x^2}\right)_{\text{exact}} = \int -k^2 U(k) e^{ikx_j} \, dk \tag{3.56}$$

$$u_j'' = \left(\frac{\partial^2 u_j}{\partial x^2}\right)_{\text{num}} = \int -k_{\text{eq}}^2 U(k) e^{ikx_j} \, dk \tag{3.57}$$

where $-k_{\text{eq}}^2$ is the equivalent wavenumber for the numerical second derivative. The spectral resolution of the scheme of Eq. (3.55) provides the equivalent wavenumber $k_{\text{eq}}$ as

$$\left(\frac{k_{\text{eq}}}{k}\right)^2 = \frac{\sin^2(kh/2)}{(kh/2)^2} \tag{3.58}$$

### 3.3.3  Compact Schemes for First Derivatives

Previously, the resolution of explicit schemes to discretized first and second derivatives is described. Next, compact schemes are introduced, where the derivatives are obtained implicitly by posing auxiliary relations. The main features of compact scheme are higher spectral resolution, while a compact stencil is used in the auxiliary equations. These have been originally used for centered discretization using Padé schemes for ODE described in [36]. Application of such schemes to PDEs can be found specifically for fluid dynamics in [39, 59].

The derivatives in compact schemes are obtained by solving linear algebraic equations relating functions with coupled derivatives, which require solving either tridiagonal or pentadiagonal matrix equations. A general compact scheme for $n$th derivative of the variable $u_j$ (at the $j$th-node) involves solving coupled equations, in a uniform grid, given by

$$\sum_{k=-N_1}^{N_1} \alpha_k u_{j+k}^{(n)} = \frac{1}{h^n} \sum_{l=-M_1}^{M_1} \beta_l u_{j+l} \tag{3.59}$$

where $h$ is the uniform spacing and $u_j^{(n)}$ is the $n$th derivative of $u_j$. It is seen that the resultant matrix is band-limited with bandwidth defined by $N_1$ for the derivative evaluation. The value of $M_1$ decides the needed boundary closure schemes [61].

A tridiagonal compact scheme is shown here to evaluate first derivatives [39]:

$$\alpha_1 u_{j+1}' + u_j' + \alpha_1 u_{j-1}' = c_1 \frac{u_{j+3} - u_{j-3}}{6h} +$$
$$b_1 \frac{u_{j+2} - u_{j-2}}{4h} + a_1 \frac{u_{j+1} - u_{j-1}}{2h} \tag{3.60}$$

For $c_1 = 0$, one can obtain a fourth-order accurate scheme if

$$a_1 = \frac{2}{3}(\alpha_1 + 2), \quad b_1 = \frac{1}{3}(4\alpha_1 - 1)$$

A choice of $\alpha_1 = 1/3$ gives a sixth-order accurate scheme.
For $c_1 \neq 0$, a sixth-order accurate scheme is obtained if

$$a_1 = \frac{1}{6}(\alpha_1 + 9), \quad b_1 = \frac{1}{15}(32\alpha_1 - 9), \quad c_1 = \frac{1}{10}(-3\alpha_1 + 1)$$

If $\alpha_1 = 3/8$, then accuracy further increases to eighth order. The scheme given by Eq. (3.60) can be applied for all nodes, if we have periodic problems, and its resolution is given by

$$\left(\frac{k_{eq}}{k}\right) = \frac{1}{kh} \frac{a_1 \sin(kh) + (b_1/2)\sin(2kh) + (c_1/3)\sin(3kh)}{1 + 2\alpha_1 \cos(kh)} \tag{3.61}$$

In [69, 72], an optimized high-accuracy upwind compact scheme $OUCS3$ has been described, for which the interior stencil is given by

$$p_{j-1} u'_{j-1} + u'_j + p_{j+1} u'_{j+1} = \frac{1}{h} \sum_{n=-2}^{2} q_n u_{j+n} \tag{3.62}$$

where $p_{j\pm1} = D \pm \frac{\eta_*}{60}, q_{\pm2} = \pm\frac{F}{4} + \frac{\eta_*}{300}, q_{\pm1} = \pm\frac{E}{2} + \frac{\eta_*}{30}, q_0 = -\frac{11\eta_*}{150}$, where $D = 0.3793894912$, $E = 1.57557379$, and $F = 0.183205192$. Here, $\eta_*$ is the upwind parameter and a choice of zero makes the scheme central. This is formally second-order accurate as noted by Taylor series. However, $D$, $E$, and $F$ are obtained by maximizing the resolution in the spectral plane. The method of optimization for these coefficients is obtained for non-periodic problem and are given in [61, 69]. For periodic problems, this essentially implies that one maximizes the objective function

$$I_\gamma = \int_0^\gamma \left| 1 - \left(\frac{k_{eq}}{k}\right) \right| d(kh) \tag{3.63}$$

where $0 \leq \gamma \leq \pi$. Despite second-order formal accuracy, the scheme has high spectral resolution, higher than explicit schemes of higher order. Even in Fig. 3.4, $(k_{eq}/k)$ for eighth-order compact schemes proposed by Lele [39] and OUCS3 scheme (Eq. (3.62)) are compared for an interior stencil, and the figure shows that OUCS3 scheme to be slightly superior to the eighth-order compact scheme of Lele [39]. Here OUCS3 scheme is fixed to be central by choosing $\eta_* = 0$ in Eq. (3.62). The figure also shows that both the schemes have $(k_{eq}/k) \simeq 1$, up to larger value of $kh$. For OUCS3 scheme, near-spectral accuracy is retained up to $kh \approx 2.3$, after which it falls off and becomes zero at $kh = \pi$.

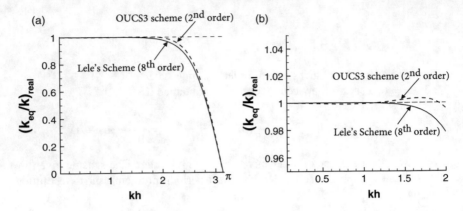

**Fig. 3.4** $(k_{eq}/k)$ plotted as a function of $kh$ for the interior stencil of OUCS3 (Eq. (3.62)) with $\eta_* = 0$ and the eighth-order compact scheme in Eq. (3.60). The interior stencil is considered only. This analysis is also valid for periodic problems

### 3.3.4 Boundary Closure for Compact Schemes for First Derivative

We have so far discussed about the interior stencils for the compact schemes, which can be used at all nodes for periodic problems. For non-periodic problems, one requires different stencils for near-boundary points ($j = 1, 2, (N-1)$ and $N$ for five-point right-hand side in the basic compact scheme) for closure of the linear algebraic equations, given by Eq. (3.60) or (3.62).

For boundary closure Adams [1] proposed, for $j = 1$ and 2, the following for use with non-periodic problems. The third stencil is for the interior nodes. Also, one should note that there would be the need to write boundary closure schemes for $j = (N-1)$ and $N$, which is left as an exercise.

$$2u'_1 + 4u'_2 = \frac{1}{h}(-5u_1 + 4u_2 + u_3) \quad \text{at } j = 1 \qquad (3.64)$$

$$u'_1 + 4u'_2 + u'_3 = \frac{3}{h}(u_3 - u_1) \quad \text{at } j = 2 \qquad (3.65)$$

$$u'_{j-1} + 3u'_j + u'_{j+1} = \frac{1}{12h}(-u_{j-2} - 28u_{j-1}$$
$$+ 28u_{j+1} + u_{j+2}) \quad \text{for } 3 \le j \le (N-2) \qquad (3.66)$$

The stencil at $j = 2$ has fourth-order accuracy, while for the interior stencil has sixth-order formal accuracy.

The above equations, Eqs. (3.64)–(3.66), written for using compact scheme for non-periodic problems in evaluating first derivatives at all nodes simultaneously would require solving the implicitly coupled unknown derivatives from the linear algebraic equation [59, 61, 72]

$$[A]\{u'\} = \frac{1}{h}[B]\{u\} \tag{3.67}$$

where both $[A]$ and $[B]$ are matrices of size $(N \times N)$. By defining $[C] = [A]^{-1}[B]$, one can convert the implicit method of evaluating the derivatives, by an equivalent explicit method given by

$$\{u'\} = \frac{1}{h}[C]\{u\} \tag{3.68}$$

While $[A]$ and $[B]$ are sparse matrices (with $[A]$ being a tridiagonal matrix and $[B]$ contains five sets of entries along diagonals), the matrix $[C]$ is a non-sparse matrix. Being an explicit relation in Eq. (3.68), the derivative at the $j$th node can be obtained from

$$u'_j = \frac{1}{h} \sum_{l=1}^{N} C_{jl} u_l$$

with $u_l = u(x_l) = \int U(k) e^{ikx_l} dk$ denoting the function at the $l$th node. Noting that in spectral representation the evaluation of derivative at the $j$th node requires the phase information at that point only, one rewrites the derivative at that node by

$$u'_j = \int \frac{1}{h} \sum C_{jl} U(k) e^{ik(x_l - x_j)} e^{ikx_j} dk \tag{3.69}$$

Equating Eqs. (3.69) and (3.51), one obtains

$$\left(\frac{[k_{eq}]_j}{k}\right) = -\frac{i}{kh} \sum_{l=1}^{N} C_{jl} e^{ik(x_l - x_j)} \tag{3.70}$$

While $C_{jl}$'s are real, $([k_{eq}]_j / k)$ is complex, with real and imaginary parts denoting numerical phase and added numerical diffusion, respectively, as noted before. Both of these are fixed by the elements of $[C]$ matrix. From Eq. (3.70), real and imaginary parts of $(k_{eq}/k)$ are obtained at all the nodes simultaneously. This in essence is the global spectral analysis (GSA) introduced in [59, 61, 72], a major innovation in numerical analysis of implicit schemes. As stated this GSA incorporates interior and boundary node information simultaneously, thereby justifying the nomenclature global in GSA. Also, it shows that futility of classifying numerical schemes by Taylor series expansion alone, due to the fact that every implicit scheme has an unique explicit equivalence given in Eq. (3.69). Thus, despite the compact appearance of the scheme in Eq. (3.67), the bandwidth of the $j$th row of $[C]$ matrix characterizes the equivalent explicit scheme, which is significantly higher than what Taylor series expansion can reveal. We have already noted that OUCS3 scheme with its tag of second-order accuracy is superior to the eighth-order Lele's compact scheme. These aspects are clearly revealed by GSA. However, implicit schemes also have problem areas, as noted next.

This can be shown by plotting $(k_{eq}/k)$ that implicit boundary closure schemes can create numerical instability, which can be noted by GSA with presence of anti-diffusion at near-boundary points for many early compact schemes, as revealed in [72].

The anti-diffusion for near-boundary points using compact schemes arises due to the use of implicit boundary closures [59, 61, 72]. In [72], it has been established that by using explicit boundary closure schemes, one can solve these problems. Thus, one uses explicit closure at $j = 1, 2, (N - 1)$ and $N$. Explicit boundary closures were introduced in proposing $OUCS3$ scheme, for which the interior stencil is as in Eq. (3.62). The explicit boundary closures used in [72] are given by

$$u_1' = \frac{1}{2h}\left[-3u_1 + 4u_2 - u_3\right] \tag{3.71}$$

$$u_2' = \frac{1}{h}\left[\left(\frac{2\beta_*}{3} - \frac{1}{3}\right)u_1 - \left(\frac{8\beta_*}{3} + \frac{1}{2}\right)u_2 + \left(4\beta_* + 1\right)u_3 \right.$$
$$\left. - \left(\frac{8\beta_*}{3} + \frac{1}{6}\right)u_4 + \frac{2\beta_*}{3}u_5\right] \tag{3.72}$$

with the help of the parameter $\beta_*$. Corresponding closures have to used for $j = (N - 1)$ and $N$. The parameter $\beta_*$ is associated with added fourth-order diffusion for numerical stabilization by removing effects of anti-diffusion at near-boundary points. In [72], better global numerical properties were obtained by using $\beta_* = -0.025$ for $j = 2$ and $\beta_* = 0.09$ for $j = N - 1$. Figure 3.5 shows real and imaginary parts of $(k_{eq}/k)$ for OUCS3 scheme with explicit boundary closures used as given in Eqs. (3.71) and (3.72). It is noted that use of such boundary closures reduces the

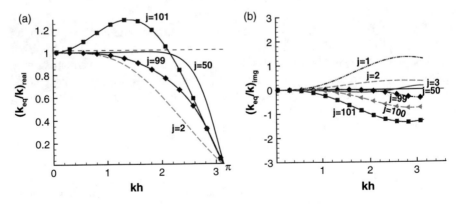

**Fig. 3.5** **a** Real and **b** imaginary parts of $(k_{eq}/k)$ plotted for OUCS3 scheme with explicit boundary closure schemes given by Eqs. (3.71) and (3.72). For the interior OUCS3 stencil here the value upwind parameter is chosen with $\eta_* = 0$ in Eq. (3.62)

problem of anti-diffusion for near-boundary points significantly. One also notices that the problem associated with $j = 1$ is not a matter of concern, as the boundary condition will be used for this node.

### 3.3.5 Compact Schemes to Evaluate Second Derivatives

One can use a tridiagonal compact scheme for the evaluation of second derivative from

$$\bar{\alpha}u''_{j+1} + u''_j + \bar{\alpha}u''_{j-1} = b_2 \frac{u_{j+2} - 2u_j + u_{j-2}}{4h^2} + a_2 \frac{u_{j+1} - 2u_j + u_{j-1}}{h^2} \tag{3.73}$$

To achieve sixth-order accuracy for Eq. (3.73), one must choose

$$(a_2 + b_2) = (1 + 2\bar{\alpha}) \quad \text{for second-order scheme}$$
$$(a_2 + 4b_2) = 12\bar{\alpha} \quad \text{additional relation for fourth-order scheme}$$
$$(a_2 + 16b_2) = 30\bar{\alpha} \quad \text{additional relation for sixth-order scheme} \tag{3.74}$$

where the first relation is also the consistency condition. The meaning of these relations is that for second-order scheme, we cannot have $a_2$ and $b_2$ as unknowns, even if we take $\bar{\alpha}$ as the free parameter. However, solving Eqs. (3.74), one can uniquely fix all the three unknowns to obtain $\bar{\alpha} = 2/11$, $a_2 = 12/11$, and $b_2 = 3/11$. The eventual scheme in Eq. (3.73) defines sixth-order accuracy for the above values of $a_2$, $b$, and $\alpha_2$. This is true for a periodic problem. Not using the last relation, we can propose a single parameter, fourth-order scheme for

$$a_2 = \frac{4}{3}(1 - \bar{\alpha}) \tag{3.75}$$

$$b_2 = \frac{1}{3}(10\bar{\alpha} - 1) \tag{3.76}$$

Similarly following the expression in Eq. (3.56), one can represent the resolution of the scheme for the second derivative given by Eq. (3.73) for periodic problems by $k_{eq}$ as

$$\left(\frac{k_{eq}}{k}\right)^2 = \frac{1}{(1 + 2\bar{\alpha}\cos(kh))} \frac{(b_2 \sin^2(kh) + 4a_2 \sin^2(kh/2))}{(kh)^2} \tag{3.77}$$

Following the laid our procedure for first derivative, one can propose an optimized compact scheme for second derivative for periodic problems, by choosing $\bar{\alpha}$ from the minimization of the objective function

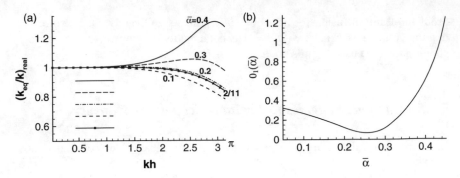

**Fig. 3.6** **a** $(k_{eq}/k)$ for the compact stencil, Eq. (3.73) plotted for different values of $\bar{\alpha}$. $\bar{\alpha} = 2/11$ is for a sixth-order scheme. **b** Integral objective function $O_I$ given by Eq. (3.78) as a function of $\bar{\alpha}$ for $\gamma = \pi$

$$O_I = \int_0^{\gamma_2} \left| 1 - \left( \frac{k_{eq}}{k} \right)^2 \right| d(kh) \qquad (3.78)$$

with $\gamma_2 \leq \pi$. In Fig. 3.6a, $(k_{eq}/k)^2$ for the periodic compact scheme of Eq. (3.73) is plotted as function of $\bar{\alpha}$. We have seen that, $\bar{\alpha} = 2/11$ is needed for a sixth-order scheme, with corresponding value of $a_2$ and $b_2$. In Fig. 3.6b, the integrated objective function $O_I$ given above in Eq. (3.78) is plotted as a function of $\bar{\alpha}$ for $\gamma_2 = \pi$. In frame (b), $O_I$ is seen to attain the minimum for $\alpha_2 = 0.26$. Thus, with this choice an optimal fourth-order scheme stencil with this optimal value of $\bar{\alpha} = 0.26$ will provide higher accurate scheme, as compared to the sixth-order scheme with $\bar{\alpha} = 2/11$.

In considering non-periodic problems, one needs at least two boundary closures for near-boundary points. Lele [39] suggested the following boundary closures for the second derivative:

$$u_1'' + 11u_2'' = \frac{1}{h^2}(13u_1 - 27u_2 + 15u_3 - u_4) \qquad (3.79)$$

$$\frac{1}{10}u_1'' + u_2'' + \frac{1}{10}u_3'' = \frac{12}{10h^2}(u_3 - 2u_2 + u_1) \qquad (3.80)$$

Stencils have to be provided for $j = (N - 1)$ and $j = N$. We note that Lele [39] proposed implicit closures and that can lead to numerical problem. As we have solved the problems for first derivatives by using explicit closures for near-boundary points, here also we propose the following composite stencils for the evaluation of second derivative for non-periodic problem as

$$\bar{\alpha} u''_{j+1} + u''_j + \bar{\alpha} u''_{j-1} = \frac{b_2}{4h^2} (u_{j+2} - 2u_j + u_{j-2})$$

$$+ \frac{a_2}{h^2} (u_{j+1} - 2u_j + u_{j-1}) \quad \text{for } 3 \le j \le (N-3) \tag{3.81}$$

$$u''_j = \frac{1}{h^2} (u_{j+1} - 2u_j + u_{j-1}) \quad \text{for } j = 2 \text{ and } j = (N-1) \tag{3.82}$$

$$u''_1 = \frac{1}{h^2} (2u_1 - 5u_2 + 4u_3 - u_4) \quad \text{for } j = 1 \tag{3.83}$$

$$u''_N = \frac{1}{h^2} (2u_N - 5u_{N-1} + 4u_{N-2} - u_{N-3}) \quad \text{for } j = N \tag{3.84}$$

Similarly GSA can be performed for the non-periodic compact scheme for second derivatives as

$$[A_2]\{u''\} = \frac{1}{h^2} [B_2]\{u\} \tag{3.85}$$

with $[A_2]$ and $[B_2]$ are $(N \times N)$ matrices. Defining $[D] = [A_2]^{-1}[B_2]$, one can write

$$\{u''\} = \frac{1}{h^2} [D]\{u\} \tag{3.86}$$

Although $[A_2]$ and $[B_2]$ are sparse matrices, with $[A_2]$ as tridiagonal and $[B_2]$ contains five diagonal entries, the matrix $[D]$ is not simply banded or sparse. Using Eq. (3.86) the second derivative at the $j$th node is obtained from

$$u''_j = \frac{1}{h^2} \sum_{l=1}^{N} D_{jl} u_l$$

For GSA, one alternatively writes the second derivative as [59]

$$u''_j = \int \frac{1}{h^2} \sum D_{jl} U(k) e^{ik(x_l - x_j)} e^{ikx_j} dk \tag{3.87}$$

Using Eq. (3.87), one performs GSA of non-periodic compact scheme to express $(k_{eq}/k)^2$ at $j$th node by

$$\left( \frac{[k_{eq}]_j}{k} \right)^2 = -\frac{1}{(kh)^2} \sum_{l=1}^{N} D_{jl} e^{ik(x_l - x_j)} \tag{3.88}$$

Once again, $([k_{eq}]_j/k)^2$ is complex, with real part representing phase representation for the second derivative and imaginary part representing numerical dispersion.

### 3.3.6  Combined Compact Difference Scheme

Consider a computational domain constructed using equi-spaced grid points with a grid spacing $h$. Simultaneous evaluation of the first, the second, and the fourth derivative terms using coupled compact difference scheme is expressed as [48]

$$a_1(u'_{i+1} + u'_{i-1}) + u'_i + b_1 h(u''_{i+1} - u''_{i-1}) + c_1 h^3(u''''_{i+1} - u''''_{i-1})$$
$$= \frac{1}{h} d_1(u_{i+1} - u_{i-1}) \qquad (3.89)$$

$$\frac{a_2}{h}(u'_{i+1} - u'_{i-1}) + b_2(u''_{i+1} + u''_{i-1}) + u''_i + c_2 h^2(u''''_{i+1} + u''''_{i-1})$$
$$= \frac{1}{h^2}\left[ d_2(u_{i+1} + u_{i-1}) + e_2 u_i \right] \qquad (3.90)$$

$$\frac{a_3}{h^3}(u'_{i+1} - u'_{i-1}) + b_3 h^2(u''_{i+1} + u''_{i-1}) + c_3(u''''_{i+1} + u''''_{i-1}) + u''''_i$$
$$= \frac{1}{h^4}\left[ d_3(u_{i+1} + u_{i-1}) + e_3 u_i \right] \qquad (3.91)$$

Various coefficients present in Eqs. (3.89)–(3.91) are found using Taylor series approximations as given next [48].

$$a_1 = \frac{31}{64}; \quad b_1 = \frac{-5}{64}; \quad c_1 = \frac{1}{960}; \quad d_1 = \frac{63}{64}$$

$$a_2 = \frac{313}{224}; \quad b_2 = \frac{-41}{224}; \quad c_2 = \frac{1}{672}; \quad d_2 = \frac{24}{7}; \quad e_2 = \frac{-48}{7}$$

$$a_3 = \frac{-1725}{56}; \quad b_3 = \frac{285}{56}; \quad c_3 = \frac{-3}{56}; \quad d_3 = \frac{-360}{7}; \quad e_3 = \frac{720}{7}$$

Equations (3.89)–(3.91) are solved in a coupled and iterative manner for the evaluation of the first, the second, and the fourth derivative terms. As an initial guess, the various derivative terms are evaluated using explicit central difference schemes. Equations (3.89)–(3.91) are solved iteratively until the maximum residue falls below the prescribed tolerance value. Iterations can be performed by using either the traditional Gauss–Seidel iterative algorithm or a bi-conjugate gradient stabilized method (*Bi-CGSTAB*) [97].

For the non-periodic problems, we suggest the following stencils for the starting boundary and the near-boundary nodes. The stencils for the boundary and the near-boundary points on the other end can simply be obtained by reverting the stencils and adding a minus sign [61]. Note that the second and the fourth derivative terms are not evaluated at the boundary nodes while the fourth derivative is not evaluated at the second and second last nodes.

$$u_1' = (-1.5u_1 + 2u_2 - 0.5u_3)/h, \ u_1'' = 0, \ u_1'''' = 0 \tag{3.92}$$

$$u_2' = (u_3 - u_1)/(2h), \ u_2'' = (u_1 - 2u_2 + u_3)/h^2, \ u_2'''' = 0 \tag{3.93}$$

$$u_3' = (-u_5 + u_1 + 8(u_4 - u_2))/(12h), \ u_3'' = (u_1 - 2u_2 + u_3)/h^2,$$
$$u_3'''' = (u_1 + u_5 - 4(u_2 + u_4) + 6u_3)/h^4 \tag{3.94}$$

Finite difference schemes are derived using Taylor series approximation in which one matches the required lower order derivative terms and truncates the higher order derivatives based on number of unknown coefficients. This results in difference between the numerically estimated and the exact derivative values. Thus, the use of difference scheme to evaluate a derivative of a function involves implicit filtering [61]. High-accuracy schemes are designed specifically to obtain significant spectral resolution by minimizing implicit filtering involved in difference approximations [13, 61, 72]. Spectral resolution of a numerical scheme is a key design parameter for the construction of high-accuracy scheme (Fig. 3.7).

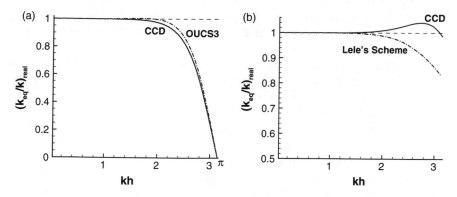

**Fig. 3.7** **a** $(k_{1eq}/k)$ and **b** $(k_{2eq}/k)^2$ plotted for first and second derivatives for CCD scheme, for periodic problems. Stencils of OUCS3-scheme Eq. (3.62) (for 1st derivative) and the Lele's scheme in Eq. (3.73) (for 2nd derivative) are included in these frames, respectively

Consider a physical domain to be divided into $N$ equi-spaced grid points, with grid spacing $h$. The physical variable $u_j$ at the $j$th grid point can be represented in the spectral plane as [61], $u_j = \int U e^{ikx_j} dk$, where $k$ is a wavenumber. Expression for the corresponding exact derivative is represented in the spectral plane as $(u'_j)_{\text{exact}} = \int ik U e^{ikx_j} dk$. However, when the same derivative has been evaluated numerically, loss of higher derivative terms in the form of truncation error brings deviation in the numerically estimated derivative value as compared to the exact value. Thus, one can represent the numerically estimated derivative value in the spectral plane as [61], $(u'_j)_{\text{numerical}} = \int ik_{\text{eq}} U e^{ikx_j} dk$. The discretization effectiveness of a numerical scheme in evaluating the first derivative term can be given by variation of the ratio of $k_{\text{eq}}/k$ across a complete wavenumber range. Ideally numerical scheme should have discretization effectiveness as unity across a complete resolved wavenumber range; however, for various difference schemes, this ratio starts deviating from the ideal value in higher wavenumber range.

This aspect has been shown in Fig. 3.8a for the $CCD$ scheme of [14] which has a central stencil, spectrally optimized upwind $CCD$ scheme of [13] and the proposed coupled compact difference scheme in evaluating the first derivative term. We have purposefully compared numerical properties of the proposed scheme with $CCD$ schemes of [13, 14], as these schemes also simultaneously evaluate the first and the second derivative terms. In addition, these scheme also use the information available at the next and previous grid points only as in the proposed scheme.

Here, we have followed the GSA technique as in [57, 61, 72, 80] to obtain the spectral resolutions of different spatial discretization schemes as given by variation of real part of $k_{\text{eq}}/k$ with respect to a non-dimensional wavenumber $kh$. Figure 3.8a shows that among the three schemes, discretization effectiveness of the coupled compact difference scheme lies in between that of $CCD$ scheme of [13] and [73].

Although spectrally optimized $CCD$ scheme of [13] displays higher resolving ability than the present scheme, it has a disadvantage of implicit addition of numerical diffusion to the solution. Thus, although upwind $CCD$ scheme in [13] shows higher resolving ability, it numerically attenuates the resolved higher wavenumber components. This aspect is clear from the variation of the imaginary part of $k_{\text{eq}}/k$ as shown in Fig. 3.8b.

For the derived numerical scheme, estimation of the discretization effectiveness for the second derivative terms $(-k_{\text{eq}}^2/k^2)$ associated with viscous diffusion process in the Navier–Stokes equation is also important. Figure 3.8c compares the spectral resolutions for the indicated difference schemes in estimating second derivative. All the schemes have a near-spectral resolution. One observes a mild overshoot of the discretization effectiveness above the ideal value of one near the Nyquist limit. Authors in [82] concluded that over-prediction of the spectral amplitude for the second derivative in the high wavenumber region helps in de-aliasing the numerical solution, as the diffusion terms are over-predicted. Overshoot associated with the proposed scheme is closer to the Nyquist limit, and hence has desirable de-aliasing nature.

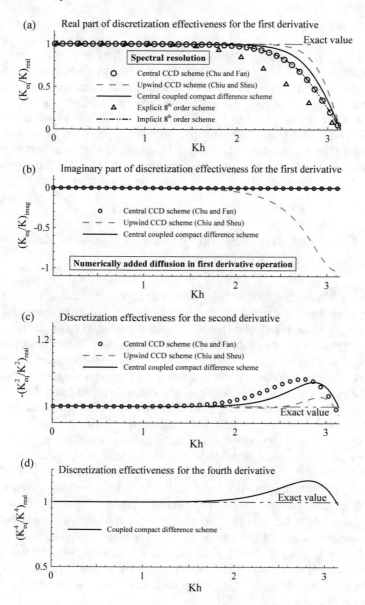

**Fig. 3.8** Comparison of the discretization effectiveness in evaluating the first, the second, and the fourth derivative terms for the indicated spatial discretization schemes at the central node in the spectral plane has been shown [48]. Here, $k$ and $k_{eq}$ denote the exact and the numerically obtained wavenumber, respectively

In addition to the estimation of the second-order derivative terms, computation of higher order even derivative term is essential to obtain numerical diffusion as shown in Eq. (7.1). Addition of the higher order even derivative term corresponds to introduction of numerical diffusion. Here a fourth-order derivative term is added for numerical diffusion. Figure 3.8d shows the discretization effectiveness for evaluating the fourth derivative term $(k_{eq}^4/k^4)$ using CCD scheme. Similar to the effectiveness variation of the second derivative term, effectiveness of fourth derivative term is overestimated in the high wavenumber range, which provides further control on aliasing error, as discussed in [82]. Numerical instabilities are usually observed due to amplification of high wavenumber components [38, 58, 61]. Thus, addition of numerical diffusion is aimed at attenuating high wavenumber components.

### 3.3.7  Computation of Unsteady Flow by Dispersion Relation Preserving (DRP) Methods

Every space–time-dependent problems are characterized by unique dispersion relation. For transitional and turbulent flows, length and time scales have been shown via their spectra. LES and DNS of such flows require numerically following the physical dispersion relation, for which the group velocity plays the role for the transportation of energy. Primary aim here is relating $V_g$ with the numerical dispersion property. Matching $V_g$ with numerical group velocity $(V_{gN})$ ensures removal of dispersion error. The overall accuracy depends on many other sources of error.

For calibrating space–time-dependent problems, Eq. (3.34) provides a tough challenge for any methods of discretization. This equation being non-dispersive and non-dissipative, initial condition convects undistorted and without attenuation to the right with the phase speed $c$, which happens to be both the phase speed and the group velocity.

The study of spatial discretization separately from temporal discretization is thus totally unwarranted and erroneous. Having chosen any combination of methods, one should still analyze the numerical scheme's dispersion and dissipation properties with the help of standard models, as we are proposing here. The latter has been variously studied in [37, 86, 91]. Vichnevetsky and Bowles [94] have in an ad hoc manner attempted to explain dispersion property of $CD_2$ finite difference and a linear Galerkin FEM incorrectly. A correct approach in estimating numerical dispersion relation was started in [69, 71, 76]. The following is based on GSA for the full computational domain in [72].

Thus, before simulation of unsteady problem for complex geometries, it is recommended that the important properties of numerical amplification factor $(|G|)$ and normalized numerical group velocity $(V_{gN}/c)$ and normalized phase speed $(c_N/c)$ for the 1D wave equation are analyzed [36, 57]

$$\frac{\partial u}{\partial t} + c\frac{\partial u}{\partial x} = 0, \qquad c > 0 \tag{3.95}$$

Analytical solution of Eq. (3.95) expresses the initial condition to convect at the constant phase speed without attenuation or distortion. This equation also automatically works for acoustics, which shares the same properties in homogeneous, open domain [21]. Thus, one would first obtain $|G|$, $c_N/c$ and $V_{gN}/c$ for the solution of 1D wave equation following [36, 57]. The $G_j$ for the $j$th node is the ratio of the bilateral Laplace amplitudes of the signal at $t$ and $t + \Delta t$ given by $\left( G_j = \frac{U(k,t^{n+1})}{U(k,t^n)} \right)$. As $G_j$ has to be complex, $(G_j = G_R + iG_I)$, one defines an amplitude from $|G_j| = \sqrt{(G_R^2 + G_I^2)}$ [36, 57] and a phase shift per time step by $\beta_j$, where $\tan \beta_j = -G_I/G_R$, enabling one to write expressions for $V_{gN}/c$ and $c_N/c$, following the notations in [36, 57] for the $j$th node by

$$\left[ \frac{c_N}{c} \right]_j = \frac{\beta_j}{\omega \Delta t} \tag{3.96}$$

$$\left[ \frac{V_{gN}}{c} \right]_j = \frac{1}{N_c} \frac{d\beta_j}{d(kh)} \tag{3.97}$$

where $N_c$ is the Courant–Friedrichs–Lewy (CFL) number ($N_c = \frac{c\Delta t}{h}$) with $\Delta t$ as the time step. These have been described in details for evaluating $|G_j|$, $(V_{gN}/c)_j$, and $(c_N/c)_j$ in [36, 61]. In Fig. 3.9 $|G|$ and $V_{gN}/c$ contours are compared for the central node, using various combined compact schemes and $RK4$ time integration scheme for the numerical solution of Eq. (3.95). The left and the right frames of Fig. 3.9a show $|G|$ and $V_{gN}/c$ contours in $(N_c, (kh))$ plane. The $|G|$ contours in Fig. 3.9a and c display necessary neutral stability only for very small $N_c$ values, for full range of $kh$. However, neutral stability is absent for $CCD$ scheme in frame (b) for the scheme due to [10], as upwinding adds diffusion. However, $|G|$ contours alone do not indicate the utility of a numerical method only. One must factor in the dispersion property indicated by $V_{gN}/c$ contours.

One percent tolerance has been marked by the region bounded by $V_{gN}/c = 0.99$ and $V_{gN}/c = 1.01$ is the $DRP$ region [50, 57, 61, 81], where one must also ensure to have $|G| = 1$. For example, $(V_{gN}/c)$ contours indicate in frame (a) that for up to $N_c = 0.15$, the $DRP$ region can be said to exist up to $kh = 1.40$. Solution components above this wavenumber value will be spurious, as these components will propagate in the non-physical upstream direction, and such spurious components are known as the $q$-wave, as distinct from the components moving in the physical direction. However, outside the zone of tolerance, even these physical components are known to suffer dispersion. All the CCD schemes in Fig. 3.9 show the extreme unphysical attribute of dispersion, indicated by the presence of $q$-waves [57, 61, 75]. For the correct prediction of the acoustic field, any spurious components present in the solution must be removed by added diffusion. We suggest adding controlled numerical diffusion by fourth derivative term, whenever and wherever it is required.

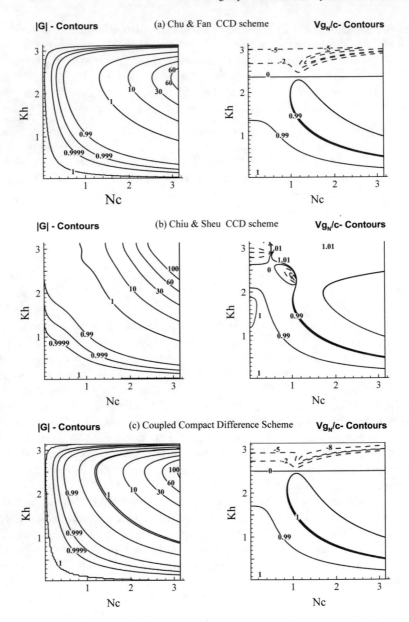

**Fig. 3.9** Contours of $|G|$ and $V_{gN}/c$ for a central node shown in $(N_c, kh)$-plane with spatial discretization are done using **a** $CCD$ scheme of [11]; **b** $CCD$ scheme of [10]; and **c** coupled compact difference scheme [32] with $RK4$ time integration scheme in solving 1D wave equation

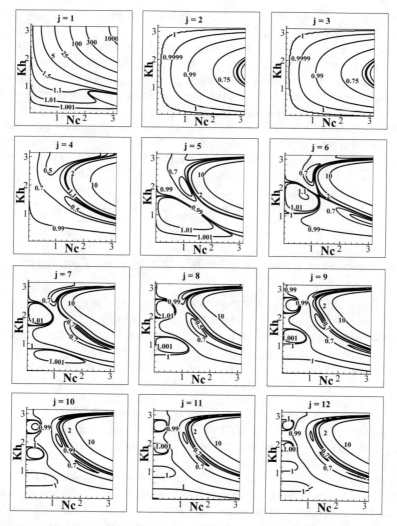

**Fig. 3.10** Comparison of $|G|$ contours for boundary and near-boundary nodes for CCD scheme [48] used with $RK4$ time integration scheme for 1D wave equation. Domain is divided into 101 equi-spaced points with $h$ as the spacing

Unlike the contours in Fig. 3.9 for an isolated central node values, $|G_j|$ contours in Fig. 3.10 are obtained by GSA to show numerical amplification factors for points near inlet boundary and near-boundary nodes. For non-periodic problems, one requires different boundary stencils at the boundary and near-boundary nodes. This causes varying properties for the near-boundary nodes, as compared to central nodes [36]. The $|G_j|$ contours for $j = 1$ display significant instability in a large region, for very small $N_c$ values. For the used central stencils for 2nd and 3th nodes, show necessary neutral stability for small $N_c$ values. However, for node-5 onward, one notices mild instability for small $N_c$ values.

Instability of boundary nodes is controlled by added numerical diffusion. Figure 3.11 shows the $|G_j|$ contours for the boundary and near-boundary nodes, with fourth-order diffusion term with diffusion coefficient $\alpha_d = 0.01$ is added with first derivative. The figure indicates that the added numerical diffusion introduces stability, as $|G_j|$ contours indicate for small $N_c$ values. Thus, one can choose controlled amount of diffusion to achieve the desired level of numerical performance. We, however, note that the unstable numerical behavior onset is due to the presence of $q$-waves, which travel upstream to reach nodes at the inlet. If one eliminates $q$-waves in the interior, subsequently there is no need to add numerical diffusion for near-boundary points. Figures 3.10 and 3.11 highlight undesirable properties of near-boundary nodes, as compared to interior nodes, as shown in Fig. 3.9c.

### 3.3.8  Further Analysis of $G_j$ with 1D Advection Equation

Many researchers have analyzed numerical schemes to follow error dynamics. To do so, one can trace backward from discrete governing equation using Taylor series expansion to an equivalent differential equation, which is an analog of the discrete equation following the process suggested in [84]. Relating the truncation error terms for space–time discretizations, one obtains the $\Gamma$-form analysis route. If the truncation error terms for spatial discretization are retained only, then the corresponding differential equation is called the $\Pi$-form equation. Unfortunately, in the literature one notices proliferation of $\Pi$-form analysis. There are only few $\Gamma$-form analyses reported with consistent results in [70, 71]. This is the approach which is also consistent with DRP route of analysis. The insignificance of $\Pi$-form analysis is exposed clearly in [71].

In the formal von Neumann approach, the error of the numerical solution for constant coefficient differential equation is assumed to follow identically the discrete equation for the signal, and the difference between exact and numerical solution is either due to roundoff and/or by the error with initial data. In this formal approach for periodic problems, error is represented by Fourier series. One investigates the normal modes one at a time, without considerations for mutual interactions of various normal modes. This approach is used for nonlinear systems by quasi-linearization and also for linear systems with variable coefficients. However, for nonlinear systems, one cannot superpose the normal modes and additionally aliasing error is a major source of misrepresentation of actual dynamics.

Numerous efforts in following error dynamics follow the method attributed to von Neumann [12, 16], which appears to be very obvious due to the linearity/quasi-linearity of the equations studied. Primary assumption used for linear problems starts with the observation that signal and associated error must follow identical dynamics due to linearity. However, the authors in [71] reported a fresh approach and established that this intuitive assumption itself is flawed for all discrete computations due

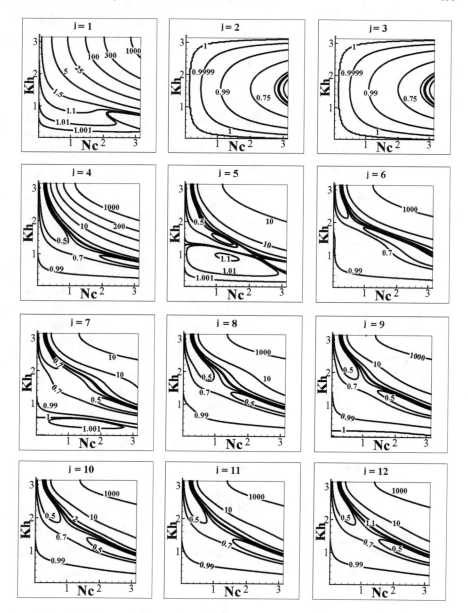

**Fig. 3.11** Comparison of effects of added diffusion with $\alpha_d = 0.01$ on $|G_j|$ contours for the boundary and near-boundary nodes for CCD scheme [32] used with $RK4$ time integration scheme for 1D wave equation. Domain has been divided into 101 equi-spaced points with $h$ as the spacing

to the coupling of space–time discretization which leads to phase and dispersion errors, in addition to the numerical method contributing error due to non-neutral $|G|$ for the model 1D convection equation.

We demonstrate the need to go beyond von Neumann analysis using the GSA results for Eq. (3.34) to analyze spatial and temporal discretization studied together. This equation forms the bedrock of analyzing numerical methods with respect to accuracy of numerical solution, error dynamics, and quantification of all these by noting the dispersion error, as attempted in [69, 71, 76, 94, 102] with different degree of success.

Performing this analysis requires representing an unknown in a hybrid form, by its bilateral Laplace transform as a function of time for the variable at the $j$th-node of a uniform grid with spacing $h$ by $u(x_j, t) = \int U(kh, t)\, e^{ikx_j}\, dk$. As noted before, the spectral amplification factor is defined by

$$G(\Delta t, kh) = \frac{U(kh, t^n + \Delta t)}{U(kh, t^n)} \tag{3.98}$$

As $G$ has to be complex, the bilateral Laplace amplitude $U(kh, t)$ also will be a complex quantity. In the continuum limit of $h$, $\Delta t \to 0$, for the 1D convection equation, there is the need for one to have

$$|G| \equiv 1 \tag{3.99}$$

The real and imaginary parts of $G$ depend not only on the equation solved, but also more importantly on the spatial–temporal discretization methods used, with other properties obtained using the hybrid representation of Eq. (3.34) with Eq. (3.69). This yields

$$\int \left[ \frac{dU}{dt} + \frac{c}{h} \sum U C_{jl}\, e^{ik(x_l - x_j)} \right] e^{ikx_j}\, dk = 0 \tag{3.100}$$

As it is true for any $k$, the integrand must itself vanish for all $k$. Implicitly, Eq. (3.100) provides the following rearrangement:

$$\frac{dU}{U} = -\left[ \frac{cdt}{h} \right] \sum_{l=1}^{N} C_{jl}\, e^{ik(x_l - x_j)} \tag{3.101}$$

The factor in square bracket on the RHS of Eq. (3.101) is $N_c$. The LHS of Eq. (3.101) is the node-dependent $G_j$ for Euler time integration given by

$$G_j = G|_{(x=x_j)} = 1 - N_c \sum_{l=1}^{N} C_{jl}\, e^{ik(x_l - x_j)} \tag{3.102}$$

From Eq. (3.70), one can use the definition of $k_{eq}$ for the summed term on the RHS as

$$G_j = G|_{(x=x_j)} = 1 - i N_c\, [k_{eq} h]_j \tag{3.103}$$

It is readily apparent that Euler time integration is unstable, as $|G_j| > 1$. Instead, we recommend any other two time level, multistage higher order methods for accurate and stable approach, as has been explained for the $RK_4$ time integration method [61, 71]. If the RHS of Eq. (3.34) is denoted as $L(u) = -c\frac{\partial u}{\partial x}$, then the following steps are used in $RK_4$ method:

Step 1: $\quad u^{(1)} = u^{(n)} + \frac{\Delta t}{2} L[u^{(n)}]$

Step 2: $\quad u^{(2)} = u^{(n)} + \frac{\Delta t}{2} L[u^{(1)}]$

Step 3: $\quad u^{(3)} = u^{(n)} + \Delta t L[u^{(2)}]$

Step 4: $\quad u^{(n+1)} = u^{(n)} + \frac{\Delta t}{6} \{L[u^{(n)}] + 2L[u^{(1)}] + 2L[u^{(2)}] + L[u^{(3)}]\}$

For the $RK_4$ time integration scheme, $G_j$ is obtained as [61, 71]

$$G_j = 1 - A_j + \frac{A_j^2}{2} - \frac{A_j^3}{6} + \frac{A_j^4}{24} \tag{3.104}$$

where

$$A_j = N_c \sum_{l=1}^{N} C_{jl} \, e^{ik(x_l - x_j)}$$

While $|G_j| \neq 1$ is surely the source of error, additionally, error is committed by phase and dispersion mismatch, a very subtle topic and has been introduced in recent times. This is explained in the following.

### 3.3.9 Quantification of Dispersion Error and Error Propagation Equation

One can obtain the numerical dispersion relation for 1D convection equation by using Fourier–Laplace transforms relating wavenumber with circular frequency. If this relation is different from the physical dispersion relation, then the difference between the two will be source of phase and dispersion errors. For the 1D convection equation, if the initial condition is

$$u(x_j, t = 0) = u_j^0 = \int A_0(k) \, e^{ikx_j} \, dk \tag{3.105}$$

then the solution for Eq. (3.34) at any time is given by

$$u_j^n = \int A_0(k) \, [|G_j|]^n \, e^{i(kx_j - n\beta_j)} \, dk \tag{3.106}$$

where $|G_j| = (G_{rj}^2 + G_{ij}^2)^{1/2}$ and $\tan(\beta_j) = -\frac{G_{ij}}{G_{rj}}$, with $G_{rj}$ and $G_{ij}$ as the real and imaginary parts of $G_j$, respectively. Comparing Eq. (3.106) with the general form of the solution, one readily notes that $n\beta_j = kc_N t$, with $c_N$ as the numerical phase speed, already defined. Above expression for phase shift per time step indicates that $c_N$ is a not a constant; instead, it is a function of $k$, making the numerical method non-dispersive. Noting that the numerical solution of Eq. (3.34) is denoted by

$$\bar{u}_N = \int A_0 \, [|G|]^{t/\Delta t} \, e^{ik(x - c_N t)} \, dk \qquad (3.107)$$

One notes that the numerical dispersion relation is, therefore, given by $\omega_N = c_N k$, which is qualitatively different from the physical dispersion relation:, $\bar{\omega} = ck$.

This simple observation has profound implications for scientific computing, as shown already through Eqs. (3.96) and (3.97). Let us define the numerical error as $e(x, t) = u(x, t) - \bar{u}_N$, and then one notices that the error dynamics is obtained by the following. Using Eq. (3.107), one obtains

$$\frac{\partial \bar{u}_N}{\partial x} = \int ik A_0 \, [|G|]^{t/\Delta t} \, e^{ik(x - c_N t)} \, dk \qquad (3.108)$$

$$\frac{\partial \bar{u}_N}{\partial t} = -\int ik \, c_N \, A_0 \, [|G|]^{t/\Delta t} \, e^{ik(x - c_N t)} \, dk$$

$$+ \int \frac{Ln \, |G|}{\Delta t} A_0 \, [|G|]^{t/\Delta t} \, e^{ik(x - c_N t)} \, dk \qquad (3.109)$$

The error propagates for Eq. (3.34) as [71]

$$\frac{\partial e}{\partial t} + c\frac{\partial e}{\partial x} = -c\left[1 - \frac{c_N}{c}\right]\frac{\partial \bar{u}_N}{\partial x} - \int \frac{dc_N}{dk}\left[\int ik' A_0[|G|]^{t/\Delta t}e^{ik'(x - c_N t)}dk'\right]dk$$

$$- \int \frac{Ln \, |G|}{\Delta t} A_0 \, [|G|]^{t/\Delta t} \, e^{ik(x - c_N t)} \, dk \qquad (3.110)$$

This correct error propagation equation is symptomatic of forced excitation, as opposed to the homogeneous equation obtained in von Neumann analysis, as a consequence of an assumption. The RHS in Eq. (3.110) provides the forcing, while it is assumed identically as zero in von Neumann analysis. Implicitly, in von Neumann analysis, one presumes $c_N \cong c$, i.e., there is no dispersion error and the numerical method is perfectly neutral, and all the terms on the RHS of Eq. (3.110) is identically zero. We also note that error grows more if the signal displays sharp spatial variation, due to the first term on the RHS.

The consequence of Eq. (3.110) for 1D convection equation shows the dependence of error on the numerical properties of the combination of space–time discretization methods. To demonstrate this, we look at OUCS3 scheme [72], which is given in

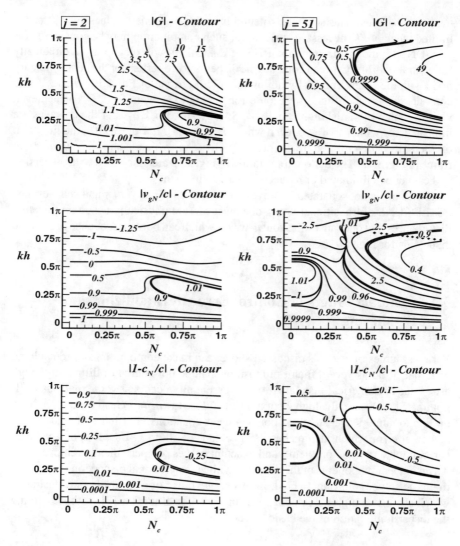

**Fig. 3.12** Contours of $|G_j|$, $V_{gN}/c$ and $(1 - c_N/c)$ for the near-boundary node $j = 2$ (left) are compared with the central node $j = 51$ (right) for the 1D advection equation solved by the OUCS3-$RK_4$ method

Eq. (3.62). For a non-periodic problem, some typical results are shown for a central and another node close to the inflow boundary, when OUCS3 scheme is used with $RK_4$ time integration scheme to solve Eq. (3.34). In Fig. 3.12, $|G|$-, $V_{gN}/c$- and $(1 - c_N/c)$-contours are plotted in $(N_c, kh)$-plane for $j = 2$ and for the central node at $j = 51$.

The properties at $j = 2$ are significantly different from those for the central node at $j = 51$. In the top frame, the combination scheme is stable for $N_c \leq 1.301$, for the central nodes. The scheme is neutrally stable for small values of $N_c$, and a limited range of $kh$, necessary attribute for such combination may be considered for LES/DNS, as the last term on the RHS of Eq. (3.110) will disappear automatically for neutral stability. One can compare the dispersion properties for the near-boundary node with the central nodes by noting the middle frame of Fig. 3.12, where $V_{gN}/c$ contours display extreme dispersion error for high $k$s, even if $G_j = 1$ is achieved via the choice of vanishingly small $N_c$. Specifically, for $kh \geq 2.4$ the $q$-waves are indicated, for which $V_{gN} \leq 0$, even when $N_c \cong 0$ [68]. In Eq. (3.110), the first term on the RHS is determined by the property $(1 - \frac{c_N}{c})$, which is shown in the bottom frame of Fig. 3.12. The bottom four frames in Fig. 3.12 are associated with dispersion and cannot be eliminated or reduced by grid refinement.

If the flow is not separated, $q$-waves are reported [47] as very small scale events. For LES/DNS of transitional flows, one must capture physically upstream traveling disturbances, but one must still avoid $q$-waves which can be a consequence of extreme dispersion for high $kh$.

## 3.4   Compact Filters to Control Error by Stabilizing and De-aliasing

Various sources of error, as discussed above, can make computations of transitional or turbulent flows very difficult proposition. The major source of difficulty though arises due to numerical instability, which essentially occurs at very small scales. If not treated with care, the high $k$ components (close to $kh = \pi$) lead to solution blow up due to numerical instability. For nonlinear equations, aliasing is the additional source of error which also makes its appearance at high $k$. Use of compact filters can control both these problems, and in actual usages one periodically removes high wavenumber components of the solution, while the numerical dispersion property of the basic solver is left unaffected. In this context, use of central filters is desirable, which attenuates high $k$ components as presented in Bhumkar [8], without altering the dispersion relation of the basic solver.

### 3.4.1   Filters for DNS and LES

Filtering is always implicitly present in all discrete computations [39, 59, 61]. In computations, the nature of filter used has to be low pass, which leaves smaller $k$ components unaffected, while removing the problematic higher $k$ components. In traditional LES, the governing equation itself is analytically filtered before discretization. In [27, 35, 42, 43, 57, 85, 93], effects of different types of filters used in classical LES have been described. The filters in traditional LES require adding additional modeled stresses (known also as SGS or Leonard stresses) which is applied on all the terms of the governing equation. This is equivalent to applying convolution, which can itself add to numerical instabilities due to aliasing caused by convolution, if not performed carefully.

In recent times, many researchers have proposed performing filtering implicitly and these are known as implicit large eddy simulation (ILES), where spatial filters [8, 21–23, 54, 55, 66, 67, 96] are applied at the end of time advancement of the governing equation. As the applied filters are post-processing operation, one can decide upon the frequency of filtering. However, the benefit comes from the fact that there is no need for any models for SGS stresses. As there are no compulsion in applying the filter at every time step, one can choose the frequency and order of filter. Significant advantages accrue if ILES is performed, allowing one to completely dispense with empirical SGS model.

To control instabilities primarily, explicit Padè filters have been proposed in [8, 21–23, 61, 66, 67, 96], to circumvent issues on instabilities due to mesh non-uniformities and the way numerical boundary conditions are applied for the use of spatial filters for ILES. The filters have central stencils applied only in the interior of the domain, while one may use one-sided stencil which have been proposed for non-periodic problems in [22, 95] for near-boundary nodes. It is now understood from the analysis in [67] using GSA that one-sided stencils can be destabilizing. Due to implicit nature of filters, such numerical instability affects in the interior.

As filters are meant for interior points only, one-dimensional filters have been proposed and developed [96], whose general stencil is given by

$$\alpha_f \hat{u}_{j-1} + \hat{u}_j + \alpha_f \hat{u}_{j-1} = \sum_{n=0}^{N} \frac{a_n}{2} (u_{j+n} + u_{j-n}) \qquad (3.111)$$

which is unfortunately defined as a $(2N)$th-order filter [22, 95]. Here, $\hat{u}_j$ represents filtered counterpart of variable $u_j$, at the $j$th grid point. It is noted that the definition of the order of filter is wrong, as in Eq. (3.111) [61]. Writing $u_j$ and $\hat{u}_j$ by hybrid Laplace transform as $u_j = \int U(k)e^{ikx_j} dk$ and $\hat{u}_j = \int \hat{U}(k)e^{ikx_j} dk$, one defines a transfer function for the periodic problem by $TF(k) = \frac{\hat{U}(k)}{U(k)}$. For the stencil given in Eq. (3.111), the transfer function for the corresponding periodic problem is

$$TF(k) = \frac{a_0 + \sum_{n=1}^{N} a_n \cos(nkh)}{1 + 2\alpha_f \cos(kh)} \qquad (3.112)$$

In designing 1D filter, one expands Eq. (3.111) by Taylor series on both the sides and matches coefficients of the identical ordered derivatives. In Eq. (3.111), if one does so up to $2N$th-order derivatives to get $N$ equations, for the $(N + 1)$ unknown coefficients, with $\alpha_f$ used as a parameter in fixing the filter performance. The rest of the unknowns, $a_0$ to $a_N$, can be solved from these $N$ equations. However, then the filter becomes a trivial identity. One of the key features of all such filter designs, therefore, rests upon matching up to $(2N - 2)$th order, which should be referred to as the order of the filter. (However, to avoid confusion, we will follow the convention of other

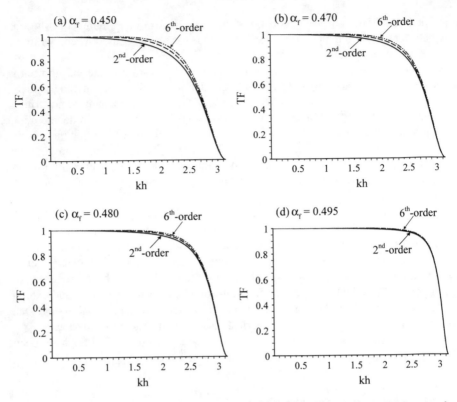

**Fig. 3.13** Transfer functions ($TF(kh)$) of periodic 1D filters plotted as a function of $kh$ for second-, fourth-, and sixth-order filters for the indicated $\alpha_f$. Note the text for determining the correct orders of filters

researchers. Thus, what has been referred to as the second-order filter in the literature is in reality a zeroth-order filter!) This, therefore, leaves the number of equations less by one, as compared to the number of unknowns. An additional equation is found by imposing a condition that $TF(k) = 0$ at $kh = \pi$. These coefficients $a_0$ to $a_N$ are given as a function of the filter coefficient $\alpha_f$ for second- to tenth-order filters in [96]. Application of filter by design then involves solving a tridiagonal system by Thomas algorithm. To be able to do so, one requires the diagonal dominance of the associated LHS matrix. This is achieved by the value of $\alpha_f$ ranging from $+0.5$ to $-0.5$. The specific choice of $\alpha_f = 0.5$ indicates no filtering of the unknowns [61], a fact that has been used in proposing an adaptive filter in [9]. In Fig. 3.13, the transfer function of the filter, $TF(k)$, is shown as a function of $kh$ for second-, fourth-, and sixth-order filters for the indicated values of $\alpha_f$. One notices in Fig. 3.13 that the second-order filter affects both low and high $kh$ components, if $\alpha_f$ is chosen below 0.47.

To design a so-called fourth- and higher order filters for non-periodic problems, authors in [96] given sets of one-sided boundary filters, to be used along with the central interior filters. One can and should use GSA to characterize the compact filters, as has been done in [61]. However, the one-sided filters in [96] have been designed by local analysis, and not by a global analysis. For example, at a near-boundary point $j$, the authors have suggested filters of the following type to retain the tridiagonal form of the resultant linear algebraic equation for filtering.

$$\alpha \hat{u}_{j-1} + \hat{u}_j + \alpha \hat{u}_{j+1} = \sum_{n=0}^{N} a_{N-n,j} u_{N-n} \tag{3.113}$$

For a so-called "fourth"-order filter, one chooses $N = 5$ for $j = 2$, for the coefficients in Eq. (3.113). These coefficients are listed as

$$a_{1,2} = \frac{1}{16} + \frac{7\alpha}{8}; \ a_{2,2} = \frac{3}{4} + \frac{\alpha}{2}; \ a_{3,2} = \frac{3}{8} + \frac{\alpha}{4}; \ a_{4,2} = -\frac{1}{4} + \frac{\alpha}{2} \ \text{and} \ a_{5,2} = \frac{1}{16} - \frac{\alpha}{8}$$

If one wants to extend it to so-called "sixth"-order representation, then one choose $N = 7$ for $j = 2$, to obtain the coefficients as

$$a_{1,2} = \frac{1}{64} + \frac{31\alpha}{32}; \ a_{2,2} = \frac{29}{32} + \frac{3\alpha}{16}; \ a_{3,2} = \frac{15}{64} + \frac{17\alpha}{32}; \ a_{4,2} = -\frac{5}{16} + \frac{5\alpha}{8};$$

$$a_{5,2} = \frac{15}{64} - \frac{15\alpha}{32}; \ a_{6,2} = -\frac{3}{32} + \frac{3\alpha}{16} \ \text{and} \ a_{7,2} = \frac{1}{64} - \frac{\alpha}{32}$$

For $j = 3$, the coefficients in Eq. (3.113) for "sixth"-order accuracy ($N = 7$) are given by

$$a_{1,3} = -\frac{1}{64} + \frac{\alpha}{32}; \ a_{2,3} = \frac{3}{32} + \frac{13\alpha}{16}; \ a_{3,3} = \frac{49}{64} + \frac{15\alpha}{32}; \ a_{4,3} = \frac{5}{16} + \frac{3\alpha}{8};$$

$$a_{5,3} = -\frac{15}{64} + \frac{15\alpha}{32}; \ a_{6,3} = \frac{3}{32} - \frac{3\alpha}{16} \ \text{and} \ a_{7,3} = -\frac{1}{64} + \frac{\alpha}{32}$$

To analyze the filter performance globally along with one-sided boundary stencils, one can write the coupled set of filter equation as

$$[\bar{A}]\{\hat{u}_j\} = [\bar{B}]\{u_j\} \tag{3.114}$$

with the boundary filters given by Eq. (3.113) are used in the first and last few rows of $[\bar{A}]$ and $[\bar{B}]$ matrices. For $\hat{u}_j$ and $u_j$, if one uses the Fourier–Laplace transforms and then equates the spectral weights for the $j$th-node, one obtains the generalized transfer function as

$$\sum_{l=1}^{N} \bar{a}_{jl} e^{ik(x_l - x_j)} \hat{U}(k) = \sum_{l=1}^{N} \bar{b}_{jl} e^{ik(x_l - x_j)} U(k) \tag{3.115}$$

with $\hat{U}$ and $U$ as the Laplace transform of $\hat{u}_j$ and $u$, respectively. Therefore, the transfer function of the filter at the $j$th-node can be expressed in the $k$-plane as

$$TF_j(k) = \frac{\sum_{l=1}^{N} \bar{a}_{jl} e^{ik(x_l - x_j)}}{\sum_{l=1}^{N} \bar{b}_{jl} e^{ik(x_l - x_j)}} \qquad (3.116)$$

As with the compact scheme for spatial discretization, one can characterize also the filters using the model 1D advection equation. If $u_j^n = u(x_j, t^n)$ is the solution of the model equation for the $j$th-node at the $n$th time step and then the solution is filtered on the advanced time solution $u_j^{n+1}$ to obtain the filtered variable $\hat{u}_j^{n+1}$, then the combined time integration and filtering would give rise to the equivalent amplification factor given by

$$\hat{G}_j(kh, N_c) = TF_j(kh)G_j(kh, N_c) \qquad (3.117)$$

Figure 3.14a and b shows the real and imaginary parts of the transfer function $(TF_j)$ for a so-called sixth-order filter using $\alpha_f = 0.47$. This figure is shown to highlight boundary stencils in Eq. (3.113) and how it affects the filter performance. In the frame (a), we note a wider bandwidth of filters for the near-boundary points and

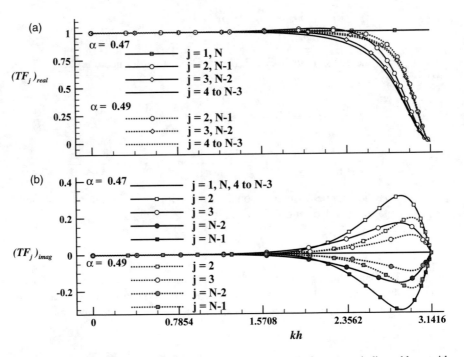

**Fig. 3.14  a** Real and **b** imaginary parts of transfer function $TF_j$ for non-periodic problems with indicated $\alpha$ at the indicated nodes for the sixth-order interior filter with near-boundary filters given in [96]

a slight overshoot of $TF_j$ for $j = 2$ and $N - 1$. We also note the stable property near the boundaries to be more for higher order filters, with more number of points becoming affected near the boundaries. For the interior, the imaginary part of $TF_j$ is identically zero, implying absent dispersive effects. To understand dispersive effects of a filter, one should look at the space–time-dependent model equations, such as 1D advection equation as reported in [61, 67].

The properties of the transfer function of the filter shown in Fig. 3.14 will attenuate $G_j(kh, N_c)$ strongly near the Nyquist limit $(kh = \pi)$. Since by design, $TF_j$ is equal to zero for $kh = \pi$. The near-boundary points of non-periodic problems, $TF_j(kh)$ being complex, will change $G_j(kh, N_c)$ due to attenuation/amplification and dispersion. Following Eq. (3.107), the filtered solution of 1D advection equation is noted as

$$\hat{u}_N = \int A_0 \left[ |\hat{G}| \right]^{t/\Delta t} e^{ik(x - c_N t)} \, dk \qquad (3.118)$$

with numerical phase speed and group velocities are obtained from

$$\left( \frac{c_N}{c} \right)_j = \frac{\hat{\beta}_j}{N_c kh} \qquad (3.119)$$

$$\left( \frac{V_{gN}}{c} \right)_j = \frac{1}{N_c} \frac{\partial \hat{\beta}_j}{\partial (kh)} \qquad (3.120)$$

where $\hat{\beta}_j = -tan^{-1} \left( [\hat{G}_j]_{\text{imag}} / [\hat{G}_j]_{\text{real}} \right)$. For central stencil for the filtering operation of a periodic problem, dispersion properties are unaffected, as $TF_j$ is strictly real (See Eq. (3.111)). Thus, $[\hat{G}_j]_{\text{imag}} / [\hat{G}_j]_{\text{real}} = [G_j]_{\text{imag}} / [G_j]_{\text{real}}$, implying no change in dispersion relation. The same is true for the interior points of a non-periodic problem. This is irrespective of any choice for the admissible value of $\alpha_f$. For these situations, central filters do not change the numerical dispersion relation of the solution for the model 1D advection equation. This is a desirable feature of a filter.

For non-periodic problems, if filter order is increased, more problems are noted near the inflow boundary points, while outflow boundary points will show excessive filtering. If one wants to change dispersion properties on purpose, then one can design an interior filter with complex $TF_j$. For this purpose, 1D upwinded filters have been proposed in [67], for which the general stencil in the interior is given as [67]

$$\bar{\alpha}_f \hat{u}_{j-1} + \hat{u}_j + \bar{\alpha}_f \hat{u}_{j-1} = \bar{\eta} u_{j+4} + (a_3/2 - 5\bar{\eta}) u_{j+3} + (a_2/2 + 10\bar{\eta}) u_{j+2}$$
$$+ (a_1/2 - 10\bar{\eta}) u_{j+1} + (a_0 + 5\bar{\eta}) u_j + (a_1/2 - \bar{\eta}) u_{j-1} + (a_2/2) u_{j-2}$$
$$+ (a_3/2) u_{j-3} \qquad (3.121)$$

This is the stencil for a fifth-order upwinded filter for interior points, with coefficients obtained as function of $\bar{\alpha}_f$, with an upwinding constant $\bar{\eta}$ as $a_0 = (22 + 20\bar{\alpha}_f)/32 - 10\bar{\eta}$, $a_1 = (15 + 34\bar{\alpha}_f)/32 + 15\bar{\eta}$, $a_2 = (-6 + 12\bar{\alpha}_f)/32 - 6\bar{\eta}$, and

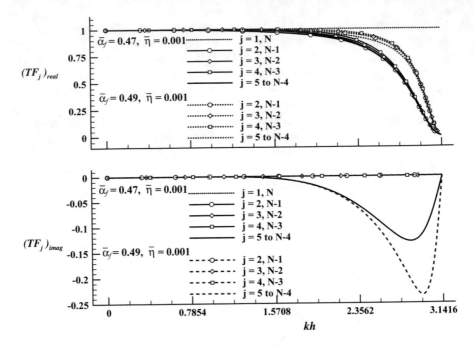

**Fig. 3.15** Real and imaginary parts of $TF(kh)$ shown for the fifth-order upwinded filter with indicated $\bar{\alpha}_f$ and $\bar{\eta}$ [67]

$a_3 = (1 - 2\bar{\alpha}_f)/32 + \bar{\eta}$. For $\bar{\eta} = 0$, one recovers the so-called sixth-order central filter, while for $\bar{\eta} \neq 0$, one obtains a complex $TF(k)$, which alters the dispersion relation.

In Fig. 3.15, real and imaginary parts of $TF(kh)$ are shown for the fifth-order upwinded filter given in Eq. (3.121) [67]. To use this filter, central filters are used for near-boundary nodes, whereas Eq. (3.121) is used from $j = 5$ to $j = (N - 4)$. The "second"-order central filter is used for $j = 2$ and $(N - 1)$, the "fourth"-order central filter is used for $j = 3$ and $(N - 2)$, and finally a "sixth"-order central filter is used for $j = 4$ and $(N - 3)$. The upwind filter removes the numerical instability problem near the inflow and excessive filtering near the outflow, as shown in Fig. 3.14 for sixth-order central filter. Hardly any difference is noted among $(TF_j)_{real}$ values for the nodes shown in Fig. 3.15a, while for near-boundary points, $(TF_j)_{imag}$ is zero in Fig. 3.15b, as central filters are used for these nodes. In Fig. 3.15b, transfer function is shown for $j = 5$ to $(N - 4)$ nodes, and the transfer function has identical imaginary value marked by solid line. More details about upwind filters and applications are given in [67], with the following features:

1. The upwind filter helps to add necessary dissipation in the interior to control instability without excessive attenuation.
2. Total control on imaginary part of $TF(k)$ allows to create "hyper-viscosity" which SGS stress models require in classical LES.

3. The upwind filter with the imaginary part of $TF(k)$ alters the dispersion relation in controlling $V_{gN}$ and hence $q$-waves.
4. The upwind filter bears resemblance with models used in detached eddy simulation [4, 92] without requiring different equations in different parts of the domain.

### 3.4.2 Two-Dimensional Filters of Different Orders

One-dimensional filters display strong directionality of the filtered solution for 2D flows [61]. In proposing upwind filter [67], the authors have shown that azimuthal filter for 2D flows past aerofoil or for rotary oscillation of cylinder in uniform flow displays stretching shear layer or vortices in azimuthal direction. Also, while solving Navier–Stokes equation, researchers noted that aliasing error as a major problem not working on too fine a grid. Aliasing originates when product terms are evaluated. In solving NSE, convective acceleration terms are the usual source of aliasing. To avoid both of the above problems, authors in [66] proposed 2D filter as solution. 2D filters more physically remove aliasing than 1D filters, and no unphysical stretching of shear layers is caused. The 2D filters are qualitatively different from the 1D filters described in Sect. 3.4.1.

Briefly, we explain aliasing error for discrete computations, which occurs in evaluating product terms numerically. Let us look at the product of two terms in the physical plane

$$w(x_j) = u(x_j)\, v(x_j) \tag{3.122}$$

These terms are written by bilateral Fourier–Laplace transform as

$$w(x_j) = \int W(k)e^{ikx_j}\, dk$$

$$u(x_j) = \int U(k_1)e^{ik_1x_j}\, dk_1$$

$$v(x_j) = \int V(k_2)e^{ik_2x_j}\, dk_2 \tag{3.123}$$

Then one obtains

$$w(x_j) = \int_{-k_m}^{k_m} \int_{-k_m}^{k_m} U(k_1)V(k_2)e^{i(k_1+k_2)x_j}\, dk_1 dk_2 \tag{3.124}$$

In discrete computing, maximum resolvable wavenumber $k_m = \pi/h$ is fixed by the Nyquist limit, where $h$ is the uniform grid spacing. Thus, for $k_1$ and $k_2$ in Eq. (3.124), the limits are same: $-k_m$ to $k_m$. This is indicated in Fig. 3.16 by $PQRS$. Equation (3.124) shows that phase for $w(x_j)$ is determined by $k = k_1 + k_2$. Due to the ranges of $k_1$ and $k_2$, range of $k$ is apparently from $2k_m$ to $-2k_m$. For example, inside $LMQ$ at the top in Fig. 3.16, $k_m \leq k \leq 2k_m$. Similarly, inside $HSG$: $-2k_m \leq k_1 + k_2 \leq -k_m$. Due to the Nyquist limit being valid for $k$, for $LMQ$,

**Fig. 3.16** Wavenumber plane domain of interest for discrete computation. Both the areas, $EBF$ and $HDG$, where aliasing causes evaluating product terms to zones $HOG$ and $EOF$

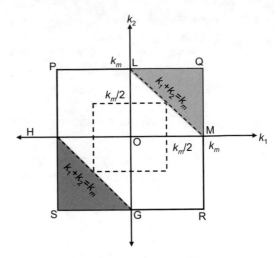

where $(k_1 + k_2) \geq |k_m|$, the phase must fold back inside the Nyquist limit. Defining a modified wavenumber $\bar{k} = (k_1 + k_2) - 2k_m = k - 2k_m$, one notes that for the aliased region,

$$e^{ikx_j} = e^{i\bar{k}x_j}$$

Thus, for $k_m \leq k \leq 2k_m$, $LMQ$ maps to $-k_m \leq \bar{k} \leq 0$, implying the transformation in wavenumber retains the phase relation for the modified wavenumber $\bar{k}$ staying inside the Nyquist limit. Thus, the region $LMQ$ will be aliased to $HOG$. Similarly, one can describe the aliasing for the zone $HSG$, which will migrate to $LOM$. Therefore, mapping the integral in Eq. (3.124) from $(k_1, k_2)$-plane to $(k, k_2)$-plane, one can write

$$w(x_j) = \int_{-k_m}^{0} \int_{-k_m}^{k+k_m} U(k - k_2)V(k_2)e^{ikx_j}\,dk_2dk$$

$$+ \int_{0}^{k_m} \int_{k-k_m}^{k_m} U(k - k_2)V(k_2)e^{ikx_j}\,dk_2dk$$

$$+ \int_{-2k_m}^{-k_m} \int_{-k_m}^{k+k_m} U(k - k_2)V(k_2)e^{ikx_j}\,dk_2dk$$

$$+ \int_{k_m}^{2k_m} \int_{k-k_m}^{k_m} U(k - k_2)V(k_2)e^{ikx_j}\,dk_2dk \qquad (3.125)$$

From Eq. (3.125), we note that the first and the last terms fall within the Nyquist limit for $k$, while the last two integrals lie outside of it to cause aliasing. Using the transformations $\bar{k} = k + 2k_m$ and $\bar{k} = k - 2k_m$ in the third and fourth integrals, respectively, we rewrite the above equation as

$$w(x_j) = \int_{-k_m}^{0} \int_{-k_m}^{k+k_m} U(k - k_2)V(k_2)e^{ikx_j} dk_2 dk$$

$$+ \int_{0}^{k_m} \int_{k-k_m}^{k_m} U(k - k_2)V(k_2)e^{ikx_j} dk_2 dk$$

$$+ \int_{0}^{k_m} \int_{-k_m}^{k-k_m} U(k - k_2 - 2k_m)V(k_2)e^{ikx_j} dk_2 dk$$

$$+ \int_{-k_m}^{0} \int_{k+k_m}^{k_m} U(k - k_2 + 2k_m)V(k_2)e^{ikx_j} dk_2 dk \tag{3.126}$$

In Eq. (3.126), we have replaced $\bar{k}$ by $k$ without changing implication for the integrals. We also note that aliasing is associated with the last two integrals of Eq. (3.126). Expressing $w(x_j)$ by its Laplace transform and equating it with the RHS of Eq. (3.126), the Laplace amplitude is given by

$$W(k) = \int_{-k_m}^{k+k_m} U(k - k_2)V(k_2)dk_2 + \int_{k-k_m}^{k_m}$$

$$U(k - k_2)V(k_2)dk_2 + W_{\text{alias}}(k) \tag{3.127}$$

with the aliased contribution $W_{\text{alias}}(k)$ given by

$$W_{\text{alias}}(k) = \int_{-k_m}^{k-k_m} U(k - k_2 - 2k_m)V(k_2)dk_2$$

$$+ \int_{k+k_m}^{k_m} U(k - k_2 + 2k_m)V(k_2)dk_2 \tag{3.128}$$

Equation (3.128) reveals aliased component to be shifted to a lower $k$, within the Nyquist limit. This is noted in solving NSE for the nonlinear convective acceleration terms. A comprehensive discussion on effects of aliasing error is given in [61], and established here that the issue can be handled effectively by 2D filter.

It will be shown that 2D filters remove high $kh$ components, those responsible for aliasing. General form of 2D filters in [66] has been proposed as

$$\hat{u}_{i,j} + \alpha_{2f}(\hat{u}_{i-1,j} + \hat{u}_{i+1,j} + \hat{u}_{i,j-1} + \hat{u}_{i,j+1})$$

$$= \sum_{n=0}^{\hat{M}} \frac{a_n}{2}(u_{i\pm n,j} + u_{i,j\pm n}) \tag{3.129}$$

In Eq. (3.129), the filtered variable is given by $\hat{u}_{i,j}$ and $u_{ij}$ is the unfiltered variable. In Eq. (3.129), $\hat{M}$ can be again associated with the order of filter, with implications as noted before for the meaning of order. This is equal to zero for second-order filter and every increase of $\hat{M}$ by 1 increments the order by two. If $N_i$ and $N_j$ are the number of points in $i$- and $j$-directions, then the values of $i$ and $j$ range from $i = 2$ to $(N_i - 1)$ and $j = 2$ to $(N_j - 1)$, for the interior points to be filtered, which can be handled by a "second"-order filter. Similarly, for the "fourth"-order filter, Eq. (3.129) is used from $i = 3$ to $(N_i - 2)$ and $j = 3$ to $(N_j - 2)$. Higher order filters are applied for lower ranges of points similarly. Equation (3.129) implies that the 2D filter requires solving a linear algebra problem with the LHS given by a pentadiagonal matrix. Any iterative solver can be employed to solve this linear algebra problem, without violating the diagonal dominance of the matrix. This is ensured by keeping $|\alpha_{2f}| \leq 0.25$ [66].

Once again using $\hat{u}$ and $u$ for filtered and unfiltered quantities in the physical plane, the 2D filter's performance is evaluated in the spectral plane by bilateral Laplace transform as

$$u_{j,l} = \int \int U(k_x, k_y)e^{i(k_x x_j + k_y y_l)} \, dk_x \, dk_y$$

$$\hat{u}_{j,l} = \int \int \hat{U}(k_x, k_y)e^{i(k_x x_j + k_y y_l)} \, dk_x \, dk_y$$

One obtains the transfer function, $TF(k_x h_x, k_y h_y)$, of the 2D filter of Eq. (3.129) by uniform grid spacing in $x$- and $y$-directions given by $h_x$ and $h_y$ by

$$TF(k_x h_x, k_y h_y) = \frac{a_0 + \sum_{n=1}^{\hat{M}} a_n[\cos(nk_x h_x) + \cos(nk_y h_y)]}{1 + 2\alpha_{2f}[\cos(k_x h_x) + \cos(k_y h_y)]} \qquad (3.130)$$

Here $k_x$ and $k_y$ are the wavenumber components along $i$- and $j$-directions, respectively. As noted already that the choice of filter coefficient is restricted to $\alpha_{2f} < 0.25$, with the value of 1/4 does no filtering of the variable. The coefficients $a_k$ of the 2D filter are fixed as functions of $\alpha_{2f}$ by expanding the RHS and LHS of Eq. (3.129) by Taylor series and equating coefficients of appropriate terms of same order on both sides. For second order, one just equates the coefficients of unfiltered and filtered quantities at the $(i, j)$th node, which is the consistency condition given by

$$a_0 + 2a_1 = 1 + 4\alpha_{2f} \qquad (3.131)$$

To obtain $a_0$ and $a_1$ for "second"-order, 2D filter, the supplementary condition is the additional condition requiring the transfer function to vanish for $k_x h_x = k_y h_y = \pi$, i.e., at point $Q$ in Fig. 3.16. This is imposed to effectively control the aliasing error, without distorting the mean flow and unphysical attenuation of the variable [66]. Thus, the additional relation is

$$a_0 = 2a_1 \qquad (3.132)$$

Solving Eqs. (3.131) and (3.132), one obtains

$$a_0 = \frac{1}{2} + 2\alpha_{2f}$$

$$a_1 = \frac{1}{4} + \alpha_{2f} \qquad (3.133)$$

For the "fourth"-order, 2D-filter, the consistency condition alters to

$$a_0 + 2a_1 + 2a_2 = 1 + 4\alpha_{2f} \qquad (3.134)$$

For this filter, one also equates the coefficients of second-order derivatives (including the mixed derivatives) at the $(i, j)$th-node to yield the following:

$$a_1 + 4a_2 = 2\alpha_{2f} \qquad (3.135)$$

Additionally, Eqs. (3.134) and (3.135) are supplemented by the vanishing of the transfer function at $Q$ in Fig. 3.16, which provides

$$a_0 - 2a_1 + 2a_2 = 0 \qquad (3.136)$$

Solving Eqs. (3.134)–(3.136), one obtains $a_0$, $a_1$, and $a_2$ for the "fourth"-order 2D filter as

$$a_0 = \frac{(5 + 12\alpha_{2f})}{8}, \quad a_1 = \frac{(1 + 4\alpha_{2f})}{4}, \quad a_2 = \frac{(-1 + 4\alpha_{2f})}{16} \qquad (3.137)$$

Similarly, for the "sixth" order, the coefficients of 2D filter are obtained as [66]

$$a_0 = \frac{(11 + 20\alpha_{2f})}{16}, \quad a_1 = \frac{(15 + 68\alpha_{2f})}{64}$$
$$a_2 = \frac{(-3 + 12\alpha_{2f})}{32}, \quad a_3 = \frac{(1 - 4\alpha_{2f})}{64} \qquad (3.138)$$

In Fig. 3.17, the transfer functions of a second-order, 2D filter are shown, for the indicated values of $\alpha_{2f}$, as a function of $k_x h_x$ and $k_y h_y$. The region corresponding to $(k_x h_x + k_y h_y) > \pi$ contributes to aliasing error, which have been shaded here for ease of understanding. Clearly, from Fig. 3.17, the transfer function of 2D filters more effectively reduces aliasing error, without unphysically attenuating low and mid-wavenumber ranges in both directions [66]. It has been shown in [9] that 2D filters can be used adaptively on designated small region of the domain to locally affect the region, where high wavenumber fluctuations can lead to solution divergence. Such adaptive filters are in use for performing receptivity calculations in [5, 62, 64] to de-alias solution effectively.

## Problems

**P3.1:** Derive governing equation for velocity potential ($\phi$) for unsteady irrotational compressible flow. Is the equation linear or nonlinear?

**P3.2:** From the solution of Eq. (3.1), obtain governing equation for compressible steady flow. Linearize this equation.

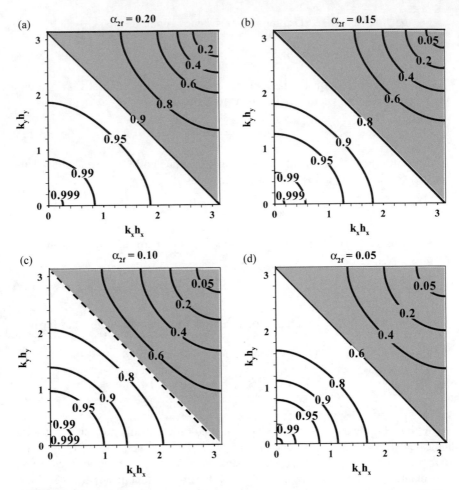

**Fig. 3.17** The transfer functions of the second-order 2D filter shown for indicated values of filter coefficient $\alpha_{2f}$. The aliased part is shaded for ease of understanding how this is avoided using 2D filter

**P3.3:** Obtain the expression for $|G|$, $|V_{gN}/c|$, and $c_N/c$ for Fourier spectral method with two-stage Runge–Kutta (RK2) method for the one-dimensional convection equation.

**P3.4:** Why should one be interested in one-dimensional convection equation to calibrate numerical methods for solving time-accurate Euler or Navier–Stokes equation?

# References

1. Y. Adams, Highly accurate compact implicit method and boundary conditions. Comput. Phys. **24**, 10–22 (1977)
2. R.L. Ash, A.J. Zuckerwar, Z. Zheng, *Second Coefficient of Viscosity in Air.* NASA-CR-187783 (1991)
3. B. Auld, *Acoustic Fields and Waves in Solids* (Wiley-Interscience, New York, 1973)
4. M.F. Barone, C.J. Roy, Evaluation of detached eddy simulation for turbulent wake applications. AIAA J. **44**(12), 3062–3071 (2006)
5. S. Bhaumik, *Direct Numerical Simulation of Inhomogeneous Transitional and Turbulent Flows.* Ph.D. Thesis, I. I. T. Kanpur (2013)
6. S. Bhaumik, T.K. Sengupta, Precursor of transition to turbulence: spatio-temporal wave front. Phys. Rev. E **89**(4), 043018 (2014)
7. S. Bhaumik, T.K. Sengupta, A new velocity-vorticity formulation for direct numerical simulation of 3D transitional and turbulent flows. J. Comput. Phys. **284**, 230–260 (2015)
8. Y.G. Bhumkar, *High Performance Computing of Bypass Transition.* Ph.D. Thesis, I. I. T. Kanpur (2011)
9. Y.G. Bhumkar, T.K. Sengupta, Adaptive multi-dimensional filters. Comput. Fluids **49**(1), 128–140 (2011)
10. L. Brillouin, *Wave Propagation and Group Velocity* (Academic Press, New York, 1960)
11. G. Buresti, A note on Stokes hypothesis. Acta Mech. **226**, 3555–9 (2015)
12. J.G. Charney, R. Fjørtoft, J. Von Neumann, Numerical integration of the barotropic vorticity equation. Tellus **2**(4), 237–254 (1950)
13. P.H. Chiu, T.W.H. Sheu, On the development of a dispersion-relation-preserving dual-compact upwind scheme for convection-diffusion equation. J. Comput. Phys. **228**, 3640–3655 (2009)
14. P.C. Chu, C. Fan, A three-point combined compact difference scheme. J. Comput. Phys. **140**, 370–399 (1998)
15. M.S. Cramer, Numerical estimates for the bulk viscosity of ideal gases. Phys. Fluids **24**(066102), 1–23 (2012)
16. J. Crank, P. Nicolson, A practical method for numerical evaluation of solutions of partial differential equations of the heat conduction type. Proc. Camb. Phil. Soc. **43**(50), 50–67 (1947)
17. J.D. Crouch, A. Garbaruk, D. Magidov, A. Travin, Origin of transonic buffet on aerofoils. J. Fluid Mech. **628**, 357–369 (2009)
18. P.A. Davidson, *Turbulence: An Introduction for Scientists and Engineers* (Oxford University Press, UK, 2004)
19. G. Emanuel, Bulk viscosity of a dilute polyatomic gas. Phys. Fluids A **2**, 2252–2254 (1990)
20. M. Gad-el-Hak, Questions in fluid mechanics. Stokes' hypothesis for a Newtonian, isotropic fluid. J. Fluids Eng. **117**, 3–5 (Technical Forum) (1995)
21. D.V. Gaitonde, J.S. Shang, J.L. Young, Practical aspects of higher-order numerical schemes for wave propagation phenomena. Int. J. Numer. Meth. Eng. **45**, 1849 (1999)
22. D.V. Gaitonde, M.R. Visbal, Further development of a Navier-Stokes solution procedure based on higher order formulas, in *37th Aerospace Sciences Meeting and Exhibition, AIAA 99–0557, Reno, NV* (1999)
23. D.V. Gaitonde, M.R. Visbal, Padé-type higher-order boundary filters for the Navier-Stokes equations. AIAA J. **38**(11), 2103 (2000)
24. T.B. Gatski, Review of incompressible fluid flow computations using the vorticity-velocity formulation. Appl. Numer. Math. **7**, 227–239 (1991)
25. T.B. Gatski, C.E. Grosch, M.E. Rose, A numerical study of the two-dimensional Navier-Stokes equations in vorticity-velocity variables. J. Comput. Phys. **48**, 1–22 (1982)
26. T.B. Gatski, C.E. Grosch, M.E. Rose, A numerical solution of the Navier-Stokes equations for three-dimensional, unsteady, incompressible flows by compact schemes. J. Comput. Phys. **82**, 298–329 (1989)

27. S. Ghosal, P. Moin, The basic equations for the large eddy simulation of turbulent flows in complex geometry. J. Comput. Phys. **118**, 24 (1995)
28. R.E. Graves, B.M. Argrow, Bulk viscosity: past to present. J. Thermophys. Heat Trans. **13**(3), 337–342 (1999)
29. W.R. Hamilton, *The Collected Mathematical Papers*, vol. 4 (Cambridge University Press, 1839)
30. G.J. Hirasaki, J.D. Hellums, A general formulation of the boundary conditions on the vector potential in three-dimensional hydrodynamics. Quart. Appl. Maths. **26**(3), 331–342 (1968)
31. G.J. Hirasaki, J.D. Hellums, Boundary conditions on the vector and scalar potentials in viscous three-dimensional hydrodynamics. Quart. Appl. Maths. **28**(2), 293–296 (1970)
32. J.T. Holdeman, A velocity-stream function method for three-dimensional incompressible fluid flow. Comput. Methods Appl. Mech. Eng. **209**, 66–73 (2012)
33. F. Jenkins, H. White, *Fundamentals of Physical Optics* (McGraw-Hill, New York, 1973)
34. S.M. Karim, L. Rosenhead, Rev. Modern Phys. **24**, 108–16 (1952)
35. C.A. Kennedy, M.H. Carpenter, Several new numerical methods for compressible shear-layer simulations. Appl. Numer. Math. **14**, 397 (1994)
36. Z. Kopal, *Numerical Analysis* (Springer, New York, USA, 1966)
37. H. Kreiss, J. Oliger, Comparison of accurate methods for the integration of hyperbolic equations. Tellus **24**, 199–215 (1972)
38. L.D. Landau, E.M. Lifshitz, *Fluid Mechanics* (Pergamon Press, New York, 1959)
39. S.K. Lele, Compact finite difference schemes with spectral-like resolution. J. Comput. Phys. **103**(1), 16–42 (1992)
40. L.N. Liebermann, The second viscosity of liquids. Phys. Rev. **75**(9), 1415 (1949)
41. M.J. Lighthill, *Fourier Analysis and Generalized Functions* (Cambridge University Press, Cambridge, UK, 1978)
42. J. Mathew, R. Lechner, H. Foysi, J. Sesterhenn, R. Friedrich, An explicit filtering ethod for large eddy simulation of compressible flows. J. Comput. Phys. **15**(8), 2279 (2003)
43. F.M. Najjar, D.K. Tafti, Study of discrete test filters and finite difference approximations for the dynamic subgrid-scale stress model. J. Comput. Phys. **8**(4), 1076 (1996)
44. J. Ortega-Casanova, R. Fernandez-Feria, A numerical method for the study of nonlinear stability of axisymmetric flows based on the vector potential. J. Comput. Phys. **227**(6), 3307–3321 (2008)
45. G.A. Oswald, K.N. Ghia, U. Ghia, Direct solution methodologies for the unsteady dynamics of an incompressible fluid, in *International Conference of Computing in Engineering, Science*, vol. 2, Atlanta, ed. by S.N. Atluri, G. Yagawa (Springer, Berlin, 1988)
46. K. Oberleithner, M. Sieber, C.N. Nayeri, C.O. Paschereit, C. Petz, H.-C. Hege, B.R. Noack, I. Wygnanski, Three-dimensional coherent structures in a swirling jet undergoing vortex breakdown: stability analysis and empirical mode construction. J. Fluid Mech. **679**, 383–414 (2011)
47. T. Poinsot, D. Veynante, *Theoretical and Numerical Combustion*, 2nd edn. (Edwards, PA, 2005)
48. J. Pradhan, B. Mahato, S.D. Dhandole, Y.G. Bhumkar, Construction, analysis and application of coupled compact difference scheme in computational acoustics and fluid flow problems. Commun. Comput. Phys. **18**(4), 957–984 (2015)
49. K.R. Rajagopal, A new development and interpretation of the Navier-Stokes fluid which reveals why the "Stokes assumption" is inapt. Int. J. Non-Linear Mech. **50**, 141–151 (2013)
50. M.K. Rajpoot, T.K. Sengupta, P.K. Dutt, Optimal time advancing dispersion relation preserving schemes. J. Comput. Phys. **229**(10), 3623–3651 (2010)
51. L. Rayleigh, *Scientific Papers 1* (Cambridge University Press, Cambridge, 1889)
52. L. Rayleigh, *Scientific Papers 2* (Cambridge University Press, Cambridge, 1890)
53. W.C. Reynolds, A.K.M.F. Hussain, The mechanics of an organized wave in turbulent shear flow. Part 3. Theoretical models and comparisons with experiments. J. Fluid Mech. **54**(2), 263–288 (1972)

54. D.P. Rizzetta, M.R. Visbal, G.A. Blaisddell, A time-implicit high-order compact differencing and filtering scheme for large-eddy simulation. Int. J. Numer. Meth. Fluids **42**, 655 (2003)

55. D.P. Rizzetta, M.R. Visbal, P.E. Morgan, A high-order compact finite-difference scheme for large-eddy simulation of active flow control, in *46th Aerospace Sciences Meeting and Exhibition, AIAA 2008-526, Reno, NV* (2008)

56. L. Rosenhead, Proc. R. Soc. London A **226**, 1–6 (1954)

57. P. Sagaut, *Large Eddy Simulation for Incompressible Flows* (Springer, Berlin, 2002)

58. A. Sengupta, V.K. Suman, T.K. Sengupta, S. Bhaumik, An enstrophy-based linear and non-linear receptivity theory. Phys. Fluids **30**, 054016 (2018)

59. T.K. Sengupta, *Fundamentals of Computational Fluid Dynamics* (University Press, Hyderabad, India, 2004)

60. T.K. Sengupta, *Instabilities of Flows and Transition to Turbulence* (CRC Press, Taylor & Francis Group, Florida, USA, 2012)

61. T.K. Sengupta, *High Accuracy Computing Methods: Fluid Flows and Wave Phenomenon* (Cambridge University Press, USA, 2013)

62. T.K. Sengupta, S. Bhaumik, Onset of turbulence from the receptivity stage of fluid flows. Phys. Rev. Lett. **154501**, 1–5 (2011)

63. T.K. Sengupta, S. Bhaumik, *DNS of Wall-Bounded Turbulent Flows: A First Principle Approach* (Springer-Nature, Singapore, 2019)

64. T.K. Sengupta, S. Bhaumik, Y.G. Bhumkar, Direct numerical simulation of two-dimensional wall-bounded turbulent flows from receptivity stage. Phys. Rev. E **85**(2), 026308 (2012)

65. T.K. Sengupta, S. Bhaumik, R. Bose, Direct numerical simulation of transitional mixed convection flows: viscous and inviscid instability mechanisms. Phys. Fluids **25**, 094102 (2013)

66. T.K. Sengupta, Y.G. Bhumkar, New explicit two-dimensional higher order filters. Comput. Fluids **39**, 1848–1863 (2010)

67. T.K. Sengupta, Y. Bhumkar, V. Lakshmanan, Design and analysis of a new filter for LES and DES. Comput. Struct. **87**, 735–750 (2009)

68. T.K. Sengupta, Y. Bhumkar, M.K. Rajpoot, V.K. Suman, S. Saurabh, Spurious waves in discrete computation of wave phenomena and flow problems. Appl. Math. Comput. **218**, 9035–9065 (2012)

69. T.K. Sengupta, S. Dey, Proper orthogonal decomposition of direct numerical simulation data of by-pass transition. Comput. Struct. **82**, 2693–2703 (2004)

70. T.K. Sengupta, A. Dipankar, A comparative study of time advancement methods for solving Navier-Stokes equations. J. Sci. Comput. **21**(2), 225–250 (2004)

71. T.K. Sengupta, A. Dipankar, P. Sagaut, Error dynamics: beyond von Neumann analysis. J. Comput. Phys. **226**, 1211–1218 (2007)

72. T.K. Sengupta, G. Ganeriwal, S. De, Analysis of central and upwind compact schemes. J. Comput. Phys. **192**, 677–694 (2003)

73. T.K. Sengupta, V. Lakshmanan, V.V.S.N. Vijay, A new combined stable and dispersion relation preserving compact scheme for non-periodic problems. J. Comput. Phys. **228**(8), 3048–3071 (2009)

74. T.K. Sengupta, M.T. Nair, A new class of waves for Blasius boundary layer, in *Proceeding of 7th Asian Congress of Fluid Mechanics* (Allied Publishers, Chennai, India, 1997), pp. 785–788

75. T.K. Sengupta, Y.G. Bhumkar, M. Rajpoot, V.K. Suman, S. Saurabh, Spurious waves in discrete computation of wave phenomena and flow problems. Appl. Math. Comput. **218**, 9035–9065 (2012)

76. T.K. Sengupta, A.K. Rao, K. Venkatasubbaiah, Spatio-temporal growing wave fronts in spatially stable boundary layers. Phys. Rev. Lett. **96**(22), 224504 (2006)

77. T.K. Sengupta, A. Sengupta, S. Sengupta, A. Bhole, K.S. Shruti, Non-equilibrium thermodynamics of Rayleigh-Taylor instability. Int. J. Thermophys. **37**(4), 1–25 (2016)

78. T.K. Sengupta, A. Sengupta, N. Sharma, S. Sengupta, A. Bhole, K.S. Shruti, Roles of bulk viscosity on Rayleigh-Taylor instability: non-equilibrium thermodynamics due to spatio-temporal pressure fronts. Phys. Fluids **28**, 094102 (2016)

79. T.K. Sengupta, H. Singh, S. Bhaumik, R. Roy Chowdhury, Diffusion in inhomogeneous flows: unique equilibrium state in an internal flow. Comp. Fluids **88**, 440–451 (2013)
80. T.K. Sengupta, S.K. Sircar, A. Dipankar, High accuracy schemes for DNS and acoustics. J. Sci. Comput. **26**, 151–193 (2006)
81. T.K. Sengupta, S.K. Sircar, A. Dipankar, High accuracy compact schemes for DNS and acoustics. J. Sci. Comput. **26**(2), 151–193 (2006)
82. T.K. Sengupta, V.V.S.N. Vijay, S. Bhaumilk, Further improvement and analysis of CCD scheme: dissipation discretization and de-aliasing properties. J. Comput. Phys. **228**(17), 6150–6168 (2009)
83. N. Sharma, T.K. Sengupta, Vorticity dynamics of the three-dimensional Taylor-Green vortex problem. Phys. Fluids **31**, 035106 (2019)
84. Y.I. Shokin, *The Method of Differential Approximation* (Springer, Berlin, 1983)
85. S. Stoltz, N.A. Adams, L. Kleiser, An approximate deconvolution model for large-eddy simulation with application to incompressible wall-bounded flows. Phys. Fluids **13**(4), 997 (2001)
86. B. Swartz, B. Wendroff, The relation between the Galerkin and collocation methods using smooth splines. SIAM J. Num. Anal. **11**(5), 994–996 (1974)
87. C.K.W. Tam, K. Viswanathan, K.K. Ahuja, J. Panda, The sources of jet noise: experimental evidence. J. Fluid Mech. **615**, 253–292 (2008)
88. H. Tennekes, J.L. Lumley, *First Course in Turbulence* (MIT Press, Cambridge, MA, 1971)
89. L. Tisza, Supersonic absorption and Stokes' viscosity relation. Phys. Rev. **61**, 531 (1942)
90. E. Touber, N.D. Sandham, Large-eddy simulation of low-frequency unsteadiness in a turbulent shock-induced separation bubble. Theo. Comput. Fluid Dyns. **23**(2), 79–107 (2009)
91. L.N. Trefethen, Group velocity in finite difference schemes. SIAM Rev. **24**(2), 113–136 (1982)
92. P.G. Tucker, Differential equation-based wall distance computation for DES and RANS. J. Comput. Phys. **190**, 229–248 (2003)
93. O.V. Vasilyev, T.S. Lund, P. Moin, A general class of commutative filters for LES in complex geometries. J. Comput. Phys. **146**, 82 (1998)
94. R. Vichnevetsky, J.B. Bowles, *Fourier Analysis of Numerical Approximations of Hyperbolic Equations* (SIAM Studies of Applied Mathematics, Philadelphia, USA, 1982)
95. M.R. Visbal, D.V. Gaitonde, High-order-accurate methods for complex unsteady subsonic flows. AIAA J. **37**(10), 1231 (1999)
96. M.R. Visbal, D.V. Gaitonde, On the use of higher-order finite-difference schemes on curvilinear and deforming meshes. J. Comput. Phys. **181**, 155 (2002)
97. H.A.V. der Vorst, Bi-CGSTAB: a fast and smoothly converging variant of Bi-CG for the solution of non-symmetric linear systems. SIAM J. Sci. Stat. Comput. **13**, 631–644 (1992)
98. E. Weinan, J.-G. Liu, Vorticity boundary condition and related issues for finite difference schemes. J. Comput. Phys. **124**(2), 368–382 (1996)
99. G.B. Whitham, *Linear and Nonlinear Waves* (Wiley-Interscience, New York, 1974)
100. A.K. Wong, J.A. Reizes, An effective vorticity-vector potential formulation for the numerical solution of three-dimensional duct flow problems. J. Comput. Phys. **55**(1), 98–114 (1984)
101. A.K. Wong, J.A. Reizes, The vector potential in the numerical solution of three-dimensional fluid dynamics problems in multiply connected regions. J. Comput. Phys. **62**(1), 124–142 (1986)
102. D.W. Zingg, Comparison of high-accuracy finite-difference schemes for linear wave propagation. SIAM J. Sci. Comput. **22**(2), 476–502 (2000)

# Chapter 4
# Computational Incompressible Aerodynamics

## 4.1 Introduction

Analysis of aerodynamic properties of aerospace vehicles has now matured to a stage, where high accuracy computing can, to a large extent, replace the design data book's sectional properties. In this chapter, we focus on main issues which make computational aerodynamics a practical tool. We emphasize the high accuracy aspects of such computing techniques, which have been developed over the last two decades. We will only highlight the fundamental aerodynamic properties in understanding the basic principles of flight in cruise configuration. Apart from this, one does note that there are model-based tools available for computing flows past complete complex geometries often employed for flying vehicles in industry or for preliminary design of aerospace vehicles. These are based on approximate form of the governing Navier–Stokes equation, for example, by solving Reynolds-averaged Navier–Stokes (RANS) equations or unsteady RANS equation.

The main emphasis here is on the design condition, namely, in the simplest symmetric steady flight of aircraft in longitudinal plane and involves with lift, drag, and pitching moment acting on the vehicle. Lift is the most fundamental property that enables a heavier-than-air vehicle to lift not only its own weight in steady and level flight, but the lifting surface should be able to generate additional lift to perform steady and unsteady maneuvers and that is required during landing and takeoff. While any airborne object would experience drag due to pressure distribution and shear stresses, lifting surfaces experience additional drag due to lift generation, which is known as the induced drag. Apart from sustaining the weight of the aerospace vehicle, the concomitant pressure distribution causes pitching moment and affects the flight dynamics of the vehicle in symmetric flight condition. Readers can consult a book on flight dynamics [1] or other books on aerodynamics [33], which will convince the readers that knowledge of pitching moment is essential to ensure even the static stability of the vehicle.

© Springer Nature Singapore Pte Ltd. 2020
T. K. Sengupta and Y. G. Bhumkar, *Computational Aerodynamics and Aeroacoustics*,
https://doi.org/10.1007/978-981-15-4284-8_4

In Chap. 3, we have described the hierarchy of different models employed in calculating forces and moments acting on aerospace vehicles starting with the Navier–Stokes equations. Other simplification follows this equation, as in RANS calculations, which provides a time-averaged picture of the flow field. At a lower hierarchy, one computes the inviscid flow field by solving Euler equation, and the solution of inviscid flow field past a complete Jumbo jet was considered a major milestone in the development of computational aerodynamics in 1980s. At the elementary level, one solves the Laplace equation by the panel method, which now forms the basis of preliminary design of all aircraft. However, problem of aerodynamics still revolves around evaluation of lift, drag, and pitching moment coefficients of the aerofoil. For this reason, airfoil is treated as the quintessential element of the lifting surface and studied in great detail, with theory and experiment complimenting each other. Other aerodynamic properties of the integral aircraft are equally relevant to understand the off-design condition. For example, flight with one engine operating, for a twin-engine aircraft, is a rudimentary requirement for certification of commercial aircraft. Such operation of aircraft also requires proper sizing and design of control surfaces.

Theoretically, one can start from the most rudimentary potential flow models characterized by panel methods described in [33], which allow one to obtain the lift and pitching moment acting on an aerodynamic surface for pressure distribution. As viscous effects are not considered in potential flows, one cannot obtain shear stresses and hence the skin friction. Not only that the additional phase shift caused on pressure distribution due to viscous action gives pitching moment, which is also significantly in error from the experimental observation. In early days, this has been attempted to be circumvented with solution of boundary layer equation to provide skin friction, only if the flow remains attached to the surface. It is explained in [33] that how one solves the potential flow for inviscid pressure distribution on the surface, followed by solution of the boundary layer equation. For compressible flows, one solves the Euler equation for inviscid pressure distribution which is followed by solution of compressible boundary layer equation. This is the weak coupling between viscous and inviscid effects on lifting surfaces with flow attached. Difficulties in coupling viscous and inviscid dynamics are experienced for separated flows and in explaining instability and transition, which are often attempted in a semi-empirical manner.

Restrictions of studying flow past aerodynamic surfaces by simplified model equation at off-design conditions are removed by solving the Navier–Stokes equation. In solving Navier–Stokes equation, carefully designed methodology works across all speed regimes for continuum flows and is advocated again, i.e., no special efforts are needed to distinguish between incompressible and compressible flows, with the flow fields computed by same methods. Thus, the same methodology used for incompressible flows past airfoils can be directly employed to study time-accurate transonic and supersonic flow past airfoils [37]. Thus, it is apparent that the methodology for solving Navier–Stokes equation for flow past airfoils have to be performed very accurately and an approach is followed which uses structured grid with finite difference methods. For this purpose, we need to develop good-quality grid around airfoil. We emphasize the use of orthogonal grids for airfoils of a different thickness and camber by describing a method to do so. Generated orthogonal grid with desirable properties helps perform DNS of flow past aerofoils, provided one takes adequate care in choosing the numerical method. Choice of numerical method completely depends upon the global spectral analysis (GSA), which helps identify properties of the adopted method. While one can calibrate methods for Navier–Stokes equation for special cases, here we propose to use a model equation which will highlight the strength and weaknesses of the chosen method for different spatio-temporal discretizations. In this respect, a more detailed text can be used as given in [32].

Basic equations in aerodynamics are the statements of conservation principles for mass, translational momentum, and energy. It is interesting, however, to note that in these equations, rotational component of energy and conservation of angular momentum is never considered. In the following, we will focus upon the discussion of numerical methods which can be used to solve Navier–Stokes equation. This is the model convection–diffusion equation, which is an appropriate linear model of Navier–Stokes equation [53]. The property shown by a numerical method for convection–diffusion equation is truly reflected for NSE and that is the reason for highlighting this here.

## 4.2 Spectral Analysis of Finite Difference Schemes for Convection–Diffusion Equation

Simulating various natural phenomena, convection, diffusion, and convection–diffusion processes are most fundamental, as they occur in a wide range of disciplines in applied physics. The governing equations which describe these are in general nonlinear or quasi-linear, and hence difficult to solve analytically, requiring numerical techniques for solution. This is not to say that all linear differential equations are amenable to closed-form analytic solution, as we note for Laplace equation for complex geometry problem in aerodynamics. The numerical solution being an approximation to the exact solution, the accuracy of the former depends on our ability to follow the physical processes. This shows the central role of numerical analysis in quantifying errors and thereby measure the accuracy of schemes. Due to the complex nature of most governing equations, nonlinearities which introduce difficulties and non-availability of exact/reference solutions needed for comparison, one uses prototype model convection–diffusion equations to calibrate space–time discretizations in the numerical method.

A full-domain spectral analysis [43, 53] with appropriate error metrics [32] reveals information about stability/instability, dispersion, and dissipation errors for all length and time scales. These information are critical in designing very high accuracy dispersion relation preserving (DRP) schemes for DNS and LES. The analysis in [43] demonstrated that signal and error follow different dynamics for the convection equation. In [36], error dynamics of a general finite-difference-based scheme for linear diffusion equation is derived, and errors are quantified for some popular schemes. In both these references, we note that the phase speed in convection equation and the coefficient of diffusion in the heat equation do not remain constant numerically. Such a situation also prevails in convection–diffusion equation as is shown below [53].

We chose the linear 1D convection–diffusion equation, a linearized model of the governing Navier–Stokes equations of fluid dynamics. It is appropriate to analyze this equation in the spectral framework for any choice of space–time discretization, which can be also used for solving the time-accurate Navier–Stokes equation. Consider the linear convection–diffusion equation given by

$$\frac{\partial u}{\partial t} + c\frac{\partial u}{\partial x} = \alpha \frac{\partial^2 u}{\partial x^2} \tag{4.1}$$

where $c$ and $\alpha$ are constants denoting the convection speed and coefficient of diffusion, respectively. This is the model 1D equation that mimics the Navier–Stokes equation, where in the corresponding vector equation for momentum conservation equation, one would replace $c$ by the corresponding velocity component and $\alpha$ would be replaced by the kinematic viscosity, $\nu$. In the pressure–velocity formulation, additionally there would be pressure gradient term. However, for the vorticity transport equation there is no pressure term and the above equation imitates the rotational aspect of the Navier–Stokes equation.

In performing GSA to represent the unknown, $u(x, t)$ of the model equation in the hybrid spectral plane is given by [32]

$$u(x, t) = \int \hat{U}(k, t)e^{ikx} dk \tag{4.2}$$

where $\hat{U}$ is the Fourier amplitude and $k$ is the wavenumber. Substituting this in the convection–diffusion equation, we obtain the transformed equation given by

$$\frac{d\hat{U}}{dt} + ick\hat{U} = -\alpha k^2 \hat{U} \tag{4.3}$$

The above equation is solved for a general initial condition $u(x, 0) = f(x) = \int \hat{F}(k)e^{ikx} dk$ to obtain the exact solution

$$\hat{U}(k, t) = \hat{F}(k)\, e^{-\alpha k^2 t}\, e^{-ikct} \tag{4.4}$$

To obtain the dispersion relation, represent the unknown by Fourier–Laplace transform, $u(x, t) = \iint \hat{U}(k, \omega)\, e^{i(kx-\omega t)}\, dk\, d\omega$, to get the dispersion relation

$$\omega = ck - i\alpha k^2 \tag{4.5}$$

Any numerical scheme used to solve problems must satisfy physical dispersion relation for accuracy [32, 43, 53] and are said to be DRP schemes. A very concise and clear explanation of the correct dispersion relation in the context of multiple time level scheme is given in [51] for the convection equation, which is extended here for the convection–diffusion equation. The above relation is used to obtain the phase speed as

$$c_{phys} = \frac{\omega}{k} = c - i\alpha k \qquad (4.6)$$

The physical group velocity is defined by

$$V_{g,phys} = \frac{\partial \omega}{\partial k} = c - 2i\alpha k \qquad (4.7)$$

Therefore, $\alpha = \frac{i}{2k}(V_{g,phys} - c)$. Further expanding the real and imaginary parts of the group velocity, we obtain

$$\alpha = \frac{i}{2k}\left[(V_{g,phys})_{real} - c)\right] - \frac{(V_{g,phys})_{imag}}{2k} \qquad (4.8)$$

Since $\alpha$ is real, $(V_{g,phys})_{real} = c$ is the condition for a physical system. As the right-hand side of Eq. (4.1) is physically diffusive and stable, hence one must have $\frac{(V_{g,phys})_{imag}}{2k} < 0$.

The physical amplification factor $G_{phys} = \hat{U}(k, t + \Delta t)/\hat{U}(k, t)$ is obtained from Eq. (4.4) as

$$G_{phys} = \frac{\hat{U}(k, t + \Delta t)}{\hat{U}(k, t)} = e^{-\alpha k^2 \Delta t} e^{-ikc\Delta t} = e^{-i\omega \Delta t} = e^{-Pe\,(kh)^2} e^{-iN_c(kh)} \qquad (4.9)$$

Next introduce $N_c \, (= \frac{c\Delta t}{h})$, as the CFL number and $Pe \, (= \frac{\alpha \Delta t}{h^2})$, as the Peclet number in the above, as the non-dimensional parameters for this equation, with $h$ as the constant grid spacing and $\Delta t$ as the time step. The $G_{phys}$ for $Pe = 0.05, 0.25$, and $0.5$ are shown in Fig. 4.1. As $Pe$ increases, the rate of diffusion also increases, as noted from Eq. (4.9).

Every numerical scheme has a numerical amplification factor $G_{num}$, associated with its numerical dispersion relation. It is essential that a DRP scheme reproduces the physics of the governing equation, $G_{num}$ should be very close to $G_{phys}$. In the case of the 1D linear convection–diffusion equation, the numerical dispersion relation is directly obtained as in [36, 43], for pure convection and pure diffusion equations from [53]:

$$\omega_{num} = kc_{num} - i\alpha_{num}k^2 \qquad (4.10)$$

**Fig. 4.1** Physical amplification factor $|G_{\mathrm{phys}}|$ contours for the 1D linear convection–diffusion equation. The top, middle, and bottom panels represent contours for the Peclet numbers 0.01, 0.25, and 0.5, respectively. Reproduced from V. K. Suman, et al., *Spectral analysis of finite difference schemes for convection diffusion equation. Computers and Fluids*, **150**, 95–114 (2017), with the permission of Elsevier Publishing

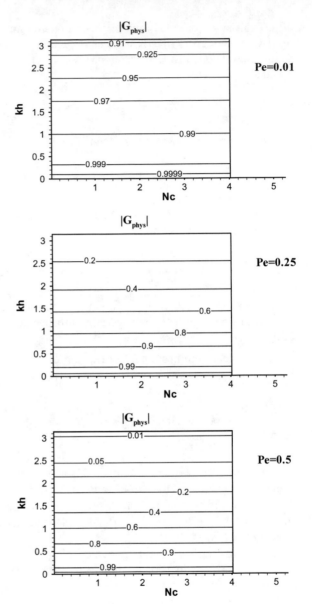

We note that $c_{\mathrm{num}}$ and $\alpha_{\mathrm{num}}$ are generally not constants in numerical simulation, for any arbitrary numerical parameter combinations. Having obtained the numerical dispersion relation, one can readily represent $G_{\mathrm{num}}$ as

$$G_{\mathrm{num}} = e^{-i\omega_{\mathrm{num}}\Delta t} = e^{-\alpha_{\mathrm{num}}k^2\Delta t}e^{-ikc_{\mathrm{num}}\Delta t} \tag{4.11}$$

The numerical phase shift in one time step is given by

$$\tan(\beta) = -\left(\frac{(G_{num})_{Imag}}{(G_{num})_{Real}}\right) \implies \beta = k c_{num} \Delta t \qquad (4.12)$$

and the non-dimensional numerical phase speed is

$$\frac{c_{num}}{c} = \frac{\beta}{k c \Delta t} = -\frac{1}{(kh) N_c} \tan^{-1}\left(\frac{(G_{num})_{Imag}}{(G_{num})_{Real}}\right) \qquad (4.13)$$

The numerical group velocity can be calculated from the numerical dispersion relation as $V_{g,num} = \frac{d}{dk}(\omega_{num})$, which on further simplification yields the following expression:

$$\frac{V_{g,num}}{V_{g,phys}} = \frac{1}{N_c} \frac{d\beta}{d(kh)} \qquad (4.14)$$

where $V_{g,phys}$ is the physical group velocity and here it is equal to $c$.

The numerical diffusion coefficient ($\alpha_{num}$) can be evaluated by noting from Eq. (4.11) that $|G_{num}| = e^{-\alpha_{num} k^2 \Delta t}$. Thus

$$\frac{\alpha_{num}}{\alpha} = -\left(\frac{\ln|G_{num}|}{Pe\ (kh)^2}\right) \qquad (4.15)$$

The above quantity indicates that a numerical scheme has exact diffusion when the ratio is unity, higher diffusion for values greater than one and lower diffusion for values between zero and one. Negative values indicate anti-diffusion, and hence are unphysical and numerically unstable.

We stress here that to obtain accurate solutions for convection–diffusion equation, the most important properties to satisfy are the physical diffusion, phase speed, and group velocity. This is achieved by a numerical scheme when all the quantities $\alpha_{num}/\alpha$, $c_{num}/c$, and $V_{g,num}/V_{g,phys}$ become unity. Next, we will show how $G_{num}$ is obtained for different schemes and also present information of $\alpha_{num}/\alpha$, $c_{num}/c$, and $V_{g,num}/V_{g,phys}$.

### 4.2.1 Spectral Analysis of Numerical Schemes

The principal step in performing spectral analysis is to transform the information from physical to the spectral plane, achieved by adopting the representation in Eq. (4.2) for the unknown at the nodes and substituting this in discretized equation to obtain the respective amplification factor [32, 44]. Different numerical schemes are studied next, and analysis results are presented.

### 4.2.2  Euler-CD$_2$-CD$_2$ Scheme

Here, the Euler method is used for the time derivative, and second-order central differencing (CD$_2$) for the spatial derivatives is employed. The discretized equation is

$$\frac{u_j^{n+1} - u_j^n}{\Delta t} + c\frac{u_{j+1}^n - u_{j-1}^n}{2h} = \alpha\frac{u_{j+1}^n - 2u_j^n + u_{j-1}^n}{h^2} \tag{4.16}$$

where $j$ denotes the nodal index and $n$ is the time level. Substituting spectral representation in the above to obtain the numerical amplification factor at the $j$th node as

$$G_{\text{num},j} = 1 - \frac{N_c}{2}(e^{ikh} - e^{-ikh}) + Pe(e^{ikh} - 2 + e^{-ikh}) \tag{4.17}$$

The ratios $c_N/c$, $V_{g,\text{num}}/V_{g,\text{phys}}$, and $\alpha_{\text{num}}/\alpha$ are calculated from Eqs. (4.13), (4.14) to (4.15), respectively.

Figures 4.2 and 4.3 show the ratios $\frac{|G_{\text{num}}|}{|G_{\text{phys}}|}$, $\alpha_{\text{num}}/\alpha$, $c_{\text{num}}/c$, and $\frac{V_{g,\text{num}}}{V_{g,\text{phys}}}$ for $Pe = 0.01$ and 0.5 in the ($N_c$, $kh$)-plane. If the ratios are unity, then the numerical solution is identical to the analytical solution. The region of instability ($|G_{\text{num}}| \geq 1$) is marked by gray region, and it exactly matches the limits obtained in [16], i.e., the solution is stable when ($2Pe \leq 1$) and ($N_c^2 \leq 2Pe$). An increase in the stable region with increase in $Pe$ is observed. We observe that $\frac{|G_{\text{num}}|}{|G_{\text{phys}}|}$ is always greater than 1 in the stable region, which implies that the numerical damping is lower compared to the physical damping. This can be reaffirmed from the plotted $\alpha_{\text{num}}/\alpha$—contours where values less than one indicate lower numerical diffusion. It is interesting to note from Fig. 4.3 that the numerical solution corresponding to $Pe = 0.5$ and $N_c = 1$ shows neutral stability for all wavenumbers.

The ratio—$c_{\text{num}}/c$, on the other hand—shows different characters for the two $Pe'$s. In Fig. 4.2, it is seen that the value is less than one, implying that $c_{\text{num}}$ is always less than $c$ for the $N_c$ values. However, for $Pe = 0.5$, the ratio is greater than 1 for $N_c < 1$ and is less than 1 for $N_c > 1$ implying faster and slower propagation speeds. At $N_c = 1$, $c_{\text{num}}$ is identical to $c$ indicating that the wave propagation is correctly captured.

One can see that this scheme cannot achieve exact solution for whatever value of $N_c$ at the given $Pe$ values, though one may find region(s) near the origin, where error is sufficiently small leading to satisfactory results. Although it may appear beneficial to work with $N_c = 1$ and $Pe = 0.5$ for the convection–diffusion equation, these parameters cannot be used to solve Navier–Stokes equation, due to the fact that for $Pe = 0.5$, the line $N_c = 1$ lies on the boundary of stability and instability. For safer operation, one would be advised to work with $N_c$ slightly lower than one. The more important concern for this combination of $N_c$ and $Pe$ is the vanishing of $\alpha_{\text{num}}$, for which one ends up solving a convection equation, instead of the convection–diffusion equation.

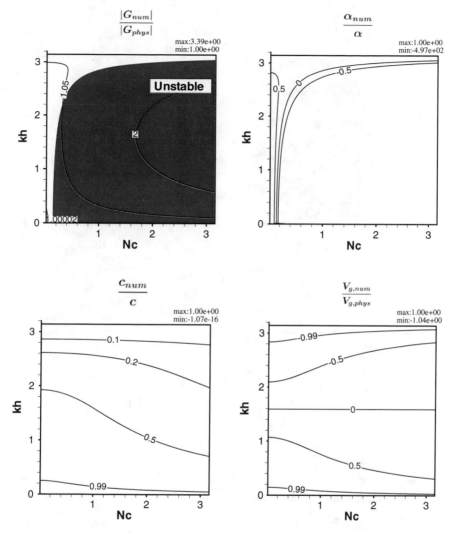

**Fig. 4.2** Contours of $\frac{|G_{num}|}{|G_{phys}|}$ (top left), $\alpha_{num}/\alpha$ (top right), $c_{num}/c$ (bottom left), and $\frac{V_{g,num}}{V_{g,phys}}$ (bottom right) for Euler-CD$_2$-CD$_2$ scheme for $Pe = 0.01$. Reproduced from V. K. Suman, et al., *Spectral analysis of finite difference schemes for convection diffusion equation. Computers and Fluids*, **150**, 95–114 (2017), with the permission of Elsevier Publishing

### 4.2.3 RK$_4$-OUCS3-CD$_2$ Scheme

Next, the impact of using compact finite differences for the convection term and four-stage Runge–Kutta method for time integration in investigated. Compact finite difference schemes are highly accurate and have spectral-like resolution in evalu-

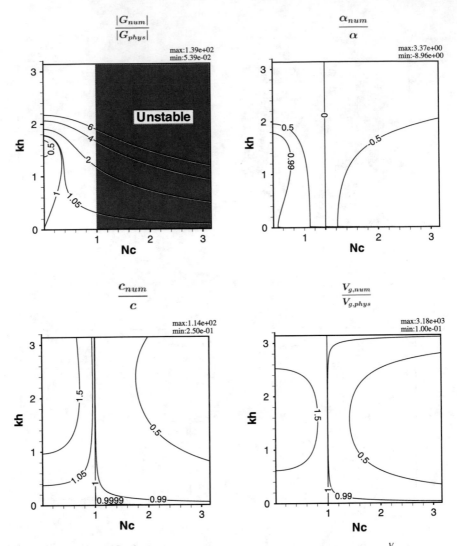

**Fig. 4.3** Contours of $\frac{|G_{num}|}{|G_{phys}|}$ (top left), $\alpha_{num}/\alpha$ (top right), $c_{num}/c$ (bottom left), and $\frac{V_{g,num}}{V_{g,phys}}$ (bottom right) for Euler-CD$_2$-CD$_2$ scheme for $Pe = 0.5$. Reproduced from V. K. Suman, et al., *Spectral analysis of finite difference schemes for convection diffusion equation. Computers and Fluids*, **150**, 95–114 (2017), with the permission of Elsevier Publishing

ating derivatives. These are used in DNS/LES simulations [3, 4, 17, 18, 25, 28, 37, 38, 56]. Here, the optimized upwind compact scheme (OUCS3) is used, which is superior in terms of resolution and accuracy for convection-dominated problems [32]. The stencil is given below, and readers are referred to [44] for further details for its development.

$$u'_1 = \frac{1}{2h}[-3u_1 + 4u_2 - u_3]$$

$$u'_2 = \frac{1}{h}\left[\left(\frac{2\beta}{3} - \frac{1}{3}\right)u_1 - \left(\frac{8\beta}{3} + \frac{1}{2}\right)u_2 + (4\beta + 1)u_3\right.$$
$$\left. - \left(\frac{8\beta}{3} + \frac{1}{6}\right)u_4 + \frac{2\beta}{3}u_5\right]$$

$$p_{j-1}u'_{j-1} + u'_j + p_{j+1}u'_{j+1} = \frac{1}{h}\sum_{k=-2}^{2} q_k u_{j+k}$$

$$j = 3, \cdots, N - 2u'_{N-1} = -\frac{1}{h}\left[\left(\frac{2\beta}{3} - \frac{1}{3}\right)u_N - \left(\frac{8\beta}{3} + \frac{1}{2}\right)u_{N-1} + (4\beta + 1)u_{N-2}\right.$$
$$\left. - \left(\frac{8\beta}{3} + \frac{1}{6}\right)u_{N-3} + \frac{2\beta}{3}u_{N-4}\right]$$

$$\text{with } u'_N = \frac{1}{2h}[3u_N - 4u_{N-1} + u_{N-2}]$$

$$\beta = -0.025 \text{ (for } j = 2) \text{ (and } \beta = 0.09 \text{ ( for } j = N - 1)$$

Also,

$$p_{j\pm1} = D \pm \frac{\eta}{60}; \quad q_{j\pm2} = \pm\frac{F}{4} + \frac{\eta}{300}; \quad q_{j\pm1} = \pm\frac{E}{2} + \frac{\eta}{30} \text{ and } q_0 = -\frac{11\eta}{150} \text{ with}$$
$$D = 0.3793894912; \quad E = 1.57557379; \quad F = 0.183205192 \text{ and } \eta = -2.0 \quad (4.18)$$

Following [32], any compact scheme can be represented in the matrix notation as

$$[A]\{u'\} = [B]\{u\}$$
$$\text{So that } \{u'\} = [C]\{u\} \tag{4.19}$$

where $[C] = [A]^{-1}[B]$, with $[A]$, $[B]$, and $[C]$ as constant matrices. The numerical amplification for four-stage Runge–Kutta time integration method is given by [32],

$$G_j = 1 - A_j + \frac{A_j^2}{2} - \frac{A_j^3}{6} + \frac{A_j^4}{24}$$

For the present discretizations, $A_j$ is given by using Eq. (4.19) as [53],

$$A_j = N_c\, C_{ij}\, e^{ik(x_j - x_i)} + 2\,Pe\,[1 - \cos(kh)] \tag{4.20}$$

which is used to determine $G_j$. The performance of the scheme at an interior node of the computational domain is investigated. Figures 4.4 and 4.5 show contours of properties for $Pe = 0.01$ and 0.5, respectively. For $Pe = 0.01$ and small values of $N_c$, the numerical diffusion is equal to the physical diffusion for a broad range of wavenumbers, as noted from Fig. 4.4. For $Pe = 0.5$, lower wavenumbers have higher diffusion, while higher wavenumbers have lower diffusion.

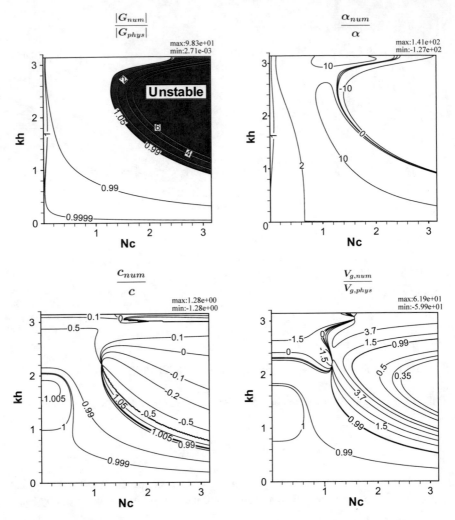

**Fig. 4.4** Contours of $\frac{|G_{\text{num}}|}{|G_{\text{phys}}|}$ (top left), $\alpha_{\text{num}}/\alpha$ (top right), $c_{\text{num}}/c$ (bottom left) and $\frac{V_{g,\text{num}}}{V_{g,\text{phys}}}$ (bottom right) for RK$_4$-OUCS3-CD$_2$ scheme for $Pe = 0.01$. Reproduced from V. K. Suman, et al., *Computers and Fluids*, **150**, 95–114 (2017), with the permission of Elsevier Publishing

The scheme displays better properties for phase speed and group velocity than the previous scheme with lower $Pe$ values, showing excellent results.

While $RK_4$ time integration and spatial discretization of convection by OUCS3 scheme are highly accurate, the discretization of diffusion term by $CD_2$-scheme is not so accurate, due to its lower spectral accuracy. This last aspect can be rectified by an implicit method for spatial discretization of first and second derivatives that is described next.

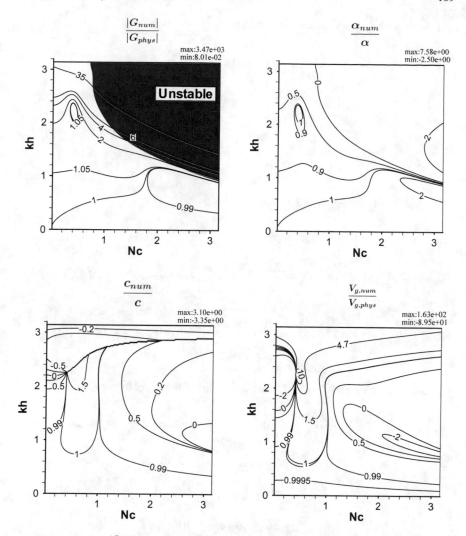

**Fig. 4.5** Contours of $\frac{|G_{num}|}{|G_{phys}|}$ (top left), $\alpha_{num}/\alpha$ (top right), $c_{num}/c$ (bottom left), and $\frac{V_{g,num}}{V_{g,phys}}$ (bottom right) for RK$_4$-OUCS3-CD$_2$ scheme for $Pe = 0.5$. Reproduced from V. K. Suman, et al., *Computers and Fluids*, **150**, 95–114 (2017), with the permission of Elsevier Publishing

## 4.2.4 RK$_4$-NCCD Scheme

Here, one investigates a variant of compact difference schemes, called the combined compact difference (CCD) scheme, which discretizes both first- and second-order derivatives simultaneously [7, 62]. Here, the improved NCCD scheme is taken from [46, 52], for the analysis. The stencil of NCCD scheme is given below. The prime indicates that a derivative in the chosen direction is considered.

One set of boundary closure scheme is shown here first

$$u'_1 = \frac{1}{2h}[-3u_1 + 4u_2 - u_3]$$

$$u''_1 = \frac{1}{h^2}[u_1 - 2u_2 + u_3]$$

$$u'_2 = \frac{1}{h}\left[\left(\frac{2\beta}{3} - \frac{1}{3}\right)u_1 - \left(\frac{8\beta}{3} + \frac{1}{2}\right)u_2 + (4\beta + 1)u_3\right.$$
$$\left. - \left(\frac{8\beta}{3} + \frac{1}{6}\right)u_4 + \frac{2\beta}{3}u_5\right]$$

$$u''_2 = \frac{1}{h^2}[u_1 - 2u_2 + u_3]$$

The interior stencils are given by

$$\frac{7}{16}(u'_{j+1} + u'_{j-1}) + u'_j - \frac{h}{16}(u''_{j+1} - u''_{j-1}) = \frac{15}{16h}(u_{j+1} - u_{j-1}),$$
$$\text{for } j = 3, \cdots, N-2$$

$$\frac{9}{8h}(u'_{j+1} - u'_{j-1}) - \frac{1}{8}(u''_{j+1} + u''_{j-1}) + u''_j = \frac{3}{h^2}(u_{j+1} - 2u_j + u_{j-1}),$$
$$j = 3, \cdots, N-2$$

The boundary closure schemes for the other end of the domain are given by

$$u'_{N-1} = -\frac{1}{h}\left[\left(\frac{2\beta}{3} - \frac{1}{3}\right)u_N - \left(\frac{8\beta}{3} + \frac{1}{2}\right)u_{N-1} + (4\beta + 1)u_{N-2} - \left(\frac{8\beta}{3} + \frac{1}{6}\right)u_{N-3} + \frac{2\beta}{3}u_{N-4}\right]$$

$$u''_{N-1} = \frac{1}{h^2}[u_N - 2u_{N-1} + u_{N-2}]$$

$$u'_N = \frac{1}{2h}[3u_N - 4u_{N-1} + u_{N-2}]$$

$$u''_N = \frac{1}{h^2}[u_N - 2u_{N-1} + u_{N-2}]$$

with    $\beta = -0.025$ (for $j = 2$) and $\beta = 0.09$ (for $j = N - 1$)        (4.21)

Following [52, 53], one can write Eq. (4.21) in a compact form as

$$[A]\{du\} = \{b\}$$

where the matrix $[A]$ and the vectors $\{du\}$, $\{b\}$ are as shown in the reference. The solution of these simultaneous equations can be expressed as

$$\{u'\} = \frac{1}{h}[D_1]\{u\}$$

$$\{u''\} = \frac{1}{h^2}[D_2]\{u\}$$

(4.22)

where the matrices $[D_1]$ and $[D_2]$ are presented in [46]. Block tridiagonal matrix algorithm (TDMA) is employed to calculate the inverses to obtain $[D_1]$ and $[D_2]$. After this step, the same approach described for the compact scheme is adopted. The function at the node $l$ is evaluated as

$$A_l = N_c \sum_{j=1}^{N} [D_1]_{lj}\, e^{ik(x_j-x_l)} - Pe \sum_{j=1}^{N} [D_2]_{lj}\, e^{ik(x_j-x_l)}$$

An interior node is considered again for investigating the accuracy of the NCCD scheme. Figures 4.6 and 4.7 show the contours of $\frac{|G_{num}|}{|G_{phys}|}$, $\frac{\alpha_{num}}{\alpha}$, $\frac{c_{num}}{c}$ and $\frac{V_{g,num}}{V_{g,phys}}$ for $Pe = 0.01$ and $0.25$, respectively. Excellent error metric behaviors are noted for this scheme for the two $Pe$'s, with the lower $Pe$ giving better results.

## 4.2.5  Wave-Packet Propagation

We next consider a wave packet, as an initial solution, and compare its evolution using numerical simulations based on RK$_4$-NCCD and $RK_4$-OUCS3-CD$_2$ schemes with the corresponding exact solution. The initial wave-packet solution is given as

$$u(x) = \exp^{-\alpha_1(x-x_o)^2}\, \sin(k_o x)$$

(4.23)

where $\alpha_1$ controls the width of the packet in the physical plane, and $k_o$ is the wavenumber around which the packet is centered in the spectral plane. The exact solution is obtained using an analytic technique, in which the initial solution is first converted in spectral space using FFT and is advanced at any later time $t$ using Eq. (4.9), and the resulting solution is converted back into the physical space using the inverse Fourier transform. The exact solution is used to provide boundary conditions for the simulation, to avoid complications brought in by these.

For the simulations, $(N_c, Pe) = (0.2, 0.01)$ is taken for the choice of $k_o \Delta x = 0.5$ with $\Delta t = 8 \times 10^{-3}$. The initial solution is shown in Fig. 4.8, and it is noted that the packet is centered at $x_0 = 20$. The properties of the schemes for these parameters are shown in Fig. 4.9. According to this, the numerical results should show minimal error for wavenumbers up to $kh = 1.5$ for the RK$_4$-NCCD scheme, and one expects this behavior in the test. This is confirmed by the numerical results presented in Fig. 4.10, where the computed and exact solutions are in excellent agreement. The information in Fig. 4.9 is interesting with respect to the group velocity, which indicates negative

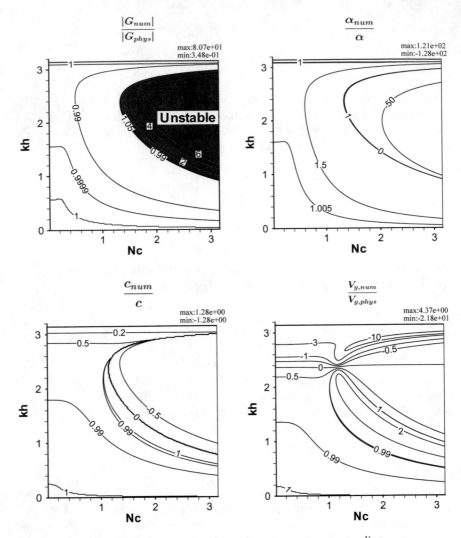

**Fig. 4.6** Contours of $\frac{|G_{num}|}{|G_{phys}|}$ (top left), $\frac{\alpha_{num}}{\alpha}$ (top right), $\frac{c_{num}}{c}$ (bottom left), and $\frac{V_{g,num}}{V_{g,phys}}$ (bottom right) for RK$_4$-NCCD scheme for $Pe = 0.01$. Reproduced from V. K. Suman, et al., *Spectral analysis of finite difference schemes for convection diffusion equation. Computers and Fluids,* **150**, 95–114 (2017), with the permission of Elsevier Publishing

value for $kh > 2.411$ for $RK_4$-OUCS3-CD$_2$ scheme and $kh > 2.365$ for RK$_4$-NCCD scheme for this combination of $N_c$ and $Pe$.

Extreme form of dispersion error leads to negative numerical group velocity, which causes the computed solution to propagate in a direction, which is opposite to the actual physical direction. In the literature and textbook [32], these events have been noted as $q$-wave propagation.

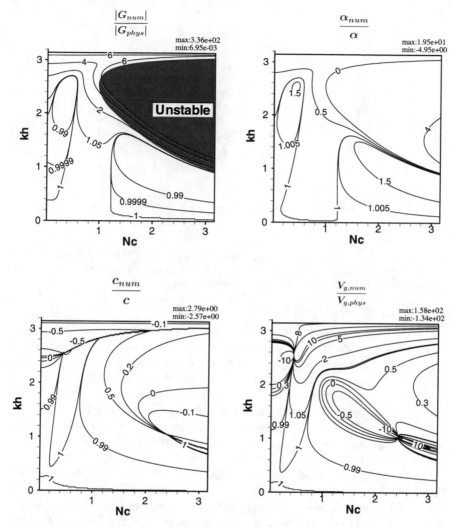

**Fig. 4.7** Contours of $\frac{|G_{num}|}{|G_{phys}|}$ (top left), $\alpha_{num}/\alpha$ (top right), $c_{num}/c$ (bottom left) and $\frac{V_{g,num}}{V_{g,phys}}$ (bottom right) for RK$_4$-NCCD scheme for $Pe = 0.25$. Reproduced from V. K. Suman, et al., *Spectral analysis of finite difference schemes for convection diffusion equation. Computers and Fluids*, **150**, 95–114 (2017), with the permission of Elsevier Publishing

## 4.2.6   Simulation of q-Waves

The main purpose of this test is to show $q$-waves, which arise for the high $kh$ components in the computations. For this demonstration, the RK$_4$-NCCD scheme is chosen and we use the initial wave-packet solution given by Eq. (4.23) keeping $(N_c, Pe)$ and $\Delta t$ the same, as in the previous case, but increase $k_o \Delta x = 2.6$, with the initial

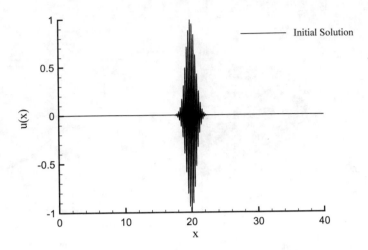

**Fig. 4.8**  Initial wave-packet solution ($k_o \Delta x = 0.5$). Reproduced from V. K. Suman, et al., *Spectral analysis of finite difference schemes for convection diffusion equation. Computers and Fluids*, **150**, 95–114 (2017), with the permission of Elsevier Publishing

packet centered at $x_0 = 20$. The initial solution is shown in Fig. 4.11. Spectral analysis for $V_{gN}$ predicts upstream propagating packet or $q$-waves, for the given initial condition from the properties shown in Fig. 4.9. It is seen that $q$-waves appear, when $kh > 2.365$, as the group velocity becomes negative above this $kh$. This prediction is confirmed by numerical results shown in Fig. 4.12, which shows the numerical solution propagating upstream. We note that $q$-waves are undesirable for high-fidelity computations, and though their amplitudes become negligible with time, a constant forcing in the form of boundary conditions will leave their presence at all times. This can be avoided by employing a finer grid such that the $kh$ value of the wave packet decreases, resulting in positive values of group velocity, or by employing numerical filtering algorithms to remove the high wavenumber components from the solution. However, that will distort the initial signal beyond recognition and is not suggested.

The results in Fig. 4.12 also show that the amplitude of the $q$-waves attenuates very rapidly, with attenuation by a factor of $10^{-13}$ in just 350 time steps for the numerical simulation. It can also be seen that the numerical solution attenuates rapidly compared to the exact solution, with the FFT plotted in Fig. 4.13 confirming this. This is because $\frac{|G_{num}|}{|G_{phys}|} = 0.99491$ and $\frac{\alpha_{num}}{\alpha} = 1.07548$ for $kh = 2.6$, which implies higher attenuation for this wavenumber. The analysis predicts that the amplitude of the wavenumber corresponding to $kh = 2.6$ attenuates by a factor of $(0.99491)^{350} = 0.1676$, compared to the exact amplitude for the same wavenumber and time, i.e., $\frac{\hat{F}_{num}(2.6, 350\Delta t)}{\hat{F}_{ex}(2.6, 350\Delta t)} = 0.1676$, where $\hat{F}(kh, t)$ denotes the FFT amplitude for the $kh$ at time $t$; the subscripts "num" and "ex" denote numerical and exact solutions, respectively. This can be verified from the FFT of numerical and exact solutions at $t = 350\Delta t$, by dividing the corresponding FFT amplitudes for a wavenumber $kh$, to exactly

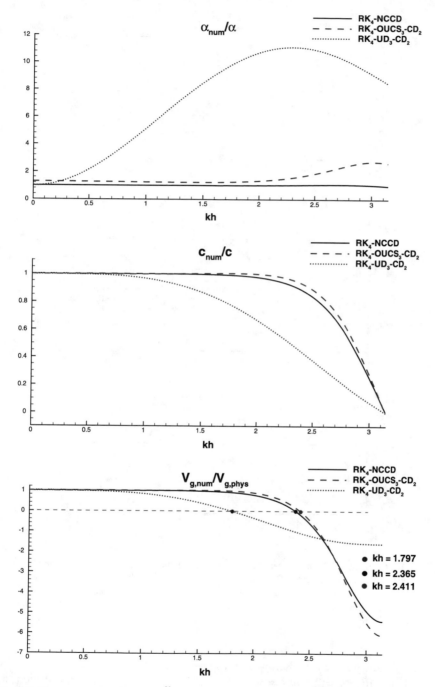

**Fig. 4.9** Plot of $\alpha_{num}/\alpha$, $\frac{c_{num}}{c}$ and $\frac{V_{g,num}}{V_{g,phys}}$ for RK$_4$-NCCD scheme for $(N_c, Pe) = (0.2, 0.01)$. Reproduced from V. K. Suman, et al., *Spectral analysis of finite difference schemes for convection diffusion equation. Computers and Fluids*, **150**, 95–114 (2017), with the permission of Elsevier Publishing

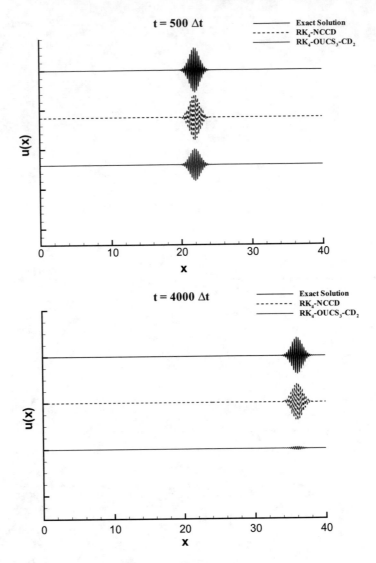

**Fig. 4.10** Comparison of numerical solution obtained using the RK₄-NCCD scheme and exact solution for $N_c = 0.2$ and $Pe = 0.01$ at the indicated times for the wave-packet problem with $k_o \Delta x = 0.5$. Reproduced from V. K. Suman, et al., *Spectral analysis of finite difference schemes for convection diffusion equation. Computers and Fluids*, **150**, 95–114 (2017), with the permission of Elsevier Publishing

determine the attenuation. Following this procedure, the attenuation is determined as 0.17166 for $kh = 2.6$ at $t = 350\Delta t$ which is in very good agreement with the predicted value.

**Fig. 4.11** Initial wave-packet solution $(k_o \Delta x = 2.6)$ for the RK$_4$-NCCD scheme for $N_c = 0.2$ and $Pe = 0.01$. Reproduced from V. K. Suman, et al., *Spectral analysis of finite difference schemes for convection diffusion equation. Computers and Fluids*, **150**, 95–114 (2017), with the permission of Elsevier Publishing

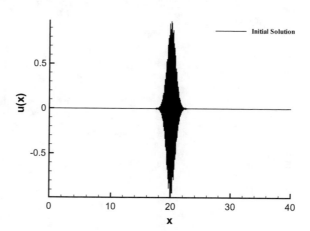

**Fig. 4.12** Comparison of numerical solution obtained using the RK$_4$-NCCD scheme and exact solution for $N_c = 0.2$ and $Pe = 0.01$ at the indicated times for the wave-packet problem with $k_o \Delta x = 2.6$. Reproduced from V. K. Suman, et al., *Spectral analysis of finite difference schemes for convection diffusion equation. Computers and Fluids*, **150**, 95–114 (2017), with the permission of Elsevier Publishing

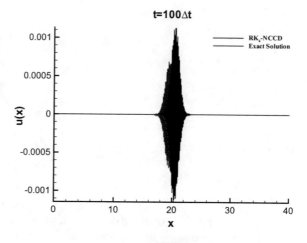

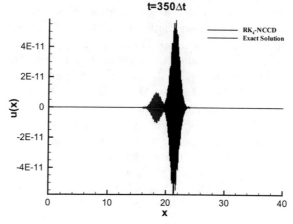

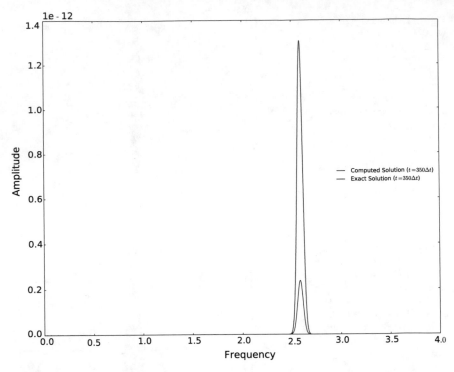

**Fig. 4.13** FFT of numerical solution obtained using $RK_4$-CCD scheme at $t = 350\Delta t$ and exact solution at $t = 350\Delta t$ for the wave-packet problem ($k_o\Delta x = 2.6$) for $N_c = 0.2$ and $Pe = 0.01$

In the previous section, we emphasized the importance of choosing the appropriate space–time discretization method that provides very accurate results, as demonstrated from the solution of a model 1D equation. While this analysis can be extended to higher dimensional convection–diffusion equation, there can be still many more sources that can contaminate the accuracy of the solution. This is specifically so for the flow past an aerofoil. The flow past an aerofoil experiences varying degree of pressure gradient and that makes the high accuracy computation difficult. Added to this problem, one has to solve the governing equation, in the finite difference context, in the transformed plane. This implies that the external domain of the aerofoil is mapped into the interior of a unit square in the transformed plane. For an arbitrary aerofoil, one performs this by grid generation. Among various possibilities of grid generation techniques, we recommend creation of an orthogonal grid, due to its superior performance in producing high accuracy solution. This is discussed in the following, with specially constructed grid for highly cambered, thick aerofoils, as in turbomachines or in natural laminar flow (NLF) wing sections. We also discuss orthogonal grid generation around NLF aerofoil with multiple roughness elements.

## 4.3   Improved Orthogonal Grid Generation Method for Highly Cambered Aerofoils

Time-accurate solution of Navier–Stokes equation for flow past an aerofoil at high angles of attack or an aerofoil with embedded surface roughness is not a trivial exercise. For such simulations, use of a good-quality structured grid with smooth grid metrics [54, 61] is necessary, as grid shocks or discontinuities in grid metrics give rise to mesh-related flow distortions and additional sources of high wavenumber errors contributing to aliasing [31]. Inadequate grid resolution results in causing spurious signal propagation [6, 9, 21, 60]. The grid aspect ratio should match the propagation direction, which otherwise results in spurious $q$-waves [41, 50]. It is necessary that grid points be clustered near the leading and the trailing edges, which are regions of flow acceleration and wake formation, respectively.

Strict orthogonal grid generation is the main focus here, so that one uses orthogonal formulation of Navier–Stokes equation, resulting in simpler formulation in the transformed plane, as compared to non-orthogonal formulation. This results in fewer operations and hence faster computations, along with lower numerical errors. Additionally, with orthogonal body-fitted grid, boundary conditions are easy and accurate to implement. While solving Navier–Stokes equation using finite difference formulation, one maps non-uniformly spaced grid points in physical $(x, y)$-plane to equi-spaced points in the computational $(\xi, \eta)$-plane, by relating the two using, $x = x(\xi, \eta)$ and $y = y(\xi, \eta)$. Such mapping must be unique, for which any point $(x, y)$ in the physical space must be a linear scalar functional of the coordinate $(\xi, \eta)$ in the computational plane [31, 54]. Thus, $\nabla x$ and $\nabla y$ are constant-valued vector fields, which result in

$$\nabla^2 x = 0, \ \nabla^2 y = 0 \tag{4.24}$$

Included angle between $\xi = $ constant and $\eta = $ constant lines in the physical plane is constrained to be orthogonal, if above numerical transformations satisfy

$$x_\xi x_\eta + y_\xi y_\eta = 0 \tag{4.25}$$

where a subscript indicates partial derivative with respect to it. One introduces a distortion function $f$ as a ratio of scale factors, $h_1 = \sqrt{x_\xi^2 + y_\xi^2}$ and $h_2 = \sqrt{x_\eta^2 + y_\eta^2}$ in $\eta$- and $\xi$-directions [23, 27, 31] by

$$f = \frac{h_2}{h_1} \tag{4.26}$$

Using Eq. (4.26) in (4.25), one obtains the following two relations:

$$x_\eta = f y_\xi$$
$$y_\xi = -f y_\eta \tag{4.27}$$

These are the Beltrami equations for general orthogonal mapping [10, 11, 27] and are nothing but the generalization of Cauchy–Riemann relations stated for conformal mapping. Here, the grid generation procedure of [23] is improved to compute flow past highly cambered aerofoil, with and without leading edge roughness elements [2].

As a demonstration, O-type grids are produced using finite difference method. Here, Eq. (4.25) and one of Eq. (4.27) are solved along with prescribed arc length in the hyperbolic direction, by taking the following steps:

(a) Points on the body are distributed along $\eta = 0$ line which fixes the scale factor $h_1$-distribution on the line.
(b) The grid line increment $\Delta S_\eta$ in the $\eta$-direction is prescribed, and this fixes the other scale factor $h_2$ given by

$$h_2 = \frac{\Delta S_\eta}{\Delta \eta} \tag{4.28}$$

Thus, the distortion function $f$ can be evaluated on $\eta = 0$ line.

(c) Equation (4.25) and either of Eq. (4.27)) are then integrated in the $\eta$-direction by one step. This produces $(x, y)$ coordinates of the $\eta = \Delta \eta$ line.
(d) Steps (a)–(c) can be repeated to obtain successive $\eta$ grid lines.

Present procedure for the evaluation of $f$ not only makes the grid generation process very fast, but also gives a rational basis for the choice of the function $f$. This process of fixing the distortion function makes the governing grid generation equations linear, and hence the grid generation process is also very fast. Furthermore, the ability to prescribe the grid line spacing in the normal direction is a very useful feature of the generated grids for viscous flow calculations.

Above-mentioned method has been implemented on various aerofoils. In Fig. 4.14, physical and computational planes are illustrated. For a typical O-type grid, physical body is represented as OPQ, where the cuts OU and QT are introduced starting from the trailing edge so that the computational domain is simply connected. Dirichlet conditions are prescribed on OPQ, and the grid spacing in the wake of the aerofoil is achieved by distributing points on the cuts OU and QT. The first grid line is located as close as 0.000001 times the chord of the aerofoil. The successive grid lines are expanded smoothly by increasing the spacing by a constant value.

We show here a grid generated [23] for NACA 0015 in Fig. 4.15. First 100 $\eta$-lines in the hyperbolic direction of the grid are shown. The slope discontinuity at the trailing edge is avoided by fitting a parabola in the open trailing edge, with continuous slope everywhere.

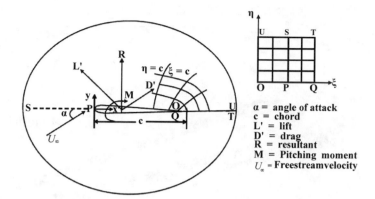

**Fig. 4.14** Physical and computational domains for an O-type grid

**Fig. 4.15** Orthogonal grid
generated around NACA
0015 aerofoil. Only first
hundred $\eta$-lines are shown
here

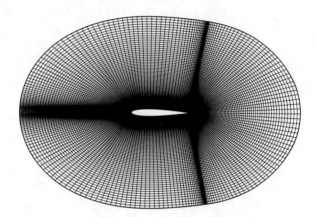

There are 233 points in $\xi$-direction, 301 lines in $\eta$-direction, and outer boundary
is 14.76 chords away. Included angles between constant $\xi$- and $\eta$-lines are virtually
orthogonal everywhere except at three points along the second $\eta$-line, where the
worst departure at a point with included angle is given by 89.476° [24]. The included
angle is calculated from

$$\theta = cos^{-1} \left[ \frac{g_{12}}{h_1 h_2} \right] \tag{4.29}$$

The grid metric $g_{12}$ is as given on the left hand side in Eq. (4.25). In calculating
the metrics, second-order central differences are used. The major advantage of the
method is that the addition and deletion of points only have local effects. The orthogonality property of the generated grids is better than efforts in [10, 47]. In [10], 1.11°
has been reported as maximum departure from orthogonality. Next, NACA 8504
aerofoil is considered, which has an 8% camber and 4% thickness with respect to

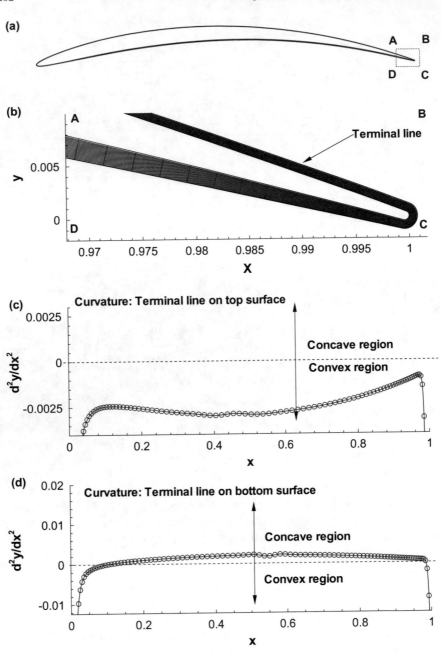

**Fig. 4.16  a** Orthogonal grid near the surface of the NACA 8504 aerofoil has been generated in [23]; **b** zoomed view of ABCD region with a highlighted terminal line; curvature variations of the terminal line on top and bottom sides are shown in (**c**) and (**d**), respectively. Reproduced from Bagade, et al., *An improved orthogonal grid generation method for solving flows past highly cambered aerofoils with and without roughness elements. Computers and Fluids*, **103**, 275–289 (2014), with the permission of Elsevier Publishing

chord, as shown in Fig. 4.16a. In this figure, orthogonal grid around NACA 8504 aerofoil with 456 points in the azimuthal direction and 250 points in the wall-normal direction is generated, but only the first 50 $\eta$-lines are shown. This part of the grid is generated following the procedure in [23], which extends only to a small distance from the aerofoil. This is generated by prescribing the gap between successive $\eta$=constant lines in the wall-normal direction following the tangent hyperbolic distribution given by

$$h_2 = H \left[ 1 - \frac{\tanh[\beta(1 - \eta)]}{\tanh[\beta]} \right]$$

where $\beta$ represents the grid clustering parameter. The grid line increment in $\eta$-direction is fixed by grid scale parameter $h_2$, and $H$ is the non-dimensional distance (in terms of chord ($c$) of the aerofoil) of the outer boundary. The $h_2$-distribution is calculated for 250 points in the wall-normal direction, with the boundary at a distance of $5c$ from the aerofoil. A region ABCD near the trailing edge of the aerofoil is shown zoomed in Fig. 4.16b. In this figure, the outermost azimuthal terminal line ($H = 5c$) is marked as a solid line. It is evident that the generated grid farther away from the aerofoil with wall-normal grid lines from the bottom surface near the leading and trailing edge will eventually intersect, due to concavity of the bottom surface. This requires modification of the algorithm in [23], in order to generate orthogonal grid around highly concave-shaped bodies. Distinction between the concave and convex nature of the surface is visible from the variation of the curvature, $\frac{d^2 y}{dx^2}$ of the terminal line on the top and the bottom surfaces, as shown in Fig. 4.16c, d, respectively.

Thus, one needs to obtain $h_2$-distribution as a function of both $\xi$ and $\eta$ to avoid grid intersection. The prescription of target in the wall-normal direction renders the modified method to be viewed as hybrid elliptic–hyperbolic method, as compared to purely hyperbolic method in [23]. Estimation of curvature on the terminal line helps one to locate the concave region and is an important parameter in the grid generation algorithm proposed here, which is enumerated next. Proposed algorithm is explained here with the help of Figs. 4.17 and 4.18.

### 4.3.1 Grid Generation Algorithm

1. Generate an orthogonal grid around a given aerofoil following the algorithm in [23] up to a small distance from aerofoil surface, so that the grid lines emerge smoothly without intersection, as shown in Fig. 4.16a.
2. Next, compute the curvature on the terminal line and identify the concave region(s), as shown in Fig. 4.16c, d.
3. Change value of $h_2$-distribution appropriately in the concave region by small amount ($\Delta h_2$) to regenerate a grid with the modified $h_2$-distribution.
4. One should compute the curvature on the terminal line again for the new grid and repeat steps 3 and 4, till the concavity has been removed completely at the

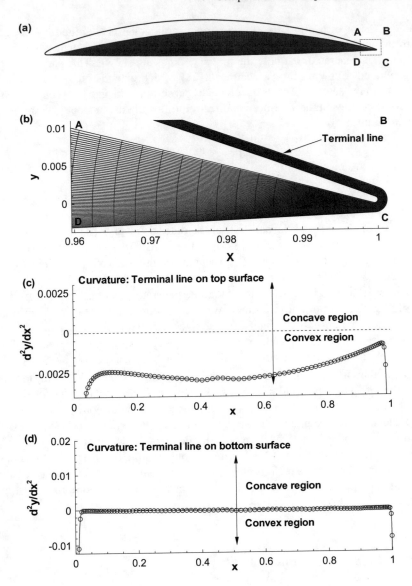

**Fig. 4.17  a** Generated orthogonal grid at the end of step 4 of the proposed algorithm in Sect. 4.3.1;
**b** zoomed view of the new OPQS region, **c** curvature variations of the new terminal line on top side,
**d** curvature variations on bottom side.  Reproduced from Bagade, et al., *An improved orthogonal
grid generation method for solving flows past highly cambered aerofoils with and without roughness
elements. Computers and Fluids*, **103**, 275–289 (2014), with the permission of Elsevier Publishing

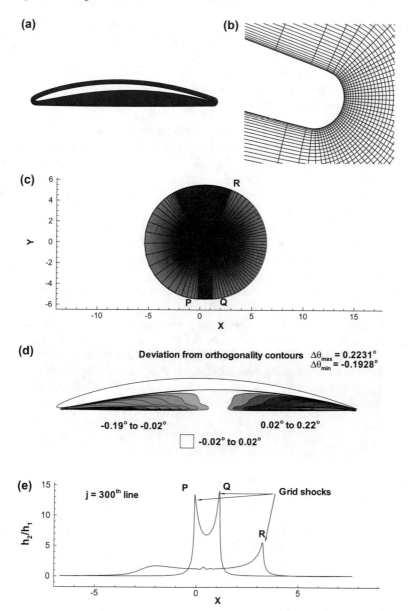

**Fig. 4.18** **a** Zoomed view of the orthogonal grid generated around NACA8504 aerofoil; **b** complete grid; **c** deviation from orthogonality contours for the grid shown in (**b**); **d** variation of grid scale parameter combination $h_2/h_1$ on the last azimuthal line. Note grid shocks at P, Q, and R in frame (**e**). Reproduced from Bagade, et al., *An improved orthogonal grid generation method for solving flows past highly cambered aerofoils with and without roughness elements. Computers and Fluids*, **103**, 275-289 (2014), with the permission of Elsevier Publishing

terminal line. At the end of this step, one obtains a grid, as shown in Fig. 4.17a, b, with a zero curvature on the bottom surface of a terminal line as shown in Fig. 4.17b.
5. Once concavity is removed from the terminal line, further hyperbolic grid generation farther away from the surface following the algorithm in [23] is straightforward. Such an interim grid is shown in Fig. 4.18.

While performing steps 3 and 4, we have prescribed a fixed limiting curvature on the terminal line as $\frac{d^2 y}{dx^2} = 0$. So the spacing in the wall-normal direction for a concave region is modified till we obtain a grid with a flat-bottom terminal line, as shown in Fig. 4.17a. Increase of spacing in the wall-normal direction causes $\xi$ = constant lines to bend as shown in Fig. 4.17b. Thus, the problem of possible intersection of $\xi$ = constant lines from the leading and the trailing edges of the bottom surface has been removed. Figure 4.18b shows a zoomed view of the constructed grid, while Fig. 4.18c shows a complete orthogonal grid around NACA 8504 aerofoil with $456 \times 250$ points. Deviation of grid lines from orthogonality is shown in Fig. 4.18d by contour plots. Maximum positive deviation $\Delta\theta_{max}$ is $0.1484°$, while maximum negative deviation $\Delta\theta_{min}$ is $0.1138°$. This is significantly lesser, as compared to $0.524°$ reported for the case of NACA 0015 symmetric aerofoil in [23].

While least deviation from orthogonality is a major requirement for use with orthogonal formulation of Navier–Stokes equations, one has to also ensure smooth variation of grid metrics, which appear explicitly in the governing equations. See Fig. 4.18e, which identifies the three locations at $P$, $Q$, and $R$ displaying grid shock. This is discussed here with respect to the orthogonal formulation of incompressible Navier–Stokes equations using stream function and vorticity, given by

$$\frac{\partial}{\partial \xi}\left(\frac{h_2}{h_1}\frac{\partial \psi}{\partial \xi}\right) + \frac{\partial}{\partial \eta}\left(\frac{h_1}{h_2}\frac{\partial \psi}{\partial \eta}\right) = -h_1 h_2 \omega \tag{4.30}$$

$$h_1 h_2 \frac{\partial \omega}{\partial t} + h_2 u \frac{\partial \omega}{\partial \xi} + h_1 v \frac{\partial \omega}{\partial \eta} = \frac{1}{Re}\left[\frac{\partial}{\partial \xi}\left(\frac{h_2}{h_1}\frac{\partial \omega}{\partial \xi}\right) + \frac{\partial}{\partial \eta}\left(\frac{h_1}{h_2}\frac{\partial \omega}{\partial \eta}\right)\right] \tag{4.31}$$

Following initial and boundary conditions are used in solving the governing equations. On the aerofoil surface, no-slip conditions are used

$$\psi = \text{constant}; \quad \frac{\partial \psi}{\partial \eta} = 0 \tag{4.32}$$

These conditions also help fix the wall vorticity, which is required as the boundary condition for the vorticity transport equation, Eq. (4.31). In O-grid topology, one introduces a cut starting from the leading edge of the aerofoil to the outer boundary. Periodic boundary conditions apply along the cut, which is introduced to make the computational domain simply connected. For the stream function equation, Eq. (4.30), at the outer boundary, Sommerfeld boundary condition is used on the $\eta$ component of the velocity field. Resultant value of stream function is used to calculate the vorticity value at the outer boundary.

From the stream function equation, the wall vorticity is calculated as

$$\omega|_{body} = -\frac{1}{h_2^2}\frac{\partial^2\psi}{\partial\eta^2}\bigg|_{body} \tag{4.33}$$

The three combinations of scale factors, namely, $h_1h_2, h_1/h_2$, and $h_2/h_1$, play important roles while solving Navier–Stokes equation. In solving the stream function equation (Eq. (4.30)), one uses iterative procedure, and the total number of iterations required to achieve convergence directly depends on these metrics. Whenever one comes across a sudden change in $h_1$ and $\xi$ = constant lines are clustered, a grid-shock form. In the presence of such grid shocks, diagonal dominance of the associated linear algebraic equation obtained from the discretized self-adjoint operator can be adversely affected, leading to poor convergence. One notices very high values of $h_2/h_1$ on the last azimuthal line in the concave part of the aerofoil, as shown in Fig. 4.18e and the formation of grid shocks near P, Q, and R, as marked in Fig. 4.18c.

To remove grid shock, we state a procedure used for a completely new grid for NACA 8504 in Figs. 4.19a, b, by prescribing criteria on minimum limiting curvature on the terminal line as $\frac{d^2y}{dx^2} = -0.003$. Zoomed view of the grid in Fig. 4.19a shows instead a convex shape of the terminal line, and the complete grid in Fig. 4.19b shows significant reduction in grid-shock intensity at points P, Q, and R, as compared to that shown in Fig. 4.18e. This is also evident from variation of $h_2/h_1$ on the last azimuthal line shown in Fig. 4.19e as compared to that in Fig. 4.18e. However, for such smooth variation of grid metrics, we also have higher deviation from orthogonality, as noted in Fig. 4.19d as compared to that shown in Fig. 4.18d.

### 4.3.2  Orthogonal Grid Generation for the SHM-1 Aerofoil with Roughness Element

Orthogonal grid for the NLF, SHM-1 aerofoil [14] in the presence of multiple roughness elements located on the top surface of the aerofoil near the leading edge has been generated in [2]. In Fig. 4.20a, the basic profile of SHM-1 aerofoil with surface roughness elements near the leading edge is shown, with the zoomed view of the region ABCD enclosing three successive roughness elements near the leading edge shown in Fig. 4.20b. These kinds of surface irregularities may be present on the aircraft wing due to manufacturing problems, ice accretion, or the presence of wing parting line. It is important to study the change in the flow field due to the presence of roughness elements.

For the presented case, generated grid has 601 points in the azimuthal direction and 700 points in the wall-normal direction up to 15c. Figure 4.20c shows a zoomed view of the grid near the roughness elements. Each surface roughness element creates additional concave regions, as shown in Fig. 4.20d. Altogether there are three such concave regions. These concave regions are filled up using the hybrid approach described before. Concavity near roughness elements is much more localized as

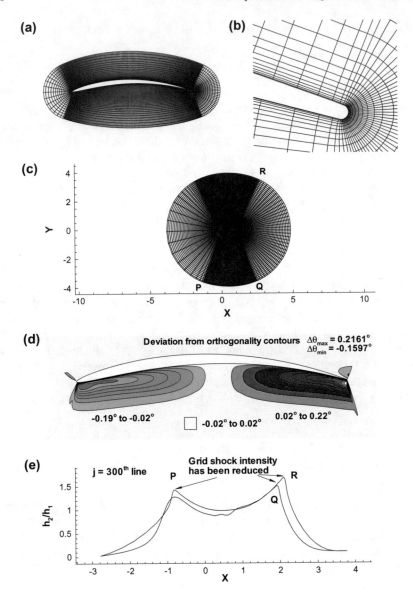

**Fig. 4.19** **a** Zoomed view of the orthogonal grid generated around NACA8504 aerofoil; **b** complete grid; **c** deviation from orthogonality contours; **d** variation of grid scale parameter combination $h_2/h_1$ on the last azimuthal line. Note intensity of grid shocks at P, Q, and R has been reduced. Reproduced from Bagade, et al., *An improved orthogonal grid generation method for solving flows past highly cambered aerofoils with and without roughness elements. Computers and Fluids*, **103**, 275–289 (2014), with the permission of Elsevier Publishing

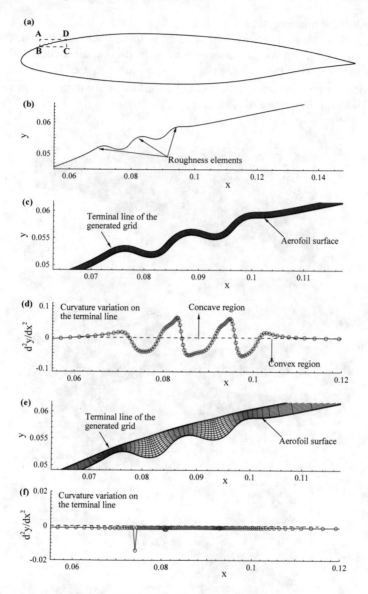

**Fig. 4.20** **a** Schematic of SHM-1 airfoil [14] with multiple surface roughness near leading edge; **b** zoomed view of surface roughness elements; **c** orthogonal grid generated following [23]; **d** curvature variations of the terminal line shown in (**c**); **e** orthogonal grid generated using proposed algorithm and **f** curvature variation of the terminal line shown in (**e**). Reproduced from Bagade, et al., *An improved orthogonal grid generation method for solving flows past highly cambered aerofoils with and without roughness elements. Computers and Fluids*, **103**, 275–289 (2014), with the permission of Elsevier Publishing

compared to the concavity of the bottom surface of the SHM-1 aerofoil. For each roughness element, 20 points have been taken in the azimuthal direction. Concave portions on the top and bottom surfaces are removed using the modified algorithm discussed above. After the execution of step 4, one obtains a grid as shown in Fig. 4.20e, with a criterion on minimum limiting curvature on the terminal line as $\frac{d^2y}{dx^2} = -0.003$, which is also noted in Fig. 4.20f.

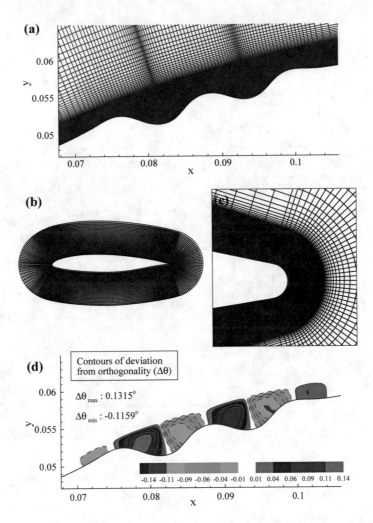

**Fig. 4.21** **a** Zoomed view of the grid near roughness elements; **b** zoomed view of grid around complete airfoil; **c** deviation from orthogonality contours for the generated grid in (**b**). Reproduced from Bagade, et al., *An improved orthogonal grid generation method for solving flows past highly cambered aerofoils with and without roughness elements. Computers and Fluids,* **103**, 275–289 (2014), with the permission of Elsevier Publishing

Close-up view of the grid near the roughness elements is shown in Fig. 4.21a, while the close-up of the complete grid is shown in Fig. 4.21b, and a zoomed view of the trailing edge of the airfoil is shown in Fig. 4.21c. As noted before, maximum deviation

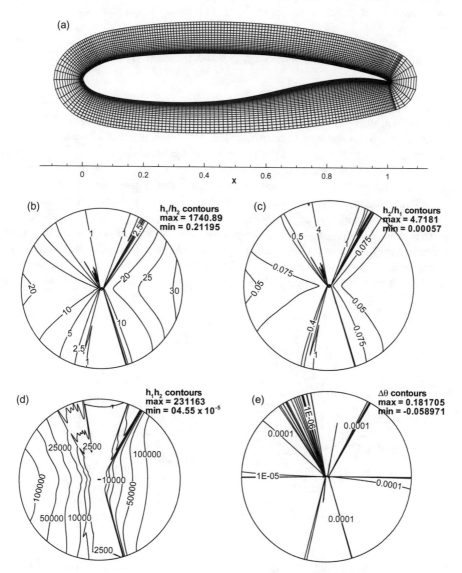

**Fig. 4.22** **a** Zoomed view of grid over WTEA-TE1 airfoil (The complete grid is generated with 1069 points in $\xi$-direction and 900 points in $\eta$-direction). However, in figure alternate fifth $\xi$-lines and alternate third $\eta$-lines are shown, **b** $h_1/h_2$ contours, **c** $h_2/h_1$ contours, **d** $h_1h_2$ contours, **e** deviation from orthogonality contours. Reproduced from Bagade, et al., *An improved orthogonal grid generation method for solving flows past highly cambered aerofoils with and without roughness elements. Computers and Fluids,* **103**, 275–289 (2014), with the permission of Elsevier Publishing

from orthogonality occurs in the concave portion and is mainly restricted to the region near the roughness elements. Figure 4.21d shows deviation from orthogonality contours with $\Delta\theta_{max} = 0.1361°$ and $\Delta\theta_{min} = -0.1347°$. As shown in Fig. 4.19, one controls grid-shock intensity by prescribing a higher convex curvature on the terminal line at the cost of increased deviation from orthogonality. For this case, other grid metrics are

$$(h_1 h_2)_{max} = 28559.3, \ (h_1/h_2)_{max} = 1180.73 \text{ and } (h_2/h_1)_{max} = 24.39$$

Following the present method, grid is also generated over WTEA-TE1 aerofoil. This NLF aerofoil has been considered for morphing wing applications [20]. The thickness-to-chord ratio for this aerofoil is 16%, as shown in Fig. 4.22 with localized camber in the last $0.30c$, near the trailing edge. Grid generation over such an aerofoil is equally challenging due to this localized camber. One needs to fill up concave region without grid line intersection. The increment in $\eta$-direction has to be very small initially, and one takes more number of points in this direction to avoid grid line intersection and low deviation from orthogonality. As the grid is generated iteratively, while satisfying orthogonality condition, the process is slow initially, till the concavity is removed. Generated grid with 1069 points in $\xi$-direction and 900 points in the $\eta$-direction takes one up to $25c$ from the aerofoil. Figure 4.22a shows the zoomed view near the aerofoil with alternate fifth $\xi$-lines, and alternate third

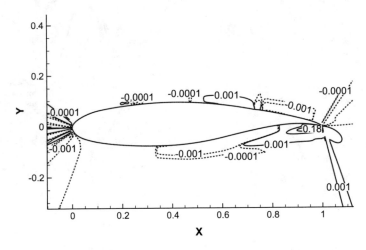

**Fig. 4.23** Deviation from orthogonality contours near WTEA-TE1 airfoil surface. Note maximum deviation from orthogonality (0.18°) is observed in a very small region near trailing edge extending up to 0.018 chord in $\eta$ direction. Dashed lines indicate negative values. Reproduced from Bagade, et al., *An improved orthogonal grid generation method for solving flows past highly cambered aerofoils with and without roughness elements. Computers and Fluids*, **103**, 275–289 (2014), with the permission of Elsevier Publishing

$\eta$-lines are shown for clarity. Figure 4.22b–e shows various grid metrics obtained for this grid, for those quantities which are directly needed in solving Navier–Stokes equation. The maximum deviation from orthogonality for the grid generated is found to be 0.165°, which is localized in a very small region near the trailing edge, as shown in Fig. 4.23. This region extends up to 0.018c in the $\eta$-direction. It is noted that except for this region near the trailing edge (0.0002c), the deviation from orthogonality is very low, of the order 0.05° to 0.007°, as shown in Fig. 4.23.

## 4.4 Low Reynolds Number Solution of Navier–Stokes Equation for AG24 Aerofoil

In Fig. 4.24, a schematic of the flow is shown, identifying different boundary segments of the computational domain. On the airfoil surface (AOC), no-slip boundary condition is applied by using $\frac{\partial \psi}{\partial \eta} = 0$, $\psi_{\text{wall}} = $ constant. These are used in solving stream function equation and obtaining the wall vorticity from $\omega_{\text{wall}} = -\frac{1}{h_2^2} \frac{\partial^2 \psi}{\partial \eta^2}$.

At the inflow (PBQ), free-stream condition obtained from potential flow solution is used $\frac{\partial \psi}{\partial y} = U_\infty(t)$. Convective boundary condition is enforced on radial velocity at the outflow (PRQ) on the outflow boundary, which is the Sommerfeld boundary condition given by [12],

$$\frac{\partial v}{\partial t} + u_c(\xi, \eta_{\max}, t) \frac{\partial v}{\partial \eta} = 0 \tag{4.34}$$

where the convection speed, $u_c$, at the outflow ($\eta = \eta_{\max}$) is taken as the wall-normal component speed evaluated at the previous time step, at the same location. Periodic boundary condition is applied for all variables on $\xi = $ constant lines, AB and CD.

To solve the pressure Poisson equation,

$$\frac{\partial}{\partial \xi}\left(\frac{h_2}{h_1} \frac{\partial P_t}{\partial \xi}\right) + \frac{\partial}{\partial \eta}\left(\frac{h_1}{h_2} \frac{\partial P_t}{\partial \eta}\right) = \frac{\partial(h_2 v\omega)}{\partial \xi} - \frac{\partial(h_1 u\omega)}{\partial \eta} \tag{4.35}$$

The required Neumann boundary conditions on the physical surface and at the far field are obtained from the normal ($\eta$)-momentum equation given by

$$\frac{h_2}{h_1} \frac{\partial p}{\partial \eta} = -h_1 u\omega + \frac{1}{Re} \frac{\partial \omega}{\partial \xi} - h_1 \frac{\partial v}{\partial t} \tag{4.36}$$

From the total pressure, $P_t = p + \rho \frac{(u^2 + v^2)}{2}$, one obtains static pressure ($p$) on the surface directly. The coefficients of lift and drag are derived from the normal and shear stresses acting on the aerofoil surface. Lift and drag coefficients are obtained from these stresses and presented as the lift coefficient variation with angle of attack

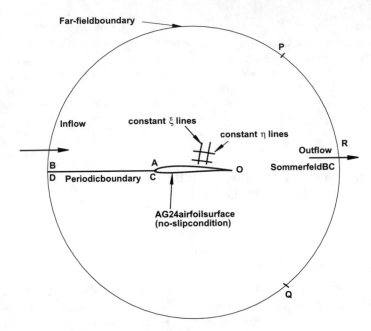

**Fig. 4.24** Schematic of the computational domain with grid lines and boundaries. Reproduced from Bagade, et al., *An improved orthogonal grid generation method for solving flows past highly cambered aerofoils with and without roughness elements. Computers and Fluids,* **103**, 275–289 (2014), with the permission of Elsevier Publishing

(lift curve) and plotting variation of the drag coefficient with the lift coefficient (drag polar). The details of these are as given in [33].

The governing kinematic and kinetic equations require different numerical methods given in the following. For the SFE, the LHS terms are discretized using second-order central difference scheme (CD2). This results in a system of linear algebraic equations given by, $[M]\psi = -h_1 h_2\{\omega\}$, which is solved iteratively using Bi-CGSTAB method [57]. Use of CD2 scheme preserves the isotropic nature of physical diffusion and ensures diagonal dominance of the $[M]$ matrix above.

For the VTE, governed by convection and diffusion processes, one requires very high accuracy methods. The convection terms in VTE are discretized by using high-resolution OUCS3 scheme [32, 44], which includes special boundary closure schemes. The diffusion terms of VTE are discretized by CD2 scheme to maintain isotropy of diffusion process. VTE is time advanced using three-stage optimized Runge–Kutta (OCRK3) technique [49], which has better dispersion relation preservation properties for unsteady flow computations.

Use of numerically generated orthogonal grid is demonstrated here by studying flow past AG24 aerofoil for $Re = 6 \times 10^4$ and $Re = 4 \times 10^5$, values commonly observed for gliders and micro air vehicles (MAVs). A grid with 528 points in the azimuthal direction and 497 points in the wall-normal direction has been used. This

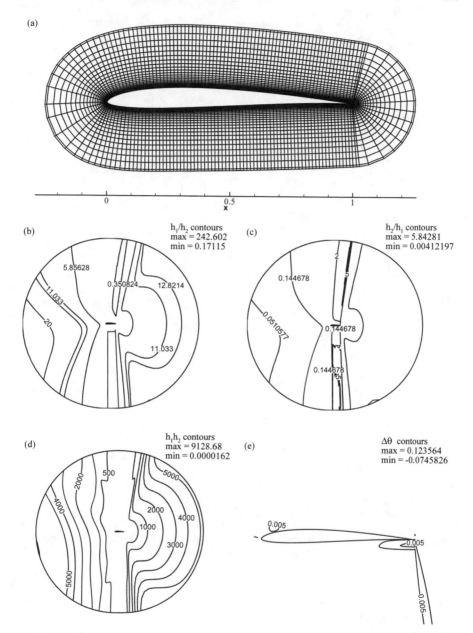

**Fig. 4.25 a** Zoomed view of grid around AG24 airfoil surface (The grid is generated using 528 points in $\xi$-direction and 497 points in $\eta$-direction), alternate fifth $\xi$-lines and alternate third $\eta$-lines are shown here; **b** $h_1/h_2$ contours; **c** $h_2/h_1$ contours; **d** $h_1h_2$ contours; **e** deviation from orthogonality contours for the generated grid. Reproduced from Bagade, et al., *An improved orthogonal grid generation method for solving flows past highly cambered aerofoils with and without roughness elements. Computers and Fluids*, **103**, 275–289 (2014), with the permission of Elsevier Publishing

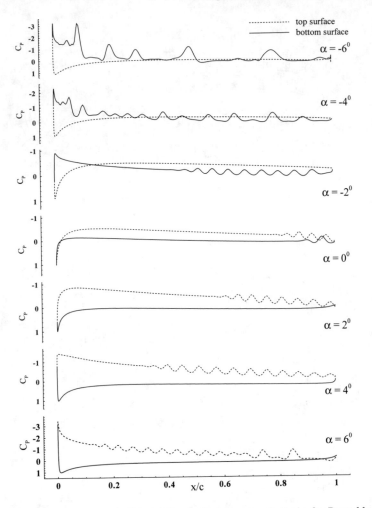

**Fig. 4.26** $C_p$ distribution over AG24 airfoil at indicated angle of attacks for Reynolds number of 60,000. Reproduced from Bagade, et al., *An improved orthogonal grid generation method for solving flows past highly cambered aerofoils with and without roughness elements. Computers and Fluids*, **103**, 275–289 (2014), with the permission of Elsevier Publishing

is shown in different frames in Fig. 4.25, with zoomed view near the aerofoil shown in frame 4.25a. Alternate fifth $\xi$-lines and alternate third $\eta$-lines are shown for clarity. The grid metrics such as $h_1/h_2$, $h_2/h_1$ and $h_1 h_2$ are shown in Fig. 4.25b–d, respectively, along with the ranges of the metrics obtained. The iso-contours of deviation from orthogonality are shown in frame Fig. 4.25e.

Figures 4.26 and 4.27 show the $C_p$-distribution for the considered angles of attack of $-6°$, $-4°$, $-2°$, $0°$, $2°$, $4°$, and $6°$ for the Reynolds numbers of 60,000 and 400,000, respectively. Figures 4.28 and 4.29 show comparison between numerical and experi-

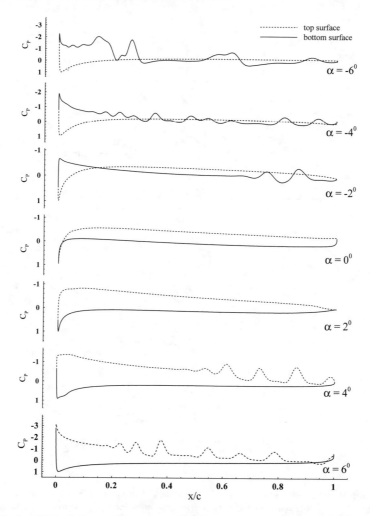

**Fig. 4.27** $C_p$ distribution over AG24 airfoil at indicated angle of attacks for Reynolds number of 400,000. Reproduced from Bagade, et al., *An improved orthogonal grid generation method for solving flows past highly cambered aerofoils with and without roughness elements. Computers and Fluids*, **103**, 275–289 (2014), with the permission of Elsevier Publishing

mental results for lift curve and drag polar for $Re = 60,000$ and $400,000$, respectively. The numerical results are found to be in qualitative agreement with experimental results [59]. It is noted that these experimental results are obtained in the presence of free-stream turbulence (FST), which is always present in any experimental setup. The computations are performed without considering any FST.

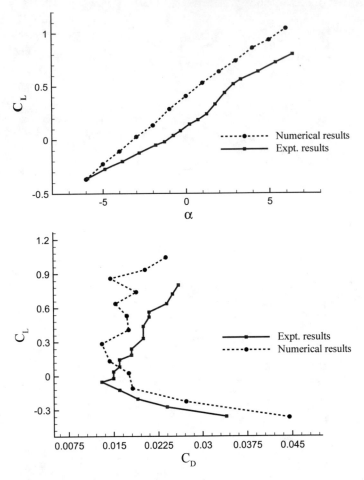

**Fig. 4.28** Lift curve and drag polar for $AG24$ airfoil at Reynolds number of 60,000. Reproduced from Bagade, et al., *An improved orthogonal grid generation method for solving flows past highly cambered aerofoils with and without roughness elements. Computers and Fluids,* **103**, 275–289 (2014), with the permission of Elsevier Publishing

## 4.5  Sensitivity to Orthogonality

Sensitivity to orthogonality of the generated grids has been analyzed with Navier–Stokes solution of flow past AG24 airfoil at zero angle of attack. Table 4.1 shows different cases, indicating grid metrics composed of $h_1$ and $h_2$, domain size, minimum and maximum deviation from orthogonality ($\Delta\theta_{\min}$ and $\Delta\theta_{\max}$, respectively), along with results for $C_L$ and $C_D$, in each case.

Here, all computations are reported for the AG24 airfoil at $\alpha = 0°$ for $Re = 60,000$. For the case-1, a grid with 528 points in azimuthal, $\xi$-direction and 497

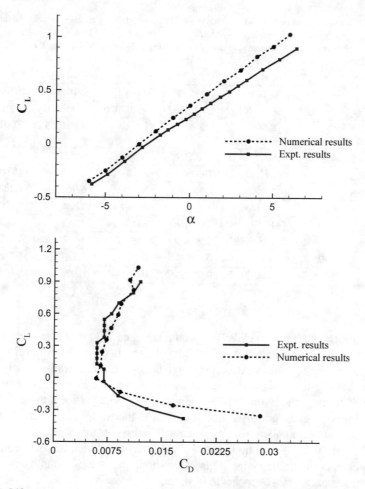

**Fig. 4.29** Lift curve and drag polar for $AG24$ airfoil at Reynolds number of 400,000. Reproduced from Bagade, et al., *An improved orthogonal grid generation method for solving flows past highly cambered aerofoils with and without roughness elements. Computers and Fluids,* **103**, 275–289 (2014), with the permission of Elsevier Publishing

points in wall-normal, $\eta$-direction is used in a domain, whose outer boundary is at $10c$. Case-2 corresponds to similar grid, but defined over a larger computational domain, with outer boundary at $15c$. Finally, case-3 is considered where the grid is generated with 1479 points in $\xi$-direction and 301 points in $\eta$-direction, while the outer boundary is placed at $15c$. Corresponding experimental results, in terms of $C_L$ and $C_D$, are given in the second column of Table 4.1.

The grid properties included in the table are of primary importance in solving Navier–Stokes equation correctly. For example, $h_1 h_2$ represents cell size, and the maximum value in the computational domain indicates the ability of computing the

**Table 4.1** Sensitivity to orthogonality analysis. Reproduced from Bagade, et al., *An improved orthogonal grid generation method for solving flows past highly cambered aerofoils with and without roughness elements. Computers and Fluids,* **103**, 275–289 (2014), with the permission of Elsevier Publishing

| Parameters | Expt.[59] | Case-1 | Case-2 | Case-3 | Case-4 | Case-5 |
|---|---|---|---|---|---|---|
| $\Delta\theta_{max}$ | – | 0.12356° | 0.19989° | 0.18128° | 0.1013° | 0.1686° |
| $\Delta\theta_{min}$ | – | −0.074582° | −0.098077° | −0.07026° | −0.0845° | −0.1561° |
| $\Delta\theta_{avg}$ | – | $8.546 \times 10^{-4}$ | $3.5546 \times 10^{-3}$ | $1.1765 \times 10^{-3}$ | 0.001055 | 0.001294 |
| $(h_1 h_2)_{max}$ | – | 9235.83 | 42829.9 | 90922 | 66321.3 | 58972.1 |
| $(h_1/h_2)_{max}$ | – | 241.628 | 185.602 | 7437.47 | 51.0538 | 157.186 |
| $(h_2/h_1)_{max}$ | – | 5.8663 | 6.4579 | 7.0302 | 6.96 | 4.33416 |
| $Stretch_{max.}$ | – | 2.3335 | 2.5925 | 2.03 | 1.886 | 1.80 |
| Domain | – | 10 Chord | 15 Chord | 15 Chord | 15 Chord | 15 Chord |
| Grid size | – | 528 × 497 | 528 × 500 | 1497 × 301 | 1497 × 300 | 1497 × 300 |
| $C_L$ | 0.125 | 0.41 | 0.26824 | 0.209875 | 0.1812 | 0.1881 |
| $C_D$ | 0.0155 | 0.0672 | 0.01026 | 0.0204 | 0.0238 | 0.024 |

flow with such cell sizes. Relatively from the table, case-3 has a maximum value of $h_1 h_2$ near the outer boundary, due to stretching the grid in the wall-normal direction up to 15c, using only 301 points. Also, it is noted that $h_2$ represents wall-normal spacing, which is usually lowest near the wall. Case-3 has smaller $h_1$ value, due to largest number of points in the azimuthal direction, yet $(h_1/h_2)_{max}$ is noted to be highest for case-3, implying finest grid used that is closest to the wall. In contrast, $(h_2/h_1)_{max}$ appears to be almost the same for all the cases. We also note that Stretch$_{max}$ in the table relates to spacing in $\xi$-direction, on the airfoil surface, which helps in controlling grid shock near the trailing edge. This value should be as close to unity as possible and in this respect, case-3 appears to have the best property.

While deviation from orthogonality is an important parameter, the maximum and minimum values in Table 4.1 occur at isolated points. In fact for different grids with different domain sizes and number of points, a more practical parameter is the averaged deviation from orthogonality, on per-point basis. In this regard, case-1 appears to have better orthogonality property (given in degrees), while the other two cases also have similar order of magnitude deviations.

Results shown in the table clearly indicate that case-2, as compared to case-1, shows better match with the experimental values. This implies that for low Reynolds numbers, the effects of the airfoil are felt at larger distances from the surface and imposing free-stream conditions at 10c results in poorer computed flow field. Thus, placing the outflow boundary at 15c improves the results significantly. Also for case-2, with wall-normal spacing increasing, adds more numerical diffusion. To test this hypothesis, in case-3 we have kept the outer boundary at 15c, while reducing the number of grid point in $\eta$-direction to 301. Case-3 shows the closest match with the experimental results.

Figure 4.28 shows the time series of (a) $C_L$ and (b) $C_D$ for the three cases of Table 4.1. The experimental time-averaged values are marked in these figures by a dotted line. It is noted that the used high accuracy methods, cases-1 and 2, show low-frequency oscillations for both $C_L$ and $C_D$, while the computed results for case-3 shows stable solution. Also for this case, the results indicate that the solution reaches the desired quasi-steady state even before $t = 30$, and that is the reason that the solution is reported up to that time only.

## 4.6  Direct Numerical Simulation of Transition Over a Natural Laminar Flow Aerofoil

Here, the main focus is in obtaining time-accurate solutions for transitional flows past an NLF aerofoil, which is designed in a way that transition from laminar to turbulent flow is delayed on the suction surface, as far aft as possible. Transition begins via flow instability and in design of NLF aerofoils, the flow instability onset is delayed. One of the most commonly studied and understood instability of wall-bounded shear layer is the classical route associated with the formation of Tollmien–Schlichting (TS) wave packets.

The solution of Orr–Sommerfeld equation reported in [15, 29, 55] for stability of flow over a flat plate was investigated as an example of zero pressure gradient Blasius boundary layer. It was shown that if small-amplitude monochromatic time-harmonic wall excitation is imposed in a localized manner, then the response field appears as TS wave packet. Results in [29, 55] predicted onset of criticality with the presence of wave packets that show initial growth and followed by decay, giving the progressive wave an appearance of wave packet. This was verified experimentally via the pioneering effort of Schubauer and Skramstad [30]. Since then spatial growth of TS wave packets is held responsible for flow transition. While different frequencies suffer differential growth rates, appearance of the wave packet is considered as a precursor to transition. However, it has been shown in recent times that additional spatio-temporal wavefront (STWF) is the true precursor of transition [4, 34] for low levels of background disturbance. However, for large-amplitude wall excitation or free-stream excitation for zero pressure gradient boundary layer, one often notices the absence of TS wave packet. One can also create transition without the presence of TS wave packet.

According to Morkovin [22], instances of flow transitions without the presence of TS wave packets are collectively categorized as *bypass transition*. We note that such terminology does not imply that there is only one bypass transition route. For example, one such bypass transition has been shown theoretically and experimentally in [19, 42]. The physical mechanism has been identified with the help of an energy-based receptivity equation obtained from Navier–Stokes equations, without making any approximation, as vortex-induced instability. An example of this route of bypass transition is noted on the attachment line, along the leading edge of an infinite swept wing via convecting aperiodic vortices in the free stream as subcritical instability [45], generated at the wing–body junction.

The classical or the bypass route of transition has been stated above for canonical zero pressure gradient flows only, while it is known that adverse pressure gradient has strong destabilizing effects on flow transition. As NLF aerofoils are of larger thickness, the aft portion of the aerofoil experiences very severe varying pressure gradients, on both the top and bottom surfaces for different angles of attack. Existence of such varying adverse pressure gradient is capable of amplifying even very small background disturbances, and no systematic criterion exists which theoretically correlates background disturbances with transition to turbulence, for either the classical or bypass route. In this context, it is worthwhile to investigate the process of transition over an aerofoil by DNS of Navier–Stokes equation [39].

In this reference, flow past an NLF SHM-1 aerofoil [14] is reported for a cruise Reynolds number of 10.3 million. NLF feature is achieved by contouring the section, with the resultant pressure gradient delaying flow transition on the suction surface, as far as possible. However, when such an aerofoil is employed for general aviation purpose, the flow eventually suffers transition in the rear half of the section due to the prevalent adverse pressure gradient, which accelerates the growth rate of the trace amount of background disturbances. In a computational framework, the background disturbances can arise from various sources of numerical errors. In nature, disturbances are omnipresent and their role in destabilizing fluid flow is a pacing item of research in fluid dynamics. These disturbances arise from surface perturbations, free-stream noise, surface vibrations, etc. We emphasize that even when explicit excitation is not employed in numerical computations, due to hypersensitivity of the flow field to background disturbances at higher Reynolds numbers, transition is observed. In an ideal scenario, one would like to remove all sources of numerical errors, which contribute to flow transition, and then apply an explicit excitation, which mimics the physical background disturbances. This is possible for Blasius boundary layer, and results are reported in [4, 13, 34, 35]. It is imperative to understand various sources of numerical errors and the way these can be eliminated or reduced [43]. It is important to choose an appropriate formulation and its accurate discretization, along with high-quality grid generation.

The other sources of errors while computing are (i) aliasing error in computing product terms, as explained in [32, 52] and (ii) spurious dispersive waves (whose extreme form is known as $q$-waves), as described in [26, 41, 58] and due to numerical stability, phase, and dispersion error, as explained in [43].

An algorithm to generate a truly orthogonal grid for thick and cambered aerofoil is developed by removing grid shocks arising from the concave part of the aerofoil near the trailing edge. The resultant orthogonal grid for SHM-1 NLF aerofoil is shown in Fig. 4.30, with highest departure from orthogonality noted for a single point as $0.0748°$ in Fig. 4.30b. Efficiency of the proposed grid generation algorithm in constructing orthogonal grid around cambered aerofoils is a significant element of the present procedure.

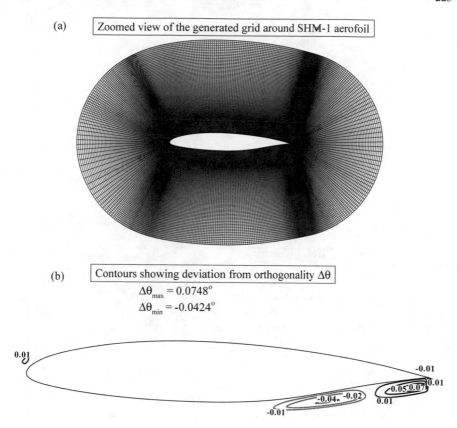

**Fig. 4.30** **a** Zoomed view of the generated grid after Step III is shown; **b** contours of deviation from orthogonality are shown for the generated grid. Reproduced from T. K. Sengupta, et al., *Direct numerical simulation of transition over a NLF aerofoil: Methods and validation. Frontiers in Aerospace Engineering (FAE)*, **2**(1), (2013), with the permission of Science and Engineering Publishing Company.

Above-mentioned orthogonal grid used with orthogonal formulation is further aided by adopting high accuracy method in solving vorticity transport equation. High-resolution ability of the used SOUCS3 [8] scheme in discretizing convection terms is demonstrated by comparing the equivalent numerical wavenumber in the spectral plane, with conventional second- and fourth-order discretization methods in Fig. 4.31a. The compact scheme has higher resolution by almost an order of magnitude, as compared to explicit second-order discretization. This spatial discretization is used to develop a DRP scheme for time integration, as the OCRK3 scheme in [49], which is used here. These combined schemes are calibrated with respect to the model 1D convection equation, and the contours of numerical amplification factor $|G|$ and numerical group velocity $V_{gN}/c$ are shown in Fig. 4.31b, c, respectively. Despite providing superior discretization properties, one notices a range of high wavenumber

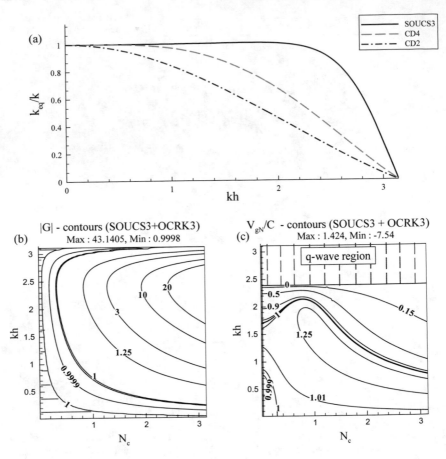

**Fig. 4.31  a** Comparison of scale resolution between second-order, fourth-order central difference scheme and the compact scheme SOUCS3 used here; **b** $|G|$ and **c** $V_{gN}/C$ contours for the solution of 1D convection equation using the SOUCS3-OCRK3 scheme. Note the presence of $q$-wave region at higher wavenumber range in (**c**). Reproduced from T. K. Sengupta, et al., *Direct numerical simulation of transition over a NLF aerofoil: Methods and validation. Frontiers in Aerospace Engineering (FAE)*, **2**(1), (2013), with the permission of Science and Engineering Publishing Company

(greater than $kh = 2.40$) displaying spurious dispersion, with $V_{gN}/c$ contours taking negative values. Such spurious upstream propagation gives rise to $q$-waves, which has been removed for a small $N_c$ range by using a fifth-order upwind filter [40], whose transfer function and effects on numerical properties are shown in Fig. 4.32.

It is known that high accuracy methods also suffer from aliasing error for high Reynolds number computations. This has been controlled by using an adaptive multi-dimensional filter [5], whose transfer function is shown in Fig. 4.33 for different filter coefficient values.

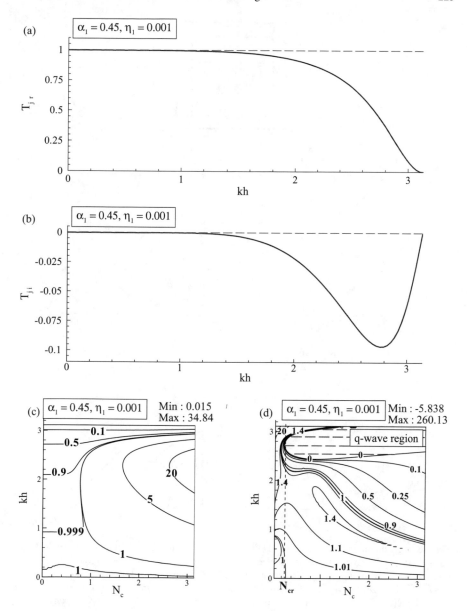

**Fig. 4.32** Variations of **a** real part and **b** imaginary part of the transfer functions of the fifth-order upwind filter [40] for filter coefficient $\alpha_1 = 0.45$ and $\eta_1 = 0.001$ are shown along with modifications in (**c**) $|G|$ and **d** $V_{gN}/C$ contours for the case of Fig. 4.31. Note the $q$-wave region has been removed over a small range of $N_{cr}$ in (**d**). Reproduced from T. K. Sengupta, et al., *Direct numerical simulation of transition over a NLF aerofoil: Methods and validation. Frontiers in Aerospace Engineering (FAE)*, **2**(1), (2013), with the permission of Science and Engineering Publishing Company

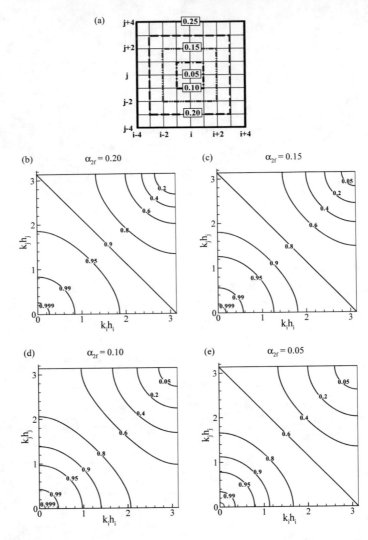

**Fig. 4.33  a** Schematic of the adaptive 2D filter [5], with the filter sub-domain shown in the transformed plane. For the 2D filters, the transfer functions are shown in frames (**b**) to (**e**) for different values of the filter coefficient $\alpha_{2f}$. Reproduced from T. K. Sengupta, et al., *Direct numerical simulation of transition over a NLF aerofoil: Methods and validation. Frontiers in Aerospace Engineering (FAE)*, **2**(1), (2013), with the permission of Science and Engineering Publishing Company

## 4.7   Flow Field Past SHM1 Aerofoil in Cruise Configuration

Validation of numerical results with the experimental results provided in [14] is presented next. The reported results are for two Reynolds numbers, $Re = 2.8 \times 10^6$ and $Re = 10.3 \times 10^6$ obtained using a ($597 \times 397$) grid. In Fig. 4.34a, variation of numerically obtained mean lift coefficient ($C_L$) with angle of attack for $Re = 2.8 \times 10^6$ is compared with experimental results provided in [14], which shows a good match between the two. The calculations are strictly 2D for $Re = 2.8 \times 10^6$ without any explicit forcing for the case shown in Fig. 4.34a, while the experimental results involve three-dimensionality and background tunnel noise. Time variation of lift and drag coefficient for $Re = 10.3 \times 10^6$ and zero angle of attack is shown in frames (b) and (c), respectively. Time-averaged values of lift and drag coefficients are shown with the horizontal dashed lines in these frames. Results show a good match between the experimental and numerically obtained values.

Although computations reported with a ($597 \times 397$) grid show a good comparison with the experimental results, next we report the numerical results obtained using a ($5169 \times 577$) grid specifically for a careful study of flow transition over a NLF aerofoil. Variation of RMS value of azimuthal component of disturbance velocity over top and bottom surfaces is shown in Fig. 4.34d at a distance of 0.000056 from the aerofoil surface. Variation of RMS component over the aerofoil surface shows that after 60% of chord there is sharp rise in fluctuations on the top surface indicating flow transition similar to the experimental results shown in [14]. On the bottom surface, one notes even a sharper variation of the RMS component of azimuthal velocity at $x = 0.70c$. Correct prediction of transition location highlights the importance of highly space–time accurate numerical solutions obtained.

Flow past SHM-1 aerofoil for $Re = 10.3 \times 10^6$ has been at zero angle of attack, and Fig. 4.35a shows stream function contours at $t = 4.50$. Top frame shows the flow field around the complete aerofoil, while in the bottom frame, flow field near the trailing edge of the aerofoil is shown at the same instant. Flow near the leading edge of the aerofoil experiences acceleration due to the favorable pressure gradient. As flow moves toward trailing edge, it experiences progressively adverse pressure gradient. Due to varying adverse pressure gradient, small separation bubbles are formed on the aerofoil surface near trailing edge, which move downstream along with the flow. These separation bubbles are observed in the zoomed view of a trailing edge, as shown in the bottom frame of Fig. 4.35a. As these bubbles move toward the trailing edge, these further excite the flow field. As shown earlier in Fig. 4.34d, flow transition on top surface starts after $0.60c$, these separation bubbles also first appear around the same location. In Fig. 4.35b, variations of displacement thickness ($\delta^*$) on top and bottom surfaces of the aerofoil are shown in frame (i) up to 60 % of chord of SHM-1 aerofoil at $t = 4.50$ . Additionally, variation of a steady flow separation parameter used in Falkner–Skan analysis, $m = \frac{x}{U_e} \frac{dU_e}{dx}$ of Fig. 4.35b, is shown for the top and bottom surfaces of SHM-1 aerofoil in frame (ii) at $t = 4.50$. A horizontal dashed line corresponding to steady separation flow criterion ($m = -0.09$) is marked for comparison purpose. Figure shows that on the top surface near the leading edge, flow

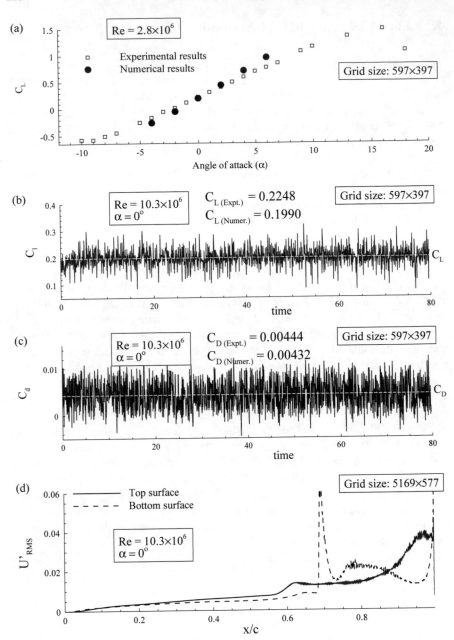

**Fig. 4.34** **a** Numerically obtained $C_L$ with angles of attack is shown along with experimental results of [14]; variation of instantaneous $C_l$ and $C_d$ is shown in frames (**b**) and (**c**), respectively; **d** variation of RMS value of the azimuthal component of the velocity over aerofoil for $Re = 10.3 \times 10^6$ on a line close to the aerofoil. Reproduced from T. K. Sengupta, et al., *Direct numerical simulation of transition over a NLF aerofoil: Methods and validation. Frontiers in Aerospace Engineering (FAE)*, **2**(1), (2013), with the permission of Science and Engineering Publishing Company

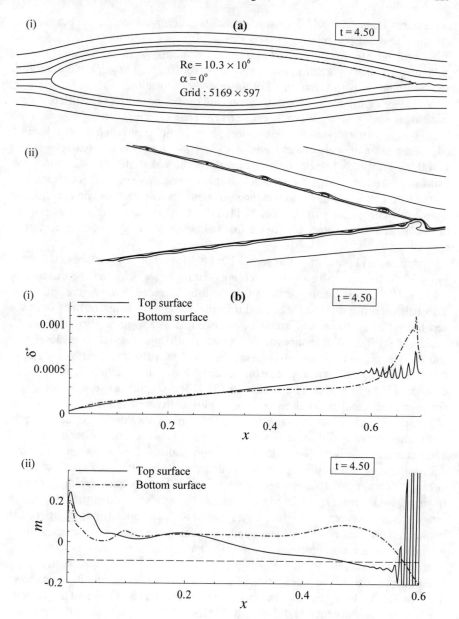

**Fig. 4.35 a** Stream function contours over the complete aerofoil as well as near the trailing edge are shown for the indicated parameters; **b** variation of the displacement thickness ($\delta^*$) and steady flow separation parameter $m$, on top and bottom surfaces of SHM-1 aerofoil, is shown in frames (i) and (ii), respectively. Reproduced from T. K. Sengupta, et al., *Direct numerical simulation of transition over a NLF aerofoil: Methods and validation. Frontiers in Aerospace Engineering (FAE),* **2**(1), (2013), with the permission of Science and Engineering Publishing Company

experiences higher favorable pressure gradient, as compared to the bottom surface. Pressure gradient parameter ($m$) smoothly decreases on the top surface after the initial peak near the leading edge. However, on the bottom surface, pressure gradient parameter remains favorable to a larger distance from the leading edge of the aerofoil, and then it has a sharp variation aft of the mid-chord location.

Variations of surface vorticity on the top and the bottom surfaces at different time instants are shown in Fig. 4.36a, b, respectively. Due to acceleration of flow near the leading edge on the top surface, one notices sharp vorticity variation near the leading edge. Formation of wavy disturbances on the top surface starts approximately from $x = 0.55c$ onward, while on the bottom surface, these disturbances start appearing from $x = 0.70c$ onward. Time variation of surface vorticity shows that disturbances on the bottom surface have similar range of vorticity variation as compared to the disturbances on the top surface. Flow field in the aft part of the aerofoil shows strong unsteady behavior, as compared to the first half, even when no explicit excitation is applied.

Formation and propagation of separation bubbles on the top surface is shown in Fig. 4.37. Variation of the azimuthal component of the velocity ($u$) at a distance of $1.242 \times 10^{-6}c$ from the top surface of the aerofoil is traced in this figure with time. This height corresponds to the second azimuthal line ($\eta = $ constant line) from the aerofoil surface. At $t = 0.40$, small wavy disturbances appear originating from the trailing edge. These disturbances move upstream with time as noted in the subsequent frames. The presence of adverse pressure gradient near the trailing edge magnifies these disturbances, which propagate further upstream as shown in this figure. The calculations are performed on a finer grid with $(5169 \times 577)$ points, to resolve the length scales correctly and avoid aliasing error. Additionally, we have used upwind filter [40] to prevent any $q$-waves contaminating the numerical solution. So the upstream propagation of the disturbances suggests physical bypass transition phenomenon [48] and not due to $q$-waves. Figure 4.37 shows how additional wave packets are introduced upstream of this propagating main disturbance, marked as $T_v$ in the frames, at $t = 1.25$ and later. This induced wave packet ($T_v$) grows quite rapidly, as noted from the frames from $t = 1.25$ to $t = 1.31$, while convecting downstream. As time progresses, this sequence of upstream migration of disturbance fronts fill up the top surface, up to about $x = 0.55c$, at around $t = 1.50$. Thereafter, disturbances formed on the top surface stop upstream propagation and continue to convect downstream, as shown till $t = 4.50$.

Similar formation and propagation of disturbances on the bottom surface is shown in Fig. 4.37b. In this figure also, the variation of azimuthal component of the velocity ($u$) is shown at a distance of $1.242 \times 10^{-6}c$ for the bottom surface. Similar to top surface, wavy disturbances originate from the trailing edge and can be observed as early as at $t = 0.50$. By $t = 0.67$, an induced small disturbance packet is noted close to $x = 0.8c$, which is marked as $B_v$. Its rapid growth is traced in the subsequent time frames, and one can note the nonlinear distortion by $t = 0.71$. The trailing edge position near $x = 0.8c$ of the aerofoil has large concavity imposing strong adverse pressure gradient, which results in such drastic amplification of disturbances. Due to nonlinear action, unsteady separation bubbles are formed in this part of the aerofoil,

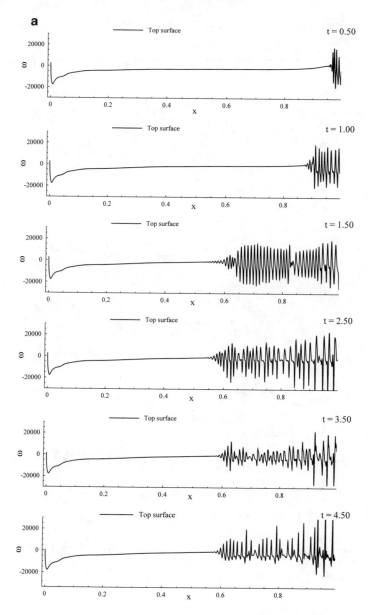

**Fig. 4.36** Variation of wall vorticity on the top surface of SHM-1 aerofoil has been shown for the case of $Re = 10.3 \times 10^6$ and $\alpha = 0°$, at the indicated times. Reproduced from T. K. Sengupta, et al., *Direct numerical simulation of transition over a NLF aerofoil: Methods and validation. Frontiers in Aerospace Engineering (FAE)*, **2**(1), (2013), with the permission of Science and Engineering Publishing Company

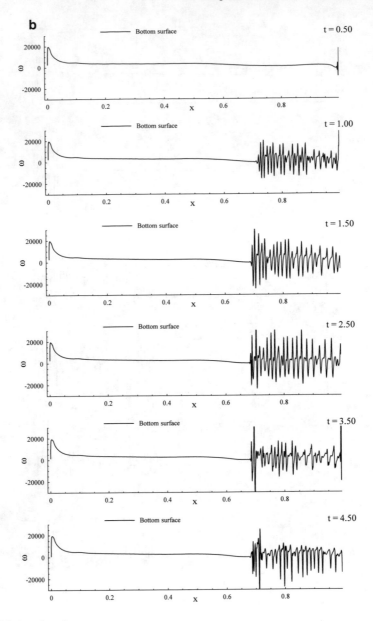

**Fig. 4.36** (continued)

resulting in increased drag. As compared to the top surface, here the upstream location of the disturbance is restricted to $x = 0.70c$, as noted in the bottom frames of Fig. 4.37b.

**Fig. 4.37** Variation of the azimuthal component of the velocity ($u$) at a distance of $1.242 \times 10^{-6}c$ from the top surface of SHM-1 aerofoil has been shown for the case of $Re = 10.3 \times 10^{6}$ and $\alpha = 0°$, at the indicated times. Reproduced from T. K. Sengupta, et al., *Direct numerical simulation of transition over a NLF aerofoil: Methods and validation. Frontiers in Aerospace Engineering (FAE)*, **2**(1), (2013), with the permission of Science and Engineering Publishing Company

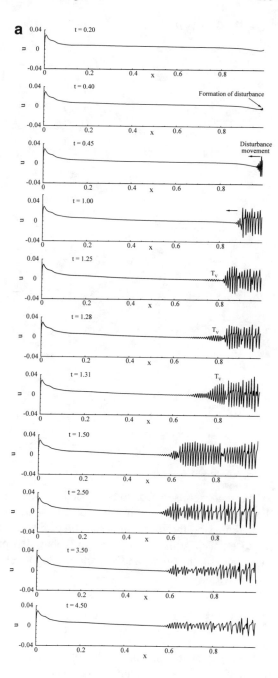

**Fig. 4.37** (continued)

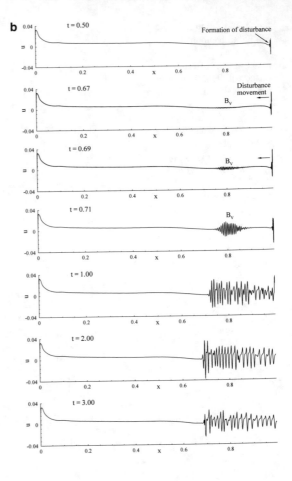

Present exercise needs to be carried out with the application of realistic explicit background disturbances in wind tunnels to highlight the role of free-stream turbulence in fixing the loads on aerofoil, specifically the drag coefficient. This would require not only the use of fine grids, but also proper description of free-stream turbulence in the tunnel and its appropriate model.

## Problems

**P4.1**: (a) Show that the satisfaction of Beltrami equations, Eq. ((4.27)), is equivalent to satisfying the orthogonality condition of grid generation given by Eq. ((4.25)). (b) Despite the condition of (a), why don't we solve the Beltrami equations only, for orthogonal grid generation? (c) Also show that the satisfaction of Beltrami equations is equivalent to satisfying covariant Laplace equations given by Equations

$$\frac{\partial}{\partial \xi}\left(f \frac{\partial x}{\partial \xi}\right) + \frac{\partial}{\partial \eta}\left(\frac{1}{f} \frac{\partial x}{\partial \eta}\right) = 0$$

$$\frac{\partial}{\partial \xi}\left(f \frac{\partial y}{\partial \xi}\right) + \frac{\partial}{\partial \eta}\left(\frac{1}{f} \frac{\partial y}{\partial \eta}\right) = 0$$

Can these Equations be useful in generating orthogonal grids?

**P4.2**: (a) Explain why solving a problem of flow past an airfoil in the presence of free stream turbulence (FST) is much tougher than solving uniform flow past the same airfoil. What are the grid metrics relevant for solving flow past bodies using orthogonal grid?

**P4.3**: This is the case of flow past AG24 airfoil at $\alpha = -2°$ for $Re = 60000$ and comparison of results are shown for simulation of the flows, with and without free stream turbulence (FST). While we do not discuss about how the FST model is designed and used, it is shown here to emphasize upon the point that all low Reynolds number airfoils operate at very low altitudes where the background disturbances are significant. This is very apparent in the shown streamlines for the cases of with and without FST.

**P4.4**: This is the case of flow past the same airfoil at same $Re$ for $\alpha = -5°$. For this angle of attack, one can see the front stagnation point anchored steadily on top surface. In fact the flow over the top surface is very steady. Whereas on the lower surface, one notices unsteady separation right from the leading edge itself. Because of the large negative AOA, the separation bubbles are appreciable in size and the flow is very unsteady. This airfoils drag polar indicates operation within $\pm 6°$ and thus this is an extreme angle of attack case and flow is hence very unsteady. This simulation is also for a case without FST.

# References

1. J.D. Anderson, *Fundamentals of Aerodynamics*, 5th edn. (Tata McGraw-Hill, India, 2010)
2. P.M. Bagade, Y.G. Bhumkar, T.K. Sengupta, An improved orthogonal grid generation method for solving flows past highly cambered aerofoils with and without roughness elements. Comput. Fluids **103**, 275–289 (2014)
3. P.M. Bagade, S.B. Krishnan, T.K. Sengupta, DNS of low Reynolds number aerodynamics in the presence of free stream turbulence. Front. Aero. Engg. **4**(1), 20–34 (2015)
4. S. Bhaumik, T.K. Sengupta, Precursor of transition to turbulence: spatiotemporal wave front. Phys. Rev. E **89**(4), 043018 (2014)
5. Y.G. Bhumkar, T.K. Sengupta, Adaptive multi-dimensional filters. Comput. Fluids **49**(1), 128–140 (2011)
6. D.L. Brown, M.L. Minion, Performance of under-resolved two-dimensional incompressible flow simulations. J. Comput. Phys. **122**, 165–83 (1995)
7. P.C. Chu, C. Fan, A three-point combined compact difference scheme. J. Comput. Phys. **140**(2), 370–399 (1998)

8.  A. Dipankar, T.K. Sengupta, Symmetrized compact scheme for receptivity study of 2D transitional channel flow. J. Comput. Phys. **215**, 245–273 (2006)
9.  D. Drikakis, P.K. Smolarkiewicz, On spurious vortical structures. J. Comput. Phys. **172**, 309–25 (2001)
10. R. Duraiswami, A. Prosperetti, Orthogonal mapping in two dimensions. J. Comput. Phys. **98**, 254–268 (1992)
11. L. Eca, 2D orthogonal grid generation with boundary point distribution control. J. Comput. Phys. **125**, 440–453 (1996)
12. P.G. Esposito, R. Verzicco, P. Orlandi, *Boundary condition influence on the flow around a circular cylinder*, in *The Proceedings of IUTAM Symposium on Bluff-Body Wakes, Dynamics and Instabilities* (Springer, Berlin, Germany, 1993), pp. 47–50
13. H. Fasel, U. Konzelmann, Non-parallel stability of a flat plate boundary layer using the complete Navier-Stokes equation. J. Fluid Mech. **221**, 331–347 (1990)
14. M. Fujino, Y. Yoshizaki, Y. Kawamura, Natural-laminar-flow airfoil development for a lightweight business jet. J. Aircraft **40**(4), 609–615 (2003)
15. W. Heisenberg, Über stabilität und turbulenz von flüssigkeitsströmen. Ann. Phys. Lpz., **379**, 577–627 (1924) (Translated as ' On stability and turbulence of fluid flows'. NACA Tech. Memo. Wash. No 1291 1951)
16. A.C. Hindmarsh, P.M. Gresho, D.F. Griffiths, The stability of explicit euler time-integration for certain finite difference approximations of the multi-dimensional advection-diffusion equation. Int. J. Numer. Methods Fluids **4**(9), 853–97 (1984)
17. S. Kawai, K. Fujii, Compact scheme with filtering for large-eddy simulation of transitional boundary layer. AIAA J. **46**, 690–700 (2008)
18. S. Laizet, E. Lamballais, High-order compact schemes for incompressible flows: a simple and efficient method with quasi-spectral accuracy. J. Comput. Phys. **228**(16), 5989–6015 (2009)
19. T.T. Lim, T.K. Sengupta, M. Chattopadhyay, A visual study of vortex-induced subcritical instability on a flat plate laminar boundary layer. Expts. Fluids **37**, 47–55 (2004)
20. M. Mamou, Y. Mébarki, M. Khalid, M. Genest, D. Coutu, A.V. Popov, Aerodynamic performance optimization of a wind tunnel morphing wing model subject to various cruise flow conditions, in *27th International Congress of the Aeronautical Sciences*, vol. 230, pp. 27–60 (2011)
21. M.L. Minion, D.L. Brown, Performance of under-resolved two-dimensional incompressible flow simulations-II. J. Comput. Phys. **138**, 734–65 (1997)
22. M.V. Morkovin, On the many faces of transition, in *Viscous Drag Reduction 1–31*, (Ed.), by C.S. Wells (Plenum Press, New York, 1969)
23. M.T. Nair, T.K. Sengupta, Orthogonal grid generation for Navier-Stokes computations. Int. J. Num. Meth. Fluids **28**, 215–224 (1998)
24. L. Prandtl, *Essentials of Fluid Dynamics* (Hafner Publishing, New York, USA, 1952)
25. S. Pirozzoli, Numerical methods for high-speed flows. Annu. Rev. Fluid Mech. **43**(1), 163–94 (2011)
26. T. Poinsot, D. Veynante, *Theo. and Numer. Combus.* Edwards, PA, 2nd edn (2005)
27. G. Ryskin, L.G. Leal, Orthogonal mapping. J. Comput. Phys. **50**, 71–100 (1983)
28. R. Sandberg, Direct numerical simulations for flow and noise studies. Procedia Eng. **61**, 356–62 (2013)
29. H. Schlichting, Zur entstehung der turbulenz bei der plattenströmung. Nach. Gesell. d. Wiss. z. Gött., *MPK*, **42**, 181–208 (1933)
30. G.B. Schubauer, H.K. Skramstad, Laminar boundary layer oscillations and the stability of laminar flow. J. Aero. Sci. **14**(2), 69–78 (1947)
31. T.K. Sengupta, *Fundamentals of Computational Fluid Dynamics* (University Press, Hyderabad, India, 2004)
32. T.K. Sengupta, *High Accuracy Computing Methods: Fluid Flows and Wave Phenomenon* (Cambridge University Press, USA, 2013)
33. T.K. Sengupta, *Theoretical and Computational Aerodynamics* (Wiley, UK, 2015)

34. T.K. Sengupta, S. Bhaumik, Onset of turbulence from the receptivity stage of fluid flows. Phys. Rev. Lett. **154501**, 1–5 (2011)
35. T.K. Sengupta, S. Bhaumik, V. Singh, S. Shukl, Nonlinear and nonparallel receptivity of zero-pressure gradient boundary layer. Int. J. Emerg. Mult. Fluid Sci. **1**(1), 19–35 (2009)
36. T.K. Sengupta, A. Bhole, Error dynamics of diffusion equation: effects of numerical diffusion and dispersive diffusion. J. Comput. Phys. **226**(1), 240–251 (2014)
37. T.K. Sengupta, A. Bhole, N.A. Sreejith, Direct numerical simulation of 2D transonic flows around airfoils. Comput. Fluids **88**, 19–37 (2013)
38. T.K. Sengupta, Y.G. Bhumkar, S. Sengupta, Dynamics and instability of a shielded vortex in close proximity of a wall. Comput. Fluids **70**, 166–175 (2012)
39. T.K. Sengupta, Y.G. Bhumkar, Direct numerical simulation of transition over a NLF aerofoil: methods and validation frontiers in aerospace. Engineering **2**, 1 (2013)
40. T.K. Sengupta, Y. Bhumkar, V. Lakshmanan, Design and analysis of a new filter for LES and DES. Comput. Struct. **87**, 735–750 (2009)
41. T.K. Sengupta, Y. Bhumkar, M. Rajpoot, V.K. Suman, S. Saurabh, Spurious waves in discrete computation of wave phenomena and flow problems. Appl. Math. Comput. **218**, 9035–9065 (2012)
42. T.K. Sengupta, S. De, S. Sarkar, Vortex-induced instability of an incompressible wall-bounded shear layer. J. Fluid Mech. **493**, 277–286 (2003)
43. T.K. Sengupta, A. Dipankar, P. Sagaut, Error dynamics: beyond von Neumann analysis. J. Comput. Phys. **226**, 1211–1218 (2007)
44. T.K. Sengupta, G. Ganeriwal, S. De, Analysis of central and upwind compact schemes. J. Comput. Phys. **192**, 677–694 (2003)
45. T.K. Sengupta, R. Jain, A. Dipankar, A new flux-vector splitting compact finite volume scheme. J. Comput. Phys. **207**, 261–81 (2005)
46. T.K. Sengupta, V. Lakshmanan, V.V.S.N. Vijay, A new combined stable and dispersion relation preserving compact scheme for non-periodic problems. J. Comput. Phys. **228**(8), 3048–3071 (2009)
47. J.L. Steger, D.S. Chaussee, Generation of body-fitted coordinates using hyperbolic partial differential equations. SIAM J. Sci. Stat. Comput. **1**, 431–437 (1980)
48. T.K. Sengupta, T. Poinsot, *Instabilities of Flows: With and Without Heat Transfer and Chemical Reaction* (Springer Wien, New York, 2010)
49. T.K. Sengupta, M.K. Rajpoot, Y.G. Bhumkar, Space-time discretizing optimal DRP schemes for flow and wave propagation problems. Comput. Fluids **47**(1), 144–154 (2011)
50. T.K. Sengupta, M.K. Rajpoot, S. Saurabh, V.V.S.N. Vijay, Analysis of anisotropy of numerical wave solutions by high accuracy finite difference methods. J. Comput. Phys. **230**, 27–60 (2011)
51. T.K. Sengupta, A. Sengupta, S. Kumar, Global spectral analysis of multi-level time integration schemes: numerical properties for error analysis. Appl. Maths. Comput. **304**, 41–57 (2017)
52. T.K. Sengupta, V.V.S.N. Vijay, S. Bhaumilk, Further improvement and analysis of CCD scheme: dissipation discretization and de-aliasing properties. J. Comput. Phys. **228**(17), 6150–6168 (2009)
53. V.K. Suman, T.K. Sengupta, C. Jyothi Durga Prasad, K. Mohan, D. Surya and Sanwalia, Spectral analysis of finite difference schemes for convection diffusion equation. Comput. Fluids **150**, 95–114 (2017)
54. J.F. Thompson, Z.U.A. Warsi, C.W.J. Mastin, Boundary-fitted coordinate systems for numerical solution of partial differential equations—A review. J. Comput. Phys. **47**, 1–108 (1982)
55. W. Tollmien, Über die enstehung der turbulenz. I, English translation. *NACA TM 609* (1931)
56. A. Uzun, G.A. Blaisdell, A.S. Lyrintzis, Application of compact schemes to large eddy simulation of turbulent jets. J. Sci. Comput. **21**(3), 283–319 (2004)
57. H.A. Van der Vorst, Bi-CGSTAB: a fast and smoothly converging variant of Bi-CG for the solution of non-symmetric linear systems. SIAM J. Sci. Stat. Comput. **12**, 631–644 (1992)
58. R. Vichnevetsky, J. Bowles, *Fourier Analysis of Numerical Approximations of Hyperbolic Equations* (Society for Industrial and Applied Mathematics, Philadelphia, 1982)

59. G.A. Williamson, B.D. McGranahan, B.A. Broughton, R.W. Deters, J.B. Brandt, M.S. Selig, *Summary of Low Speed Airfoil Data. Dept. of Aerospace Engineering*, University of Illionois at Urbana-Champaign, vol. 5 (2007)
60. H.C. Yee, R.K. Sweby, Dynamical approach study of spurious steady-state numerical solutions for nonlinear differential equations, part II: global asymptotic behaviour of time discretizations. Int. J. Comput. Fluid Dynam. **4**, 219–83 (1995)
61. Y. Zhang, Y. Jia, S.S.Y. Wang, 2D nearly orthogonal mesh generation with controls on distortion function. J. Comput. Phys. **218**, 549–71 (2006)
62. Q. Zhou, Z. Yao, F. He, M. Shen, A new family of high-order compact upwind difference schemes with good spectral resolution. J. Comput. Phys. **227**(2), 1306–39 (2007)

# Chapter 5
# Computational Compressible Aerodynamics

## 5.1 Introduction

In the previous chapter, we kept our attention focused on incompressible flow problems in aerodynamics for low and high Reynolds numbers. One of the distinctive features of incompressible flows is that the orthogonal grid formulation takes more compact form and hence, produces more accurate results, as compared to results obtained from non-orthogonal grid or unstructured grid with different formulations. It has been shown in Hoffmann and Chiang [41] and Sengupta [93] that for compressible flows, there are no distinct differences in formulations for orthogonal and non-orthogonal grids. However, the accuracy of solution will still be better if one uses orthogonal grids. Hence, in this chapter while we discuss about compressible aerodynamics, we will prefer the usage of orthogonal grids, as may be obtained by methods given in Chap. 4.

In this chapter, we will first of all show the differences between the results obtained for transitional flow past an airfoil at low speed, which will be obtained from both incompressible and compressible Navier–Stokes equations (NSE). This will be followed by obtaining time-accurate solutions for flow past airfoils at transonic Mach numbers, with shock present in the solution. Finally, we will show results obtained for deep dynamic stall past airfoils using compressible NSE.

### Navier–Stokes Equation in the Transformed Plane

Here, we provide briefly, derivation of the steps for the governing conservation of mass, momentum, and energy equations, termed as the Navier–Stokes equation. For details, readers are referred to the texts by Hoffmann and Chiang [41] and Sengupta [93]. In the Cartesian frame, the Navier–Stokes equation is given by the non-dimensional form as

$$\frac{\partial Q}{\partial t} + \frac{\partial E}{\partial x} + \frac{\partial F}{\partial y} + \frac{\partial G}{\partial z} = \frac{1}{Re}\left\{\frac{\partial E_v}{\partial x} + \frac{\partial F_v}{\partial y} + \frac{\partial G_v}{\partial z}\right\} \quad (5.1)$$

© Springer Nature Singapore Pte Ltd. 2020
T. K. Sengupta and Y. G. Bhumkar, *Computational Aerodynamics and Aeroacoustics*,
https://doi.org/10.1007/978-981-15-4284-8_5

where the state vector and non-dimensional flux terms are given in terms of density, total internal energy, pressure, and velocity components in the Cartesian frame as

$$Q = \lfloor \rho \ \rho u \ \rho v \ \rho w \ \rho e_t \rfloor^T$$

$$E = \lfloor \rho u \ (\rho u^2 + p) \ \rho u v \ \rho u w \ (\rho e_t + p)u \rfloor^T$$

$$F = \lfloor \rho v \ \rho u v \ (\rho v^2 + p) \ \rho v w \ (\rho e_t + p)v \rfloor^T$$

$$G = \lfloor \rho w \ \rho u w \ \rho v w \ (\rho w^2 + p) \ (\rho e_t + p)w \rfloor^T$$

$$E_v = \lfloor 0 \ \tau_{xx} \ \tau_{xy} \ \tau_{xz} \ \tilde{R} \rfloor^T$$

$$F_v = \lfloor 0 \ \tau_{xy} \ \tau_{yy} \ \tau_{yz} \ \tilde{S} \rfloor^T$$

$$G_v = \lfloor 0 \ \tau_{xz} \ \tau_{yz} \ \tau_{zz} \ \tilde{T} \rfloor^T$$

$$\tilde{R} = u\tau_{xx} + v\tau_{xy} + w\tau_{xz} + \frac{\kappa}{Pr(\gamma - 1)} \frac{\partial}{\partial x}(a^2)$$

$$\tilde{S} = u\tau_{xy} + v\tau_{yy} + w\tau_{yz} + \frac{\kappa}{Pr(\gamma - 1)} \frac{\partial}{\partial y}(a^2)$$

$$\tilde{T} = u\tau_{xz} + v\tau_{zy} + w\tau_{zz} + \frac{\kappa}{Pr(\gamma - 1)} \frac{\partial}{\partial z}(a^2)$$

The non-dimensional parameters, Reynolds number $(Re_\infty)$, Prandtl number $(Pr)$, and Mach number $(M)$ are defined as

$$Re_\infty = \frac{\rho_\infty U_\infty L}{\mu_\infty}; \quad Pr = \frac{\mu C_p}{\kappa}; \quad M_\infty = \frac{U_\infty}{a_\infty}$$

where $\kappa$ is the coefficient of thermal conductivity; $\gamma = c_p/c_v$, the ratio of specific heats which for air is given as 1.4; $a = \sqrt{\gamma R T}$ is the speed of sound, with the universal gas constant is used as $R = 287 J/(kg K)$, and $e_t = e_{int} + (u^2 + v^2 + w^2)/2$ is the total internal energy. The internal energy is given by $e_{int} = c_v T$.

For accurate implementation of boundary conditions, one solves the NSE by transforming the equations of motion from the physical to the transformed plane. Thus, the physical space $(x, y, z)$ is mapped to the computational space $(\xi, \eta, \zeta)$ by the following relations:

$$\tau = t \qquad\qquad\qquad\qquad (5.2a)$$

$$\xi = \xi(t, x, y, z) \qquad\qquad\qquad\qquad (5.2b)$$

$$\eta = \eta(t, x, y, z) \tag{5.2c}$$

$$\zeta = \zeta(t, x, y, z) \tag{5.2d}$$

These transformation relations can be numerically evaluated and for time-dependent deforming bodies, one would require to evaluate the following:

$$\xi_t = J\left[x_\tau(y_\zeta z_\eta - y_\eta z_\zeta) + y_\tau(x_\eta z_\zeta - x_\zeta z_\eta) + z_\tau(x_\zeta y_\eta - x_\eta y_\zeta)\right]$$

$$\eta_t = J\left[x_\tau(y_\eta z_\zeta - y_\zeta z_\xi) + y_\tau(x_\zeta z_\xi - x_\xi z_\zeta) + z_\tau(x_\xi y_\zeta - x_\zeta y_\xi)\right]$$

$$\zeta_t = J\left[x_\tau(y_\eta z_\xi - y_\xi z_\eta) + y_\tau(x_\xi z_\eta - x_\eta z_\xi) + z_\tau(x_\eta y_\xi - x_\xi y_\eta)\right]$$

where $J$ is the Jacobian of the transformation defined by

$$J = \frac{\partial(\xi, \eta, \zeta)}{\partial(x, y, z)} = \frac{1}{x_\xi(y_\eta z_\zeta - y_\zeta z_\eta) - x_\eta(y_\xi z_\zeta - y_\zeta z_\xi) + x_\zeta(y_\xi z_\eta - y_\eta z_\xi)} \tag{5.3}$$

The Cartesian derivatives are replaced by the grid metrics of the transformation given by Eqs. (5.2a) through (5.2d) to yield strong conservation law form as

$$\{\partial_\tau \hat{Q} + \partial_\xi \hat{E} + \partial_\eta \hat{F} + \partial_\eta \hat{G}\} = \frac{1}{Re}\{\partial_\xi(\hat{E}_v) + \partial_\eta(\hat{F}_v) + \partial_\zeta(\hat{G}_v)\} \tag{5.4}$$

where,

$$\hat{Q} = \frac{Q}{J}$$

$$\hat{E} = \frac{1}{J}\lfloor \rho U \quad \rho u U + \xi_x p \quad \rho v U + \xi_y p \quad \rho w U + \xi_z p \quad (\rho e_t + p)U - \xi_t p\rfloor^T$$

$$\hat{F} = \frac{1}{J}\lfloor \rho V \quad \rho u V + \eta_x p \quad \rho v V + \eta_y p \quad \rho w V + \eta_z p \quad (\rho e_t + p)V - \eta_t p\rfloor^T$$

$$\hat{G} = \frac{1}{J}\lfloor \rho W \quad \rho u W + \zeta_x p \quad \rho v W + \zeta_y p \quad \rho w W + \zeta_z p \quad (\rho e_t + p)W - \zeta_t p\rfloor^T$$

$$\hat{E}_v = \frac{1}{J}(\xi_x E_v + \xi_y F_v + \xi_x G_v) = \frac{1}{J}\lfloor 0 \quad (\xi_x \tau_{xx} + \xi_y \tau_{xy} + \xi_z \tau_{xz})$$

$$(\xi_x \tau_{xy} + \xi_y \tau_{yy} + \xi_z \tau_{zy}) \quad (\xi_x \tau_{xz} + \xi_y \tau_{yz} + \xi_z \tau_{zz}) \quad (\xi_x \tilde{R} + \xi_y \tilde{S} + \xi_z \tilde{T})\rfloor^T$$

$$\hat{F}_v = \frac{1}{J}(\eta_x E_v + \eta_y F_v + \eta_z G_v) = \frac{1}{J} \lfloor 0 \ (\eta_x \tau_{xx} + \eta_y \tau_{yx} + \eta_z \tau_{zx})$$

$$(\eta_x \tau_{xy} + \eta_y \tau_{yy} + \eta_z \tau_{zy}) \ (\eta_x \tau_{xz} + \eta_y \tau_{yz} + \eta_z \tau_{zz}) \ (\eta_x \tilde{R} + \eta_y \tilde{S} + \eta_z \tilde{T})\rfloor^T$$

$$\hat{G}_v = \frac{1}{J} \ (\zeta_x E_v + \zeta_y F_v + \zeta_z G_v)$$

$$= \frac{1}{J} \lfloor 0 \ (\zeta_x \tau_{xx} + \zeta_y \tau_{yx} + \zeta_z \tau_{zx}) \ (\zeta_x \tau_{xy} + \zeta_y \tau_{yy} + \zeta_z \tau_{zy})$$

$$(\zeta_x \tau_{xz} + \zeta_y \tau_{yz} + \zeta_z \tau_{zz}) \ (\zeta_x \tilde{R} + \zeta_y \tilde{S} + \zeta_x \tilde{T})\rfloor^T$$

The contravariant components of velocity $(U, V, W)$ are defined for the transformed plane as

$$U = \xi_t + \xi_x u + \xi_y v + \xi_z w; \ V = \eta_t + \eta_x u + \eta_y v + \eta_z w; \ W = \zeta_t + \zeta_x u + \zeta_y v + \zeta_z w$$

Although we have shown the compressible 3D NSE in the transformed plane, in the following some applications are shown for 2D flow past some typical aerofoils.

## 5.2 Governing Equations, Numerical Methods for 2D Compressible Flow Past Airfoils

The 2D, unsteady, compressible NSE are a set of four coupled, nonlinear PDEs. These are solved in the divergence (also called as strong conservation) form, in a finite difference framework. The divergence formulation is preferred, as it effectively captures correct jumps at discontinuities in the solution [50].

The 2D, unsteady, compressible NSE without body force is written in vector form in the physical plane as

$$\frac{\partial \hat{Q}}{\partial t} + \frac{\partial \hat{E}}{\partial x} + \frac{\partial \hat{F}}{\partial y} = \frac{\partial \hat{E}_v}{\partial x} + \frac{\partial \hat{F}_v}{\partial y} \tag{5.5}$$

where the conservative variables are given by

$$Q = \lfloor \rho \ \rho u \ \rho v \ \rho e_t \rfloor^T$$

$$E = \lfloor \rho u \ \rho u^2 + p \ \rho u v \ (\rho e_t + p)u \rfloor^T$$

$$F = \lfloor \rho v \ \rho u v \ \rho v^2 + p \ (\rho e_t + p)v \rfloor^T$$

and the viscous flux vectors $\hat{E}_v$ and $\hat{F}_v$ are given as

$$E_v = \lfloor 0 \; \tau_{xx} \; \tau_{xy} \; \tilde{R} \rfloor^T$$

$$F_v = \lfloor 0 \; \tau_{xy} \; \tau_{yy} \; \tilde{S} \rfloor^T$$

In Eq. (5.5), the variables $\rho$, $p$, $u$, $v$, $T$, and $e_t$ represent the density, the fluid pressure, Cartesian components of fluid velocity, the absolute temperature, and the specific internal energy of the fluid, respectively, and $\tau_{xx}$, $\tau_{xy}$, $\tau_{yx}$, and $\tau_{yy}$ are the components of the symmetric viscous stress tensor and are related to the gradients of velocity as

$$\tau_{xx} = \frac{1}{Re_\infty} \left( 2\mu \frac{\partial u}{\partial x} + \lambda \nabla.V \right)$$

$$\tau_{yy} = \frac{1}{Re_\infty} \left( 2\mu \frac{\partial v}{\partial y} + \lambda \nabla.V \right)$$

$$\tau_{xy} = \tau_{yx} = \frac{\mu}{Re_\infty} \left( \frac{\partial u}{\partial y} + \frac{\partial v}{\partial x} \right)$$

The heat conduction terms $q_x$ and $q_y$ are given by

$$q_x = \kappa \frac{\partial T}{\partial x} \text{ and } q_y = \kappa \frac{\partial T}{\partial y}$$

Additional information for Eq. (5.5) is provided by the Sutherland's law relating coefficient of viscosity $\mu$, with $T$ by

$$\mu = \frac{1.45 \times 10^{-6} T^{\frac{3}{2}}}{T + 110}$$

The closure of system of equations is provided by the ideal gas equation given by $p = \rho R T$, and is used to define $e_t$ as, $e_t = C_v T + \frac{(u^2 + v^2)}{2}$, where $C_v$ is the specific heat at constant volume and is given by

$$C_v = \frac{R}{\gamma - 1}$$

Here, the governing equations are solved on a body-conforming grid, transforming the governing equations into $(\xi\text{-}\eta)$ coordinate system with the following motivations. Also, in order to capture shocks and discontinuities in the solution, the equations of motion in the transformed plane are written in the conservative form. Here, the equations are transformed from the $(x, y)$-plane to the computational $(\xi, \eta)$-plane, using the following transformation:

$$\xi = \xi(x, \ y) \ \text{and} \ \eta = \eta(x, \ y)$$

The governing equations are transformed to the computational plane as

$$\frac{\partial Q}{\partial t} + \frac{\partial E}{\partial \xi} + \frac{\partial F}{\partial \eta} = \frac{\partial E_v}{\partial \xi} + \frac{\partial F_v}{\partial \eta} \tag{5.6}$$

where corresponding conservative variables and flux variables are given as $Q = \hat{Q}/J$, and

$$E = \frac{1}{J} \begin{bmatrix} \xi_x(\rho u) + \xi_y(\rho v) \\ \xi_x(\rho u^2 + p) + \xi_y(\rho uv) \\ \xi_x(\rho uv) + \xi_y(\rho v^2 + p) \\ \xi_x u(\rho e_t + p) + \xi_y v(\rho e_t + p) \end{bmatrix}$$

$$F = \frac{1}{J} \begin{bmatrix} \eta_x(\rho u) + \eta_y(\rho v) \\ \eta_x(\rho u^2 + p) + \eta_y(\rho uv) \\ \eta_x(\rho uv) + \eta_y(\rho v^2 + p) \\ \eta_x u(\rho e_t + p) + \eta_y v(\rho e_t + p) \end{bmatrix}$$

$$E_v = \frac{1}{J} \begin{bmatrix} 0 \\ \xi_x \tau_{xx} + \xi_y \tau_{xy} \\ \xi_x \tau_{xy} + \xi_y \tau_{yy} \\ \xi_x(u\tau_{xx} + v\tau_{xy} - q_x) + \xi_y(u\tau_{yx} + v\tau_{yy} - q_y) \end{bmatrix}$$

$$F_v = \frac{1}{J} \begin{bmatrix} 0 \\ \eta_x \tau_{xx} + \eta_y \tau_{xy} \\ \eta_x \tau_{xy} + \eta_y \tau_{yy} \\ \eta_x(u\tau_{xx} + v\tau_{xy} - q_x) + \eta_y(u\tau_{yx} + v\tau_{yy} - q_y) \end{bmatrix}$$

The grid metrics of the transformation are defined as

$$\xi_x = J y_\eta; \ \xi_y = -J x_\eta; \ \eta_x = -J y_\xi; \ \text{and} \ \eta_y = J x_\xi$$

with the Jacobian of transformation $J$ given by

$$J = \frac{1}{(x_\xi y_\eta - x_\eta y_\xi)}$$

The values of $x_\xi$, $x_\eta$, $y_\xi$, and $y_\eta$ are calculated numerically using finite difference formula.

## 5.2.1 Non-dimensionalization of Governing Equations

The governing equations of fluid motion are non-dimensionalized to obtain certain objectives. First, non-dimensionalization provides conditions upon which kinematic and dynamic similarities can be obtained for geometrically similar situations. The independent and dependent variables are non-dimensionalized as

$$t^* = \frac{tU_\infty}{L}, \qquad \xi^* = \frac{\xi}{L}, \eta^* = \frac{\eta}{L}$$

$$\mu^* = \frac{\mu}{\mu_\infty}, u^* = \frac{u}{U_\infty}, v^* = \frac{v}{U_\infty}$$

$$\rho^* = \frac{\rho}{\rho_\infty}, T^* = \frac{T}{T_\infty}, p^* = \frac{p}{\rho_\infty U_\infty^2}, e_t^* = \frac{e_t}{U_\infty^2}$$

$L, U_\infty, T_\infty, \rho_\infty$, and $\mu_\infty$ are the characteristic length, velocity, temperature, density, and viscosity, respectively, and the starred quantities refer to the non-dimensional variables. Here, the characteristic length is taken as the airfoil chord $(c)$, and the rest of the characteristic quantities are taken as free-stream values.

For the sake of clarity, non-dimensionalized quantities are expressed without the asterisks. The non-dimensionalized governing equations in the computational space are obtained as

$$\frac{\partial Q}{\partial t} + \frac{\partial E}{\partial \xi} + \frac{\partial F}{\partial \eta} = \frac{\partial E_v}{\partial \xi} + \frac{\partial F_v}{\partial \eta} \tag{5.7}$$

The viscous shear stress components in the transformed plane are given by

$$\tau_{xx} = \frac{\mu}{Re_\infty} \left[ \frac{4}{3} \left( \xi_x u_\xi + \eta_x u_\eta \right) - \frac{2}{3} \left( \xi_y v_\xi + \eta_y v_\eta \right) \right]$$

$$\tau_{yy} = \frac{\mu}{Re_\infty} \left[ \frac{4}{3} \left( \xi_y v_\xi + \eta_y v_\eta \right) - \frac{2}{3} \left( \xi_x u_\xi + \eta_x u_\eta \right) \right]$$

$$\tau_{xy} = \tau_{yx} = \frac{\mu}{Re_\infty} \left[ \xi_y u_\xi + \eta_y u_\eta + \xi_x v_\xi + \eta_x v_\eta \right]$$

and the heat conduction terms in the transformed plane are given by

$$q_x = -\frac{\mu}{Pr Re_\infty (\gamma - 1) M_\infty^2} \left( \xi_x T_\xi + \eta_x T_\eta \right)$$

$$q_y = -\frac{\mu}{Pr Re_\infty (\gamma - 1) M_\infty^2} \left( \xi_y T_\xi + \eta_y T_\eta \right)$$

## 5.2.2 Auxiliary Conditions

The initial and boundary conditions are collectively known as the auxiliary conditions and are generally classified as Dirichlet, Neumann, and Robin conditions.

### 5.2.2.1 Initial Conditions

Although in real flows the initial conditions can take complicated descriptions, free-stream initial conditions are used here, which is a simple idealization. In implementing free-stream initial conditions, all the flow properties at all the points at $t = 0$ are initialized with their respective free-stream values, except no-slip wall condition is applied for $j = 1$. Physically, this would mean that the airfoil is introduced into the free-stream flow impulsively.

### 5.2.2.2 Boundary Conditions

Some of the flow properties on the boundary of the domain are unknowns and evolve with time. The values of these unknowns cannot be specified a priori, and these depend on the solution of the interior, as well as, on the information provided from the exterior. The most common boundary conditions in compressible flow calculations are (1) solid no-slip wall, (2) far-field boundary condition at inflow and outflow, and (3) periodic boundary condition.

**Solid, no-slip, and adiabatic wall boundary condition**: The assumption of no-slip wall boundary condition is widely accepted for viscous flows. The no-slip wall condition on any non-porous surface specifies that the relative velocity of the fluid with respect to the surface is zero

$$u - u_{wall} = 0; \quad \text{and} \quad v - v_{wall} = 0$$

Additional constraints on the fluid properties on the wall are brought in by specifying the surface temperature (isothermal wall condition) or by assuming that the heat flux through the wall is zero (adiabatic condition). Mathematically, this implies that

$$T_{wall} = T_w \text{ or } q_x = q_y = 0$$

No heat-transfer conditions imply that the gradient of temperature on the solid wall is zero and hence, for an orthogonal grid one obtains

$$T_{j=1} = T_{j=2}$$

where $u_{wall}$ and $v_{wall}$ are Cartesian components of velocity, and $T_w$ is the specified wall temperature. At a solid wall, the viscous condition is implemented by setting velocity components on the stationary wall to zero and then one gets

$$u_{j=1} = 0; \quad \text{and} \quad v_{j=1} = 0$$

At the wall, pressure can be specified by simple extrapolation which is equivalent to requirement that the normal gradient at the boundary to vanish.

**Periodic boundary condition**: The periodic boundary can represent an artificial or not a physical boundary condition. These types of boundary conditions are generally found naturally in O- and C-type grids, representing a line (plane in 3D) composed of grid points with different computational coordinate(s) but the same physical location. This means that the grid is folded such that it touches itself. The periodic boundary condition is implemented by enforcing the flow variables and all of its derivatives to be continuous across the boundary. In rotary machines, like gas turbines, the flow variables may be assumed as periodic, for the ease of computations.

**Far-field boundary condition**: Computations of external flow past bodies have to be conducted on a finite domain. Therefore, one must specify artificial, far-field boundary conditions. These must satisfy the following. First, the truncation of the domain should not affect the flow solution. Secondly, any outgoing disturbances should not be reflected back into the domain. Due to their elliptic nature, subsonic and transonic flow problems are particularly sensitive to the far-field boundary conditions. An inadequate implementation can lead to a significant slow down of convergence to the steady state and can lead to erroneous results. Furthermore, the accuracy of the solution is likely to be adversely affected. Here, characteristics-based boundary condition is used to create far-field boundary condition. The mathematical details are provided next. At any instant of time, the far-field boundary condition can only be any one of the following: a subsonic inflow or a supersonic inflow, a subsonic outflow, or a supersonic outflow. Thus, the boundary conditions at the inflow /outflow can be determined based on the characteristic variables. The inflow and outflow boundary conditions are applied at the boundary points, depending upon the sign of normal component of velocity ($V_n$). For orthogonal grid, the velocity component is same as the contravariant component of velocity ($\eta_x u + \eta_y v$), in the transformed plane. In this case, local one dimensionality is assumed, and Riemann invariants are used. In the present work, the far-field boundary is specified on $\eta = \eta_{max}$. The components of velocity, normal, and tangent to this boundary are given by

$$V_n = \frac{\eta_x u + \eta_y v}{\sqrt{\eta_x^2 + \eta_y^2}}$$

$$V_t = \frac{\eta_y u - \eta_x v}{\sqrt{\eta_x^2 + \eta_y^2}}$$

Furthermore, the Riemann invariants are defined as [41]

$$R^+ = V_n + \frac{2a}{\gamma - 1} \tag{5.8}$$

$$R^- = V_n - \frac{2a}{\gamma - 1} \tag{5.9}$$

and the characteristic velocities are the eigenvalues of the flux Jacobian matrices given as

$$\lambda_1 = V_n + a \tag{5.10}$$

$$\lambda_2 = V_n - a \tag{5.11}$$

where $a$ is the speed of sound. The procedure to calculate the far-field flow properties is as follows.

**Inflow boundary**: A boundary is treated as an inflow boundary, if $V_n < 0$. Consequently, $\lambda_2$ is also negative and $\lambda_1$ can be either positive or negative. If $\lambda_1$ is negative, it implies that the boundary is a supersonic inflow, and all waves are traveling into the computational domain. Hence, the Riemann invariants are set to their free-stream values

$$R^+ = R_\infty^+, \quad R^- = R_\infty^-, \quad s = s_\infty \quad and \quad V_t = V_{t\infty}$$

If $\lambda_1$ is positive (as in case of subsonic inflow), the Riemann invariant corresponding to the outgoing wave is extrapolated from the interior, and $R^-$, $s$ and $V_{t\infty}$ are set to the free-stream values.

$$R^- = R_\infty^-, \quad s = s_\infty \quad and \quad V_t = V_{t\infty}$$

**Outflow boundary**: A boundary is treated as an outflow boundary, if $V_n > 0$. Hence, $\lambda_1$ is always positive and $\lambda_2$ can be either positive or negative. If $\lambda_2$ is positive (in case of supersonic outflow), then all the Riemann variables are extrapolated from interior of the domain. When $\lambda_2$ is negative, then $R^-$ is set to its free-stream value, that is, $R^- = R_\infty^-$, and the remaining variables are extrapolated from the interior.

Once all the Riemann invariants at the far-field boundary are calculated, the flow properties on the boundary can be calculated as follows:

$$V_n = \frac{R^+ + R^-}{2}$$

and

$$a = \frac{1}{4} (\gamma - 1) \left( R^+ - R^- \right)$$

In addition, the speed of sound is given by

$$a^2 = \frac{\gamma p}{\rho}$$

From the above two equations, one obtains

$$\frac{p}{\rho} = \frac{1}{\gamma} \left[ \frac{1}{4} (\gamma - 1) \left( R^+ - R^- \right) \right]^2$$

Furthermore,

$$\log \left( \frac{p}{\rho^\gamma} \right) = s$$

From the above, one can determine the pressure and density on the boundary. Once the pressure and density are determined, the value of the temperature can be obtained using the equation of state.

Having discussed the governing equation and the auxiliary conditions for compressible NSE, we can evaluate the difference between incompressible and compressible formulations for flow past WTEA-TE1 airfoil at a relatively low Mach number flow for $M_\infty = 0.1815$. Even for the compressible flow formulations, we compare two sets of solutions obtained for time-accurate DNS results with the time-averaged, RANS formulation, also known as the Reynolds-averaged NSE.

### 5.2.3  Geometry and Computational Domain

The SHM-1 airfoil has a maximum thickness of 15% and exhibits two regions of concavity on the top and bottom surface of the airfoil, near the trailing edge, as shown in Fig. 5.1a. The surface of NACA 0012 airfoil is shown in Fig. 5.1b which has relatively simple and symmetric geometry. The chord length of the airfoil $(c)$ is usually taken as the length scale, and hence the non-dimensional chord will be unity.

For numerical simulation of compressible external flows, a large computational domain is preferred to ensure the accurate implementation of the far-field boundary condition and to reduce the acoustic wave reflections from the boundary. Therefore, a large computational domain around the airfoil (of approximately $10c$) is selected and is shown in Fig. 5.2.

Grid is clustered near the wall so that enough resolution is obtained to capture the boundary layer dynamics. In the remaining part of grid, lesser number of points are used, where flow displays some sort of steadiness. A tangent hyperbolic function [92] used for stretching the azimuthal grid lines, which determines the distance between successive $\eta = $ constant lines $(\Delta S_\eta = h_{22} \Delta \eta)$ in the wall-normal direction, is prescribed by the following distribution:

$$h_{22} = H \left[ 1 - \frac{tanh \left[ \beta (1 - \eta) \right]}{tanh (\beta)} \right]$$

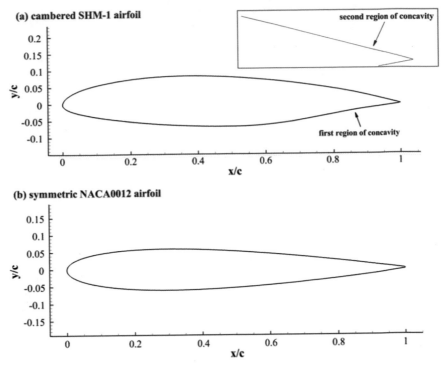

**Fig. 5.1** Profiles of **a** SHM-1 and **b** NACA 0012 airfoils

where $\beta$ is the clustering parameter and $H$ is the non-dimensional distance of the outer boundary in terms of $c$.

Grids around the SHM-1 and NACA 0012 airfoil are generated for demonstration. The grid around SHM-1 airfoil contains 663 points in the $\xi$-direction and 450 points in the $\eta$-direction. The orthogonal grid generated around the SHM-1 airfoil is shown in Fig. 5.3. Following the similar procedure, orthogonal grid is generated around the symmetrical NACA 0012 airfoil, which contains 636 points in the $\xi$-direction and 350 points in the $\eta$-direction. Close-up views of orthogonal grids around NACA 0012 and SHM-1 airfoils are shown in Fig. 5.4a and b, respectively. The near-wall resolution for grids around NACA 0012 and SHM-1 airfoils are given by $\Delta s_\eta = 0.006652$ and $\Delta s_\eta = 0.00139$, respectively. Also, dense grid (with 2658 points in $\xi$- and 360 points in $\eta$-direction) is generated around NACA 0012 airfoil to simulate high Mach number (greater than drag divergence Mach number) flows, in order to capture the flow features correctly. The near-wall resolution for the dense grid is $\Delta s_\eta = 0.003175$.

While solving Eq. (5.7), one decouples the space and time derivative terms. The space derivatives are calculated first, followed by the time integration. In the presented work [94], different numerical schemes are used for evaluating each of the derivative terms: High-resolution, high-order optimized upwind compact schemes (OUCS) are used for convective flux derivative calculation, second-order central difference (CD2)

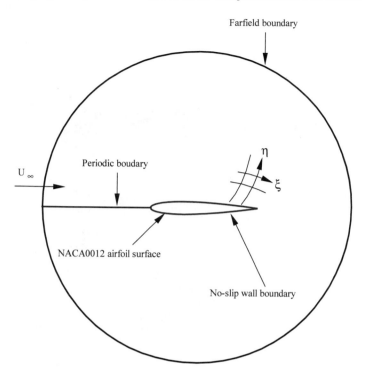

**Fig. 5.2**   Sketch of computational domain and boundary conditions for studying compressible flow past NACA 0012 airfoil [Reproduced from Sengupta et al. *Computers and Fluids*, **88**, 19–37 (2013), with the permission of Elsevier Publishing.]

scheme for viscous flux derivative calculation and classical, four-stage Runge–Kutta method for time integration. Adaptive pressure-based, artificial diffusion terms due to Jameson, Schmidt, and Turkel [45] are added to the solution to damp higher wavenumbers which arise due to nonlinearities and discontinuities.

## 5.2.4   Optimized Upwind Compact Schemes

Compact schemes are implicit and have superior stability and DRP properties [92]. In the presented work [94], high accuracy optimized DRP upwind schemes—OUCS schemes and their symmetrized versions, S-OUCS [25] schemes, have been used to evaluate the convective flux terms in $\eta$- and $\xi$-directions, respectively. These schemes are obtained by carrying out optimization in the spectral plane [93, 102], therefore ensuring the higher resolution. The procedure of full- domain spectral analysis of any numerical scheme for the DRP properties is provided in [92, 93, 101, 105].

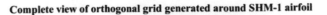

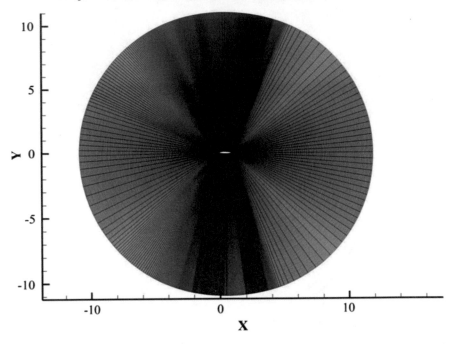

**Fig. 5.3** Orthogonal grid generated around SHM-1 airfoil

The interior stencil of OUCS2 scheme [102] used to evaluate first derivative $(u')$ is given by

$$b_1 u'_{j-1} + b_2 u'_j + b_3 u'_{j+1} = \frac{1}{h} \sum_{k=-2}^{2} (a_k u_{j+k}) \qquad (5.12)$$

where $b_1 = \frac{b_2}{3} - \frac{\alpha_1}{12}, b_3 = \frac{b_2}{3} + \frac{\alpha_1}{12}; a_{\pm 2} = \pm \frac{b_2}{36} + \frac{\alpha_1}{72}; a_{\pm 1} = \pm \frac{7b_2}{9} + \frac{\alpha_1}{9}, b_2 = 36,$
$a_0 = -\frac{\alpha_1}{4}$ and $h$ is the interval of equi-spaced grid. The parameter $\alpha_1$ is the coefficient of $(\frac{\partial^6 u}{\partial x^6})$, which is treated as the free parameter to fix the order of representation [92]. The interior stencil of the OUCS2 is a fifth-order accurate upwind scheme for $\alpha_1 < 0$ and a sixth-order accurate central scheme results for $\alpha_1 = 0$.

Another compact scheme is also used the OUCS3 scheme, whose interior stencil is given by

$$m_{j-1} u'_{j-1} + u'_j + m_{j+1} u'_{j+1} = \frac{1}{h} \sum_{k=-2}^{2} q_k u_{j+k} \qquad (5.13)$$

where $m_{j\pm 1} = D \pm \frac{\zeta}{60}; q_{\pm 2} = \pm \frac{F}{4} + \frac{\zeta}{300}; q_{\pm 1} = \pm \frac{E}{2} + \frac{\zeta}{30}$ and $q_0 = -\frac{11\zeta}{150}$ with $D = 0.3793894912, E = 1.57557379$, and $F = 0.183205192$. The parameter $\zeta$ is

**(a)**

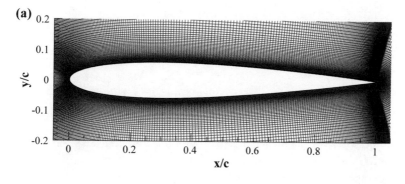

**(b)**

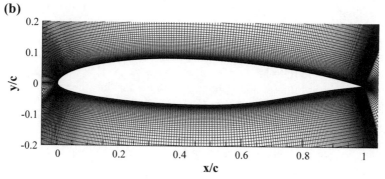

**Fig. 5.4** Close-up view of grid generated around **a** SHM-1 and **b** NACA 0012 airfoils [Reproduced from Sengupta et al., *Computers and Fluids*, **88**, 19–37 (2013), with the permission of Elsevier Publishing.]

the coefficient of $(\frac{\partial^4 u}{\partial x^4})$ which is treated as the free parameter to fix the order of representation [92]. The interior stencil of OUCS3 scheme is formally second-order accurate, as noted by Taylor's series approximation.

While the interior stencils of OUCS2 and OUCS3 schemes, given by Eqs. (5.12) and (5.13), are sufficient for periodic problem, one-sided boundary closures are required for non-periodic problems. These boundary closures require special attention, as these have to deal with the boundary conditions. Anti-diffusion due to these boundary closures can affect the whole flow field through implicitness of OUCS schemes. Solution to this problem is provided in [92, 102], by adapting explicit stencils. At the boundaries, these explicit stencils are stable at all points [92]. For both the OUCS schemes, the following explicit closure schemes for the boundary and near boundary points are used:

$$u'_1 = \frac{-3u_1 + 4u_2 - u_3}{2h} \tag{5.14}$$

$$u_2' = \frac{1}{h}\left[\left(\frac{2\beta_2}{3} - \frac{1}{3}\right)u_1 - \left(\frac{8\beta_2}{3} + \frac{1}{2}\right)u_2 + \left(4\beta_2 + 1\right)u_3\right.$$
$$\left. - \left(\frac{8\beta_2}{3} + \frac{1}{6}\right)u_4 + \frac{2\beta_2}{3}u_5\right] \tag{5.15}$$

Similar closures can be written for $j = N$ and $(N - 1)$, with the coefficients on the right-hand side changing signs. The value of $\beta$ in Eq. (5.15) is chosen as 0.02 for $j = 2$ and 0.09 for $j = N - 1$, while studying the global property of the scheme in spectral plane. The interior stencils along with boundary closures form an implicit linear system of equations, with a tridiagonal coefficient matrix. This tridiagonal system of equations is solved using Thomas algorithm [94].

Upwinding creates directional bias, by affecting the waves traveling in the direction in which upwinding is applied and dissipating the waves traveling in the opposite direction. Also, as mentioned earlier, coefficients in the boundary closure schemes for $j = 2$ and $j = (N - 1)$ are different, whose effects lead to alteration of numerical properties on the right-hand side causing asymmetry. However, one would prefer schemes with minimum added numerical diffusion in a certain bandwidth of wavenumbers, so as to not to alter the physical diffusion properties of the governing equations. It is for this reason that the symmetrized compact scheme S-OUCS is applied in the $\xi$-direction. Symmetrized scheme removes the inherent bias due to upwinding, which is otherwise present in OUCS scheme. While implementing the symmetrized versions of respective schemes (S-OUCS), derivatives at all the nodes are calculated using OUCS schemes by traveling from $j = 1$ to $N$ and then again going from $N$ to 1. Symmetrized derivative is taken as the average of the derivatives obtained by following these two directions. Procedure for full-domain analysis of any numerical scheme for stability, DRP properties, is provided in [25, 101].

### 5.2.5  Evaluation of Viscous Flux Derivatives

Viscous flux terms represent isotropic diffusion phenomenon. Hence, for evaluation of the viscous flux terms and their derivatives, a second-order central discretization (CD2) scheme is used, preserving its isotropic nature. The viscous flux derivatives at the point $(i, j)$ are calculated as

$$\left(\frac{\partial E_v}{\partial \xi}\right)_{i,j} = \frac{\left(E_{v\,i+1,j} - E_{v\,i-1,j}\right)}{2\Delta\xi} \tag{5.16}$$

$$\left(\frac{\partial F_v}{\partial \eta}\right)_{i,j} = \frac{\left(F_{v\,i,j+1} - F_{v\,i,j-1}\right)}{2\Delta\eta} \tag{5.17}$$

The viscous stress terms are also evaluated similarly, using CD2 scheme.

### 5.2.6 Numerical Properties of OUCS2 and OUCS3 Schemes

For the analysis of space–time discretization schemes, the linear convection equation as a model is used, which represents many flows and wave phenomena, given in Eq. (3.95).

Using Fourier transform, the unknown at the $j$th node of uniformly spaced discrete grid of spacing $h$ can be represented as $u(x_j, t) = \int U(k, t)e^{ikx_j}dk$, and the exact spatial derivative at the same node is given by $\left[\frac{\partial u}{\partial x}\right]_{exact} = \int ikU(k, t)e^{ikx_j}dk$. The numerical spatial derivative $u'_j$ can be shown [92] as equivalent to

$$[u'_j]_{numerical} = \int ik_{eq}Ue^{ikx_j}dk \tag{5.18}$$

In general, $[k_{eq}]$ is complex, with a real part that represents the numerical phase and an imaginary part that represents the numerical dissipation added by the choice of the numerical method.

We can express left-hand side in terms of the nodal numerical amplification factor $(G_j)$ for the four-stage Runge–Kutta time integration scheme [102] as given in Eq. (3.14) by

$$G_j = G(x = x_j) = 1 - A_j + \frac{A_j^2}{2} - \frac{A_j^3}{6} + \frac{A_j^4}{24}$$

where $A_j = Nc\sum_{l=1}^{N} C_{jl}e^{ik(x_l-x_j)}$. If the initial condition for Eq. (3.95) is represented by

$$u(x_j, t = 0) = u_j^0 = \int A_0(k)e^{ikx_j} dk \tag{5.19}$$

then the general solution at any arbitrary point can be obtained as

$$u_j^n = \int A_0(k) \left[|G_j|\right]^n e^{i(kx_j - n\beta_j)}dk] \tag{5.20}$$

where $|G_j| = \sqrt{(G_rj^2 + G_ij^2)}$ and $tan(\beta_j) = -\frac{G_{ij}}{G_{rj}}$, with $G_{rj}$ and $G_{ij}$ are real and imaginary parts of $G_j$, respectively. The phase of the solution is determined by $n\beta_j = kc_Nt$, where $c_N$ is the numerical phase speed. The general numerical solution to Eq. (3.95) is denoted as

$$u_{numerical} = \int A_0(k) \left[|G_j|\right]^n e^{i(x - c_Nt)} dk \tag{5.21}$$

Having obtained $G_j$ at any node, one can obtain numerical phase speed and group velocity at the same node using the numerical dispersion relation $\omega_N = c_Nk$ [101] as follows

$$\left[\frac{c_N}{c}\right]_j = \frac{\beta_j}{\omega \Delta t} \tag{5.22}$$

$$\left[\frac{V_{gN}}{c}\right]_j = \frac{1}{hN_c}\frac{d\beta_j}{dk} \tag{5.23}$$

For the optimized three-stage Runge–Kutta (ORK3) time integration scheme, the numerical amplification factor is given by

$$G_j = 1 - A_j + bA_j^2 - aA_j^3 \tag{5.24}$$

where variations of $a$ and $b$ are given in Fig. 14.34 of [93] as a function of $N_c$.

Thus, $G_j$, $\frac{c_N}{c}$, and $\frac{V_{gN}}{c}$ at any node for any method can be obtained using Eqs. (3.109) and (5.23). These numerical properties for OUCS3, OUCS4, and S-OUCS4 schemes are already reported in [25, 101, 102]. Here, numerical properties for OUCS2 scheme are presented and compared with the properties of the OUCS3 scheme. Figures 5.5, 5.6, 5.7, 5.8, 5.9, and 5.10 show the $|G_j|$, $\frac{c_N}{c}$, and $\frac{V_{gN}}{c}$ contours for OUCS2 (for $\alpha_1 = 0$) and OUCS3 (for $\zeta = 0$) schemes plotted side by side for the sake of comparison. These results are shown for a domain with $N = 101$ equi-spaced points [112]. Figures 5.5 and 5.6 show $|G_j|$-contours plotted for near boundary and interior nodes. Comparing $|G_j|$-contours for $j = 51$ in Fig. 5.5, one notices that for OUCS2 scheme stable region is obtained for $N_c < 0.82$ whereas, for OUCS3 scheme, stable region is obtained for $N_c < 1.00$. From Fig. 5.10, it can be seen that positive values of $\frac{V_{gN}}{c}$ for interior stencil of OUCS2 scheme are obtained for $kh < 2.28$, whereas for interior stencil of OUCS3 scheme positive values of $\frac{V_{gN}}{c}$ are obtained for $kh < 2.43$. In Fig. 5.7, for OUCS3 scheme there is $\frac{c_N}{c} = 1.0$ contour for low $N_c$ values, which is also the case for OUCS2 schemes. Thus, OUCS2 and OUCS3 schemes show comparable properties. Though accuracy noted by Taylor's series approximation for OUCS2 central scheme is six and OUCS3 scheme is two, one gets comparable DRP properties for both the methods.

### 5.2.7  Artificial Diffusion Model of JST

Artificial diffusion terms are added to the numerically evaluated convective flux terms, to damp the high wavenumber waves arising from nonlinearities and discontinuities in the solution. Adaptive pressure-based, nonlinear artificial diffusion terms due to Jameson, Schmidt and Turkel (JST) [45] are added explicitly to accurately capture discontinuities. In JST scheme, the artificial diffusion term in $\xi$-direction is $D_\xi$ and in $\eta$-direction this is $D_\eta$ defined as

$$D_\xi = \sigma_\xi J^{-1}\{(d_{\xi 2}) - (d_{\xi 4})\}$$

$$D_\eta = \sigma_\eta J^{-1}\{(d_{\eta 2}) - (d_{\eta 4})\}$$

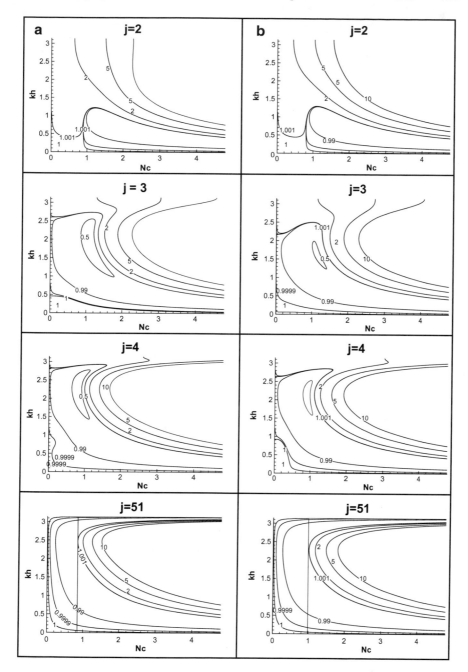

**Fig. 5.5** $|G|$ contours at indicated nodes (domain: $j = 1$ to $j = 101$) for **a** $OUCS3 + ORK_3$ on left with $\zeta = 0$, and **b** $OUCS2 + ORK_3$ on right with $\alpha_1 = 0$. Areas on the left of vertical lines in bottom two frames represent regions of numerical stability. Areas on the left of leftmost contour (labeled 1) in bottom two frames represent regions of neutral stability and should be used for numerical computations

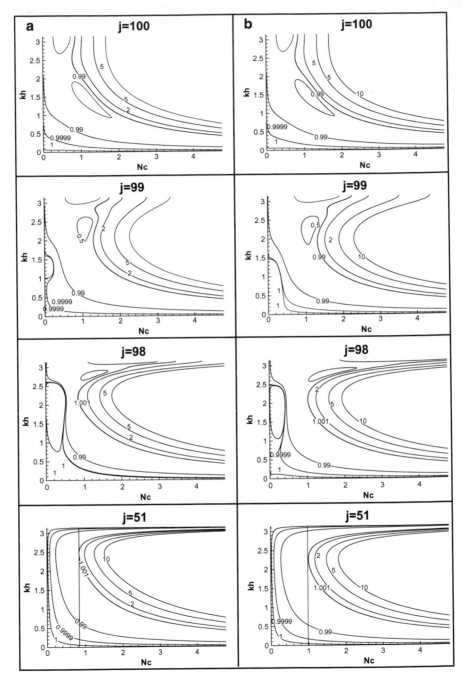

**Fig. 5.6** $|G|$ contours at indicated nodes (domain: $j = 1$ to $j = 101$) for **a** $OUCS3 + ORK_3$ on left with $\zeta = 0$, and **b** $OUCS2 + ORK_3$ on right with $\alpha_1 = 0$. Areas on the left of vertical lines in bottom two frames represent regions of numerical stability. Areas on the left of leftmost contour (labeled 1) in bottom two frames represent regions of neutral stability and should be used for numerical computations

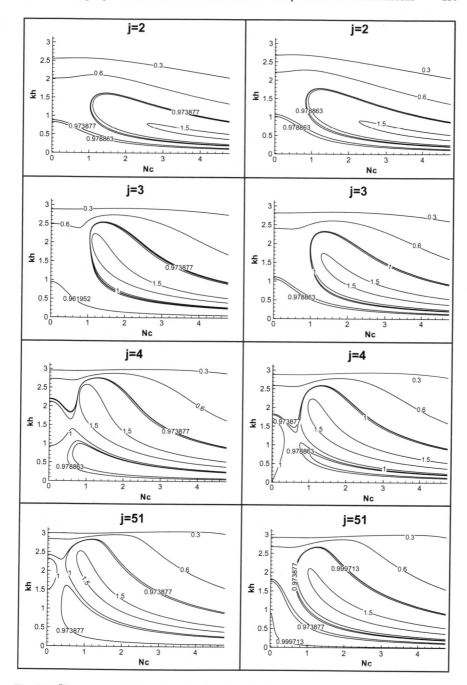

**Fig. 5.7**  $\frac{c_N}{c}$ contours at indicated nodes (domain: $j = 1$ to $j = 101$) for **a** $OUCS3 + ORK_3$ on left with $\zeta = 0$, and **b** $OUCS2 + ORK_3$ on right with $\alpha_1 = 0$

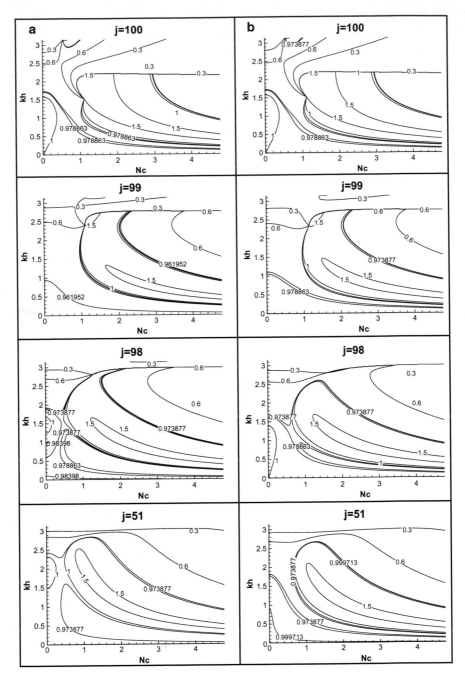

**Fig. 5.8** $\frac{c_N}{c}$ contours at indicated nodes (domain: $j = 1$ to $j = 101$) for **a** $OUCS3 + ORK_3$ on left with $\zeta = 0$, and **b** $OUCS2 + ORK_3$ on right with $\alpha_1 = 0$

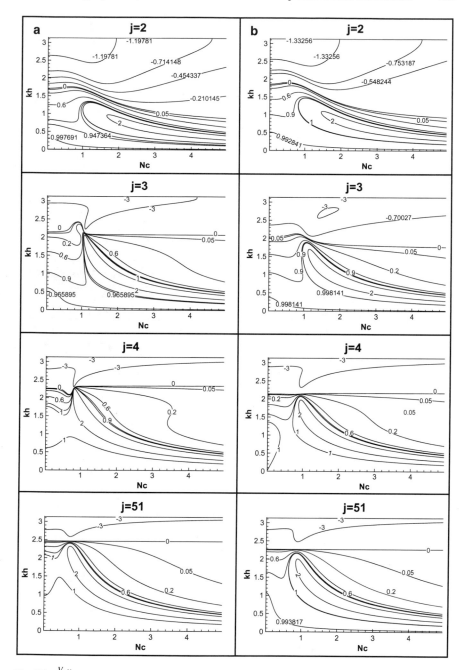

**Fig. 5.9** $\frac{V_{gN}}{c}$ contours at indicated nodes (domain: $j = 1$ to $j = 101$) for **a** $OUCS3 + ORK_3$ on left with $\zeta = 0$, and **b** $OUCS2 + ORK_3$ on right with $\alpha_1 = 0$

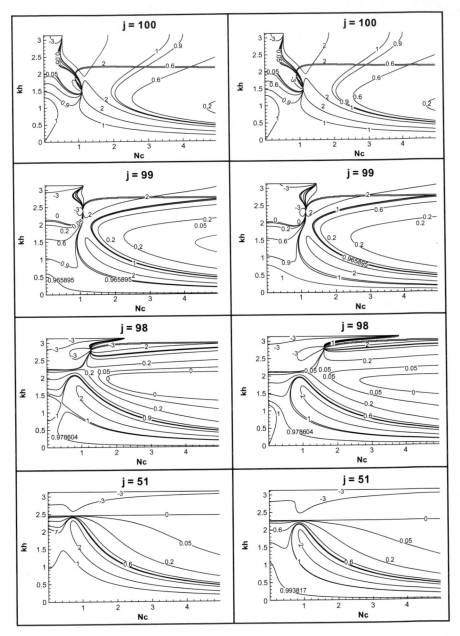

**Fig. 5.10** $\frac{V_{gN}}{c}$ contours at indicated nodes (domain: $j = 1$ to $j = 101$) for **a** $OUCS3 + ORK_3$ on left with $\zeta = 0$, and **b** $OUCS2 + ORK_3$ on right with $\alpha_1 = 0$

where $\sigma_\xi$ and $\sigma_\eta$ are the spectral radii of the flux Jacobian matrices [41], respectively, and the terms $d_{\xi 2}$, $d_{\eta 2}$, $d_{\xi 4}$, and $d_{\eta 4}$ are defined as

$$d_{\xi 2} = \epsilon_{2\xi} \nabla_\xi \Delta_\xi (J Q)$$

$$d_{\xi 4} = \epsilon_{4\xi} \left(\nabla_\xi \Delta_\xi\right)^2 (J Q)$$

$$d_{\eta 2} = \epsilon_{2\xi} \nabla_\eta \Delta_\eta (J Q)$$

$$d_{\eta 4} = \epsilon_{4\xi} \left(\nabla_\eta \Delta_\eta\right)^2 (J Q)$$

where $\epsilon_{2\xi}$, $\epsilon_{4\xi}$, $\epsilon_{2\eta}$, and $\epsilon_{4\eta}$ are defined at points $i$, $j$ in the computational domain as

$$\epsilon_{2\xi} = \kappa_2 \, max \left(\Upsilon_{i+1,j}, \Upsilon_{i,j}, \Upsilon_{i-1,j}\right)$$

$$\epsilon_{2\eta} = \kappa_2 \, max \left(\Upsilon_{i,j+1}, \Upsilon_{i,j}, \Upsilon_{i,j-1}\right)$$

$$\epsilon_{4\xi} = \kappa_4 \, max \left(0, \kappa_4 - \epsilon_{2\xi}\right)$$

$$\epsilon_{4\eta} = \kappa_4 \, max \left(0, \kappa_4 - \epsilon_{2\eta}\right)$$

where the values of $\kappa_2$ and $\kappa_4$ parameter are used as 0.25 and 0.15, respectively, for the transonic flow problem. The term $\Upsilon_{i,j}$ is defined at the point $(i, j)$ as

$$\Upsilon_{i,j} = \frac{|p_{i+1,j} - 2p_{i,j} + p_{i-1,j}|}{|p_{i+1,j}| + |2p_{i,j}| + |p_{i-1,j}|} \tag{5.25}$$

The term $\Upsilon_{i,j}$ is a measure of the second derivative of pressure and has very high value near a shock. Hence, it acts as a pressure switch activating the second difference terms, wherever there are discontinuities in pressure. Fourth difference terms are implicitly added in the solution and are switched to zero, near shocks. The diffusion terms calculated in this way are then subtracted from the numerically evaluated convective flux derivative terms, and these modified derivative terms are used in time integration.

We note in passing that JST model is used for its robustness with RANS procedure. In the present chapter, we show that this can be used easily with solvers written for ILES and DNS of flows, as it was shown in [94] for flow past airfoils with shock. At the same time, we would like to highlight that compact filters can also be used for solving compressible flows, as post-processing step in solution process. This will be exhibited with the solution for the problem of dynamical stall.

## 5.2.8  Application of Two-Dimensional Filters

Explicit two-dimensional filters introduced in [98] are presented next. Such filters perform the roles of band-limiting the unknowns as needed in ILES, and more importantly avoiding numerical instability often faced at high wavenumbers. This is aided by improved understanding of filters by spectral analyses [32, 34, 99, 102, 105]. The transfer function of the central, 1D filters obtained in the spectral plane, explains control of instabilities by these Padè filters. Such filters make use of solution of tridiagonal matrix equation for the filtered quantities, expressed as quantities with caret evaluated in terms of the unfiltered variables $u$ on the right-hand side of the equation given by

$$\alpha_1 \hat{u}_{j-1} + \hat{u}_j + \alpha_1 \hat{u}_{j+1} = \sum_{n=0}^{M} \frac{a_n}{2} (u_{j+n} + u_{j-n}) \qquad (5.26)$$

with $M$ defining the size of the stencil. One aspect is always wrongly stated in the literature and is about the so-called order of filters. The above equation clearly demonstrates that the filtered and unfiltered quantities do not even match at the lowest order. Thus, the talk of order of filters in literature is clearly a mistake [13, 32, 34, 88]. This legacy of error arises due to similarity of stencils of filters, with the stencils used for evaluating first derivative used in central differencing. For example, $M = 1$ is stated to provide a second-order filter, and successive increase of $M$ by one increases the order of the filter by two. In Eq. (5.26), $\alpha_1$ is a free parameter restricted between $-0.5$ and $+0.5$ to maintain diagonal dominance of the associated linear algebraic system. Although boundary filters have been proposed in [33, 34, 118] separately, without any coupling between these two sets of filter equations, the transfer function of the filters is provided by expressing Eq. (5.26) as $[A] \{\hat{u}_j\} = [B] \{u_j\}$ and obtaining it in the spectral plane following the methodology given in [93] as

$$T_j(k) = \frac{\sum_{l=1}^{N} b_{jl} \, e^{ik(x_l - x_j)}}{\sum_{l=1}^{N} a_{jl} \, e^{ik(x_l - x_j)}} \qquad (5.27)$$

The role of $T_j(k)$ in stabilizing an unstable method is noted by the alteration of the original amplification factor, at the $j$th node, from $G_j(k)$ to $|\hat{G}_j(k)| = |G_j(k)||T_j(k)|$. With a central filter, the role of $T_j(k)$ is to filter the unknowns differentially for different $k$. It has also been noted [34, 99] that filtering can be reduced in two ways for 1D filters: either by increasing the value of $M$ or increasing the value of $\alpha_1$ toward its limiting value of 0.5.

One-sided filters used in non-periodic problems [34] and analyzed [99], show such filters make the transfer function complex, thereby providing additional phase shift and causing numerical dispersion relation to change apart from attenuating the function. This aspect of filtering can affect computations adversely. At the same time, proper analysis also allows one to use filters constructively. For example, for most of the numerical methods in imposing the numerical dispersion relation, show an extreme form of dispersion error for a range of $kh$, for which waves move spuriously in opposite to physical direction. Existence of this is shown [101] using 1D convection equation and called as $q$-waves. The authors in [99] have introduced a spatial one-sided or upwind composite filter, and such filters allow one to add a controlled amount of dissipation, while changing the basic dispersion relation of the numerical scheme in such a way that for a small range of $N_c$ values, one prevents the formation of $q$-waves.

Despite advantages of such filters, it is preferable to use central filters, even when very high-order filters are needed. While filters with large $M$ have large stencil, these cannot be used at points close to the boundary, and an alternative was proposed [34] through the use of least order centered (LOC) filters. It is noted from Eq. (5.26) that one can use the lowest stencil size filter near the boundary, and one can progressively increase the stencil size of the filter, as one moves inward from the boundary. This is the essence of LOC filters, which do not alter the dispersion relation that arises due to use of the basic numerical method.

There is another aspect of filters that remained unattended for long in using 1D filters. This is related to directionality introduced in filtering by 1D filters used sequentially for multi-dimensional problems. This is circumvented here, by using 2D filters for multi-dimensional flows. Such multi-dimensional filters can be used in conjunction with JST model of diffusion, or completely used by itself. Some basics of 2D filters are provided next.

Despite advantages, 1D filters show strong directional behavior of the solution as noted in [99, 103] for some 2D flows. Filtering in azimuthal or wall-normal direction causes unphysical vortex smearing in the respective direction. This is removed, when one uses 2D filters instead, which has a general stencil given by

$$\hat{u}_{i,j} + \alpha_2(\hat{u}_{i-1,j} + \hat{u}_{i+1,j} + \hat{u}_{i,j-1} + \hat{u}_{i,j+1}) = \sum_{n=0}^{M} \frac{a_n}{2}(u_{i\pm n,j} + u_{i,j\pm n}) \qquad (5.28)$$

where $M$ has similar role as before and is equal to one for a lowest stencil size filter. Every increase of $M$ by one increases the fictitious "order" of the filter by two. For the lowest "order" 2D filter, $i$ and $j$ vary from 2 to $(N_i - 1)$ and 2 to $(N_j - 1)$, respectively, in $i$ and $j$ directions. For the "fourth-order" filter, $M = 2$ and Eq. (5.28) applies from 3 to $(N_i - 2)$ and 3 to $(N_j - 2)$, respectively. Similarly, higher "order" filters can be defined.

In Eq. (5.28), filtered quantities $\hat{u}$, on the left-hand side, are evaluated in terms of the unfiltered variables $u$, on the right-hand side. Filtering coefficient $\alpha_2$ and other coefficients $a_n$'s on the right-hand side determines the transfer function. Application of the filter amounts to solving a linear algebraic equation with the left-hand side giving rise to a pentadiagonal matrix. To solve this linear algebraic equation, one must have the diagonal dominance of the matrix on the left-hand side of Eq. (5.28) and is ensured by $|\alpha_2| \leq 1/4$.

Transfer function of the filter is defined in the $k$-plane, obtained by taking Fourier transform of both sides in Eq. (5.28). Thus, the transfer function of the 2D lowest "order" filter for the $(i, j)$th-node is given by

$$T_{ij}(k_i h_i, k_j h_j) = \frac{a_o + a_1[cos(k_i h_i) + cos(k_j h_j)]}{1 + 2\alpha_2[cos(k_i h_i) + cos(k_j h_j)]} \qquad (5.29)$$

In this expression, $h_i$ and $h_j$ are the uniform grid spacings in the physical plane in $i$ and $j$ directions, respectively. The coefficients $a_0$ and $a_1$ are obtained in terms of $\alpha_2$, by equating the Taylor series expansion on both sides of Eq. (5.28) about the $(i, j)$th point. The first relation is obtained by equating the coefficients of $u_{i,j}$ with $\hat{u}_{i,j}$, which gives the consistency condition as

$$a_0 + 2\,a_1 = 1 + 4\alpha_2 \qquad (5.30)$$

For the second relation, we set the condition on $T_{ij}$ to attenuate completely at the Nyquist limit $(k_i h_i = k_j h_j = \pi)$, i.e., $T_{ij}(k_i h_i = k_j h_j = \pi) = 0$. This gives the additional relation as

$$a_0 = 2a_1$$

Solving these two last equations, one gets $a_0$ and $a_1$ in terms of $\alpha_2$ as

$$a_0 = \frac{1}{2} + 2\alpha_2 \quad a_1 = \frac{1}{4} + \alpha_2$$

For $M = 2$ the transfer function is obtained as

$$T_{ij}(k_i h_i, k_j h_j) = \frac{a_o + a_1[cos(k_i h_i) + cos(k_j h_j)] + a_2[cos(2k_i h_i) + cos(2k_j h_j)]}{1 + 2\alpha_2[cos(k_i h_i) + cos(k_j h_j)]}$$

$$(5.31)$$

And the consistency condition is given by

$$a_0 + 2a_1 + 2a_2 = 1 + 4\alpha_2$$

To evaluate three unknown $a_n$'s, additionally the coefficients of the second derivatives on either side of Eq. (5.28) are equated as

$$a_1 + 4a_2 = 2\alpha_2$$

Above two equations are augmented by the vanishing condition of the transfer function (5.31) at the Nyquist limit:

$$a_0 - 2a_1 + 2a_2 = 0$$

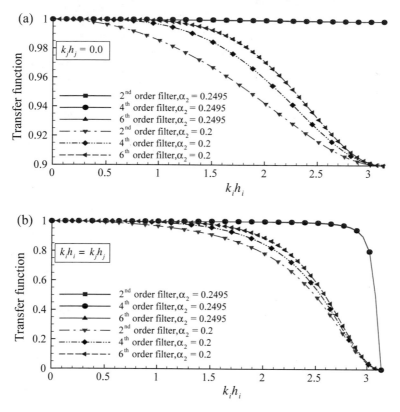

**Fig. 5.11** Transfer function variation of 2D second-, fourth-, and sixth-order filter at the indicated values of filtering coefficient $\alpha_2$, plotted as a function of $k_i h_i$ along: **a** $k_i h_i$ axis; **b** $k_j h_j = k_i h_i$ line

Solution of these three equations yields the following:

$$a_0 = \frac{(5 + 12\alpha_2)}{8}; \quad a_1 = \frac{1 + 4\alpha_2}{4}; \quad a_2 = \frac{-1 + 4\alpha_2}{16}$$

A similar exercise for the filter with $M = 3$ yields the following coefficients for the right-hand side of Eq. (5.28):

$$a_0 = \frac{20\alpha_2 + 11}{16}, \quad a_1 = \frac{68\alpha_2 + 15}{64} \quad a_2 = \frac{12\alpha_2 - 3}{32} \quad \text{and } a_3 = \frac{1 - 4\alpha_2}{64}$$

Variations of $T_{ij}$ in the spectral $(k_i h_i, k_j h_j)$-plane are plotted in Figs. 5.11a, b along specific directions for two values of $\alpha_2$. Two-dimensional filters proposed in Eq. (5.28) appear to be directional and to test it, we plot the transfer functions for "second-," "fourth-," and "sixth"-order filters along the $k_i h_i$-axis and along $k_i h_i = k_j h_j$ line. Note that the value of the transfer function along $k_j h_j$-axis will be same as obtained for the variation along the $k_i h_i$-axis. Results are plotted as a function of $k_i h_i$, from zero to its Nyquist limit in these figures. The transfer function is purely real and in Fig. 5.11a results are shown for $\alpha_2 = 0.2495$ and 0.20 along the $k_i h_i$-axis. The transfer function for $\alpha_2 = 0.2495$ varies very little between 1.0 and 0.999, with some difference among different filters noted in the fourth decimal place for intermediate wavenumbers only, whereas for $\alpha_2 = 0.20$, there is significant variation of $T_{ij}$ among different filters, where it varies from 1 to 0.90 along the $k_i h_i$-axis, again only in the intermediate range of wavenumbers.

In contrast, along the $k_i h_i = k_j h_j$ line, variation of $T_{ij}$ is from a value of one (at $k_i h_i = k_j h_j = 0$) to zero at the Nyquist limit of $(k_i h_i = k_j h_j = \pi)$ for both values of $\alpha_2$. However, one cannot distinguish among three different cases plotted for $\alpha_2 = 0.2495$ and all of them drop from one to zero for $k_i h_i \geq 2.5$. For $\alpha_2 = 0.20$, the transfer functions are distinct for different filters with different orders.

Advantage of using 2D filters over conventional 1D filters is explained with the help of Fig. 5.12. Frame (i) in this figure shows variation of transfer function in $(k_i h_i, k_j h_j)$-plane for 1D "fourth"-order filter with $\alpha_1 = 0.20$, applied sequentially. Transfer function variation for a 2D filter for $M = 2$ with $\alpha_2 = 0.20$ is shown in frame

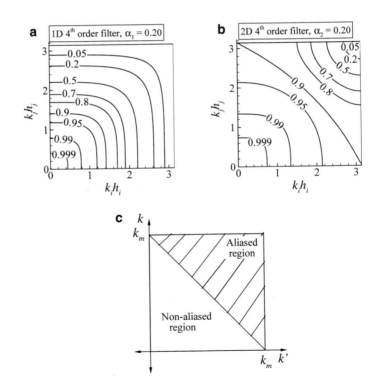

**Fig. 5.12** Transfer function contours in $(k_i h_i, k_j h_j)$-plane shown for (**a**) 1D, "fourth-order" filter with $\alpha = 0.20$ and (**b**) 2D, fourth-order filter with $\alpha_2 = 0.2$. Frame (**c**) shows non-aliased and aliased region (hatched lines) in $(k, k')$-plane for product evaluation

(ii) of the figure. Restriction on transfer function to attain zero value for $k_i h_i = \pi$ or $k_j h_j = \pi$, imposed in 1D filters, leads to excessive filtering of solution as shown in frame (i) for high values of either $k_i h_i$ or $k_j h_j$. For example, $T_{ij}$ takes a value of 0.95 at $k_i h_i \simeq 1.2$ in frame (i), while same value of transfer function is attained at $k_i h_i \simeq 2.15$ in frame (ii), for 2D filter with $\alpha_2 = 0.20$, demonstrating that 1D filters cause unnecessary large attenuation of low $kh$ components as compared to 2D filters. Ideal filters only remove high $kh$ components, which otherwise cause numerical instabilities and aliasing. Thus, in addition, 1D filter compromises accuracy of solution, as compared to 2D filtering by removing physically important low $kh$, which 2D filters will not do so, as plotted transfer function in frame (ii) of Fig. 5.12 shows.

Additionally, design of 2D filter restricts transfer function to attain zero value at $k_i h_i = k_j h_j = \pi$, and this has two important numerical consequences. First, it avoids excessive filtering, a serious drawback of all 1D filters. Secondly, 2D filters beneficially affect aliasing problem. Aliasing error occurs when one numerically evaluates nonlinear terms, or while one evaluates linear terms in transformed plane, due to grid metrics appearing in transformed plane equations. Calculations performed with a higher resolution schemes are more prone to aliasing error, as compared to low-resolution spatial schemes. This is explained briefly again with the help of frame (iii) in Fig. 5.12.

Consider a product term $f(\xi)\,\omega(\xi)$ to be evaluated in transformed plane. The Fourier spectral representations of these two terms are given as $f(\xi) = \int_{-k_m}^{k_m} F(k')$ $e^{ik'\xi}\,dk'$ and $\omega(\xi) = \int_{-k_m}^{k_m} \Omega(k)\,e^{ik\xi}\,dk$, where $k_m$ is the maximum resolved wavenumber for a given grid. Aliasing occurs when combined wavenumber $(k + k')$ is greater than $k_m$, as shown hatched in frame (iii) in Fig. 5.12. Due to aliasing, high $kh$ components near $k_i h_i = k_j h_j = \pi$ are folded back to resolved $kh$. Most problem occurs near the Nyquist limit of individual terms simultaneously. 2D filters automatically help in dealiasing the solution due to the property of transfer function, as shown in frame (ii) in Fig. 5.12. It is noted that filtering does not attenuate solution by and large, when either $k_i h_i$ or $k_j h_j$ have small values, as in 1D filters applied sequentially with transfer function shown in frame (i) in Fig. 5.12. Additionally, amount of filtering near $k_i h_i = \pi$ or $k_j h_j = \pi$ can be explicitly controlled via choice of $\alpha_2$.

## 5.2.9 Computational Issues of Transonic Flows

Till 1980s, prediction of transonic flows was based on solving nonlinear transonic small-disturbance (TSD) and full potential equations (FPE). The solution of Euler equations became important in the 1980s. Successful shock capturing algorithms for Euler equations were extended to solutions of the RANS equations. Use of software packages gained popularity [44]. Comprehensive review of the experiments and computations of transonic flows are to be found in [52, 115]. Self-sustained shock wave motions on airfoils have been investigated experimentally and numerically in [52], and other references therein. Lee [51] and Lee et al. [53] have given explanation of the mechanism for self-sustained shock oscillation and proposed a feedback model to estimate the frequency of type A shock motion. Idealized shock waves/ turbulence interactions have been reviewed by Andreopoulos *et al.* [3] indicating direct effects of the Rankine–Hugoniot relations on the turbulence amplification through shock wave. The physical mechanism of the shock motion is not completely understood, even though some work has been performed [52]. Numerical investigations have been reported for the compression corner interactions [58, 87] and incident shock interactions [82, 113].

Use of numerical diffusion and turbulence models is noted in the simulation of transonic flows. Visbal [117] simulated mass-averaged NSE with an algebraic eddy viscosity model to investigate compressibility on dynamic stall and found "good" prediction of pressure coefficient, only when strong shock-induced boundary layer separation is absent. Numerical simulations have been performed using 2D URANS with turbulence models by Marvin et al. [61]; Rumsey et al. [90]; Xiao, Tsai and Liu [119]. In [31], turbulence models of Spalart–Allmaras (S-A) and $k - \omega$ of Menter (M-SST) have been used to study wind tunnel effects on transonic flows. Liu *et. al.* [120] made use of two-equation, $k - \omega$ model to solve URANS equations to study transonic buffet on a Bauer–Garabedian–Korn supercritical airfoil. DNS of transonic flows can be found in [2, 81]. In [2], authors have reported DNS results to investigate upstream moving pressure waves over a supercritical airfoil. Pirozzoli et al. [81] have performed DNS to study shock–boundary layer interactions and identified three distinct flow regions: (i) a zero pressure gradient region upstream of the origin of the interaction; (ii) a supersonic adverse pressure gradient region, where the interaction between shock and boundary layer takes place, and (iii) a subsonic region, where the flow relaxes to an equilibrium state.

### 5.2.10   *Experimental Results Used for Validation*

Experimental results for 2D section of NACA 0012 airfoil reported in [38] are used for validation. Although the tunnel used for experiments was not designed for 2D testing, the geometry of the model (with large span-to-chord ratio, of 3.43) used is adequate for 2D testing. Also, width of the tunnel is sufficient to minimize sidewall boundary layer effects, making conditions suitable for 2D tests. The corrections for wall-induced lift interference are based on tunnel geometric characteristics, leading to incremental change in angle of attack, as given in [8]. Flow around 15 % thick natural laminar flow (NLF) SHM1 airfoil is also simulated using OUCS2 and OUCS3 scheme for discretizing convection terms. NLF airfoil has been designed with an objective to make the flow laminar over a large streamwise extent of the airfoil, by moving the maximum thickness location downstream. This extends the region of favorable pressure gradient on the suction surface. Fujino et al. [27] have tested the SHM1 airfoil that has a higher drag divergence Mach number and lower drag, and a thickness ratio provides sufficient fuel carrying capacity. Larger segments of airfoil upper (about 0.4c) and lower (about 0.6c) surfaces have favorable pressure gradient to delay transition. The experimental results available for SHM1 airfoil in [27] are used to validate the subsonic flow case.

### 5.2.11 Experimental Issues

The validation of computation is done with available experimental results, and thus knowledge of experimental conditions and error sources of the data is desirable. In [12, 31], problems related to transonic tunnel tests involving wall interference, 3D effects in 2D tests, aeroelastic effects, flow unsteadiness, etc. have been discussed, which affect the quality of the data. Correction for tunnel wall effects in experimental data comes with uncertainty and may not be directly used for CFD prediction. The wall corrections are related to incremental velocity, changed angle of incidence, altered drag, and streamline curvature [29] for subsonic wind tunnel. Supercritical flow region need not be small, with respect to tunnel dimensions, and wall corrections are more complicated, as distortions of the flow field can strongly influence the flow and shock [43].

Closed boundaries provide stronger reflections of unsteady pressure waves, and ventilated boundaries provide diffused reflection of unsteady pressure waves [63]. A completely different type of interference occurs when the shocks on the model strengthen and extend to the tunnel walls. When unsteady shock from the model reaches the tunnel walls, even a fully adaptive wall (controlled by steady flow equations) will produce significant time-dependent interference on the model. This condition sets upper limit for buffeting and oscillatory time-dependent measurements. It is known that 3D effects readily develop in the turbulent boundary layer of flows, which are nominally 2D. This 3D nature is likely to be amplified by interaction with shock wave and/or in separated flows [12]. As discussed in [64], flow unsteadiness is noted for velocity, pressure, and temperature as fluctuations. Most transonic tunnels have high level of unsteadiness, in particular, pressure fluctuations, which have significant adverse effects. The main effect on high levels of flow unsteadiness at transonic speeds is to alter the mean position of the shock. In case of natural transition, such noise has influence on the transition location [78]. Free-stream turbulence (FST) can also have an influence on attached boundary layers to cause separation onset [76]. The axial force balances are also sensitive to flow unsteadiness at transonic speeds [71]. Pressure fluctuations can be controlled by installing additional throat which demands more power and rarely incorporated in large transonic wind tunnels [63]. Some evidences of strong influence of time-dependent boundary conditions on the development of unsteady transonic flows are noted [63]. Authors in [31] concluded that available experimental data on transonic airfoils are insufficient for turbulence-model validation. In [70], authors have also concluded that comparison with wind tunnel has been limited to cases with insignificant wall effects, because of the lack of appropriate data. These issues have been discussed in [10, 112] with respect to limitations of transonic flows.

## 5.3  Transitional Flow Comparison Between RANS and DNS for Flow Over an Airfoil

We compare among various methods in computing transitional flows in this section. Relative strengths and weaknesses of methods for simulations by compressible RANS and DNS of a low Mach number flow are described. For the DNS we show the results obtained by using compressible and incompressible flow models. For the present purpose, we focus on a natural laminar flow airfoil, namely, the WTEA-TE1 airfoil for a free-stream Mach number, $M_\infty = 0.1815$, with the airfoil placed at an angle of attack of $\alpha = -0.0328°$. Computed results are compared with experimental pressure and skin friction distributions [6].

Computing transitional flows is challenging for DNS, as compared to solving fully developed turbulent flows using first principle approach. DNS of such a flow for incompressible and compressible flow models is compared with RANS results of compressible flow using a well-calibrated transition model for transitional flow over WTEA-TE1 airfoil [60]. The aerodynamic performance of this model was optimized [60] for transition delay by morphing. Review of morphing wing to delay transition using shape deformation is discussed in [107], while adaptive wing and flow control aspects of morphing wing are to be found in [83–85, 111]. The works in [60, 83] are relevant, as these studies focus on the ability to predict transition computationally [6].

RANS results with $\gamma - Re_\theta$ model of Langtry and Menter [49] have been reported also in [89]. Here, these RANS approach results are compared with DNS results obtained using a compressible flow solver developed for orthogonal grid. This DNS solver uses compact scheme [93, 94] and can be directly used for low $M_\infty$ flows. The same flow has been solved using incompressible NSE solver with the same orthogonal grid [96]. The geometry of the airfoil is shown in Fig. 5.13, with locations of Kulite transducers marked, where the unsteady pressure was measured [60]. Note that the airfoil used in the experiment had surface discontinuity at the junction of the aft part made flexible, while the front was rigid, as indicated in the inset of Fig. 5.13, which is used for grid generation used for DNS.

In the experiments, point at which transition occurs is estimated by the methods: The Kulite pressure sensors were placed on the upper surface to measure unsteady pressure. The RMS value of pressure signals can be used to indicate the state of the

**Fig. 5.13** The sketch of the WTEA-TE1 airfoil used for experiments, with the location of Kulite pressure sensors marked [6]

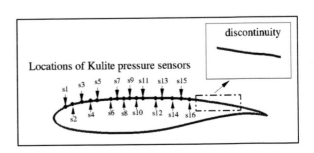

**Fig. 5.14** Infrared picture of the upper surface of WTEA-TE1 airfoil indicating transition location [6]

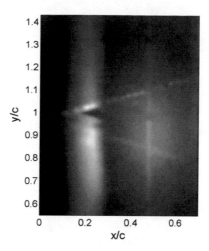

flow as turbulent or not. Also, infrared pictures have been used to visualize the top surface of the airfoil. Infrared picture in Fig. 5.14 indicates abrupt change in heat transfer to occur where the flow becomes transitional for the first time. This is based on the rapid variations of upper surface wall temperature during transition.

### 5.3.1 Numerical Methods for Compressible DNS

The governing equations for compressible DNS solver have been described with high-resolution, DRP method used with convective derivatives evaluated using OUCS3 scheme. Diffusive terms are evaluated using CD2 method. Artificial diffusion model of [45] is used here, as was also reported in [94]. An optimized method for time advancement using three-stage Runge–Kutta (OCRK3) method [93] has been used. The orthogonal grid was generated using the hyperbolic grid generation technique described in the previous chapter and in [93] for the WTEA-TE1 airfoil with a refined wall-normal spacing ($\Delta S_\eta = 3.072 \times 10^{-5}$), while the outer boundary was located at 10 chords ($c$) distance.

The solutions obtained by RANS equation with model for transition, used a 2D multi-block structured solver, that uses five-stage, Runge–Kutta ($RK_5$) time integration scheme along with JST model [89]. In addition to local time stepping, multi-grid and residual smoothing are used for the RANS solver, along with the transition model [49] on a C-grid with $624 \times 128$ points. The wall-normal spacing used is $1 \times 10^{-6}$, while outer boundary was placed at $50c$.

The same problem has been also solved using a DNS methodology for incompressible NSE with ($\psi$, $\omega$)-formulation. The spatial discretization for convective acceleration has been performed by OUCS3 scheme and time integration by an optimized four-stage, third-order Runge–Kutta method (OCRK3) described in [93].

Furthermore, 2D adaptive filter [11] has been used for the incompressible DNS solver, with the same grid used for the compressible DNS.

### 5.3.2  Flow Field over the WTEA-TE1 Airfoil

Flow over the WTEA-TE1 airfoil is simulated for $M_\infty = 0.1815$ and $\alpha = -0.0328°$. In Fig. 5.15, mean $C_p$-distribution is displayed versus $x/c$, obtained by DNS using compressible and incompressible flow solvers. These results are compared with the results obtained using the RANS solver. In this figure, $C_p$-distributions match rather well, displaying a kink for the mean $C_p$-distribution, where we noted also the slope discontinuity in Fig. 5.13. This is due to airfoil surface discontinuity at the junction between rigid and flexible parts of the airfoil tested. The prominent kink for all the results obtained by RANS and DNS solvers. However, the use of JST model in compressible DNS solver diffuses this discontinuity. This is perhaps due to addition of 2D diffusion in the JST model. In the figure, experimental data are shown for the top surface till the point of slope discontinuity. It is noted that the incompressible DNS solver over-predicts lift by significant amount, as compared to RANS and compressible DNS solvers' results. Also as would be expected, the incompressible formulation shows the suction peak at the leading edge. In contrast, the compressible DNS solver displays slow acceleration of the flow near the leading edge.

In Fig. 5.16, $C_f$ variation is shown with $x/c$ obtained using the three solvers. This helps comparison for the transition location by these competing methods. In

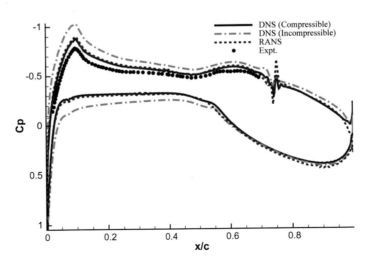

**Fig. 5.15** Mean $Cp$-distribution variation with $x/c$, as obtained by DNS compressible and DNS incompressible solver, along with the RANS solver. One notices the distinct location of slope discontinuity

**Fig. 5.16** Mean
$Cf$-distribution variation
with $x/c$, as obtained by
compressible,
incompressible DNS and
RANS solvers shown for **a**
the upper surface and **b** the
lower surface of the
WTEA-TE1 airfoil

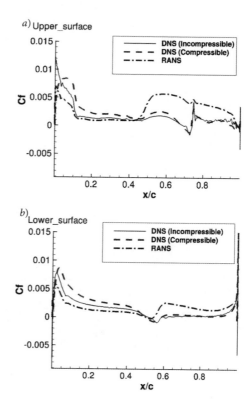

Fig. 5.16a, mean $C_f$-distribution is shown on the upper surface of this natural laminar flow airfoil. One notes that for the RANS, incompressible and compressible DNS, transition occurs at $x/c$ equal to $0.45c$, $0.48c$, and $0.46c$, respectively. The RANS solver shows that once transition "occurs" following the empirical model, the flow is thereafter considered to be fully developed turbulent, resulting in over-estimation of $C_f$. Also, this computed flow shows insensitivity for any surface discontinuity, as noted earlier for $Cp$-distribution. But the DNS solvers are capable in capturing the actual $C_f$ variation over a region, along with rapid fluctuations near the kink. This is in effect, an inadequacy of the model used in RANS, which considers transition to occur at a single point, which is not physical.

In Fig. 5.16b, mean $C_f$-distribution is shown for lower surface and it is seen that RANS shows earlier and stronger transition, while compressible and incompressible DNS show milder transition. Like the top surface, once the flow transition occurs for the RANS solver, subsequently the $C_f$ value remains higher, as fixed by the empirical model for the transitional turbulence model [49].

From Fig. 5.16, one notes that RANS predicts transitional location using the $\gamma - Re_\theta$ model [49]. Subsequent higher drag is due to the turbulence model making the drag estimates more conservative. The properties of the transitional flows displayed

**Fig. 5.17** Distribution of RMS of the streamwise component of velocity obtained by DNS solvers for **a** the upper and **b** the lower surface of the WTEA-TE1 airfoil [6]

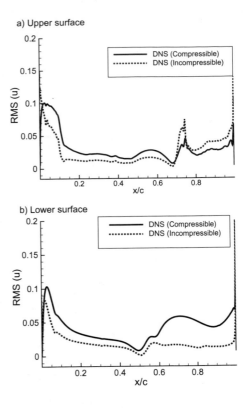

by $C_f$ distribution clearly establish that the DNS are better, with compressible flow solver showing preferred behavior for the surface pressure and skin friction.

To capture fluctuating quantities, in Fig. 5.17, root mean square (RMS) values of streamwise component of velocity obtained using compressible and incompressible DNS solvers are shown as function of $x/c$. This aids in accurately locating transitional zone, rather than a single point in RANS solver. Figure 5.17a shows RMS of streamwise component of velocity on the upper surface, with onset of transition shown by the localized peak of the RMS at identical location. The dip before transition also is noted for both the methods to be at the same location. The transitional flow, following this peak, displays the incompressible formulation to have higher RMS value, implying higher Reynolds stress. RMS of streamwise component of velocity for the lower surface of the airfoil shown in Fig. 5.17b indicates the pre-transition dip to occur at identical position for both the DNS formulations. The incompressible DNS results are characterized by higher turbulence near aft portion of the airfoil, as compared to the compressible formulation.

In Fig. 5.18, Fourier transform of the pressure time series (from $t = 22.60$ to $38.93$) obtained by compressible DNS, at the even-numbered Kulite pressure sensors' location in the experiments [60], is shown with data sampled at the rate of $0.001$. The fundamental frequency is noted as $8.4627$ kHz, and one observes wide-band

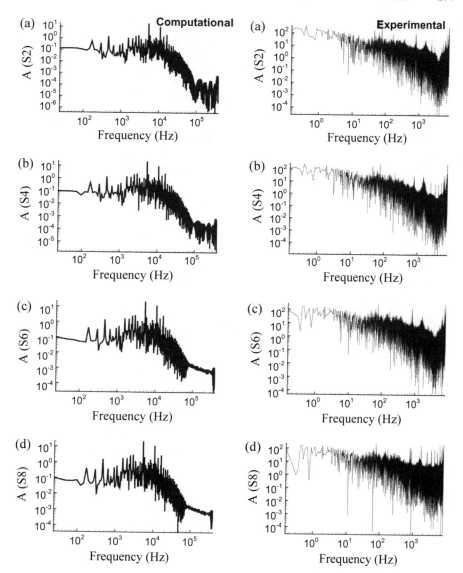

**Fig. 5.18** Fourier transform of pressure signals obtained by compressible DNS solver (computed for time series from $t = 22.60$ to $38.93$), at the even-numbered locations of Kulite pressure sensors: **a–h** correspond to 2nd−16th Kulite pressure sensors used in the experiment [60]

spectra. It is noted that with JST model (results not shown) there are only discrete higher harmonics (five). Use of high accuracy 1D filter creates a spectrum, which appears as more realistic and hence recommended.

In the RHS column, corresponding experimental pressure spectrum is shown, where the sampling rate of 15 kHz was used. Thus, one resolves only up to 7.5 kHz. Higher than this value of frequency will suffer from effects of aliasing. Direct comparison with DNS reveals that this resolution in the experiment is inadequate, with results affected due to aliasing for the higher frequencies of the experimental data.

Present section shows the relative strengths and weaknesses of some of the methods used to simulate transitional flows. If one depends upon empirical method, then undoubtedly, RANS formulation is a faster method to provide lift value effectively, as compared to the DNS using compressible or incompressible formulations for this flow at $M_\infty = 0.1815$, for which experimental results are also available. For drag prediction, the RANS method can predict a transition location, which appears satisfactory. But results obtained by DNS are able to capture better and realistically the physical nature of transitional flows. This aspect is also brought out better in simulating transonic flows in the presence of shock, as described next.

## 5.4  Transonic Aerodynamics: Flow Over Airfoils

Transonic flow covers a range of $M_\infty$, which is most economical in creating lift, and hence it is the very efficient flight regime in aerodynamics due to enhanced ability for maneuvers. For this flow, one notices rapid transition from subsonic to supersonic flow, as $M_\infty$ is increased, with shock wave formation and violent unsteadiness as the distinguishing features of the off-design performance. All higher speed flows must go through the transonic regime during its evolution [73]. If $M_\infty$ is increased continuously from zero, the transonic range begins when the highest local Mach number ($M_{loc}$) reaches unity value and ends when the lowest $M_{loc}$ reaches unity [57]. In a transonic flow, subsonic and supersonic flow co-exist, in the domain of interest [73]. Figure 5.19 illustrates the definition of transonic flow for the case of a flow over NACA 0012 airfoil with $M_\infty = 0.779$, and Reynolds number of $3 \times 10^6$, at an angle of attack $\alpha = -0.14°$, where one can see supersonic ($M > 1$) flow regions above and below airfoil surface, separated by the sonic line ($M = 1$) from the subsonic region ($M < 1$).

Transonic flows represent many complex flow phenomena such as creation of shock, shock oscillations, creation of contact discontinuities, shock–boundary layer interaction, and aeroelastic buffet onset. The flow pattern and forces exerted on the moving body in the transonic range differ significantly, from those in the low speed range. As $M_\infty$ for attached viscous flow over airfoil surface increases, one notes complex transonic phenomena. The least value of $M_\infty$, for which $M_{loc} = 1$ is seen in the flow, is the critical Mach number. With further increase of $M_\infty$, formation of supersonic pocket bounded by the sonic line is noted. Flow encounters shock jump, experiencing abrupt variation of properties, while traveling through

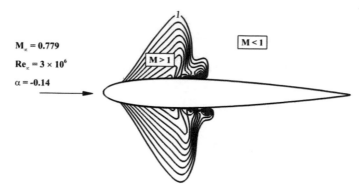

**Fig. 5.19** Definition of transonic flow

supersonic to subsonic region. When one reports shock, one is talking of a time-averaged condition. In reality, the discontinuity emanates waves in upstream and downstream directions. Shocks form on upper and/or lower surface depending on angle of attack. The shock also interacts with the boundary layer, thickening it, and depending upon shock strength can cause unsteady flow separation. Unsteady separated flow due to the shock–boundary layer interaction may reattach on the airfoil surface. As $M_\infty$ further increases, shock becomes stronger and moves in the downstream direction. At sufficiently high value of $M_\infty$, flow becomes fully separated from shock wave location to trailing edge, and shock becomes stronger moving further downstream. Such flow may have the presence of weak shock near the trailing edge. A detailed description of the effects of $M_\infty$ on the flow can be found in [10, 74, 79].

## 5.4.1 Classification of Shock Oscillation

Shock oscillation is one of the interesting phenomena associated with the transonic flows [10]. Self-sustained shock wave oscillations on airfoils at transonic flow lead to buffeting [52]. Boundary layer separation induced by moving shock and evolution of coherent vortical structures in the trailing edge region plays an important role in overall flow behavior [17]. Pressure fluctuations excite structural modes, called as buffeting [63]. Tijdeman reports [114] study of shock motions over oscillating airfoils and classifies such motions to be one of the three types: Types A, B, and C. Shock motion of type A is sinusoidal, and shock remains present on the airfoil surface during complete cycle of oscillation. In Type B motion, which is similar to type A shock wave motion, the shock wave disappears during a part of its backward motion. Shock wave motion of type C is entirely different, and periodic shock waves are formed on the airfoil surface that moves upstream, while increasing the strength. But the shock wave weakens and continues its upstream propagation after leaving the airfoil from the leading edge. This is repeated periodically and alternates between upper and lower surfaces. These types of shock motions are observed for oscillating, as well as, steady airfoils with severe flow separation downstream of the shock [115]. Under these circumstances, shocks do not remain normal but cyclically takes the appearance of lambda-type shocks.

### 5.4.2  Upstream Propagating Kutta Waves

Unsteady disturbances travel downstream within the separated flow region from shock location to trailing edge, and upon interacting with trailing edge gives rise to upstream moving Kutta waves [53, 114]. Kutta waves are generated as a result of satisfying the unsteady Kutta condition [53]. It is a model used for inviscid flows, and not at all necessarily needed in DNS [94]. However, Kutta waves are physical and can be associated with acoustic response of unsteady flow near the trailing edge. These upstream moving waves travel from the trailing edge to shock, interact with shock, and further travel upstream, leaving the leading edge of the airfoil [17]. As noted in [51, 114], period of oscillation of shock is closely related to the time taken by downstream propagating disturbance waves in separated boundary layer from shock location to the trailing edge, plus the time taken by the upstream moving disturbance waves outside the boundary layer from trailing edge to shock. The unsteady pressure fluctuations generated by the low-frequency large-amplitude shock motions are highly undesirable from the structural integrity and aircraft maneuverability point of view [52]. Transonic speed range introduces flutter problems that can occur for both attached and separated flows [46].

### 5.4.3  Need for Time-Accurate Solution of Navier–Stokes Equations

Transonic flows are highly time-dependent, and the dynamics is understood as a balance between viscous actions with the convective process [73]. For the type of pressure distributions associated with modern wings, the viscous effects are significant at full-scale Reynolds number [12]. In actual flows, changes in flow field occur irregularly, and fluctuations appear without unsteady boundary conditions [63]. The system of equations, augmented by empirical laws for the dependence of viscosity and thermal conductivity, with other flow variables, and a constitutive law defining the nature of the fluid, describes flow phenomena [40]. To obtain numerical solution with shocks and discontinuities, conservative formulation is preferred.

Whether the governing equations are elliptic, parabolic, or elliptic–parabolic in nature, the numerical method employed essentially solves it as a hyperbolic equation [92]. Hence, the method used to solve governing differential equations must preserve the correct space–time behavior, for all wavenumber–circular frequency combinations, [92]. Compressible flows for high Reynolds numbers involve wide range of spatial and temporal scales, augmented further by the presence of shocks and contact discontinuities exciting higher wavenumbers.

Large number of numerical methods and approaches have been developed, which can be broadly classified as space-centered and upwind schemes. Upwind schemes are popular due to the added numerical diffusion and are classified as flux-vector splitting, flux-difference, and total-variation diminishing schemes [40, 41, 74]. Achieving high resolution is the main goal of compact schemes. These can be used in transformed plane on uniformly distributed grid points. These are also robust and computationally economical, as it involves solving system of tridiagonal equations and hence easy to solve.

In the present book, high accuracy optimized upwind compact schemes (OUCS) are used, which were developed for any flows [92, 93, 102]. DNS of 2D wall-bounded turbulent flows is reported in [97], where OUCS3 scheme for spatial discretization has been used. In [100], symmetrized version of the OUCS scheme has been used, in parallel computing framework, to solve 3D unsteady compressible NSE for supersonic flow past a cone–cylinder for $M_\infty = 4$ and $Re = 1.12 \times 10^6$.

Results of the subsonic and transonic flow past NACA 0012 airfoil are discussed next, with numerical results compared with the experimental results given in [38]. This is followed by computational results for the subsonic flow past SHM-1 airfoil. The computations in this section are carried out using two different compact schemes, OUCS2 and OUCS3 for discretizing convection terms.

### 5.4.4  Transonic Flow Past NACA 0012 Airfoil

Compressible subsonic and transonic flows are characterized with respect to shock formation, shock–boundary interaction, and entropy generation for flow past airfoils. Viscous, transonic flows past NACA 0012 airfoil are simulated for a Reynolds number of $3 \times 10^6$ and angle of attack $\alpha = -0.14°$ for different values of free-stream Mach number. The various cases for which the simulations performed are listed in Table 5.1.

**Table 5.1**  Computational matrix for the study of viscous flow past NACA 0012 airfoil

| Case No. | Mach No., $M_\infty$ | Reynolds No., $Re_\infty$ | Angle of attack, $\alpha$ (in deg) |
|---|---|---|---|
| 1 | 0.6 | $3 \times 10^6$ | −0.14 |
| 2 | 0.758 | $3 \times 10^6$ | −0.14 |
| 3 | 0.779 | $3 \times 10^6$ | −0.14 |
| 4 | 0.8 | $3 \times 10^6$ | −0.14 |

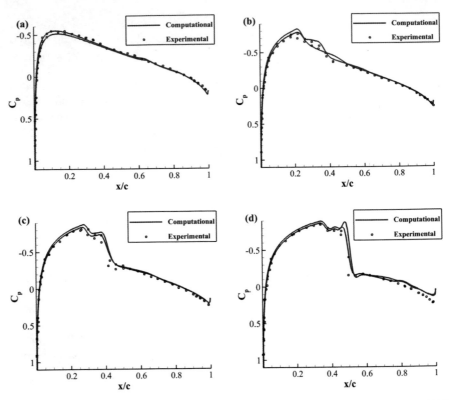

**Fig. 5.20** Comparison of computed and experimental $C_p$-distribution for flow around NACA 0012 airfoil with $Re_\infty = 3 \times 10^6$ at $\alpha = -0.14°$ and **a** $M_\infty = 0.6$; **b** $M_\infty = 0.758$; **c** $M_\infty = 0.779$ and **d** $M_\infty = 0.8$. Computed data is time-averaged over the interval of $t = 10$ to $70$

Equation (5.7) is solved using the numerical methods and boundary conditions mentioned before. The experimental correction for angle of attack ($\Delta\alpha = -1.55C_L$) is implemented in the computation, which is due to wall-induced lift interference effects [38], where $C_L$ denotes the sectional normal-force coefficient. An orthogonal grid, around NACA 0012, is generated using the grid generation method discussed, with 636 points in the azimuthal directions and 350 points in the wall-normal direction. The values of the diffusion constants appearing in the JST diffusion model are fixed as $\kappa_2 = 0.25$ and $\kappa_4 = 0.15$. Computations are performed using OUCS3-RK4 scheme, with a small non-dimensional time step of $\Delta t = 10^{-6}$, which is necessary for better DRP properties. For the time-accurate solution of NSE, definitive initial conditions are not known. Thus, a uniform flow condition is assumed at $t = 0$, and the results at early time are not relevant physically. In most of the cases, it takes about $t \simeq 10$ for the flow to develop into a meaningful viscous flow state. Hence, in the reported cases, time averages are taken from time series starting from $t = 10$.

Comparisons between computed and experimental $C_p$-distributions, for cases listed in Table 5.1, are shown in Fig. 5.20. The computed results are time-averaged,

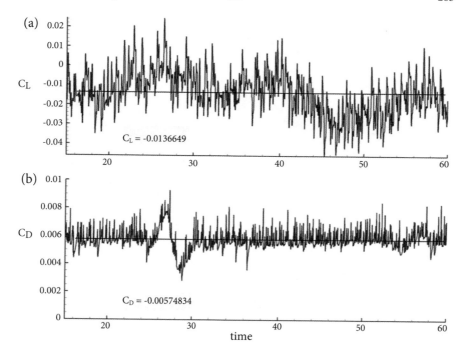

**Fig. 5.21** Variation of **a** $C_L$ and **b** $C_D$ with time for flow around NACA 0012 airfoil with $Re_\infty = 3 \times 10^6$ at $\alpha = -0.14°$ and $M_\infty = 0.779$

**Table 5.2** Computed mean values of $C_L$ and $C_D$ for the cases of viscous flow past NACA 0012 airfoil

| Case No. | Mach No., $M_\infty$ | Coefficient of drag, $C_D$ | Coefficient of lift, $C_L$ |
|---|---|---|---|
| 1 | 0.6 | 0.0019 | −0.0163 |
| 2 | 0.758 | 0.0027 | −0.0142 |
| 3 | 0.779 | 0.0057 | −0.0136 |
| 4 | 0.8 | 0.01293 | −0.0142 |

and results obtained numerically show very good agreement with experimental results. One notices that the shock has been captured with good accuracy, with the pre-shock dip in $C_p$ noted upstream of the shock. These computations do not make use of any transition and/ or turbulence model. For the Case 3, the variations of lift and drag coefficients with time are plotted in Fig. 5.21a and b, respectively, where the mean values are indicated by solid lines. The mean coefficients of lift ($C_L$) and drag ($C_D$), obtained for the cases, are listed in Table 5.2.

Figure 5.22 shows comparison between computed and experimental $C_D$ and $C_L$ values. For a constant $Re_\infty$, the coefficient of drag increases with the increase in $M_\infty$. The value of $M_\infty$, above which drag increases rapidly, is called as drag divergence

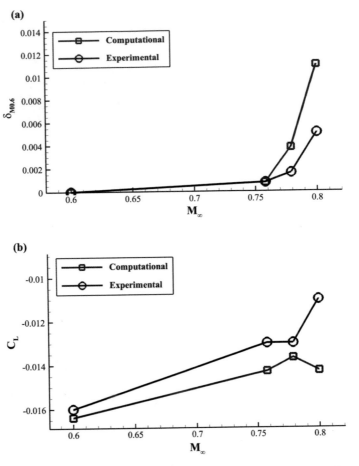

**Fig. 5.22** **a** $\delta_{M0.6}$ versus $M$ showing correct prediction of drag divergence behavior and **b** $C_L$ versus $M$ plotted with computed values

Mach number ($M_{DD}$). Similar behavior is also observed in the results obtained by present computations. The drag shows rapid increment, as the value of $M_\infty$ becomes larger than 0.779. A new variable $\delta_{C_D}$ is introduced here as

$$\delta_{C_D} = C_D - C_{D_{0.6}} \tag{5.32}$$

where $C_{D_{0.6}}$ is the value of drag coefficient at $M_\infty = 0.6$. The behavior of drag increment is shown in Fig. 5.22a, where it can be seen that $C_D$ versus $M_\infty$ curve agrees well with the experimental results [38]. Accurate prediction of drag divergence Mach number is obtained. The comparison between computed and experimental variation of $C_L$ with $M_\infty$ is plotted in Fig. 5.22b, and once again show agreement between the computed and experimental results.

## 5.4.5 Subsonic and Transonic Flow Features

The Case 1 in Table 5.1 is for subsonic (subcritical) flow around NACA 0012 airfoil. Computed $M$-contours are shown in Fig. 5.23 for this case, at indicated time instants with maximum value of $M_{Max}$ indicated in the domain. For this case, value of $M_\infty$ is 0.6, while $M_{Max}$ in the domain is variable and attains values as high as 0.79. Similarly, $p$- and $\rho$-contours are shown in Figs. 5.24 and 5.25, respectively, at indicated time instants, with their maximum and minimum values given in the domain. Figure 5.26 shows entropy ($s$)-contours for Case 1, and one can notice that creation of entropy is in the vicinity of the airfoil surface. The effectiveness of the high accuracy numerical methods employed in capturing these small bubbles is clearly demonstrated.

Other cases for which computations have been performed here are essentially transonic (supercritical) flow cases. It is observed that flow encounters a favorable pressure gradient near the leading edge of the airfoil up to some location, depending upon $M_\infty$. This location on the airfoil surface is hereafter referred to as $x_{cr}$. The flow accelerates in the favorable pressure gradient region, and a local region of supersonic flow is established. The maximum Mach number is observed in this region, whose magnitude depends upon the free-stream flow parameters listed in Table 5.1. The supersonic pocket is established near the surface of the airfoil enclosed by the sonic line. Across the part of sonic line, which is downstream of the supersonic pocket, large jumps in the flow properties are seen due to the presence of the shock. The shock strength increases as $M_\infty$ increases, when $Re_\infty$ and $\alpha$ are held constant. Beyond $x_{cr}$ flow tends to recover pressure and an adverse pressure gradient is established with flow starting to decelerate. The sudden retardation of the supersonic flow is due to the shock. The presence of shock alters the pressure distribution downstream of the shock, which can be observed in Fig. 5.20 b, c, and d. The shock interacts with the boundary layer and causes the flow to separate forming small unsteady bubbles, which travel downstream.

Figures 5.27, 5.28, and 5.29 show computed $M$-contours at indicated time instants with the maximum values of the $M_{Max}$ indicated in the domain. It is clearly seen that, with the increment in $M_\infty$, shock strength and maximum value of local Mach number in the domain increase. Figures 5.30, 5.31, and 5.32 show computed $p$-contours at indicated time instants with their local minimum and maximum values in the domain noted. As the shock strength increases with the $M_\infty$, larger jump in the $p$ values across the shock is seen. Similarly, Figs. 5.33, 5.34, and 5.35 show computed $\rho$-contours at indicated time instants with their local minimum and maximum values in the domain. Shock region is clearly seen in all these figures, and high gradients of variables are noted across the shock.

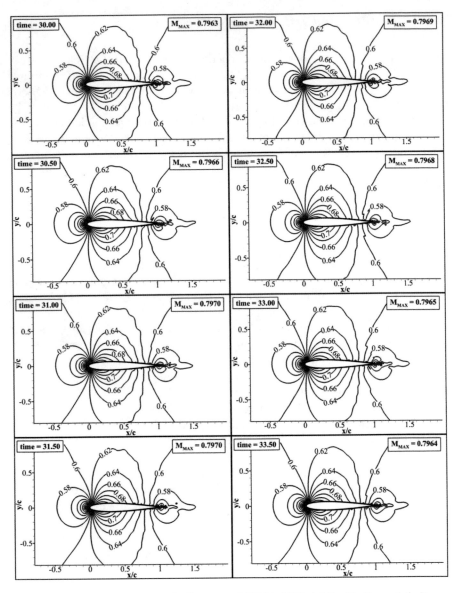

**Fig. 5.23** Computed $M$-contours for flow around NACA 0012 airfoil with $M_\infty = 0.6$, $Re_\infty = 3 \times 10^6$ at $\alpha = -0.14°$

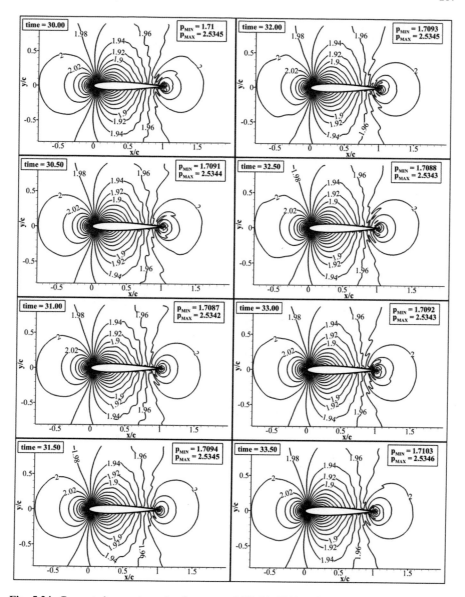

**Fig. 5.24** Computed $p$-contours for flow around NACA 0012 airfoil with $M_\infty = 0.6$, $Re_\infty = 3 \times 10^6$ at $\alpha = -0.14°$

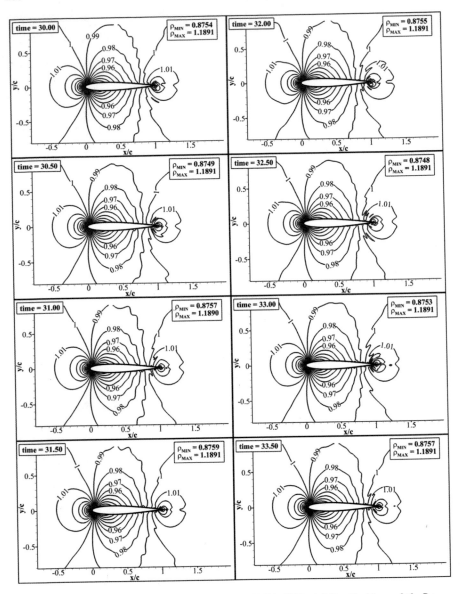

**Fig. 5.25** Computed $\rho$-contours for flow around NACA 0012 airfoil with $M_\infty = 0.6$, $Re_\infty = 3 \times 10^6$ at $\alpha = -0.14°$

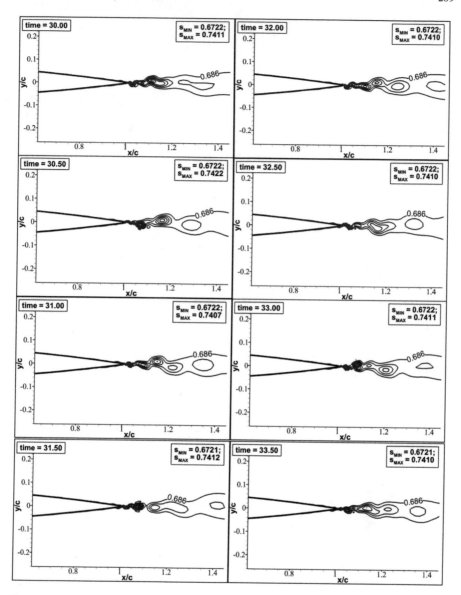

**Fig. 5.26** Computed $s$-contours for flow around NACA 0012 airfoil with $M_\infty = 0.6$, $Re_\infty = 3 \times 10^6$ at $\alpha = -0.14°$

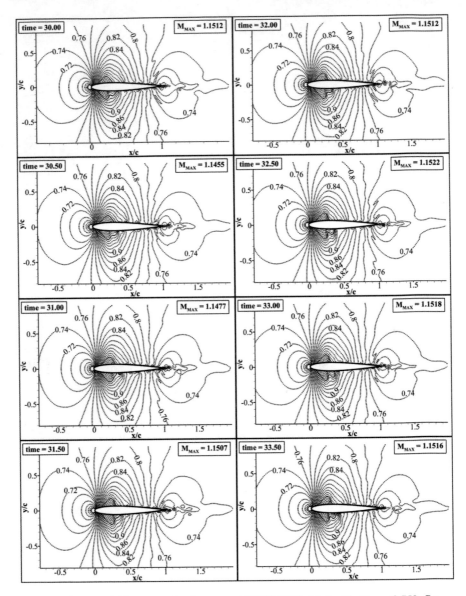

**Fig. 5.27** Computed $M$-contours for flow around NACA 0012 airfoil with $M_\infty = 0.758$, $Re_\infty = 3 \times 10^6$ at $\alpha = -0.14°$

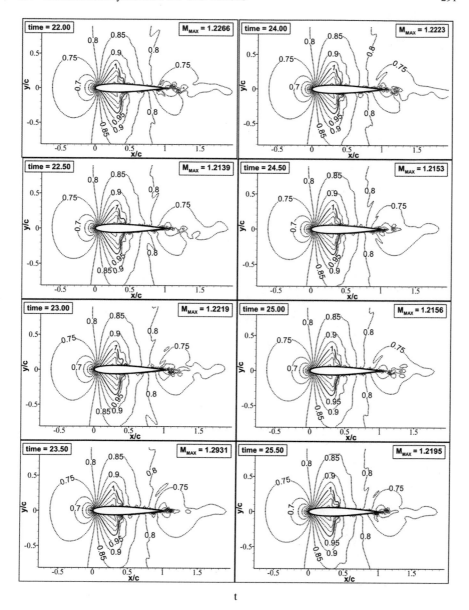

t

**Fig. 5.28** Computed $M$-contours for flow around NACA 0012 airfoil with $M_\infty = 0.779$, $Re_\infty = 3 \times 10^6$ at $\alpha = -0.14°$

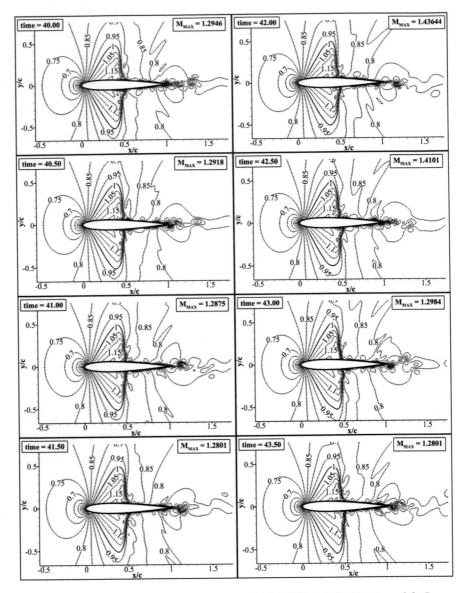

**Fig. 5.29** Computed $M$-contours for flow around NACA 0012 airfoil with $M_\infty = 0.8$, $Re_\infty = 3 \times 10^6$ at $\alpha = -0.14°$

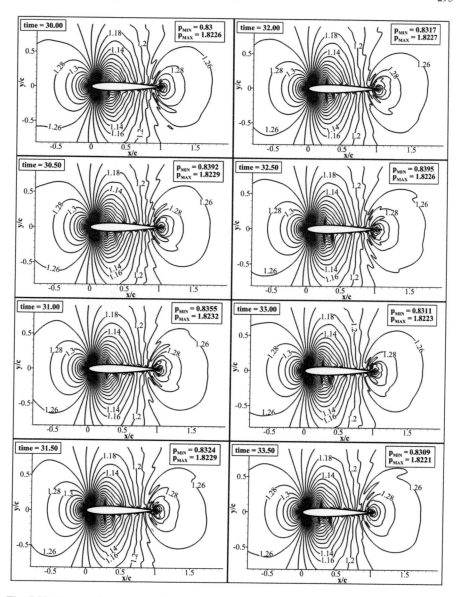

**Fig. 5.30** Computed $p$-contours for flow around NACA 0012 airfoil with $M_\infty = 0.758$, $Re_\infty = 3 \times 10^6$ at $\alpha = -0.14°$

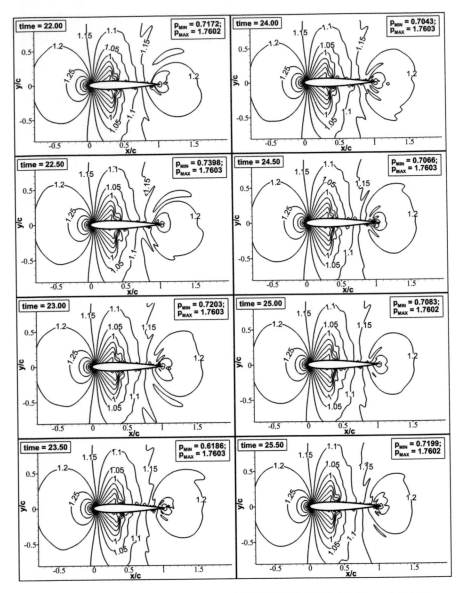

**Fig. 5.31** Computed $p$-contours for flow around NACA 0012 airfoil with $M_\infty = 0.779$, $Re_\infty = 3 \times 10^6$ at $\alpha = -0.14°$

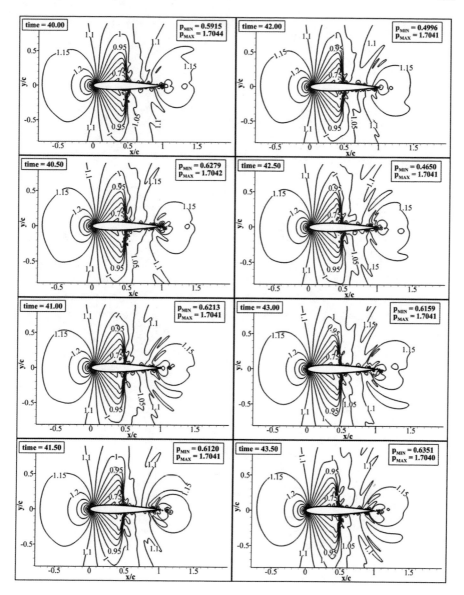

**Fig. 5.32** Computed $p$-contours for flow around NACA 0012 airfoil with $M_\infty = 0.8$, $Re_\infty = 3 \times 10^6$ at $\alpha = -0.14°$

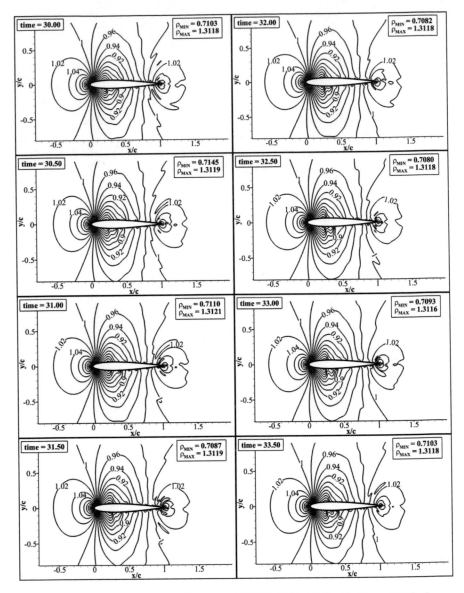

**Fig. 5.33** Computed $\rho$-contours for flow around NACA 0012 airfoil with $M_\infty = 0.758$, $Re_\infty = 3 \times 10^6$ at $\alpha = -0.14°$

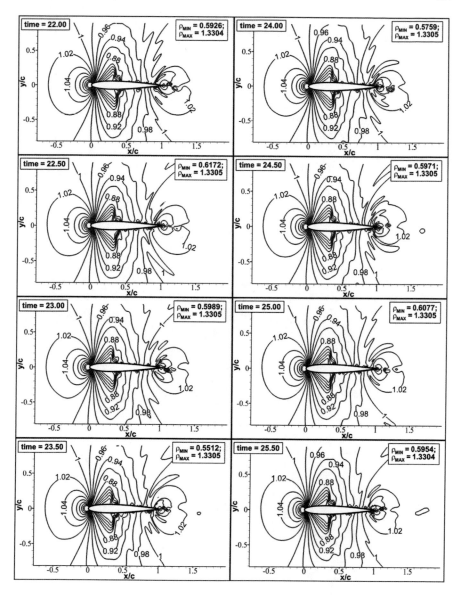

**Fig. 5.34** Computed $\rho$-contours for flow around NACA 0012 airfoil with $M_\infty = 0.779$, $Re_\infty = 3 \times 10^6$ at $\alpha = -0.14°$

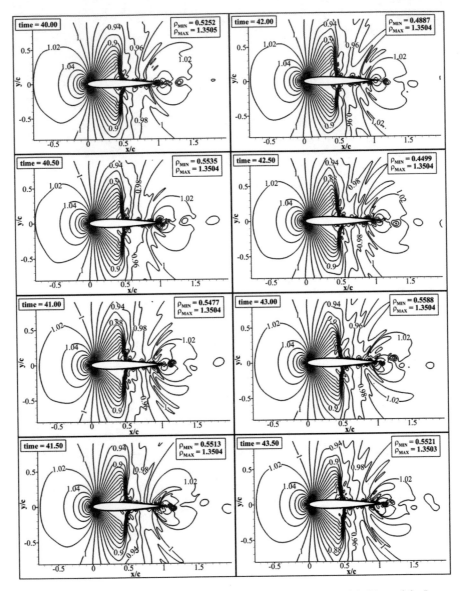

**Fig. 5.35** Computed $\rho$-contours for flow around NACA 0012 airfoil with $M_\infty = 0.8$, $Re_\infty = 3 \times 10^6$ at $\alpha = -0.14°$

## 5.5 Viscous Subsonic Flow Past SHM-1 Airfoil

To further show the capabilities of the procedure, simulations are performed for flow over SHM-1 airfoil for $M_\infty = 0.62$ and $Re_\infty = 13.6 \times 10^6$, at $\alpha = 0.27°$. For this simulation, an orthogonal grid around SHM-1 is generated, with 663 points in the azimuthal directions and 450 points in the wall-normal direction. The values of the diffusion constants appearing in the JST diffusion model are the same as before. A non-dimensional time step of $\Delta t = 10^{-6}$ is used for better DRP properties. These simulations are performed with convective flux derivatives evaluated using both OUCS2 and OUCS3 schemes separately, and results compared in Fig. 5.36, between experimental and computed $C_p$-distribution [27]. Although OUCS2 scheme has fifth-order accuracy, as compared to second-order accuracy of the OUCS3 scheme, identical results are obtained with both the schemes in capturing flow features correctly. This is due to the fact that both the schemes have comparable spectral resolution [93]. Figures 5.37a, 5.38a, and 5.39a show computed $M$-, $p$-, and $\rho$-contours, respectively, at different time instants, for flow around SHM-1 airfoil using the OUCS2 scheme, while Figs. 5.37b, 5.38b, and 5.39b show the corresponding computed $M$-, $p$-, and $\rho$-contours computed with OUCS3 scheme, at the same time instants. One can notice identical flow features captured by both the schemes.

**Fig. 5.36** Comparison of computed and experimental $C_p$-distribution for flow around SHM-1 airfoil with $Re_\infty = 13.6 \times 10^6$ at $\alpha = 0.27°$ and $M_\infty = 0.62$. Computed data is time-averaged over $t = 10$ to 20 [Reproduced from Sengupta et al., *Computers and Fluids*, **88**, 19–37 (2013), with the permission of Elsevier Publishing.]

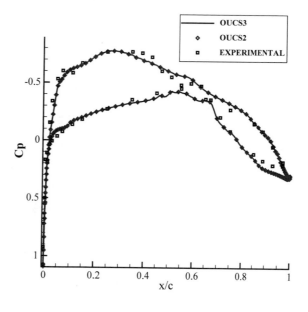

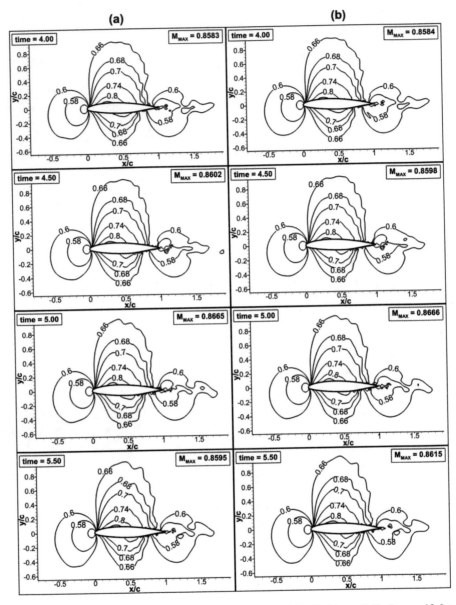

**Fig. 5.37** Computed $M$-contours for flow around SHM-1 airfoil with $M_\infty = 0.62$, $Re_\infty = 13.6 \times 10^6$ at $\alpha = 0.27°$. Numerical schemes used for evaluating convective flux terms are **a** OUCS2 and **b** OUCS3

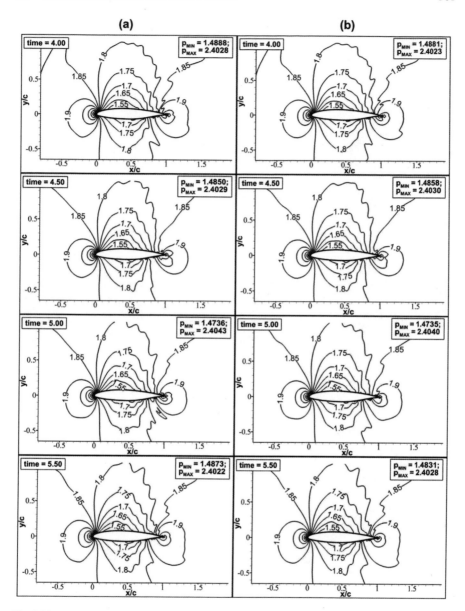

**Fig. 5.38** Computed $p$-contours for flow around SHM-1 airfoil with $M_\infty = 0.62$, $Re_\infty = 13.6 \times 10^6$ at $\alpha = 0.27°$. Numerical schemes used for evaluating convective flux terms are **a** OUCS2 and **b** OUCS3

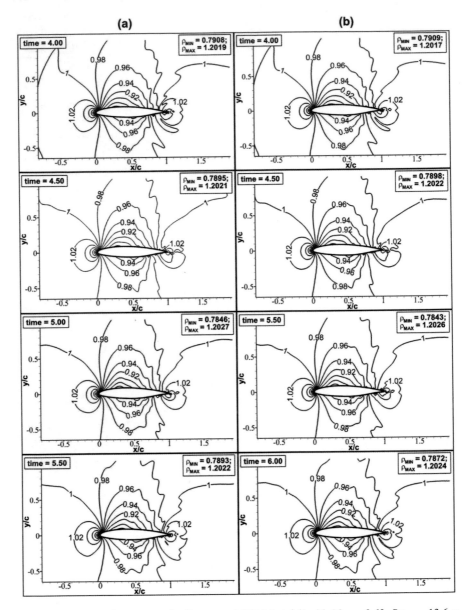

**Fig. 5.39** Computed $\rho$-contours for flow around SHM-1 airfoil with $M_\infty = 0.62$, $Re_\infty = 13.6 \times 10^6$ at $\alpha = 0.27°$. Numerical schemes used for evaluating convective flux terms are **a** OUCS2 and **b** OUCS3

### 5.5.1 Unsteady Flow Behavior of Compressible Flows

The flow over an airfoil displays time-dependent behavior, even when shock waves are not formed. When strong shocks form, such unsteadiness increases due to physical dispersion of higher wavenumber components. With a DRP scheme, such convection of high wavenumbers is naturally captured, for the subcritical and supercritical flow past NACA 0012 airfoil. In Fig. 5.40, snapshots of the $C_p$-distribution on the airfoil surface are shown for the case of $M_\infty = 0.6$. Similar results are plotted in Figs. 5.41, 5.42, and 5.43 for indicated combinations of $Re_\infty$, $\alpha$, and $M_\infty$. For supercritical flows with shock, one notices convection of pressure pulses in the downstream direction, from the near vicinity of the shock. It is associated with the sharp pressure discontinuity, which is responsible for wide spectrum of physical variables. However, such unsteadiness for high wavenumbers is not numerical.

It is well known that discontinuities are sites of $q$-waves (spurious upstream propagating waves due to extreme dispersion), as explained in [95] and in figures shown, no such upstream propagating waves in the separated flow region are noted. This type of unsteady downstream propagating waves is attributes of adverse pressure gradient in the flow, due to creation of inflection point in the velocity profile. The time average of $C_p$-distribution shows very good match with experimental value, indicating the fact that such pressure measurements are averaged over a finite time interval. Similar $C_p$-distribution results are also reported in [39], which show unsteadiness in instantaneous values, whereas time-averaged values show match with experimental values. For supercritical cases, shock is noted by the sharp discontinuity of $C_p$-distribution, and oscillatory behavior of shock is noted from the time series of the $C_p$-distribution.

### 5.5.2 Creation of Rotational Effects

In cases of NACA 0012 airfoil at small angles of attack, two different streams of flow are created, which originate from the top and bottom surfaces of the airfoil and reach the trailing edge with different values of entropy(s). Such dissimilar flows will create a mixing layer, which is unstable and roll into vortices of different length scales, as shown in Fig. 5.44, with entropy and vorticity contours plotted for the different cases studied here. Entropy values are calculated from

$$s - s_0 = c_v \ln \frac{p}{\rho^\gamma} \tag{5.33}$$

It is evident that entropy is continuously created as a consequence of vorticity being created by flow instability at the mixing layer. The relationship between vorticity, total enthalpy, and entropy gradient is as given by Crocco's theorem [40],

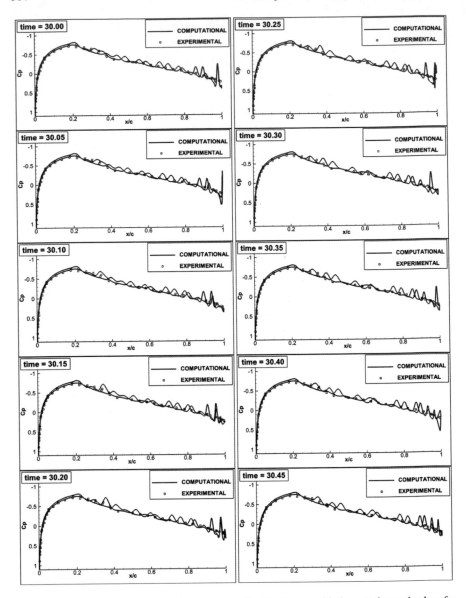

**Fig. 5.40** Comparison of computed instantaneous $C_p$-distribution with the experimental values for flow around NACA 0012 airfoil with $M_\infty = 0.6$, $Re_\infty = 3 \times 10^6$ at $\alpha = -0.14°$

$$T\nabla S + \vec{V} \times \vec{\omega} = \nabla h_0 + \frac{\partial \vec{V}}{\partial t}$$

where $h_0$ is the total entropy.

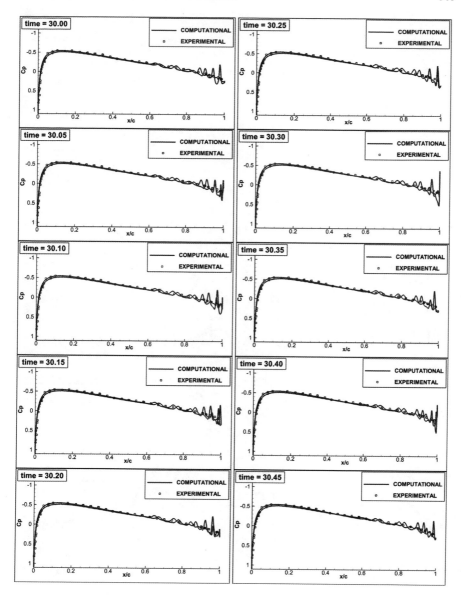

**Fig. 5.41**   Comparison of computed instantaneous $C_p$-distribution with the experimental values for flow around NACA 0012 airfoil with $M_\infty = 0.758$, $Re_\infty = 3 \times 10^6$ at $\alpha = -0.14°$

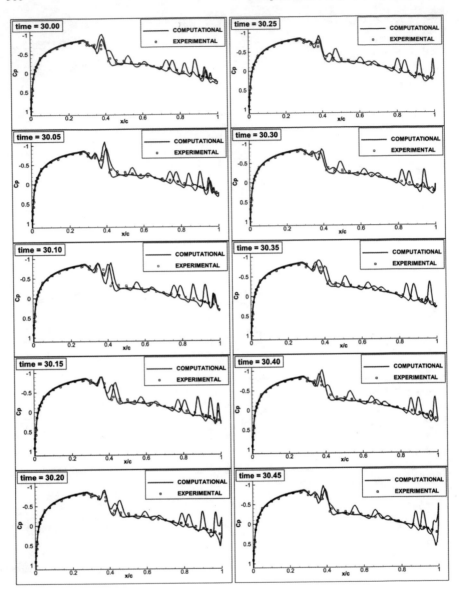

**Fig. 5.42** Comparison of computed instantaneous $C_p$-distribution with the experimental values for flow around NACA 0012 airfoil with $M_\infty = 0.779$, $Re_\infty = 3 \times 10^6$ at $\alpha = -0.14°$

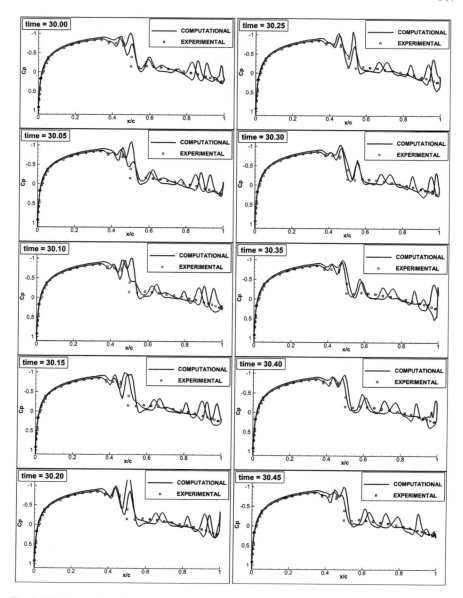

**Fig. 5.43** Comparison of computed instantaneous $C_p$-distribution with the experimental values for flow around NACA 0012 airfoil with $M_\infty = 0.8$, $Re_\infty = 3 \times 10^6$ at $\alpha = -0.14°$

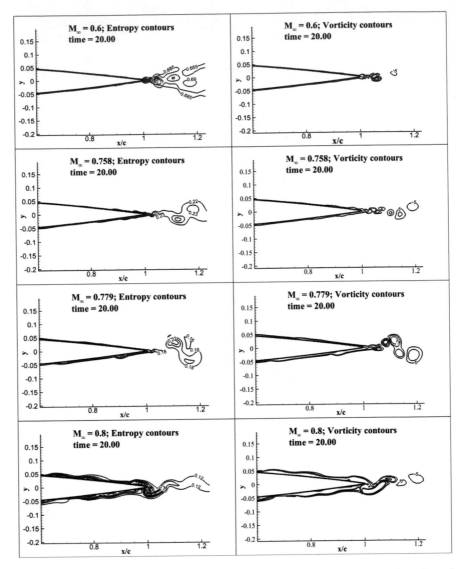

**Fig. 5.44** Entropy and vorticity contours plotted side by side for different cases as shown in each frame

### 5.5.3   Strong Shock Case and Creation of Entropy Gradient

Results obtained by numerical simulation performed for a case of $M_\infty = 0.82$, $Re_\infty = 3 \times 10^6$, and $\alpha = -0.14°$, for the flow over NACA 0012 airfoil are also reported in Figs. 5.45 to 5.50. The value of $M_\infty$ here is much higher than $M_{DD}$ for NACA 0012 airfoil. For this case, the flow gradient is expected to be much higher, and an orthogonal grid around NACA 0012 airfoil is generated with 2658 points in $\xi$- and 360 points in $\eta$-directions, respectively, to capture such gradients. Near-wall resolution of the grid is given by $\Delta s_\eta = 0.003175$. Very strong shocks with oscillatory motion are observed on upper and lower surfaces of the airfoil for this case. In Fig. 5.45, computed $C_p$-distribution is compared with the experimental value [38], showing excellent match with sharp drop in the $C_p$-distribution near the shock region. Contours of computed $M$- and $p$-contours in the vicinity of the airfoil for this case are shown in Figs. 5.46 and 5.47, respectively. Since the oncoming flow has higher value of $M_\infty$, the flow accelerates near the leading edge to higher value of $M$ than that has been observed in previous cases. For some time instants (as at $t = 15.0$), the maximum $M$ reaches a high value of 1.6872. The computed $\rho$-contours at the corresponding times are presented in Fig. 5.48. In Fig. 5.49a and b, $M$- and $s$-contours are plotted, respectively, which display the presence of very strong shock. Moreover, the strength of the shock is evident from the creation of additional entropy, aligned with the location of the shock. As usual, one notices the creation of entropy in the shear layer at the aft region of the airfoil and in the near wake.

**Fig. 5.45** Comparison of computed and experimental $C_p$-distribution for flow around NACA 0012 airfoil with $Re_\infty = 3 \times 10^6$ at $\alpha = -0.14°$ and $M_\infty = 0.82$. Computed data is time-averaged over $t = 10$ to 20 [Reproduced from Sengupta et al., *Computers and Fluids*, **88**, 19–37 (2013), with the permission of Elsevier Publishing.]

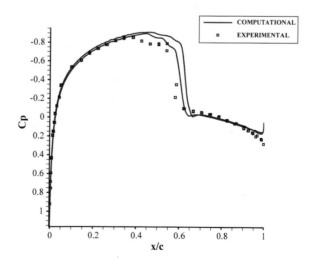

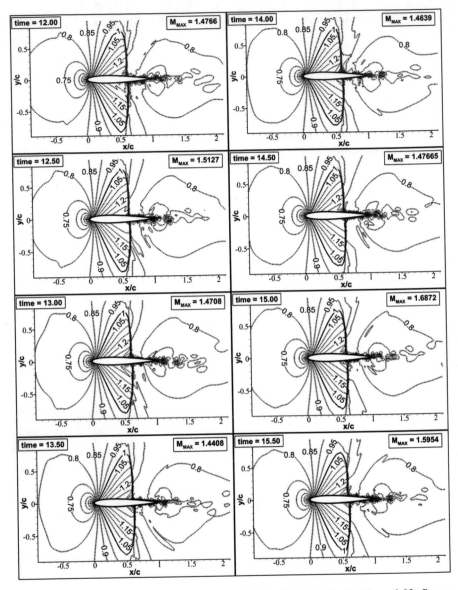

**Fig. 5.46** Computed $M$-contours for flow around NACA 0012 airfoil with $M_\infty = 0.82$, $Re_\infty = 3 \times 10^6$ at $\alpha = -0.14°$

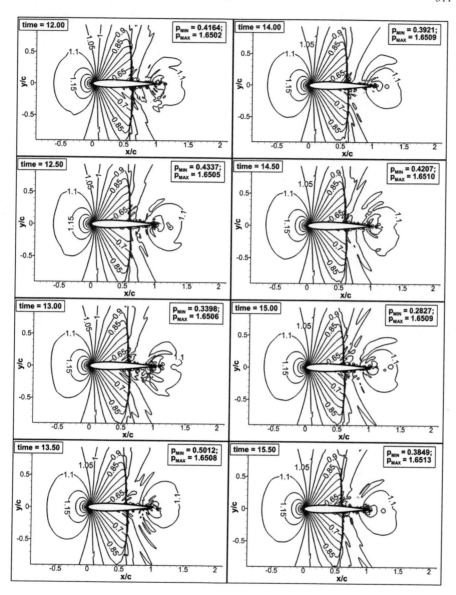

**Fig. 5.47** Computed $p$-contours for flow around NACA 0012 airfoil with $M_\infty = 0.82$, $Re_\infty = 3 \times 10^6$ at $\alpha = -0.14°$

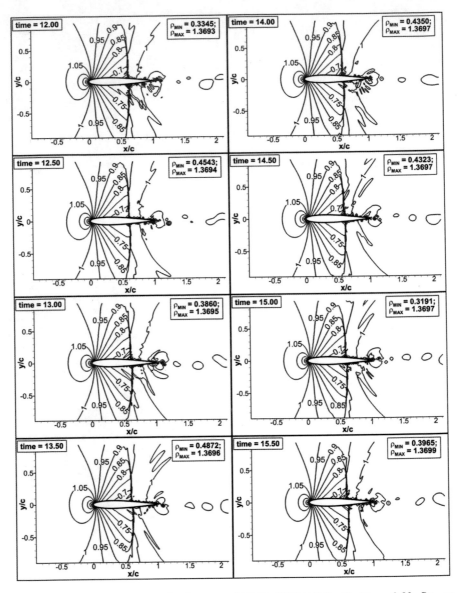

**Fig. 5.48** Computed $p$-contours for flow around NACA 0012 airfoil with $M_\infty = 0.82$, $Re_\infty = 3 \times 10^6$ at $\alpha = -0.14°$

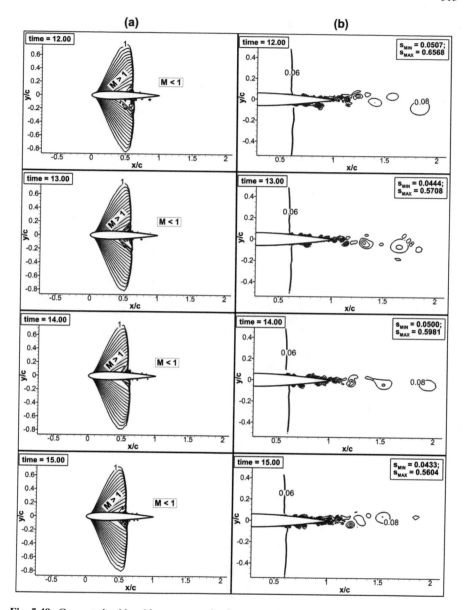

**Fig. 5.49** Computed **a** $M$ and **b** $s$-contours for flow around NACA 0012 airfoil with $M_\infty = 0.82$, $Re_\infty = 3 \times 10^6$ at $\alpha = -0.14°$ [Reproduced from Sengupta et al., *Computers and Fluids*, **88**, 19–37 (2013), with the permission of Elsevier Publishing.]

### 5.5.4  Feedback Model of Shock Wave Motion

Types of the shock wave motion suggested by Tijdeman [114] have already been discussed [10]. A possible mechanism for self-sustained shock oscillation of type A, during transonic buffeting with separated flow, was first proposed by Lee [51] and Lee et al. [53], while authors in [17] extend this mechanism to explain type C shock wave motion. From the flow evolution, type A shock wave motion is noted here for the cases of $M_\infty = 0.779$ and $M_\infty = 0.8$, in which shock wave oscillates over the airfoil surface about a mean position and exists during the complete cycle of oscillation, even though its strength varies. Due to the movement of the shock, pressure waves are formed, which propagate downstream in the separated flow region at a speed $a_p$. On reaching the trailing edge, the disturbances generate upstream moving waves at speed $a_u$, termed as 'Kutta waves' [53]. These waves interact with shock and cause oscillatory motion of the shock, closing the loop. In Fig. 5.50, contours for divergence of instantaneous velocity field are plotted, which highlights the oscillatory shock, upstream and downstream moving waves. It is noted that when shear layer interacts with sharp trailing edge, it produces a pressure wave that propagates upstream [24]. Lee suggested that the period of shock wave oscillation should agree with the time it takes for a disturbance to propagate from an upstream moving wave to reach the shock from the trailing edge. This model is applied here for the case of $M_\infty = 0.8$, as flow is separated beyond the shock and can be seen from Fig. 5.44, to calculate the period of oscillation of the shock present on the upper surface of the airfoil.

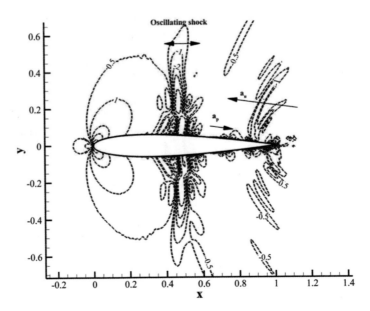

**Fig. 5.50** Divergence of instantaneous velocity field for flow around NACA 0012 airfoil with $M_\infty = 0.8$, $Re_\infty = 3 \times 10^6$ at $\alpha = -0.14°$

According to Lee [51], the time period is given by the following relation:

$$T = \int_{x_s}^{c} \frac{1}{a_p} dx + \int_{c}^{x_s} \frac{1}{a_u} dx \qquad (5.34)$$

where $x_s$ is the mean location of shock wave. This location can be obtained with a statistical analysis by plotting the streamwise evolution of the skewness ($S_p$) of pressure fluctuations ($p'$) given by [24].

$$S_p = \frac{\overline{(p'^3)}}{(\overline{p'^2})^{3/2}} \qquad (5.35)$$

The streamwise evolution of skewness is shown in Fig. 5.51. The profile exhibits sharp change from positive to negative value. One can obtain mean shock location $x_s$ corresponding to abscissa between the positive and negative peaks, where $S_p = 0$. This location is obtained as $x_s/c = 0.4734$ for the case of flow around NACA 0012 airfoil with $M_\infty = 0.8$, $Re_\infty = 3 \times 10^6$, and $\alpha = -0.14°$. The magnitude of $a_u$ is described as [51, 114]

$$a_u = (1 - M_{loc})a_{loc} \qquad (5.36)$$

where $a_{loc}$ is the local speed of sound. This equation is further simplified as [24]

$$a_u = (1 - M_\infty)a_{loc} \qquad (5.37)$$

Using above equation, speed of upstream propagating pressure waves is obtained as $0.24U_\infty$. The speed of downstream propagating pressure waves $a_p$ can be obtained from the cross-correlation analysis of pressure signals in the separated boundary layer. A cross-correlation coefficient $R(x, y, t)$ for two variables $m$ and $n$ with time delay $\tau$ can be defined as

$$R(m, n, \tau) = \frac{\overline{m'(t)n'(t+\tau)}}{\overline{m'^2 n'^2}} \qquad (5.38)$$

where $m$ and $n$ are the pressure at a point, and $m'(t)$ and $n'(t)$ are the fluctuating parts of $m$ and $n$, respectively. The bar stands for time averaging. The results are shown in Fig. 5.52, where the positive time delay is obtained, which indicates that the pressure disturbances within the separated region behind the shock waves propagate downstream toward the airfoil trailing edge. The local propagation speed of pressure disturbances is then obtained by dividing the spatial distances between the neighboring points of the pressure measurement by the time delay between the peaks of the corresponding cross-correlation shown in Fig. 5.52. As suggested by Lee [51], nearly constant speed of downstream propagation waves is obtained at different points. The speed of downstream propagating pressure disturbance is noted as $0.41U_\infty$, for the

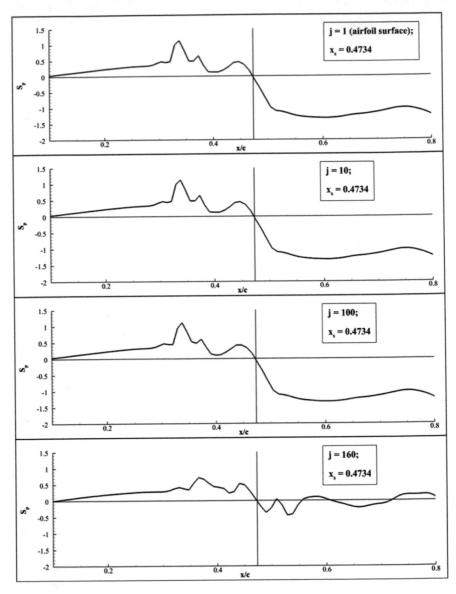

**Fig. 5.51** Skewness plotted versus non-dimensional chord to show the location mean shock wave

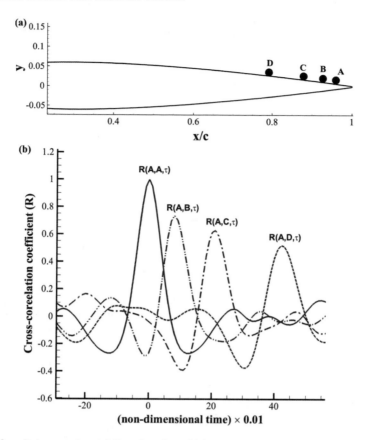

**Fig. 5.52** **a** Points on the airfoil surface for which cross-correlation is performed. **b** Cross-correlation coefficient plotted versus non-dimensional time at located points for flow around NACA 0012 airfoil with $M_\infty = 0.8$, $Re_\infty = 3 \times 10^6$ at $\alpha = -0.14°$

case of flow over NACA 0012 airfoil with $M_\infty$. Now using Eq. (5.34), one can calculate time period of shock oscillations. This is obtained as $3.501c/U_\infty$ for the flow over NACA 0012 airfoil with $M_\infty = 0.8$, $Re_\infty = 3 \times 10^6$ and $\alpha = -0.14°$.

## 5.6 Dynamic Stall: An Example of Unsteady Aerodynamics

It is noted that the lift experienced by an airfoil is steady and increases almost linearly for small steady angles of attack ($\alpha$). When this steady angle of attack is increased beyond a limit ($\alpha_{st}$), a drastic drop in the lift is observed due to flow separation; a phenomenon called the static stall, and as a consequence, the lift experienced also becomes unsteady. However, an airfoil undergoing pitching oscillation can achieve higher angles of attack instantaneously, without drastic loss of lift, as experienced

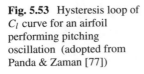
**Fig. 5.53** Hysteresis loop of $C_l$ curve for an airfoil performing pitching oscillation (adopted from Panda & Zaman [77])

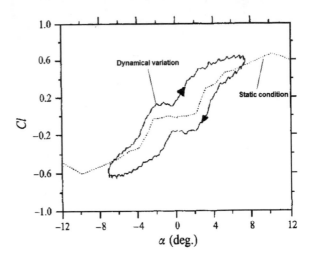

during the static stall. The instantaneous angle of attack can take values beyond $\alpha_{st}$ during the upstroke motion of the airfoil displaying a new dynamic behavior, while sustaining significantly higher lift. As $\alpha$ increases dynamically beyond $\alpha_{st}$, a situation arises when a strong vortex originates near the leading edge of the airfoil and moves downstream toward the trailing edge. This vortex creates a lower pressure on the suction surface of the airfoil, creating the extra lift observed. This is the *dynamic stall* and has been extensively discussed in [22, 55, 66, 112, 116, 117]. The instantaneous lift, drag, and pitching moment coefficient ($C_l$, $C_d$, and $C_m$, respectively) values depend on whether the airfoil is in upstroke or in downstroke motion. The overall cyclic pitching motion creates a hysteresis loop in the plots of $C_l$, $C_d$, and $C_m$ versus $\alpha$, as shown in Fig. 5.53, without experiencing massive separation at all.

The dynamic stall can be observed for rotary, flapping wings, and turbine blades. Helicopter blades undergo cyclic pitching in forward flight, as individual airfoil section encounters cyclic flow field during its sojourn as advancing and retreating section in each cycle of rotation. For a constant downwash (signifying the lift created), the relative airspeed during retreat is lower than that is during the advancing half of the cycle, that is, reflected on the angle of attack varying cyclically. However, the rotation rate of the blade has to be controlled, so that the increase in angle of attack should not expose the section to stall. If it occurs, then the periodic cycle of stall and unstall will give rise to large torsional oscillations of the rotor blades. This limits the forward speed of most modern helicopters [55, 66]. A similar sequence occurs for flapping wing aerodynamics. Thus, understanding dynamic stall is very vital for rotary and flapping wing aerodynamics, playing a crucial role in the design of airfoils and wings for unsteady aerodynamic applications.

Thus, the dynamic stall problem occurs on an airfoil during rapid, transient motion in which angle of attack surpasses the static stall limit, in its most general appellation. It has a significant influence on the performance of various engineering and natural living forms, as in helicopters, highly maneuverable aircraft, jet engines, wind turbines, flapping wing aerial vehicles, birds, and insects. The dynamic stall is a strong function of the airfoil geometry (particularly the leading edge shape), Reynolds number, Mach number, degree of unsteadiness or non-dimensional pitch rate, state of the airfoil boundary layer, airfoil initial angle of attack before pitching, three-dimensionality, type of airfoil motion, location of pitch point, etc. The dynamic stall can be of advantage or disadvantage, depending on the circumstances where it appears. While it is a limiting factor on forward speed of a helicopter due to the abrupt pitching moment variations during dynamic stall, creating adverse control loads and causing restrictions on the flight envelope, in contrast, dynamic stall helps fighter aircraft designers due to available dynamic lift that allows enhanced maneuverability.

### 5.6.1 Phenomenological Description of Dynamic Stall

To understand the dynamic stall over an oscillating airfoil, aerodynamic load coefficients ($C_l$ in Fig. 5.54b, $C_d$ in Fig. 5.54c, and $C_m$ in Fig. 5.54d) are sketched with varying angle of attack (AoA or $\alpha$, in degrees), as given in [91]. The airfoil pitches about the quarter-chord point as shown in Fig. 5.54a, and the variation of AoA is in the increasing order of the number in the upstroke. The separate dashed line denotes the static stall condition in each frame. During the upstroke motion, the condition is marked first where the flow is attached (numbered as 1), and the airfoil is well below static stall condition. The airfoil passes through the second stage, termed as the moment stall (numbered as 2, marked with diamond), where the airfoil is beyond the static stall AoA, with a strong vortex, namely, the dynamic stall vortex (DSV) present over the suction surface. This DSV originates near the leading edge of the airfoil and convects downstream toward the trailing edge, causing a low-pressure region over the airfoil, leading to an increase in lift, drag, and nose-down pitching moment, as shown in Figs. 5.54b, c and d, respectively. In the frames, number 3 (marked with circle) denotes the shedding of DSV in the near wake, which causes the abrupt drop in lift and drag shown in Figs. 5.54b and c, respectively. This abrupt loss of lift is termed as lift stall. For the depicted cyclic variation shown in the figure, the number 4 shown during the downstroke motion denotes complete stall condition, while number 5 indicates reattachment of detached shear layer from leading edge toward trailing edge. These sketches are representative to explain dynamic stall and are not actual data [91].

For sufficiently high oscillation frequency and maximum $\alpha$, the shedding of DSV is well defined, with loads showing large fluctuations, and the results do not depend on Reynolds number, type of motion and airfoil shape. This is referred to as deep dynamic stall [22, 55, 67]. For airfoils operating at large $C_l$, compressibility effects are noted at very low free-stream Mach numbers [65, 67, 68, 112], in the range $M_\infty \sim 0.2 - 0.3$. At higher $M_\infty$, transonic flow over the airfoil is noted, with occurrence of normal shock, and shock wave–boundary layer interactions are observed, resulting in change of the dynamic stall mechanism. It is apparent that $M_\infty$ plays an important

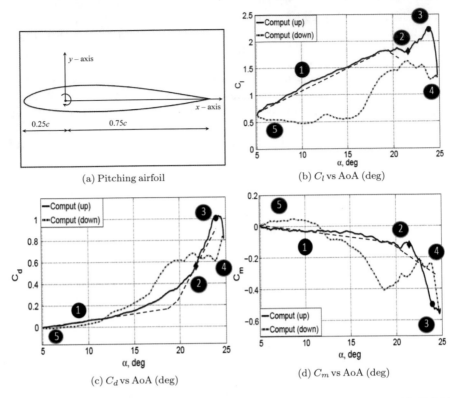

(a) Pitching airfoil

(b) $C_l$ vs AoA (deg)

(c) $C_d$ vs AoA (deg)

(d) $C_m$ vs AoA (deg)

**Fig. 5.54** Dynamic stall mechanism [Reproduced from Jyoti et al., *Physics of Fluids*, **29**, 076104 (2017), with the permission of AIP Publishing.]

role in the dynamic stall process. Therefore, a much better understanding of the formation and behavior of DSV is required at higher compressible flight speeds.

It is noted [91] with respect to experiments of Gray et al. [35] for $M_\infty \sim 0.4$, the pitching moment diverges from the attached flow value at a considerably lower $\alpha$ than for $M_\infty \sim 0.2$, with the areas enclosed by the $C_m$ loops different in these cases. As the $C_m$ loop is related to pitch damping (note in Corke et al. [20]), compressibility influences aeroelastic stability of helicopter rotors in the forward flight. It is noted that helicopter rotors experience variable Mach number in every revolution. Fukushima and Dadone [28] noted that the formation of shock waves limited the rise of velocity at the suction peak, and thereby the lift on airfoil is limited by shock-induced separation, without any delay in stall. It has been observed that as $M_\infty$ exceeds 0.6, the dynamic lift overshoot on aircraft wings is virtually eliminated [19]. This decrease in overshoot has also been observed for helicopter blades oscillating in pitch by Lorber and Carta [59], while performing oscillatory as well as constant pitch rate motion tests on airfoils.

To reduce the amount of wind tunnel testing needed to determine dynamic stall behavior, empirical as well as semi-empirical dynamic stall models have been developed. Leishman and Beddoes [56] observed that a shock appears at high Mach numbers and as $\alpha$ increases, the shock strengthens and moves toward the rear of the airfoil and leads to shock-induced separation. Visbal [117] found that the compressibility gives rise to additional mechanisms in dynamic stall processes such as shock wave–boundary layer interactions. Most of the previous attempts to computationally study flow past unsteady airfoils have used unsteady Reynolds-averaged Navier–Stokes (URANS) equation with turbulence models. Ekaterinaris et al. [26] give a comprehensive review of the computational methods that were used to study dynamic stall at the turn of the last century. Martinat et al. [62] compare various URANS and hybrid turbulence modeling approaches noting that the standard URANS methods are dissipative which quickly attenuate the instabilities and vortex structures related to dynamic stall. The inability of existing URANS-based computational fluid dynamic (CFD) analyses to accurately predict the dynamic stall phenomenon [1, 23, 55] warrants a deeper and more systematic investigation of the effect of Mach number on the physics of dynamics stall using model-free CFD analysis.

An attempt has been made without transition and turbulence models in Suchandra [112], where the author has computationally investigated and compared results with experimental results of McCroskey et al. [65, 67, 68] for $M_\infty = 0.283$. A good agreement was observed between the computations and the experiments. The fundamental investigation of dynamic stall where Mach number is used as an independent parameter has seldom been attempted. The topic is technologically significant due to its implication for the development of superior high-fidelity tools for rotor blade analysis.

## 5.6.2 Characterization of Dynamic Stall

The dynamic stall phenomenon is strongly influenced by the main factors [66]: (i) DSV and its shedding, (ii) associated shear layer eruptions, and (iii) localized compressible effects (for $M_\infty \geq 0.2$). Experimentally, DSV formation and shedding from the upper surface of the airfoil induces unsteady pressure field [66]. We have already alluded to the limiting case of *deep dynamic stall* [22, 55, 66] in this context. For lower than limiting conditions, the vortex shedding is less well defined and the development of vortices, and the loads depend on parameters of flow and geometry. This is termed as the *light dynamic stall*. The boundary layer forming over the suction surface exhibits leading edge and trailing edge separations, and the subsequent stall is termed as either the leading edge stall or the trailing edge stall.

Leading edge stall [22] is related to the formation of a small separation bubble near the LE. Trailing edge stall occurs due to the separation of boundary layer at the TE. Increasing the AoA moves the point of separation upstream over the airfoil, and sometimes the stall condition does not fall strictly in one of the categories, instead displays mixed stall behavior [66]. Additionally, thinner airfoils show a separate type of stall, referred to as *thin -airfoil stall* in the literature [22, 66].

For $M_\infty \simeq 0.2$, pitching airfoils exhibit small supersonic pocket(s) near the leading edge during the upstroke motion. This leading edge supersonic flow can show the absence of shock waves [66] or show very weak shocks. These supersonic pockets increase the tendency of leading edge stall [66]. For $M_\infty > 0.3$, similar acceleration is likely to cause shock-induced leading edge stall [47, 80]. We will further discuss lift and moment stall [66], as well as classification of vortices [106] observed during dynamic stall.

Most previous computational studies on this topic have used unsteady Reynolds-averaged Navier–Stokes (unsteady RANS or URANS) equation with turbulence models. Authors in [23] have used Transonic Unsteady Rotor Navier–Stokes (TURNS) analysis (originally developed in [109, 110]) which uses Spalart–Allmaras, one equation turbulence model [108] and algebraic Baldwin–Lomax model [7]. Authors in [116] have used RANS and shear stress transport (SST) $k - \omega$ turbulence model in combination with Langtry [48] and Menter's transition transport equations [69] to study DSV formation on OA-209 airfoil. However, authors in [106] have reported results for incompressible flow past unsteady airfoils via ILES, without any model on body-fitted orthogonal grid, formulating the problem using stream function ($\psi$) and vorticity ($\omega$), as dependent variables.

Next, compressible NSE [10, 94] results are discussed, which solved NSE in a rotating, non-inertial frame of reference. Direct simulation, without any transition or turbulence model, has been performed on body-conforming orthogonal O-type grid around NACA 0012 airfoil. Governing equations and auxiliary conditions are derived in non-inertial framework. High accuracy optimized compact schemes $OUCS3$ and $OUCS2$ [93] have been used for derivatives of convective terms and optimized three-stage Runge–Kutta ($OCRK_3$) method [86, 93, 104] has been used for time integration. The computations are validated and compared with the experimental dynamic stall data [65, 66, 68].

## 5.7  Governing Equations for Dynamic Stall

Governing equations and boundary conditions are written in body-fixed, non-inertial frame [93, 112], which makes the formulation simple and eliminates the need for multiple grid generation [117]. Also, implementation of boundary conditions in inertial frame introduces large errors. Thus, NSE for compressible flow is modified for a non-inertial frame, results of which are validated for $M_\infty \approx 0.3$ [112]. In the non-inertial frame, the continuity equation remains invariant, while the momentum equations are augmented with pseudo-forces in the non-inertial frame (by the Eulerian, centrifugal, and Coriolis acceleration terms). The work done due to these pseudo-forces is considered in the energy equation.

The pure pitching motion of the airfoil is described by instantaneous $\alpha$ changing as $\alpha = \alpha_m + \alpha_a \sin(\omega_a t)$. The angular velocity vector is given by

$$\overrightarrow{\Omega} = \alpha_a \omega_a \cos(\omega_a t)(-\hat{k})$$

The governing equations of motion are conservation laws for mass, momentum, and energy, given by unsteady, compressible NSE. The 2D, unsteady, compressible NSE is solved. Wave propagation noted from NSE is certainly like the Euler equations [54]. The NSE is solved in divergence (also called as strong conservation) form, in a finite difference framework. The divergence form is preferred, as it is effective in capturing correct jump relations across discontinuities and nonlinearities, which arise in the solution [50, 112]. The NSE in non-inertial frame, expressed in non-dimensionalized form [93, 112], is given in the physical $(x, y)$-plane as [91]

$$\frac{\partial \hat{Q}}{\partial t} + \frac{\partial \hat{E}}{\partial x} + \frac{\partial \hat{F}}{\partial y} = \frac{\partial \hat{E}_v}{\partial x} + \frac{\partial \hat{F}_v}{\partial y} + \hat{S} \tag{5.39}$$

where the conservative variables are given as

$$\hat{Q} = [\rho; \ \rho u; \ \rho v; \ \rho e_t]^T$$

The convective flux vectors $\hat{E}$ and $\hat{F}$ are given as

$$\hat{E} = [\rho u; \ \rho u^2 + p; \ \rho u v; \ (\rho e_t + p)u]^T$$

$$\hat{F} = [\rho v; \ \rho u v; \ \rho v^2 + p; \ (\rho e_t + p)v]^T$$

The viscous flux vectors $\hat{E}_v$ and $\hat{F}_v$ are given as

$$\hat{E}_v = [0; \ \tau_{xx}; \ \tau_{xy}; \ u\tau_{xx} + v\tau_{xy} - q_x]^T$$

$$\hat{F}_v = [0; \ \tau_{yx}; \ \tau_{yy}; \ u\tau_{yx} + v\tau_{yy} - q_y]^T$$

The source terms are given as

$$\hat{S} = [0; \ \rho f_x; \ \rho f_y; \ \rho \overrightarrow{f} . \overrightarrow{V}]^T$$

where $\overrightarrow{f} = f_x \hat{i} + f_y \hat{j}$; $\overrightarrow{V} = u\hat{i} + v\hat{j}$. The terms $f_x$ and $f_y$ are time-dependent pseudo-body-forces arising in the non-inertial frame and are given as

$$f_x = \underbrace{y\alpha_a \omega_a^2 \sin(\omega_a t)}_{\text{Eulerian term}} + \underbrace{x\alpha_a^2 \omega_a^2 \cos^2(\omega_a t)}_{\text{Centrifugal term}} - \underbrace{2v\alpha_a \omega_a \cos(\omega_a t)}_{\text{Coriolis term}}$$

$$f_y = \overbrace{-x\alpha_a \omega_a^2 \sin(\omega_a t)} + \overbrace{y\alpha_a^2 \omega_a^2 \cos^2(\omega_a t)} + \overbrace{2u\alpha_a \omega_a \cos(\omega_a t)}$$

The components of the symmetric viscous stress tensor are given as

$$\tau_{xx} = \frac{\mu}{Re_\infty}\left(2\frac{\partial u}{\partial x} - \frac{2}{3}\nabla.\vec{V}\right); \quad \tau_{yy} = \frac{\mu}{Re_\infty}\left(2\frac{\partial v}{\partial y} - \frac{2}{3}\nabla.\vec{V}\right); \quad \tau_{xy} = \frac{\mu}{Re_\infty}\left(\frac{\partial u}{\partial y} + \frac{\partial v}{\partial x}\right)$$

The heat conduction terms, $q_x$ and $q_y$, are given as

$$q_x = -\frac{\mu}{Pr Re_\infty(\gamma-1)M_\infty^2}\frac{\partial T}{\partial x}; \quad q_y = -\frac{\mu}{Pr Re_\infty(\gamma-1)M_\infty^2}\frac{\partial T}{\partial y}$$

All terms in Eq. (5.7) are non-dimensionalized with chord ($c$) as the length scale; the free-stream density ($\rho_\infty$) for density; the free-stream velocity ($U_\infty = M_\infty\sqrt{\gamma RT_\infty}$) for velocity; $\rho_\infty U_\infty^2$ for pressure; the free-stream temperature ($T_\infty$) for temperature; $c/U_\infty$ for time; the free-stream dynamic viscosity $\mu_\infty$, for dynamic viscosity; and $U_\infty^2$ for total specific energy $e_t$. Additionally, Sutherland's viscosity law relating the coefficient of dynamic viscosity $\mu$ with absolute temperature $T$ is used.

We note that $x$, $y$, $u$, and $v$ are in the moving frame, whereas $p$, $\mu$, $\rho$, $T$, $t$, $R$, $\gamma$, $\lambda$, and $\kappa$ are same in both inertial and non-inertial frames. Using free-stream velocity originally given in inertial frame needs to be transferred to the rotating frame value using the relationship:

$$\vec{V}_{Inertial} = \vec{V}_{Moving} + \vec{\Omega} \times \vec{r}$$

where $\vec{r} = x\hat{i} + y\hat{j}$ is the position vector.

Equation (5.7) is the transformed plane representation of the governing equation in the generalized curvilinear coordinates, which is solved [93, 112]. On the airfoil surface, no-slip condition is imposed on the (non-inertial) velocity components, and adiabatic wall condition is imposed on heat conduction terms: $u = 0$; $v = 0$; $q_x = 0$ and $q_y = 0$. The details of implementation are provided in [10, 91, 112], for the dynamic stall problem.

### 5.7.1  Numerical Methods and Grids

For the flow over the pitching airfoil with strong shocks forming at relatively higher $M_\infty$, the unsteadiness is due to physical dispersion of higher wavenumber components, and the used DRP scheme captures these components. The optimized compact $OUCS3$ scheme and its symmetrized version $S - OUCS$ [25, 93] are used to evaluate convective flux terms in $\eta$- and $\xi$-directions, respectively. Optimized three-stage Runge–Kutta ($OCRK_3$) method is used for time integration providing the necessary DRP property. Parallel computing by the Schwarz domain decomposition technique [100] is performed with message passing interface (MPI) used to parallelize the solver, with details in [91, 93, 112]. Viscous flux terms are calculated using second-order central discretization ($CD_2$) in self-adjoint form, preserving isotropic nature. Artificial diffusion terms are needed to suppress numerical instabilities arising due

**Table 5.3** JST coefficients used for the simulated cases for dynamic stall [91]

| $M_\infty$ | $\kappa_2$ | $\kappa_4$ |
|---|---|---|
| 0.283 (15 cycles) | 0 | 0.2 |
| 0.4 (15 cycles) | 0.15 | 0.15 |
| 0.5 (first 4 cycles) | 0.25 | 0.2 |
| 0.5 (next 11 cycles) | 0.20 | 0.15 |

to nonlinearities and discontinuities, and its details are given in subsection 5.2.7. The cases presented here have been simulated using the minimum possible values of the JST coefficients as given in Table 5.3 [91].

A computational domain of approximate radius as $24c$, with $c = 0.61m$, is used as in [65, 67, 68], to ensure accurate implementation of far-field boundary condition and to reduce acoustic wave reflections from the boundary. An orthogonal, body-conforming O-grid is generated using hyperbolic grid generation technique [5, 72, 93]. A finer grid around NACA 0012 airfoil consists of 2001 points in $\xi$-direction (azimuthal direction) and 350 points in $\eta$-direction (wall-normal direction) for the results. Initially, a coarser grid consisting of 636 points in the azimuthal direction was used for developing the high accuracy compact difference code. A close-up view of these grids is shown in Fig. 5.55. The cell spacing for the first grid line in wall-normal direction is approximately $2 \times 10^{-6}c$, which corresponds to $y^+ \sim 0.3$, as per

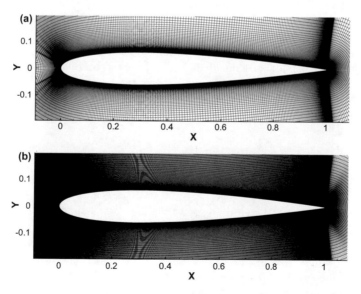

**Fig. 5.55** Close-up view of orthogonal grid generated around NACA0012: **a** 636 × 350, **b** 2001 × 350

flat plate approximation [36]. It is noted [112] that match of $C_l$ with experimental data improves with increasing number of azimuthal grid points. Also, lift stall is sharper with refined azimuthal grids with better match with experimental lift stall.

## 5.8 Computing Deep Dynamic Stall

Next, results are presented for the constant parameters given by $Re_\infty = 3.45 \times 10^6$; $p_\infty = 1$ atmosphere, mean angle of attack of $\alpha_m = 14.91°$, amplitude of pitching oscillation as $\alpha_a = 9.88°$; the chord of 0.61m, reduced frequency of $k = \frac{\omega_a c}{2U_\infty} = 0.151$, and the non-dimensional time period equal to 20.805 [65–68].

Certain features of deep dynamical stall are shown in Figs. 5.56, 5.57, and 5.58, using the finer azimuthal grid. On the left column of Figs. 5.56 and 5.57, the non-dimensional pressure contours are shown for the indicated angles of attack during upstroke and downstroke, respectively. Also shown in these frames are the representative streamlines in the near vicinity of the airfoil in body-fixed frame. On the right column frames, corresponding wall pressure distribution is shown for the airfoil. Vortices near the leading and trailing edges are seen attached, which are responsible for the visualized low pressure over the suction surface. These are the structures of evolving DSV.

In Fig. 5.58, non-dimensional vorticity contours are depicted for the upstroke and downstroke motion of the airfoil for the deep dynamic stall case shown in Figs. 5.56 and 5.57. The computed values are as obtained in the body-fixed frame. One can notice the correspondence of these large coherent vortices with the pressure contours. An interesting observation that can be made during the downstroke is the occurrence of large suction peaks in $C_p$ near the trailing edge shown in the right frames of Fig. 5.57. This peak is associated with the trailing edge separation, while such trailing edge suction peaks are not present in the experimental data, as described next. The other reason could be related to 2D computations will over-emphasize trailing edge separation.

### 5.8.1 Limitation of Dynamical Stall Experimental Data

Most of the experimental data used for validating computational results are taken from [65, 68]. According to these well-documented reports, following information helps in understanding the experimental results and the associated problem faced in computing the flow field. In these experiments, pressure was not measured at the trailing edge of the airfoil; instead, the data were extrapolated from neighboring points. As a consequence, the trailing edge suction peak created due to the trailing edge vortex is not well represented in the experimental data. The Kulite pressure transducers used were with frequency response of around 250 Hz. In contrast, computed deep dynamical stall results exhibited frequency in excess of 2kHz. Also, long air column

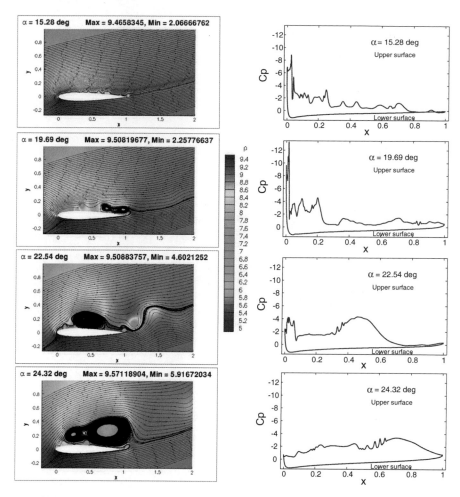

**Fig. 5.56** Upstroke airfoil motion for a case (with $OUCS3$ scheme). On the left, non-dimensional pressure contours around the airfoil along with representational streamlines at indicated angles of attack, in rotating body-fixed frame. *Non-dimensional $p = 8.92$ corresponds to 1 atm.* Maximum and minimum values of non-dimensional $p$ are indicated. Results are with 2001 grid points in azimuthal direction. On the right, instantaneous $C_p$ plots for corresponding angles of attack [112]

used in the transducers causes damping of higher frequencies in the signal. These high-frequency components are noted to have significantly high amplitude, as noted in the $C_p$-plots in Figs. 5.56 and 5.57.

The experimental data are obtained as measured, without any wind tunnel correction. Available corrections are essentially for steady flows, not applicable here. The presence of wall in the downstream of the airfoil causes flow acceleration in the

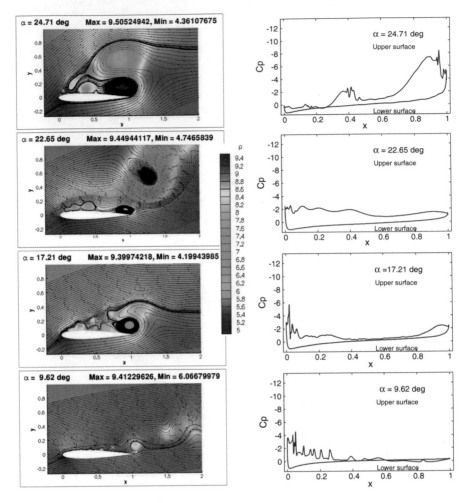

**Fig. 5.57** Downstroke airfoil motion for a case (with *OUCS*3 scheme). **Left column:** Non-dimensional pressure contours around the airfoil along with representational streamlines at indicated angles of attack, in rotating body-fixed frame. *Non-dimensional p = 8.92 corresponds to 1 atm.* Maximum and minimum values of non-dimensional *p* are indicated. Results are with 2001 grid points in azimuthal direction. **Right column:** Instantaneous $C_p$ plots for corresponding angles of attack [112]

streamwise direction, which along with additional acceleration created by tunnel wall boundary layer thickening will distort the flow measurements, including reduction of trailing edge separation. This will cause alteration of load measured.

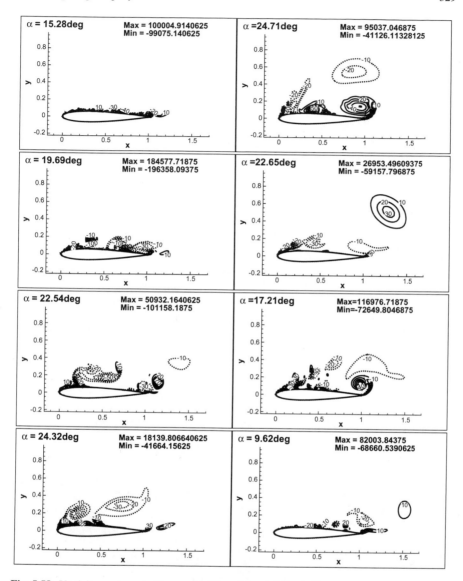

**Fig. 5.58** Vorticity contours in the near vicinity of the airfoil at indicated angles of attack for the deep dynamic stall case are shown. Maximum and minimum values of vorticity are indicated with solid lines are for positive contours and the dashed lines are for negative contours. Results are with 2001 grid points in azimuthal direction. The frames on left, from top to bottom, depict the upstroke motion of the airfoil and the frames on right depict downstroke events [112]

## 5.8.2  Why Compressible Flow Computations Necessary?

High Reynolds number unsteady flows over airfoils are known to exhibit transonic nature near the leading edge suction region, as noted by a supersonic pocket experimentally [66]. This is observed computationally as well [112], and one such snapshot is shown in Fig. 5.59, for the deep dynamic stall case, even though the free-stream Mach number is small enough to be termed as incompressible flow ($M_\infty = 0.283$). While the compressibility effects are perceptible, the flow transits from subsonic to supersonic and back to subsonic without the formation of any shock.

In the following, the results for deep dynamic stall are compared and discussed for $M_\infty = 0.283$, 0.4, and 0.5, as effects of compressibility have been observed at very low Mach numbers ($M_\infty = 0.2 - 0.3$) for high angles of attack. As we have already noted above that locally the flow can accelerate to supersonic Mach number without forming shock waves for such low $M_\infty$ flows. While such acceleration is possible due to delay of separation due to the dynamically varying AoA, a slight increase in $M_\infty$ will cause bigger supersonic pockets which will terminate into strong shock(s). Subsequent shock–boundary layer interaction will in turn nullify all the positive benefits of dynamic stall. Hence, a study of compressibility effect on dynamic stall is necessary and is presented next, as such effects are equally important for aircraft in maneuvers near static stall angles. The computational results are first compared with the experimental results of McCroskey et al. [65, 67, 68] for $M_\infty = 0.283$ for the loads and subsequently, higher Mach number cases are presented.

**Fig. 5.59** Enlarged view of upper portion of the airfoil near the leading edge, showing supersonic pockets. The figure shows Mach number $M$ contours around the airfoil at indicated angle of attack (upstroke) for deep dynamic stall case in a rotating body-fixed frame. Results are with 2001 grid points in azimuthal direction

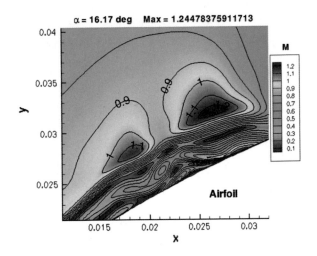

### 5.8.3 Validating Computational and Experimental Results for Aerodynamic Load Coefficients for $M_\infty = 0.283$

The dynamic stall process is not perfectly repetitive and varies from cycle to cycle, for which results are averaged over multiple cycles, thereby smoothing such variations and the mean dynamic stall behavior for all single-frequency oscillations are presented. Figure 5.60 shows hysteresis loop of pitching moment coefficient ($C_m$) for $M_\infty = 0.283$, ensemble-averaged over different number of oscillation cycles, starting from the last quarter of second cycle. The vertical lines (standard deviation bars) give the variation of the load data over the constituent cycles used in averaging. Indicated loops clearly establish the convergence of the shape of load curve using 12 or more cycles. One clearly notices that there are apparently no differences between results obtained using 11 and 12 cycles. There is minimal change in the standard deviation even with more cycles being used for averaging. Only large deviations are noted at high AoA, especially during the downstroke phase, due to unsteady separated flow. These and the following results have been presented in [91].

In Fig. 5.61, computational and experimental results [65, 67, 68] for $M_\infty = 0.283$ are compared, where the hysteresis loops for pressure-based load coefficients, i.e., lift coefficient ($C_l$), drag coefficient ($C_d$), and pitching moment coefficient ($C_m$) are shown. These load coefficients do not include viscous contributions and have to be understood as such. These depictions are useful [65] for large-amplitude unsteady motions, as the viscous components on these coefficients are negligible, as compared to pressure force component. Shown results are averaged over 15 cycles, with averaging started from the last quarter of the second cycle. This helps in avoiding the large unsteady contributions due to the flow transience at the onset of computations. Also in the absence of realistic initial data, here the computations are started with uniform flow field which is strictly not representative. Solid line in the figure corresponds to events during upstroke and dotted line denotes downstroke motion of the airfoil.

From Fig. 5.61a, it is noted that the computed $C_l$ is different from the experimental $C_l$. However, overall shape of $C_l$ curve matches better with the experimental data, as compared to $C_m$ in Fig. 5.61c and $C_d$ in Fig. 5.61b, where the match is better with experimental data at low angles of attack. Largest deviations from the experimental data are at high $\alpha$ values are seen during downstroke. This could be attributed to 3D turbulence effects [9, 62], not captured by 2D ILES computations, with computed separation stronger than that is noted experimentally. As mentioned earlier, experimentally measured pressure field in McCroskey et al. [65, 68] did not measure pressure at the trailing edge of the airfoil, which is extrapolated from the data at the neighboring points.

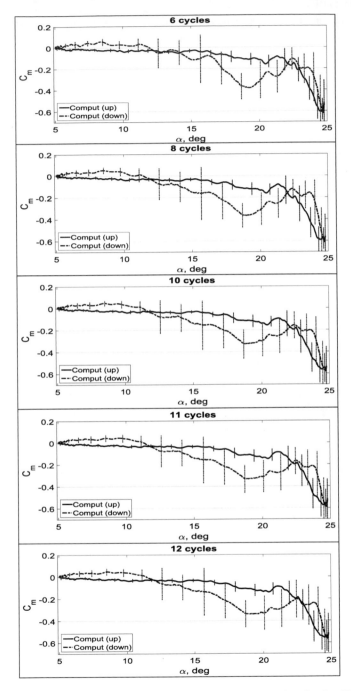

**Fig. 5.60** Hysteresis loop of $C_m$ shown for the indicated numbers of cycles for $M_\infty = 0.283$ [Reproduced from Jyoti et al., *Physics of Fluids*, **29**, 076104 (2017), with the permission of AIP Publishing.]

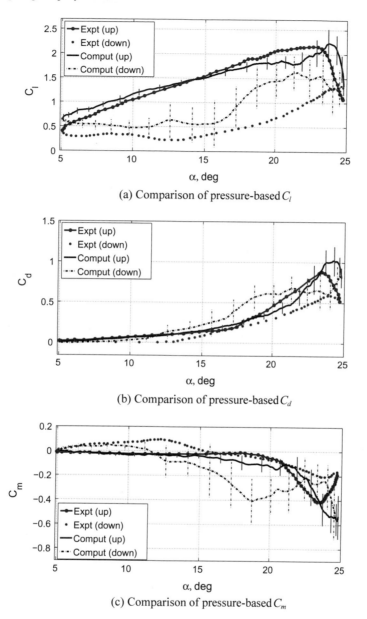

(a) Comparison of pressure-based $C_l$

(b) Comparison of pressure-based $C_d$

(c) Comparison of pressure-based $C_m$

**Fig. 5.61** Lift, drag, and pitching moment coefficients for $M_\infty = 0.283$ compared with experimental data in [65, 68] [Reproduced from Jyoti et al., *Physics of Fluids*, **29**, 076104 (2017), with the permission of AIP Publishing.]

## 5.8.4  *Compressibility Effects on Dynamic Stall*

Next, we show results of dynamic stall process with an individual cycle (cycle 7 has been chosen) investigating the compressibility effects. If the boundary layer undergoes transition before the advent of adverse pressure gradient, then there are major changes in the dynamic stall process due to compressibility [117]. Visbal [117] observed that the trailing edge stall becomes leading edge stall due to separation caused by shock wave–boundary layer interaction, which resulted in stall delay and reduction of maximum lift coefficient. Compressibility significantly alters the DSV life cycle, from stall onset, detailed DSV formation, duration of augmented dynamic lift during pitching cycle, shock—boundary layer interactions, and release of DSV in the near wake. DSV forms as a rapid succession of events in a very small range of $\alpha$, with the leading edge flow separation occurring at a very low $\alpha$. Thereafter, the vorticity developed near the leading edge region coalesces into a vortex, which further strengthens the vortex with increasing $\alpha$. The vortex develops distinctly into the DSV, which for the sake of better understanding the DSV formation are presented in Fig. 5.62, with time for the three displayed Mach numbers. Similarly in Fig. 5.63, the pressure, vorticity, and streamline contours are presented during the shedding process, where the fully developed DSV is about to convect down into the near wake. Along with the formation of leading edge vortex (LEV), small-scale vortices of anti-clockwise vorticity (which originates at the foot of DSV and convecting toward leading edge) can be observed in Figs. 5.62a, d, g.

The analysis of computed seventh pitching cycle is performed to note the non-dimensional time and $\alpha$ values at which the DSV forms (moment stall) and shedding process begins (lift stall) during the upstroke motion in Table 5.4 for the same Mach number. On comparing the compressibility effects for both the processes of lift and moment stalls, one makes the following observations:

(a) For $M_\infty = 0.283$, it is observed that the DSV forms near the leading edge, as shown in Fig. 5.62a. But in Fig. 5.62d (for $M_\infty = 0.4$), it forms near mid-chord point. It further shifts downstream toward the trailing edge for $M_\infty = 0.5$ as observed in Fig. 5.62g.

(b) The time a vortex takes to grow from the nascent stage to a fully developed DSV stage, and the residence time for which DSV remains close to and over the airfoil, decreases with increase in Mach number. From Table 5.4, it could be observed that moment and lift stalls occur at earlier times and at smaller angles of attack with increasing Mach number. From Table 5.4, it is observed that the difference between the non-dimensional time at which moment and lift stalls occur for $M_\infty = 0.283$ is 1.45; for $M_\infty = 0.4$ this is 0.70, and for $M_\infty = 0.5$ this is 0.40. Similarly, the angular difference between both the processes decreases with increase in Mach number. For instance, the difference in $\alpha$ between moment and lift stall for $M_\infty = 0.283$ is 1.65 deg; for $M_\infty = 0.4$, this difference in $\alpha$ is 1.69 deg; and for $M_\infty = 0.5$ the same difference is 1.08 deg (as noted in Table 5.4). The convection speed of DSV increases with compressibility with shock forming later for higher Mach numbers. The DSV forms later and moves

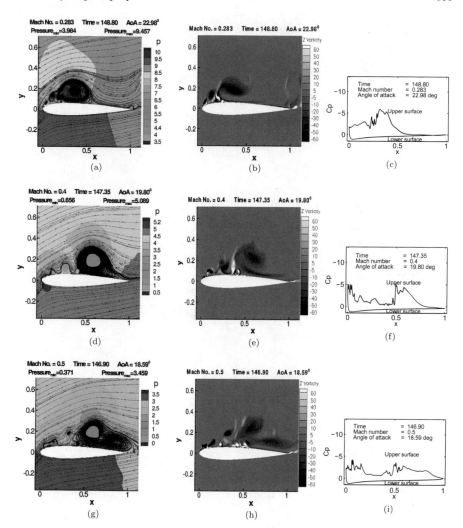

**Fig. 5.62** Formation of dynamic stall vortex (DSV). Left column shows pressure contours along with representative streamlines. Middle column shows vorticity contours where negative values show the clockwise vorticity and positive values show the anti-clockwise vorticity. Right column shows instantaneous $C_p$ plots [Reproduced from Jyoti et al., *Physics of Fluids*, **29**, 076104 (2017), with the permission of AIP Publishing.]

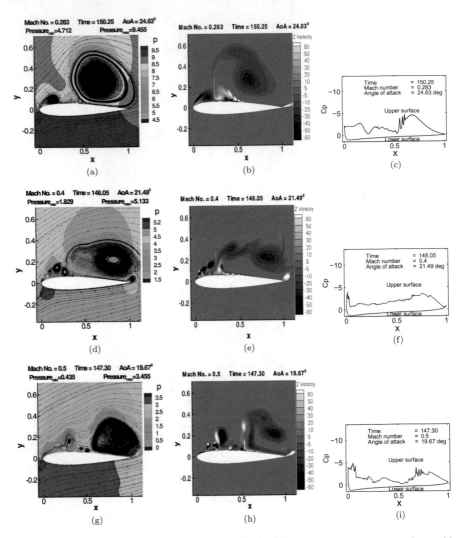

**Fig. 5.63** Shedding of dynamic stall vortex. Left column shows pressure contours along with representative streamlines. Middle column shows vorticity contours where negative values show the clockwise vorticity, and positive values show the anti-clockwise vorticity. Right column shows instantaneous $C_p$ plots [Reproduced from Jyoti et al., *Physics of Fluids*, **29**, 076104 (2017), with the permission of AIP Publishing.]

unimpeded after it crosses the shock jump, as the pressure gradient is negligible after the shock, and the DSV has easy passage.

(c) Instantaneous $C_p$ plots presented in Figs. 5.62c, f, i denote the strength of DSV forming on the suction surface, with the suction due to DSV decreases with increasing Mach number. Thus, compressibility weakens the strength of DSV. An attempt to explain the reason for weakening of strength with compressibility has been made by Chandrasekhara and Carr [15], who noted that the flow separates at lower $\alpha$ with increase in Mach number due to less adverse pressure gradient near the leading edge. It is also observed computationally in Fig. 5.64b, d, and f, the reduction in $C_p$ values near leading edge (depicting the strength of corresponding small-scale vortices) at a particular $\alpha$ is observed. In Fig. 5.64b, d, and f, $C_p$ values are averaged over 15 cycles of oscillation at a particular $\alpha$, whereas in Fig. 5.64a, c, and e, instantaneous pressure contours and streamlines are shown.

(d) With increase in Mach number, the vortex does not remain coherent and becomes disorganized as it moves over the airfoil. In Fig. 5.63b, a single DSV could be observed with a concentrated vorticity and of comparatively higher strength than the DSVs shown in Fig. 5.63e, h, where it is observed that the vortices of different strengths have combined together to form the DSV and thus, the DSV is fragmented and does not remain organized with concentrated vorticity. It is in contrast to the results shown in [15], where the vortex was noted as tightly wound. The differences can be due to changes in the parameters like $Re_\infty$, $\alpha_m$, $\alpha_a$, and $k$, which has a strong influence on DSV [15].

### 5.8.5 Post-Stall Flow Field

To understand the flow field at later stages beyond the dynamic stall, we need to know the flow field away from the airfoil. Global characteristics of the flow are needed to understand the interactions that occur as the vortex moves down the airfoil. When the vortex moves past the trailing edge, several events take place, such as trailing edge vortex (TEV) formation and the redistribution of the flow field over the airfoil. Therefore, in order to utilize and/or control dynamic stall, such events need to be investigated. Gad-el-Hak and Ho [30], and Ohmi et al. [75] have observed vortical

**Table 5.4** Details of the formation and shedding of dynamic stall vortex (DSV) [Reproduced from Jyoti et al., *Physics of Fluids*, **29**, 076104 (2017), with the permission of AIP Publishing.]

| $M_\infty$ | Moment stall (Non-dimensional time) | Moment stall ($\alpha$, in deg) | Lift stall (Non-dimensional time) | Lift stall ($\alpha$, in deg) |
|---|---|---|---|---|
| 0.283 | 148.80 | 22.98 | 150.25 | 24.63 |
| 0.4 | 147.35 | 19.80 | 148.05 | 21.49 |
| 0.5 | 146.90 | 18.59 | 147.30 | 19.67 |

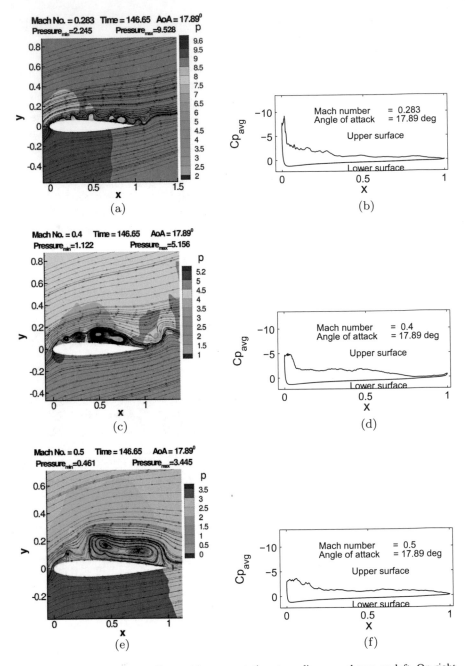

**Fig. 5.64** Pressure contours along with representative streamlines are shown on left. On right, corresponding $C_p$ plots averaged over 15 cycles are shown [Reproduced from Jyoti et al., *Physics of Fluids*, **29**, 076104 (2017), with the permission of AIP Publishing.]

structures forming near the trailing edge. The complex flow field of an oscillating airfoil results mainly due to mutual interaction between the DSV and the TEV.

This stage can be followed for different $M_\infty$, from the events noted in Fig. 5.65. As the DSV reaches the trailing edge, the fluid with anti-clockwise vorticity is being rapidly sucked from under the trailing edge, leading to the formation of TEV, as observed in Figs. 5.63d, e. TEV grows quickly beneath the DSV and lifts the DSV from the airfoil suction surface. The time the TEV travels over the suction surface is a smaller fraction of the cycle, as compared to the passage time of DSV over the suction side. Thereafter, small vortices near the leading edge combine to start forming a stronger vortex (another cycle of DSV), which combines with the TEV and resides on the suction surface of the airfoil, as shown in Fig. 5.65a. This combined DSV and TEV evolves with time, moving upward while increasing in size, as the combined structure convects downstream. This sequence of phenomena is common for $M_\infty =$ 0.283 and 0.4, as shown in Figs. 5.65b, e with the vorticity contours, and it is observed that till the DSV has lifted up clear and is convected downstream into the wake. But for the higher Mach number of $M_\infty = 0.5$, a change in the process of DSV formation is observed. Here, the combined two vortices are shown in Fig. 5.65h. First vortex grows near the leading edge and with time, it convects downstream, while gaining strength. At the same time, another vortex forms by breaking the DSV, which was formed earlier (just before LEV and TEV formation took place). This vortex lifts up and does not convect entirely in the downstream direction (as shown in Fig. 5.65h). These two vortices finally coalesce to make a single stronger DSV. The combination of this newly formed DSV along with the TEV resides near the trailing edge on the airfoil suction surface simultaneously, as shown in Fig. 5.65g. The same mechanism had also been observed experimentally by increasing the reduced frequency and keeping other parameters constant [77].

### 5.8.6  Supersonic Pocket and Shock Formation Near the Leading Edge

The rapid acceleration of flow around the leading edge of the pitching airfoil causes the flow to become locally supersonic, as described in [91]. In Fig. 5.66, such supersonic pockets are shown whose size increases with increasing $M_\infty$. The size of the supersonic region determines the shock location which also affects the shock strength. In general, the larger the supersonic region, the higher will be the Mach number immediately upstream of the shock wave [4]. Therefore, a stronger shock with higher entropy generation is observed for the higher Mach number case. The flow speed around the supersonic pocket increases from subsonic to supersonic and back to subsonic speed without any shock formation for the case of $M_\infty = 0.283$ (shown in Fig. 5.66a), as also noted in experiments of McCroskey et al. [65, 67, 68]. But a larger supersonic pocket, along with a weak shock formation, is observed for the higher Mach number case of $M_\infty = 0.4$, as shown in Fig. 5.66c. Here, a gradual

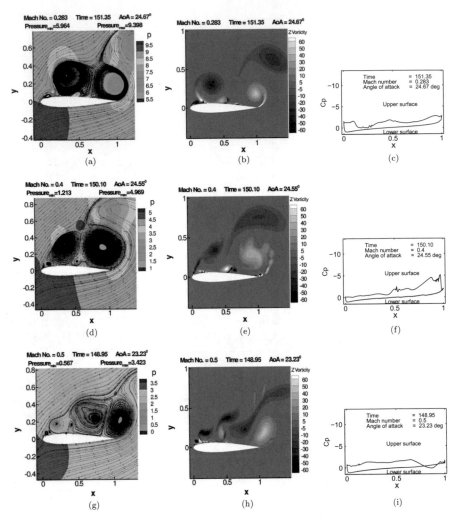

**Fig. 5.65** Post-stall flow field. Left column shows pressure contours along with representative streamlines. Middle column shows vorticity contours where negative values show the clockwise and positive values show the anti-clockwise vorticity. Right column shows instantaneous $C_p$ plots [Reproduced from Jyoti et al., *Physics of Fluids*, **29**, 076104 (2017), with the permission of AIP Publishing.]

thickening of boundary layer upstream of the shock occurs, without undergoing any transition or separation due to the diffused pressure rise (associated with the weak shock, corresponding to a local Mach number just greater than unity). For the even higher Mach number of $M_\infty = 0.5$, a lambda shock and waviness of shear layer near the leading edge are observed, as shown in Fig. 5.66e. Since the flow is supersonic in the pocket and the pressure rises as one approaches the shock, the laminar bound-

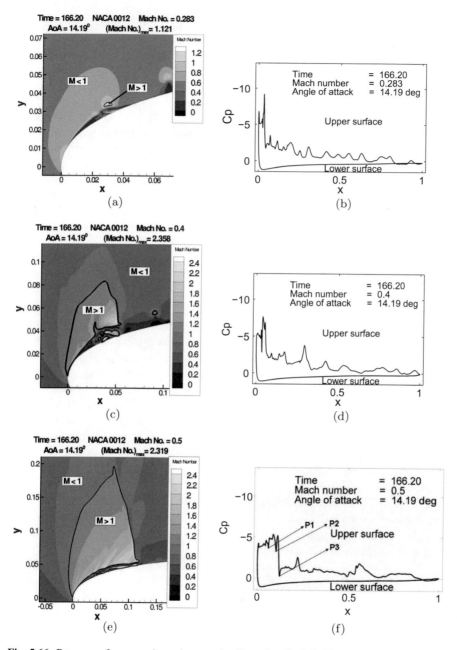

**Fig. 5.66** Presence of supersonic pocket near leading edge. On left, Mach number contours are shown and on right, corresponding instantaneous $C_p$ plots are shown [91] [Reproduced from Jyoti et al., *Physics of Fluids*, **29**, 076104 (2017), with the permission of AIP Publishing.]

ary layer separates well upstream of the main shock position, as has been reasoned also in [16]. Along with the separated shear layer which develops waviness (noted in the close-up view of the supersonic pocket) at the same non-dimensional time in Fig. 5.67a for $M_\infty = 0.5$, a sharp change in the direction of the mainstream flow also occurs. This is noted along with a well-defined oblique shock joining the primary shock at a distance over the airfoil, which is termed as the lambda shock [42]. Multiple spikes are observed as Mach number fluctuates with the separated wavy shear layer. The flow downstream of the shock undergoes through a series of accelerations and decelerations, resulting in expansion and compression waves and series of shock waves. The last shock in this series is the strongest, and the flow becomes subsonic downstream. The same pattern has also been observed during the interferometric investigations of compressible dynamic stall over a transiently pitching airfoil in [16], where Reynolds number and reduced frequency $(k)$ are smaller. The generation mechanisms of wavy structures in compressible shear layers have been discussed in literature [18].

In Fig. 5.66, results are shown for the three Mach numbers at the identical AoA of $14.19^o$ in the ninth pitch cycle. In Fig. 5.66f, P1 and P2 are the points indicating sudden jump in $C_p$ values, where weak shock forms. At P3, the jump in $C_p$ value is comparatively higher indicating stronger shock formation. As one compares Figs. 5.66b, d, and f, one notices reduction in peak pressure with increase in Mach number. Thus, suction peak and adverse pressure gradient are highest for the $M_\infty = 0.283$ case, and as a consequence, the supersonic pocket is smaller, as shown in Fig. 5.66a. Also, the location at which shock occurs (corresponding to the sudden jump in $C_p$ value) shifts toward the TE with increase in $M_\infty$. Fluctuations in the $C_p$ values (downstream of the LE) correspond to vortex formations. In Fig. 5.66c, the supersonic pocket (for $M_\infty = 0.4$) is comparatively away from the airfoil surface than for $M_\infty = 0.5$ case [as shown in Fig. 5.66e], where the supersonic pocket is closer to the airfoil surface, and this can be attributed to the reduction in $C_p$ value and induced wall-normal velocity, and because of it, the separation is restricted close to the surface.

In Figs. 5.67b–e, local Mach numbers are shown over the airfoil upper surface, along a particular azimuthal grid line with constant J-index. The wall-normal distance in these figures corresponds to the distance of the chosen grid line from the leading edge of the airfoil. For J= 5 case (at a wall-normal distance of $1.8 \times 10^{-5}$), with the grid line very close to the airfoil surface, no shock is observed, as the local Mach number is below 0.5 in Fig. 5.67b. But for J= 15 and 20 (at wall-normal distances of $2.1 \times 10^{-4}$ and $3.8 \times 10^{-4}$, respectively), multiple strong shocks are observed, where the local Mach number is greater than 1, as shown in Figs. 5.67c, d. For J = 100 (at wall-normal distance of $1.8 \times 10^{-2}$), which is comparatively far away from the surface, multiple weak shocks along with a strong shock are observed, as shown in Fig. 5.67e.

In the top frame of Fig. 5.68, an enlarged view of the boundary layer is shown at a non-dimensional time $t = 166.2$ for the case of $M_\infty = 0.5$, where the velocity profiles are shown at the indicated stations on the airfoil. In Fig. 5.68a (at location A), it is observed that the flow is still attached to the surface. In Fig. 5.68b, a small

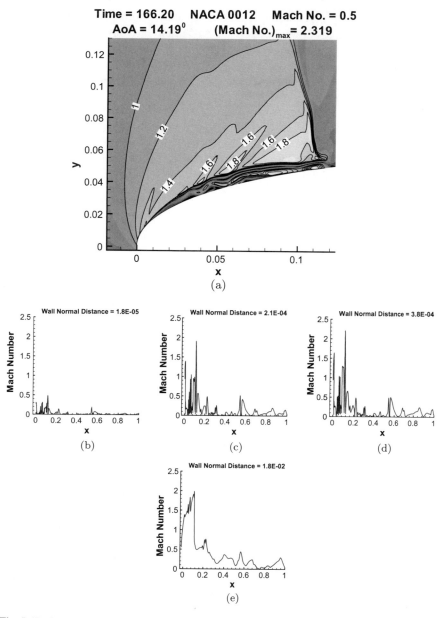

**Fig. 5.67**  Close-up view of the supersonic pocket and variation of local Mach number along the upper surface of the airfoil (for particular grid line of J-index). The wall-normal distance of the grid line from the leading edge is mentioned [91]  [Reproduced from Jyoti et al., *Physics of Fluids*, **29**, 076104 (2017), with the permission of AIP Publishing.]

reversal in flow is observed, along with the presence of inflection point. Increase in flow reversal and upward shifting of inflection point are observed in Fig. 5.68c, which implies that the shear layer is becoming thicker. At later locations, multiple inflection points are observed, as shown in Figs. 5.68d, e, and f. In Fig. 5.68d, e, it is observed that the flow closer to the airfoil surface tries to re-laminarize and as one moves away from the airfoil surface, the flow reversal is observed, where the local Mach number is greater than one. In Fig. 5.68f, three inflection points along with small zone of flow reversal are observed. As one moves downstream toward TE, separated flow could be observed due to increase in flow reversal, as shown in Figs. 5.68g, h. In Fig. 5.68h, it is observed that the separation bubble is biggest as the maximum reversal in flow direction occurs.

### 5.8.7  Ensemble-Averaged Loads and Moment

The hysteresis loops for pressure-based load coefficients, i.e., lift ($C_l$), drag ($C_d$) and pitching moment coefficients ($C_m$), are compared for different Mach numbers in Figs. 5.69 to 5.71. The results are averaged over 15 cycles for the data shown in these figures. The solid diamond and circle denote the occurrence of moment and lift stalls, respectively. Solid line corresponds to upstroke, and dotted line denotes downstroke motion. In these figures, it is observed that with increase in Mach number, the moment and lift stall occurs at smaller $\alpha$, as has also been observed in individual cycle analysis.

### 5.8.8  Comparison of Pressure-Based Aerodynamic Lift Coefficient, $C_l$

It has been noted [91] that the size of hysteresis loop and the maximum attainable lift coefficient value reduces as Mach number increases, as shown in Fig. 5.69, and is attributed to the reduction of the strength of DSV (as has already been noted for individual cycle analysis in Fig. 5.62c, f, and i). Also, $\Delta C_p$ around the airfoil reduces with increase in Mach number as shown in Fig. 5.64a, c, and e. The dynamic stall is not well developed for higher Mach numbers, as the slopes after both the moment and lift stalls are less steeper. In helicopters, to maintain a level flight, the retreating side of the rotor must produce sufficient lift to balance the lift produced by the advancing side. But the maximum dynamic pressure on retreating blade could be dramatically less than that on the advancing blade. Therefore, to produce equal amount of lift, $C_l$ value should be higher on retreating side. From Fig. 5.69c, it is observed that at higher $\alpha$ values (approximately from 21 to 24.79 degrees), $C_l$ remains almost constant for the highest Mach number investigated, and this is higher than the values observed for the lower Mach numbers of $M_\infty = 0.283$ and 0.4 cases. In Fig. 5.69a,

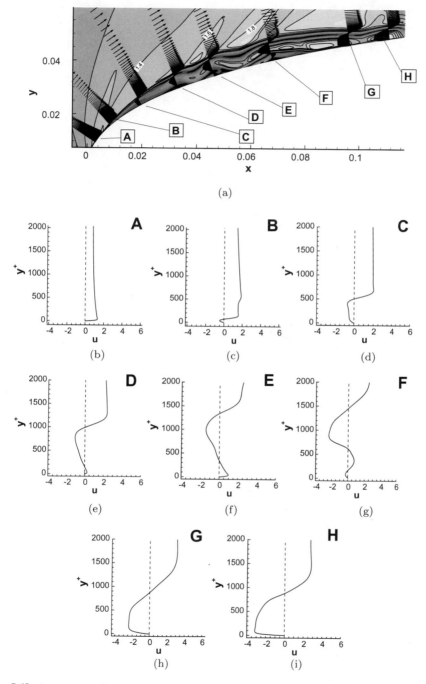

**Fig. 5.68** Appearance of separated and wavy shear layer for $M_\infty = 0.5$ [91] [Reproduced from Jyoti et al., *Physics of Fluids*, **29**, 076104 (2017), with the permission of AIP Publishing.]

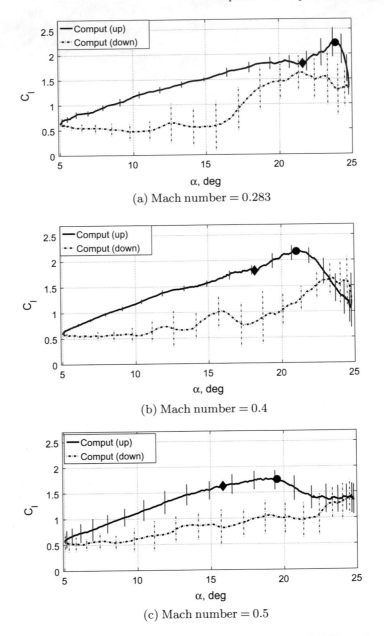

(a) Mach number $= 0.283$

(b) Mach number $= 0.4$

(c) Mach number $= 0.5$

**Fig. 5.69** Comparison of pressure-based $C_l$ for different Mach numbers [91]  [Reproduced from Jyoti et al., *Physics of Fluids*, **29**, 076104 (2017), with the permission of AIP Publishing.]

b, the continuous decrease in $C_l$ value at higher $\alpha$ is observed, while the decrease is sudden for $M_\infty = 0.283$ case. The stall mechanism for $M_\infty = 0.5$ behaves more in line with a jet or gust of air than that is for a concentrated, vortical structure, as has been observed by Bowles [14] and Chandrasekhara et al. [16] during interferometric investigations of compressibility effects on dynamic stall. The recovery of the lift near the highest $\alpha$ during downstroke is due to the shedding of the TEV which contains negative vorticity. After airfoil pitches past the dynamic stall angle, the leading edge flow becomes separated and the shear layer grows unstable, forming several smaller vortices. The undulations during the downstroke of the hysteresis loop occur due to the passage of positive and negative vortices of relatively weaker strength on the upper surface of the airfoil, following the DSV and TEV.

### 5.8.9  Comparison of Pressure-Based Aerodynamic Drag Coefficient, $C_d$

From Fig. 5.70, the reduction in hysteresis loop is observed (as in Fig. 5.69) with increasing Mach number. The maximum drag coefficient value decreases, but at maximum $\alpha$, it is noted that $C_d$ is almost same for $M_\infty = 0.4$ (in Fig. 5.70b) and 0.5 (in Fig. 5.70c) cases. Also, it is lesser than the value for $M_\infty = 0.283$ case (in Fig. 5.70a). Until $\alpha$ reaches around 20 degrees, value of $C_d$ increases with increasing $M_\infty$. For higher $\alpha$ values of the $M_\infty = 0.5$ case, it is noted that $C_d$ remains almost constant, which is lesser than the values observed for $M_\infty = 0.283$ and 0.4. Hence, it is concluded that the compressibility behaves favorably at higher $\alpha$ values, as reduction in drag coefficient values is one of the major areas for helicopter designers.

### 5.8.10  Comparison of Pressure-Based Aerodynamic Pitching Moment Coefficient, $C_m$

In Fig. 5.71, the pitching moment curves are presented for deep stall, with the DSV well developed and which convects into the wake during upstroke. The anti-clockwise contours denote positive damping and clockwise contours denote negative damping [20]. In Fig. 5.71a, three contours are observed, the first contour is anti-clockwise which is followed by a big clockwise contour (denoting the negative damping) and then again the stall-vortex-free, separated flow allows a third loop which is anti-clockwise. It has been observed in helicopter applications that the negatively damped portions in the pitching moment curves account for high-frequency stall flutter due to the rotor control input [37]. Therefore, positive damping is very important and favorable for the helicopter design from the rotor aeroelastic stability point of view. With an increase in compressibility, the mechanism for negative damping changes and it develops from the growth of lambda shock (as observed in Fig. 5.66e) during

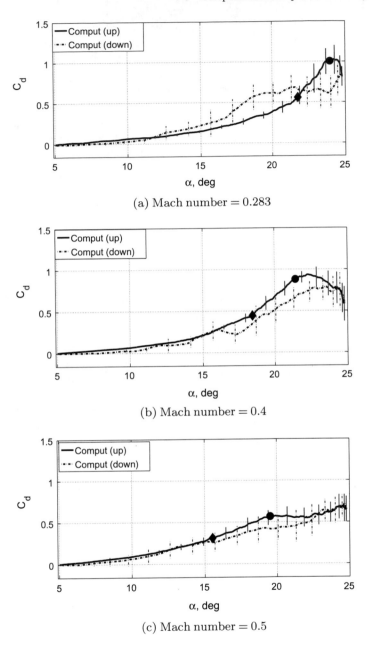

(a) Mach number = 0.283

(b) Mach number = 0.4

(c) Mach number = 0.5

**Fig. 5.70** Comparison of pressure-based $C_d$ for different Mach numbers [91] [Reproduced from Jyoti et al., *Physics of Fluids*, **29**, 076104 (2017), with the permission of AIP Publishing.]

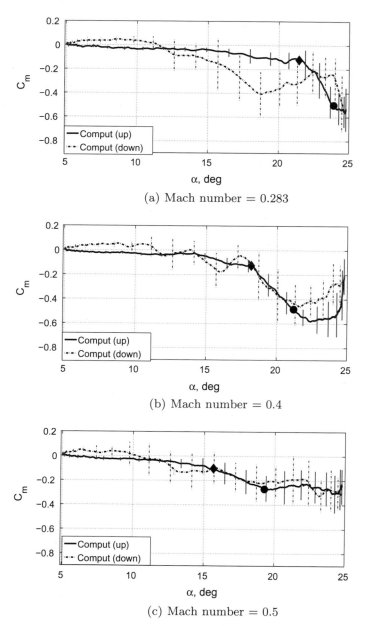

(a) Mach number $= 0.283$

(b) Mach number $= 0.4$

(c) Mach number $= 0.5$

**Fig. 5.71** Comparison of pressure-based $C_m$ for different Mach numbers [91] [Reproduced from Jyoti et al., *Physics of Fluids*, **29**, 076104 (2017), with the permission of AIP Publishing.]

the pitch up portion of the cycle [21]. There is a decrease in the pressure upstream of the shock front which in turn increases the leading edge suction. When it interacts with higher static pressure downstream of the shock, it results in a nose-up pitching moment which reduces the aerodynamic damping. The area under the positive damping increases with increase in Mach number and hence, compressibility contributes to additional aeroelastic stability. The dynamic stall process is thus considered as a problem for helicopter design due to abrupt pitching moment variations, as shown in Figs. 5.71a, b. This results in reduction of rotor and control system fatigue life. But for $M_\infty = 0.5$, the pitching moment variations are less severe as the slope between moment and lift stalls is less steeper (as shown in Fig. 5.71c). It could be due to change in DSV mechanism explained in Sect. 5.8.4, where it is noted that with increase in Mach number, strength of DSV reduces and is not well organized. In Fig. 5.71b, at higher $\alpha$ values after lift stall, it is observed that $C_m$ remains almost constant for some time, during post-stall stage, when TEV forms and resides over airfoil along with DSV, and these both convect down into the wakes, resulting in sudden change in $C_m$ and causing fully developed stall condition.

## Problems

**P5.1:** What is basis of entropy production and rotationality in transonic flows over airfoils?

**P5.2:** Show the evolving supersonic pocket in flow past the NACA 0012 airfoil for $M_\infty = 0.779$, $Re_\infty = 3 \times 10^6$ and angle of attack of $\alpha = -0.14^0$. Also discuss about upstream moving wave from the near vicinity of trailing edge, which is also known as Kutta waves.

## References

1. Abhishek, A. *Analysis, Validation, Prediction and Fundamental Understanding of Rotor Blade Loads in an Unsteady Maneuver*. PhD dissertation, Dept. of Aerospace Engg., Univ. of Maryland (2010)
2. A. Alshabu, H. Olivier, I. Klioutchnikov, Investigation of upstream moving pressure waves on a supercritical airfoil. Aero. Sci. Tech. **10**, 465–473 (2006)
3. Y. Andreopoulos, J.H. Agui, G. Briassulis, Shock-wave turbulence interactions. Ann. Rev. Fluid Mech. **32**, 309–345 (2000)
4. H. Babinsky, J.K. Harvey, *Shock Wave-Boundary-Layer Interactions* (Cambridge Univ, Press, 2011)
5. P.M. Bagade, Y.G. Bhumkar, T.K. Sengupta, An improved orthogonal grid generation method for solving flows past highly cambered aerofoils with and without roughness elements. Comput. Fluids **103**, 275–289 (2014)
6. Bagade, P. M., Laurendeau, E., Bhole, A., Sharma, N. and Sengupta, T. K. Comparison of RANS and DNS for transitional flow over WTEA-TE1 airfoil. In proc. of IUTAM Symp. on *Advances in Computation, Modeling and Control of Transitional and Turbulent Flows*,

(Editors: Profs. T. K. Sengupta, S. K. Lele, K. R. Sreenivasan and P. A. Davidson) World Scientific Publishing Company, Singapore, 349–357 (2015)

7. B.S. Baldwin, H. Lomax, Thin layer approximation and algebraic model for separated turbulent flows. AIAA Paper **78–257**, (1978)

8. R.W. Barnwell, A similarity rule for sidewall-boundary-layer effect in two-dimensional wind tunnels. AIAA Paper **79–0108**, (1979)

9. Berton, E., Allain, C., Favier, D. and Maresca, C. Experimental methods for subsonic flow measurements. *Notes on Numer. Fluid Mech. and Multidiscip. Des.*, **81**, 97–104 (2002)

10. Bhole, A. *Direct Numerical Simulation of Transonic flow over Airfoils.* M.S. thesis, Dept. of Aerospace Engg., IIT Kanpur (2013)

11. Y.G. Bhumkar, T.K. Sengupta, Adaptive multi-dimensional filters. Comput. Fluids **49**, 128–140 (2011)

12. Binion, T. W. *Limitations of Available Data.* AGARD-AR-138, May (1979)

13. C. Bogey, C. Bailly, On the application of explicit spatial filtering to the variables or fluxes of linear equations. J. Comput. Phys. **225**, 1211–7 (2007)

14. Bowles, P. O. *Wind Tunnel Experiments on the Effect of Compressibility on the Attributes of Dynamic Stall.* Univ. of Notre Dame (2012)

15. M.S. Chandrasekhara, L.W. Carr, Flow visualization studies of the Mach number effects on dynamic stall of an oscillating airfoil. J. Aircraft **27**(6), 516–522 (1990)

16. M.S. Chandrasekhara, L.W. Carr, M.C. Wilder, Interferometric investigations of compressible dynamic stall over a transiently pitching airfoil. AIAA J. **32**(3), 586–593 (1994)

17. L. Chen, C. Xu, X. Lu, Numerical investigation of the compressible flow past an airfoil. J. Fluid Mech. **643**, 97–126 (2010)

18. J. Chen, X.-T. Shi, T.-J. Wang, Z.-S. She, Wavy structures in compressible mixing layers. Acta Mech. Sin. **29**, 633–640 (2013)

19. A. Choudhry, R. Leknys, M. Arjomandi, R. Kelso, An insight into the dynamic stall lift characteristics. Exp. Therm. and Fluid Sci. **58**, 188–208 (2014)

20. T.C. Corke, P.O. Bowles, C. He, E.C. Matlis, Sensing and control of flow separation using plasma actuators. Philos. Trans. R. Soc. A **369**, 1459–1475 (2011)

21. T.C. Corke, O.F. Thomas, Dynamic stall in pitching airfoils: aerodynamic damping and compressibility effects. Annu. Rev. Fluid Mech. **47**, 479–505 (2015)

22. Crimi, P. *Dynamic stall Technical Report* No. AGARD-AG-172 (1973)

23. Datta, A. and Chopra, I. Prediction of UH-60A dynamic stall loads in high altitude level flight using CFD/CSD coupling. *In 61$^{st}$ Annual Forum Proceedings-American Helicopter Society*, Grapevine, Texas, June 1–3 (2005)

24. S. Deck, Numerical simulation of transonic buffet over a supercritical airfoil. AIAA J. **43**(7), 1556–1565 (2005)

25. A. Dipankar, T.K. Sengupta, Symmetrized compact schemes for receptivity study of 2D channel flow. J. Comput. Phys. **215**, 245–273 (2006)

26. J.A. Ekaterinaris, M.F. Platzer, Computational prediction of airfoil dynamic stall. Prog. in Aerosp. Sci. **33**(11), 759–846 (1998)

27. M. Fujino, Y. Yoshizaki, Y. Kawamura, Natural-laminar-flow airfoil development for a lightweight business jet. J. Aircraft **40**(4), 609–615 (2003)

28. Fukushima, T. and Dadone, L. U. Comparison of dynamic stall phenomena for pitching and vertical translation motions. NASA CR-2793 (1977)

29. Garner, H. C., Rogers, E., Acum, W. and Maskell, E. *Subsonic Wind Tunnel Wall corrections.* AGARDograph, AG-109 (1966)

30. Gad-el-Hak, M. and Ho, C-M. Unsteady vortical flow around three-dimensional lifting surfaces. *AIAA J.*, **24**(5), 713–721 (1986)

31. A. Garbaruk, M. Shur, M. Strelets, P. Spalart, Numerical study of wind tunnel wall effects on transonic airfoil flow. AIAA J. **41**, 1046–1054 (2003)

32. D.V. Gaitonde, J.S. Shang, J.L. Young, Practical aspects of higher-order numerical schemes for wave propagation phenomena. Int. J. Num. Methods in Engg. **45**(12), 1849–1869 (1999)

33. D.V. Gaitonde, M.R. Visbal, Further development of a Navier-Stokes solution procedure based on higher-order formulas. AIAA J. **99–0557**, (1999)
34. D.V. Gaitonde, M.R. Visbal, Padé-Type Higher-Order Boundary Filters for the Navier-Stokes equations. AIAA J. **38**(11), 2103–2112 (2000)
35. L. Gray, J. Liiva, *Wind Tunnel Tests of Thin Airfoils Oscillating Near Stall*, vol. II (Data Report. Boeing Vertol Co, Philadelphia PA, 1969)
36. G. Grötzbach, Spatial resolution requirements for direct numerical simulation of the Rayleigh-Bénard convection. J. Comput. Phys. **49**(2), 241–264 (1983)
37. Ham, N. D. and Young, M. I. Limit cycle torsional motion of helicopter blades due to stall. *J. Sound and Vibration*, **4**(3), 431IN17433–432444 (1966)
38. C.D. Harris, Two-dimensional Aerodynamic Characteristics of the NACA0012 airfoil in the Langley 8-foot Transonic Pressure Tunnel. NASA Report-TM **81927**, (1981)
39. V. Hermes, I. Klioutchnikov, H. Oliver, Numerical investigation of unsteady wave phenomena for transonic airfoil flow. Aero. Sci. Tech. **25**, 224–233 (2013)
40. Hirsch, C. *Numerical Computation of Internal and External Flows*. **1**, *Fundamentals of Numerical Discretization*, Wiley-Interscience Publication, New York, USA (1994)
41. Hoffmann, K. A. and Chiang, S. T. *Computational Fluid Dynamics*. **II**, $4^{th}$ Ed., Engineering Education System, Wichita, Kansas, USA (2000)
42. Houghton, E. L., Carpenter, P. W., Collicott, S. and Valentine, D. *Aerodynamics for Engineering Students*. $6^{th}$ Ed., Butterworth-Heinemann (2013)
43. J.L. Jacocks, *An Investigation of the Aerodynamic Characteristics of Ventilated Test Section Walls for Transonic Wind Tunnel: PhD Dissertation* (Univ, Tennessee, 1976)
44. A. Jameson, K. Ou, 50 years of transonic aircraft design. Prog. Aero. Sci. **47**(5), 308–318 (2011)
45. Jameson A., Schmidt W. and Turkel E., *Numerical Solution of the Euler Equations by Finite Volume Methods Using Runge-Kutta Time-Stepping Schemes, AIAA 1981–1259*, AIAA 14th Fluid and Plasma Dynamic Conf., June 23–25, (1981) Palo Alto, California
46. Jones, W. P. (Ed.) AGARD Manual on Aeroelasticity. *AGARD*, (1961)
47. Kemp, L. D. *An Analytic Study for the Design of Advanced Rotor Airfoils*. NASA Report **CR-112**, 297 (1973)
48. R. Langtry, F.R. Menter, Correlation based transition modeling for unstructured parallelized computational fluid dynamics codes. AIAA J. **47**(12), 2894–2906 (2009)
49. Langtry R. and Menter F. R. Transition modeling for general CFD applications in aeronautics. *AIAA*, $43^{rd}$ *AIAA Aerospace Sciences Metting and Exhibit, Reno, Nevada*, **522** (2005)
50. P.D. Lax, Weak solutions of nonlinear hyperbolic equations and their numerical computation. Commun. Pure and Appl. Math. **7**, 159–193 (1954)
51. B.H.K. Lee, Oscillatory shock motion caused by transonic shock boundary-layer interaction. AIAA J. **28**, 942–944 (1990)
52. B.H.K. Lee, Self-sustained shock oscillations on airfoils at transonic speeds. Prog. Aero. Sci. **37**, 147–196 (2001)
53. B.H.K. Lee, H. Murty, H. Jiang, Role of Kutta waves on oscillatory shock motion on an airfoil. AIAA J. **32**, 789–796 (1994)
54. S.K. Lele, T. Poinsot, Boundary conditions for direct simulations of compressible viscous flows. J. Comp. Phys. **101**, 104–129 (1992)
55. Leishman, J. G. *Principles of Helicopter Aerodynamics*. $2^{nd}$ Ed., Cambridge Univ. Press (2006)
56. J.G. Leishman, T.S. Beddoes, A Semi-Empirical Model for Dynamic Stall. J. American Helicopter society **34**(3), 3–17 (1989)
57. H. Liepmann, A. Roshko, *Elements of Gasdynamics* (Dover Publications, Inc., Mineola, New York, 1956)
58. M.S. Loginov, N.A. Adams, A.A. Zheltovodov, Large-eddy simulations of shock-wave/turbulent-boundaty-layer interaction. J. Fluid Mech. **565**, 135–169 (2006)
59. P.F. Lorber, F.O. Carta, Airfoil dynamic stall at constant pitch rate and high Reynolds number. J. Aircraft **25**(6), 548–556 (1988)

60. Mamou M., Mébarki, Y., Khalid M., Genest M., Coutu D., Popov A. V., Sainmont C., Georges T., Grigorie L., Botez R. M., Brailovski V., Terriault P., Paraschivoiu I. and Laurendeau E. *Aerodynamic Performance Optimization of a Wind Ttunnel Morphing wing Model Subject to Various Cruise Flow Conditions.* 27th *International Congress of the Aeronautical Sciences,* ICAS (2010)

61. J.G. Marvin, L.L. Lewy, H.L. Seegmiller, Turbulence modelling for unsteady transonic flows. AIAA J. **18**, 489–496 (1980)

62. G. Martinat, M. Braza, Y. Hoarau, G. Harran, Turbulence modelling of the flow past a pitching NACA0012 airfoil at $10^5$ and $10^6$ Reynolds numbers. J. Fluids and Struct. **24**(8), 1294–1303 (2008)

63. Maybey, D. G. *Physical Phenomena Associated with Unsteady Transonic Flows, in Unsteady Transonic Aerodynamics, vol. 120, Progress in Astronautics and Aeronautics, AIAA Series.* AIAA, Washington, DC, USA (1989)

64. Maybey, D. G. Some remarks on the design of transonic tunnels with low levels of flow unsteadiness. *NASA CR*-2722 (1976)

65. McAlister, K. W., Pucci, S. L., McCroskey, W. J. and Carr, L. W. An experimental study of dynamic stall on advanced airfoil sections: Vol. 2: Pressure and Force Data. *NASA Technical Memorandum*, 84245 (1982)

66. McCroskey, W. J., McAlister, K. W., Carr, L. W., Pucci, S. L., Lambert, O. and Indergrand, R. F. *Dynamic stall on advanced airfoil sections.* 36th Annual National Forum of the American Helicopter Society (1980)

67. W.J. McCroskey, K.W. McAlister, L.W. Carr, S.L. Pucci, O. Lambert, R.F. Indergrand, Dynamic stall on advanced airfoil sections. J. Am Helicopter Soc. **26**(3), 40–50 (1981)

68. McCroskey, W. J., McAlister, K. W., Carr, L. W. and Pucci, S. L. An experimental study of dynamic stall on advanced airfoil sections: Vol. 1: Summary of the Experiment. *NASA Technical Memorandum*, 84245 (1982)

69. F.R. Menter, Two-equation eddy-viscosity turbulence models for engineering applications. AIAA J. **32**(8), 1598–1605 (1994)

70. J.E. Mercer, E.W. Geller, M.L. Johnson, A. Jameson, Transonic flow calculations for a wing in a wind tunnel. AIAA J. **18**(9), 707–711 (1981)

71. Maybey, D. G. Flow unsteadiness and model vibration in wind tunnels at subsonic and transonic speeds. *Aeronautical Research Council, ARC CP*-1155 (1971)

72. M.T. Nair, T.K. Sengupta, Orthogonal grid generation for Navier-Stokes computations. Int. J. Numer. Methods Fluids **28**, 215–224 (1998)

73. T.H. Moulden, *Fundamentals of Transonic Flow* (Wiley-Interscience Publication, Canada, 1984)

74. D. Nixon, *Unsteady Transonic Aerodynamics, Progress in Astronautics and Aeronautics Series (120)* (AIAA, Washington, DC, 1989)

75. K. Ohmi, M. Coutanceau, T.P. Loc, A. Dulieu, Vortex formation around an oscillating and translating airfoil at large incidences. J. Fluid Mech. **211**, 37–60 (1990)

76. H. Otto, Systematical investigation of the influence of wind tunnel turbulence on the results of force measurements. AGARD CP **174**, (1976)

77. J. Panda, K.B.M.Q. Zaman, Experimental investigation of the flow field of an oscillating airfoil and estimation of lift from wake survey. J. Fluid Mech **265**, 65–95 (1994)

78. S.R. Pate, C.J. Schueler, Radiated aerodynamic noise effects on boundary layer transition in supersonic and hypersonic wind tunnels. AIAA J. **7**, 450–457 (1968)

79. H.H. Pearcy, *Shock induced Separation and Its Prevention by Design and Boundary Layer Control, Boundary Layer and Flow Control,* vol. 2 (Pergamon, Oxford, UK, 1961)

80. Pearcey, H. H., Wilby, P. G., Riley, M. J. and Brotherhood, P. *The Derivation and Verification of a New Rotor Profile on the Basis of flow Phenomena; Aerofoil Research and Flight Tests.* Aerodynamics of Rotary wings, AGARD CP–111 (1972)

81. S. Pirozzoli, M. Bernardini, F. Grasso, Direct numerical simulation of transonic shock/boundary layer interaction under conditions of incipient separation. J. Fluid Mech. **657**, 361–393 (2010)

82. Pirozzoli, S. and Grasso, F. Direct numerical simulation of impinging shock wave/turbulent boundary layer interaction at M = 2.25. *Phys. Fluids*, **18**, 065113 (2006)
83. Popov, A. V., Botez, R. M. and Labib, M. Transition point detection from the surface pressure distribution for controller design. *J. Aircraft*, **45**(1) (2008)
84. Popov, A. V., Grigorie, L. T., Botez, R. M., Mébarki, Y. and Mamou, M. Modeling and testing of a morphing wing in open-loop architecture. *J. Aircraft*, **47**(3) (2010)
85. Popov, A. V., Grigorie, L. T., Botez, R. M., Mébarki, Y. and Mamou, M. Closed-loop control validation of a morphing wing using wind tunnel tests. *J. Aircraft*, **47**(4) (2010)
86. M.K. Rajpoot, T.K. Sengupta, P.K. Dutt, Optimal time advancing dispersion relation preserving schemes. J. Comput. Phys **229**, 3623–3651 (2010)
87. MJWu Ringuette, M. and Martin, M. P., Coherent structures in direct numerical simulation of turbulent boundary layers at Mach 3. J. Fluid Mech. **594**, 59–69 (2008)
88. D.P. Rizzetta, M.R. Visbal, G.A. Blaisdell, A time-implicit high-order compact differencing and filtering scheme for large-eddy simulation Int. J. Num. Methods Fluids **42**, 665–693 (2003)
89. M. Robitaille, A. Mosahebi, E. Laurendeau, *Verification and Validation of the NSCODE Implementation of the $\gamma - Re_\theta$ Transition Model* (Design and Operations, conf. Royal Aero. Soc., Bristol, UK, In Advanced Aero Concepts, 2014)
90. C.L. Rumsey, M.D. Sanetrik, R.T. Biedron, N.D. Melson, E.B. Parlette, Efficiency and accuracy of time-accurate turbulent Navier-Stokes computations. Comput. Fluids **25**, 217–236 (1996)
91. J. Sangwan, T.K. Sengupta, P. Suchandra, Investigation of compressibility effects on dynamic stall of pitching airfoil. Phys. Fluids **29**, 076104 (2017)
92. T.K. Sengupta, *Fundamentals of Computational Fluid Dynamics* (Univ. Press, Hyderabad, India, 2004)
93. T.K. Sengupta, *High Accuracy Computing Methods: Fluid Flows and Wave Phenomena* (Cambridge Univ. Press, USA, 2013)
94. T.K. Sengupta, A. Bhole, N.A. Sreejith, Direct numerical simulation of 2D transonic flows around airfoils. Comput. Fluids **88**, 19–37 (2013)
95. Sengupta, T. K. and Bhumkar, Y. G. Physical and spurious disturbances in computing and bypass transition (*manuscript under preparation*). (2009)
96. T.K. Sengupta, Y.G. Bhumkar, Direct numerical simulation of transition over a natural laminar flow airfoil. Frontiers in Aerospace Engineering **2**(1), 39–52 (2013)
97. T.K. Sengupta, S. Bhaumik, Y.G. Bhumkar, Direct numerical simulation of two-dimensional wall-bounded turbulent flows from receptivity stage. Phys. Rev. E **85**(2), 026308 (2012)
98. T.K. Sengupta, Y.G. Bhumkar, New explicit two-dimensional higher order filters. Comput. Fluids **39**(10), 1848–1863 (2010)
99. T.K. Sengupta, Y. Bhumkar, V. Lakshmanan, Design and analysis of a new filter for LES and DES. Comput. Struct. **87**, 735–750 (2009)
100. T.K. Sengupta, A. Dipankar, A.K. Rao, A new compact scheme for parallel computing using domain decomposition. J. Comput. Phys. **220**, 654–677 (2007)
101. T.K. Sengupta, A. Dipankar, P. Sagaut, Error dynamics: beyond von neumann analysis. J. Comput. Phys. **226**, 1211–1218 (2007)
102. T.K. Sengupta, G. Ganeriwal, S. De, Analysis of central and upwind compact schemes. J. Comp. Phys. **192**, 677–694 (2003)
103. T.K. Sengupta, T.T. Lim, S.V. Sajjan, S. Ganesh, J. Soria, Accelerated flow past a symmetric aerofoil: experiments and computations. J. Fluid Mech. **591**, 255–288 (2007)
104. T.K. Sengupta, M.K. Rajpoot, Y.G. Bhumkar, Space-time discretizing optimal DRP schemes for flow and wave propagation problems. Comput. Fluids **47**, 144–154 (2011)
105. T.K. Sengupta, S.K. Sirkar, A. Dipankar, High accuracy schemes for DNS and acoustics. J. Sci. Comput. **26**, 151–193 (2006)
106. T.K. Sengupta, V. Vikas, A. Johri, An improved method for calculating flow past flapping and hovering airfoils. Theor. Comput. Fluid Dyn. **19**(6), 417–440 (2005)

107. A.Y.N. Sofla, S.A. Meguid, K.T. Tan, W.K. Yea, Shape morphing of aircraft wing: status and challenges. J. Mater. Desg. **31**, 1284–1292 (2010)
108. P.R. Spalart, S.R. Allmaras, A one-equation turbulence model for aerodynamic flows. AIAA J. **92–0439**, (1992)
109. G.R. Srinivasan, J.D. Baeder, S. Obayashi, W.J. McCroskey, Flowfield of a lifting rotor in hover - A Navier-Stokes simulation. AIAA J. **30**(10), 2371–2378 (1992)
110. Srinivasan, G. R. and Baeder, J. D. TURNS: A free-wake Euler/Navier-Stokes numerical method for helicopter rotors. "Technical Notes", *AIAA J.*, **31**(5), 959–962 (1993)
111. E. Stanewsky, Adaptive wing and flow control technology. Progress in Aerospace Sciences **37**, 583–667 (2001)
112. Suchandra, P. *Direct Simulation of 2D Compressible Flow around Airfoil undergoing Pitching Oscillation*. M. S. thesis, Department of Aerospace Engineering, IIT Kanpur (2015)
113. S. Teramoto, Large-eddy simulations of transitional boundary layer with impinging shock wave. AIAA J. **43**, 2354–2364 (2005)
114. Tijdeman, H. Investigation of the transonic flow around oscillating airfoils. *National Aerospace Lab. Amsterdam, Netherlands*, TR–77-090U (1977)
115. H. Tijdeman, R. Seebass, Transonic flow past oscillating airfoils. Ann. Rev. Fluid Mech. **12**, 181–222 (1980)
116. Thu, A. M., Jeon, S. E., Byun, Y. H. and Park, S. H. Dynamic stall vortex formation of OA-209 airfoil at low Reynolds number. World Academy of Science, Engineering and Technology, *Int. J. Mech., Aerospace*, Industrial and Mechatronics Engineering, **8**(2) (2014)
117. Visbal, M. R. Dynamic stall of a constant-rate pitching airfoil. *AIAA 26th Aerospace Sciences Meeting*, 88–0132 (1988)
118. M.R. Visbal, D.V. Gaitonde, On the use of higher-order finite-difference schemes on curvilinear and deforming meshes. J. Comput. Phys. **181**, 155–85 (2002)
119. Xiao, Q., Tsai, H. M. and Liu, F. Computation of shock induced separated flow with a lagged $k - \omega$ turbulence model. *AIAA 2003–3464, AIAA* (2003)
120. Q. Xiao, H.M. Tsai, F. Liu, Numerical study of transonic buffet over a supercritical airfoil. AIAA J. **44**, 620–628 (2006)

# Chapter 6
# Acoustic Wave Equation

## 6.1 Introduction

An acoustic signal or an acoustic disturbance consists of pressure oscillations traveling in an elastic medium. The variation of pressure can either be triggered by a vibrating surface (for example, diaphragm of a speaker) or by a turbulent fluid flow. In a broad sense, a sound wave is any disturbance that propagates in an elastic medium, which may be either a gas, a liquid, or a solid. Engineers working on acoustic-related problems have subdivided acoustics-related studies into two broad categories, such as the acoustic field triggered by the structure-borne and due to the fluid-borne sound. Acoustic waves can be classified on the basis of the frequency as sonic waves, ultrasonic waves, and infrasonic waves. This classification is based on the human ear perception ability, which usually can hear sound waves in the frequency range of 20 Hz to 20 kHz. Sound waves with frequencies above 20 kHz are classified as ultrasonic waves while those below 20 Hz are classified as infrasonic waves.

We communicate ideas by creating audible acoustic waves. The acoustic field is known to create positive as well as negative effects on human health. In a separate branch of science known as psycho-acoustics, psychological and physiological responses of human beings associated with the acoustic field are studied. Human ears are able to perceive sound in the form of fluctuations in pressure corresponding to audible frequency range only. For sensing acoustic signal, in addition to the frequency range, the amplitude of pressure fluctuation is also important. A young person can detect fluctuations in pressure with amplitude as low as $20 \mu$ Pa, which is around 10 order smaller compared to the background atmospheric pressure.

Although humans cannot hear sound in the ultra and infrasonic range, these waves are equally important in number of engineering applications. For example, ultrasound devices operating in between frequencies from 20 kHz up to several giga-Hertz are used to detect objects and measure distances. Ultrasound imaging is widely used in medical diagnosis. Another important application of devices operating at ultrasound frequencies is in nondestructive testing of engineering objects. Infrasonic waves

© Springer Nature Singapore Pte Ltd. 2020
T. K. Sengupta and Y. G. Bhumkar, *Computational Aerodynamics and Aeroacoustics*,
https://doi.org/10.1007/978-981-15-4284-8_6

have considerably larger wavelength compared to the sonic and ultrasonic waves and these waves can cover long distances with little change in their amplitude. Infrasonic frequency range is useful for developing instruments which are used to detect earthquakes and to gather information about rock and petroleum formations below surface of the Earth.

In addition to various engineering applications, researchers are also concerned about acoustic noise reduction. Noise is usually referred to the unwanted and objectionable audible signal. Identification of acoustic noise sources, propagation of acoustic disturbances, and methods to reduce acoustic noise levels have been important research topics for many decades. These problems are relevant to the modern city life, where the community is constantly exposed to the road traffic noise as well as other sources of noise. For example, in a vehicle's cabin, the acoustic noise is contributed by the structural vibrations of the body panels, road loads, and the aerodynamic noise. While designing modern vehicles, residential buildings, transport stations, offices, public places such as auditorium and movie theaters, acoustician and noise control engineers consider aural comfort as an important design parameter. Thus, the study of acoustics is important for us.

Here, we are going to mostly discuss about the propagation of acoustic signal in air, where the pressure fluctuations are above and below the ambient atmospheric pressure. Sound propagates in fluids in the form of longitudinal waves. Acoustic signal is a succession of compression and rarefaction in the elastic medium as shown in Fig. 6.1. In a longitudinal wave, the particle displacement is parallel to the direction of wave propagation. Particles oscillate along their mean position, as the acoustic wave propagates. Acoustic wave propagation occurs irrespective of whether the fluid is in a stationary or in a moving condition. During acoustic wave propagation, the region of a medium where distance between particles decreases one observes increase in acoustic pressure and the corresponding zone is identified as the compression zone, while in the rarefaction zone amplitude of the acoustic pressure decreases. Acoustic waves are propagated in air in the form of small perturbations in pressure and density, which are adiabatic in nature. Although the particle deviates from its mean position, corresponding change in pressure is very small. Acoustic waves propagate inside the domain in the form of wave without causing the net transfer of mass. The disturbance is only associated with the transfer of energy and momentum. Propagation of an acoustic wave needs an elastic medium, which can be of both solid and fluid. The very basic example of the wave motion can be visualized by throwing a stone in a quiet pond. The ripples propagate outward very smoothly to the end of the pond. The sound waves behave in a different manner. The elastic nature of the medium transfers oscillatory motion of the particle to its neighboring particle. It is also important to note that the transfer of momentum between two successive particles is the medium property and is equal to acoustic wave propagation speed $c$ in that medium. Some of the basic definitions related to sound propagation have been discussed in this chapter. This information is essential for the beginners in this research area before getting exposed to the essential properties of numerical methods to solve computational aeroacoustic $(CAA)$ problems and obtain solutions of different $CAA$ problems.

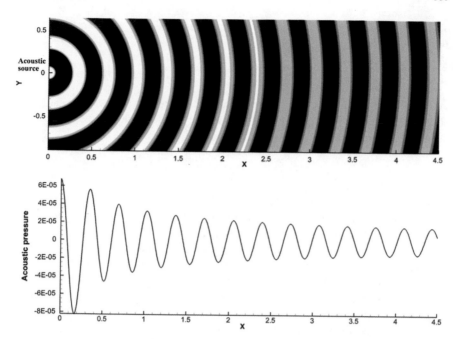

**Fig. 6.1** Top frame shows sound waves propagating outward from a spherical source while the bottom frame shows corresponding amplitude variation in space

Calculation of the acoustic field is a challenging task as the amplitude of the acoustic wave can be of equal magnitude as of the numerical error associated with the simulation. Almost all the engineering problems associated with acoustics display a wideband frequency nature, which demands use of high-resolution and high-accuracy schemes for the simulation. For accurately simulating high-frequency events, one is forced to compute on a very fine mesh (as the wavelength of acoustic disturbance becomes smaller) with a significantly lower time step. One also needs to simulate such problems for a longer duration in order to capture the effects of low-frequency events. Thus, simulation of computational acoustics problem is a challenging task.

### 6.1.1 Characterization of Acoustic Waves

Acoustic waves travel through an elastic medium and carry energy from one point in the domain to the other point through vibration of medium molecules. Acoustic waves travel much faster when the intermolecular spacing is less for a given medium as compared to the medium where intermolecular spacing is large. Thus, the speed of acoustic waves in solid is higher as compared to that in a gas or in a liquid. Speed of acoustic waves in air depends on the temperature, pressure, and density of the air and is given as

$$c = \sqrt{\frac{\gamma p}{\rho}} \tag{6.1}$$

where $\gamma$ is the ratio of specific heats, $p$ is an ambient pressure, and $\rho$ is the density of air. For majority of the engineering problems, the direct relation between the speed of sound and temperature of the air is given as

$$c = 20.05\sqrt{T} \tag{6.2}$$

If any solid object (let's say a tuning fork) is subjected to any disturbance force, the closely packed particles in the tuning fork start vibrating which excites the fluid surrounding it. Thus, vibration leads to the oscillation of the closely packed particles in a physical medium. The acoustic wave has similar nature like light waves, to display reflection, diffraction, scattering, and refraction. One can sense the reflection of acoustic waves by the echo of sound in a largely bounded area. To understand propagation of an acoustic wave, consider a circular piston oscillating with some amplitude and frequency along $x$-axis. As the piston moves forward along $x$-axis, air immediately next to the face of the piston moves along with the piston. This causes increase in pressure of the fluid element situated very close to piston surface. This element expands in the forward direction displacing the next layer of air and causing its compression. Thus, a pressure pulse is formed which travels at the speed of sound. As the piston oscillates, acoustic field composed of compression and rarefaction zones is formed.

Acoustic waves propagate through air in a form of longitudinal waves by alternate compression and expansion of air molecules. Successive compression and expansion of air molecules lead to pressure fluctuation about mean atmospheric pressure. Amplitude of this fluctuation is known as sound pressure. In general, engineering acoustic problems involve measurement of sound pressure at various points with respect to time to characterize the acoustic field. Sound pressure is measured in units of pascals (Pa). A young person with normal hearing ability can detect pressure fluctuations at micropascal ($\mu$ Pa) level corresponding to the audible frequency range. While dealing with the acoustic problems, one needs to always keep it in mind that for recognizing acoustic signal, it's not only the amplitude of acoustic disturbance that matters but the frequency associated with the acoustic signal is equally important. Any signal with frequency either less than 20 Hz or greater than 20 kHz will not be recognized by humans as audible signal, no matter how high the sound pressure is.

Acoustic field in most of the practical problems usually consists of superposition of large number of acoustic waves with different frequencies. Each of the acoustic wave can be characterized by it's own frequency and amplitude. These individual acoustic waves with monochromatic frequencies are termed as pure tone, which can be mathematically described as

$$p(t) = p_0 \cos(2\pi f t + \phi) \tag{6.3}$$

Equation (6.3) represents a pure tone sound wave. Here, $p(t)$ is instantaneous pressure, $p_0$ is the maximum amplitude of the sound disturbance, and $f$ is the frequency. Time (t) and frequency (f) are measured in seconds (s) and Hertz (Hz), respectively, while $\phi$ provides the phase lag. In addition to characterizing sound using maximum amplitude $p_0$, one can combine the effects of the time-varying amplitude and the frequency to define a new parameter as root mean square value of the pressure. Sometimes it is also denoted as $RMS$ value. It helps to obtain a quantitative measure of any fluctuating parameter. For example, if one obtains an average value over full cycle of a sine or cosine wave, it will be zero. But if we squared each instantaneous value then the average won't be zero. One can find out the square root of the corresponding averaged value and get a quantitative measure of the amplitude of the fluctuation.

So, the $RMS$ value of pressure for Eq. (6.3) is

$$p_{\text{RMS}}^2 = \frac{1}{2} p_0^2 \tag{6.4}$$

For a non-periodic signal, a large time interval has to be taken to estimate the $RMS$ value accurately. Engineers also use the term sound energy density to characterize the acoustic field which is nothing but the sound energy stored per unit volume and is given as $D = p_{av}^2/\rho c^2$. Here, $p_{av}^2$ is the squared average pressure at a point in the domain.

Apart from quantifying the amplitude of the acoustic wave, prescription of the time period or frequency of the signal is also important. Time taken to complete one full oscillation cycle is defined as period of the signal. By definition, frequency ($f$) is the number of cycles completed in one second. This gives us relation between the time period $T$ and the frequency $f$ as $T = 1/f$. Distance traveled by a wave in one full cycle is known as wavelength. Mathematically, wavelength is given by $\lambda = cT = c/f$, where $c$ is speed of sound, $T$ is time period, and $f$ is the frequency.

Acoustic signal usually consists of superposition of large number of individual acoustic waves with different frequencies as shown in Fig. 6.2. Number of frequencies present in the acoustic signal may be countably infinite. Plot of amplitude of pressure fluctuation with respect to frequency provides spectrum for an acoustic signal. So, if the number of frequencies present in the signal are infinite then the plot will show a *continuous spectrum*. In general engineering applications, acoustic waves are superposition of large number of longitudinal waves with different phases, frequencies, and amplitudes. These collection of longitudinal waves constitute the frequency spectrum of the signal. Regardless the type of the spectrum, all the waveforms remain unchanged while traveling through air for a short distance. Thus, a musical instrument heard at a distance of 10 m, sounds same as at 5 m, but amplitude of the signal will decrease due to distribution of acoustic energy over a larger area. Over a large distance, high-frequency waves (more than 1000 Hz) are prone to attenuate by viscous nature of the air.

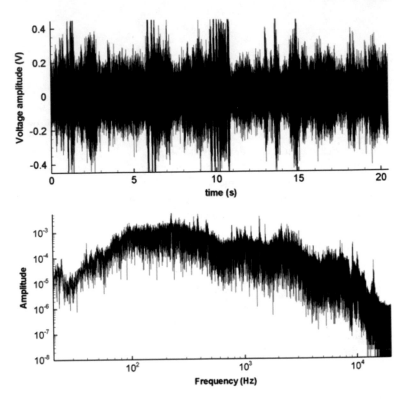

**Fig. 6.2** Measurement of road noise using a microphone is shown in the top frame and corresponding acoustic spectra have been shown in the bottom frame

### 6.1.2 Sound Intensity and Sound Power

An acoustic source radiates power resulting in the development of an acoustic field via variation in sound pressure. Humans are able to recognize an audible signal due to the sound pressure which, in turn, depends on the sound power emitted by an acoustic source. High sound pressure level can cause hearing damage. The sound pressure is dependent on the distance between the source and the receiver as well as the acoustic environment which includes the size and sound-absorbing nature of the walls of a room. By measuring sound pressure at a given location inside room, one can not quantify the acoustic or a noise source, which can either be a speaker or noise-making machine component. One needs to find out the radiated sound power to quantify various sound sources. Sound power is the rate at which sound energy is radiated into the surrounding. It is measured in watt (W). If we consider a non-directional point source, sound power can be expressed mathematically as

$$W_s = (4\pi r^2) I_s(r) \tag{6.5}$$

where $I_s(r)$ denotes maximum sound intensity on the imaginary sphere of radius $r$ in W/m$^2$. In case of line source distributed over length l, Eq. (6.5) modifies to Eq. (6.6) as shown below:

$$W_c = (2\pi r l) I_c(r) \qquad (6.6)$$

where $W_c$ is total sound power radiated.

Inside the room, we can measure sound pressure at various locations and find out whether the acoustic noise is in the permissible limit or not. If the noise level is high, then we will have to find out effective ways to reduce it. We can work on noise reduction strategies provided that we have sufficient information about noise radiated by different acoustic sources. Thus, measuring sound power of different sources can help us to prioritize our action plan as the machine that makes more noise needs more attention for effective noise reduction. However, one can relate sound power to sound pressure only under carefully controlled conditions such as in an anechoic or reverberant chamber where special assumptions about the sound field hold true. In contrast, another quantity identified as sound intensity can be measured in any sound field and provides us the necessary information similar to sound power. Sound intensity is defined as energy carried by sound waves per unit area. Sound intensity is measured in watts per square meter (W/m$^2$) and is a directional quantity. It has a maximum ($I_{max}$) value when the measuring plane is perpendicular to the direction of acoustic wave propagation. Sound intensity $I$ is obtained by the product of the sound pressure $p$ and the particle velocity $v$ and is useful while locating sound sources.

Consider an acoustic point source radiating acoustic waves outward. The intensity variation of acoustic waves in terms of acoustic power $W_s$ in the radial direction is given by

$$I(r) = \frac{W_s}{4 \times \pi \times r^2} \qquad (6.7)$$

where $r$ is a distance of the point at which intensity has been calculated from the center of the sphere. This relationship is also known as inverse square law, which states that in a free field environment, intensity of acoustic waves must decrease as one moves away from the sound source. As acoustic waves propagate spherically outward, doubling the radius causes four times increase in surface area of the sphere. So, to preserve sound energy, intensity must be reduced by a factor of four (Fig. 6.3).

An acoustic field can be classified with respect to the environment in which acoustic waves travel. The relationship between pressure and intensity is described in following two special cases, the free field and the diffuse field. The free field term corresponds to propagation of acoustic waves in an idealized free space where acoustic waves do not undergo any reflections. For this condition to hold true, source must be located far away from the ground in an open space or it has to be kept in an anechoic chamber where all the acoustic disturbances are absorbed. As per the inverse square law, one observes drop in sound pressure level by 6 dB as the distance

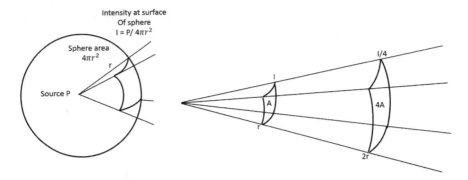

**Fig. 6.3** Sound intensity reduces one-fourth times by increase in radius two times

from the source is doubled. The relationship between the sound intensity and the pressure for free field environment is given as

$$|I| = \frac{p_{RMS}^2}{\rho c} \tag{6.8}$$

In contrast to a free field environment, a diffuse field environment corresponds to the acoustic field where acoustic waves are reflected from the surrounding solid surfaces so many times that these waves travel in all directions with equal magnitude and probability. Such acoustic field can be approximated inside a reverberant room. As the acoustic waves travel in all directions with equal magnitude and probability, net intensity $|I|$ is zero. Intensity also depends on the phase angle between the sound pressure and the particle velocity. For the case without a phase difference between the sound pressure and the particle velocity, one obtains net sound intensity. In contrast if the sound pressure and the particle velocity are $90^o$ out of phase then the product of the pressure and the particle velocity provides an instantaneous intensity signal which varies sinusoidally with a zero mean intensity.

### 6.1.3   Acoustic Measurements

Human ears are receptive to acoustic waves over a wide band of amplitude and frequency range. Various levels designed to perform acoustic measurements are described here. Sound pressure level is the most common term associated with microphone measurements. For a given acoustic field, sound pressure level is obtained by

$$L_p = 20 \log_{10} \left[ \frac{p_{RMS}}{p_{ref}} \right] \, dB \tag{6.9}$$

where $p_{RMS}$ is the $RMS$ value of the sound pressure at a point and $p_{ref}$ is the reference pressure level. For sound propagation through air medium, reference sound pressure $p_{ref}$ is considered as $2 \times 10^{-5}$ N/m$^2$. Sound pressure level is usually measured in decibel (dB). Another way of prescription of an acoustic field is by using sound power level which is given as

$$L_w = 10 \log_{10} \frac{W}{W_0} \text{ dB} \tag{6.10}$$

where $W$ is sound power in watts. Sound power level is measured with respect to the reference value, $W_0$. Value of $W_0$ is usually considered as $10^{-12}$ W. One can also describe sound intensity level as

$$L_I = 10 \log_{10} \frac{I}{I_{ref}} \text{ dB} \tag{6.11}$$

where $I$ is the intensity level. For acoustic waves traveling through air, the reference intensity $I_{ref}$ is considered as $10^{-12}$ W/m$^2$. Intensity levels are also measured in dB scale.

## 6.1.4 Reflection, Scattering, and Diffraction of Acoustic Waves

Any disturbance in the fluid caused by displacement of a fluid element from its mean position propagates away in all directions at the speed of sound $c$. For homogeneous fluid, the disturbance propagates uniformly in all directions similar to an expanding spherical shell. Acoustic waves propagate normal to the direction of local wavefront. Any complex acoustic field simultaneously displays phenomenon of reflection, interference, scattering, diffraction, and refraction associated with the wave motion.

Acoustic waves undergo reflection as soon as these encounter change in medium. The reflection of the acoustic wave is dependent on the nature of the obstructing surface. Some surfaces completely reflect back the acoustic energy while some surfaces transmit and absorb part of the acoustic energy and reflect the remaining fraction of the incident acoustic energy. Acoustic wave reflection is maximum for the fully reflecting boundary. For a pure reflection, the angle of incidence is equal to the angle of reflection as shown in Fig. 6.4. In most of the practical applications, pure reflection case is hardly observed. If the dimensions of the room are small then one is not able to recognize the difference between the direct and the reflected acoustic waves as the human ear cannot distinguish between two identical sounds which approach ear less than 1/15th part of a second. If the gap between the direct and reflected sound is more than this, then we hear echo.

While designing concert halls, auditoriums, movie theaters as well as class rooms, engineers need to take care of problems associated with reflection of acoustic waves.

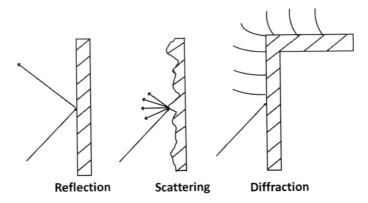

**Fig. 6.4** Schematic of reflection, scattering, and diffraction of acoustic waves

In a poorly designed hall, listeners' ability to distinguish between different words of a speech gets hampered significantly due to phenomenon of reverberation in which each word echoes repeatedly. Engineers try to minimize such problems by covering reflecting surfaces inside the room with sound-absorbing materials or acoustical tiles. Such sound-absorbing materials are usually porous. Sound waves get trapped into the tiny air-filled spaces of a porous material and undergo multiple reflections inside the small cavities until their energy is dissipated. Reflection of sound also plays important part in the design of sonar which is commonly used for underwater object detection such as submarines, lost vessels as well as sea surfaces. The reflection of sound is also used by bats and toothed whales to locate food and habitat.

The reflection of acoustic waves can be altered by changing either the solid material from which the reflection is occurring or by changing the surface finish. For a truly reflecting surface, the angle of incidence and reflection are same. However, if the acoustic waves are incident on a rough surface, then the energy of the incident wave gets distributed through large number of reflections, which is also known as scattering. Therefore, scattering can reduce the reflected acoustic energy in one direction by converting it in to many scattered reflections. Scattering is also dependent on the frequency and wavelength of the incident acoustic disturbance. If the surface roughness is comparable to the wavelength of the acoustic wave, then scattering is observed as shown in Fig. 6.4.

Diffraction of acoustic waves is usually observed when the waves either pass around an obstacle or propagate through a slit as shown in Fig. 6.4. In such cases, the edge of the obstacle or the slit acts as a secondary sound source which radiates acoustic waves at the same frequency and wavelength but with reduced acoustic intensity compared to the primary acoustic source. Corresponding spreading of acoustic waves from the secondary source is called diffraction. As acoustic waves bend around the obstacle by undergoing diffraction process, one is able to hear the sound despite of being on the other side of the obstacle. Such zone with reduced acoustic intensity is referred as a shadow zone.

Apart from acoustic reflection and diffraction, acoustic waves also display interference which occurs when multiple waves interact. Superposition of large number of acoustic waves provides the physical basis for interference. It can either happen due to superposition of acoustic waves radiated by multiple acoustic sources or due to superposition of the incident and reflected waves. In such acoustic fields, one observes constructive and destructive interference at various locations in the domain. For a closed reverberant space, a source inside a domain with reflecting boundaries also creates interference, either constructive or destructive. In auditoriums, destructive interference at certain locations leads to creation of dead spots where the amplitude and clarity of acoustic waves are inferior. One can redesign such spots by using appropriate sound-absorbing materials or sound-reflecting surfaces so as to avoid destructive interference. In addition, the reflecting surfaces inside an auditorium can be arranged in such a way that there is increased level of sound in the audience.

## 6.2  Governing Differential Equations for Acoustic Wave Propagation

In this section, we are going to derive governing differential equations for solving computational acoustics and aeroacoustics problems. In these problems, the frequency of the sound waves falls in the acoustic range and the medium for acoustic wave propagation is air. For computational acoustics problems, usually the acoustic source is a vibrating surface while for the computational aeroacoustics problems, acoustic waves are triggered due to the unsteady nature of the fluid flow.

Fluids are collection of large number of molecules. Spacings between the molecules are larger for gases as compared to the liquids. In contrast to solids, fluid molecules are not fixed in lattice but are able to move freely relative to each other. Thus, the number of molecules present inside a given volume of a fluid changes continuously with time. One needs to be careful while defining fluid properties such as density, pressure, viscosity, velocity, as these properties change with space and time. One needs to find out these properties at different points in the domain for understanding the flow field. At each point, these properties are found out corresponding to a small volume that surrounds that point. If the chosen volume is too small or too large then we will not be able to measure above properties accurately as the microscopic or macroscopic effects will be significant. For example, the density $\rho$ of a fluid can be defined as

$$\rho = \lim_{\delta v \to \delta v^*} \frac{\delta m}{\delta v} \tag{6.12}$$

where the limiting volume $\delta v^*$ is about $10^{-9}$ mm$^3$ for all liquids and gases at atmospheric pressure which contains approximately $3 \times 10^7$ molecules [3]. For such volume, change in fluid properties due to micro/macroscopic uncertainty can be neglected and variation of fluid properties can be considered smooth. Such fluid is called as continuum and we can apply differential calculus to analyze the flow properties. Laws of mechanics are written for an arbitrary quantity of mass of fixed identity. In case of fluid flow, one needs to convert the system laws to apply to a specific region of interest. Continuum hypothesis enables us to write fluid flow (conservation laws) equations in differential equation form. Various conservation laws (Mass and Momentum) are given by [3]

**Mass conservation law**:

$$\frac{\partial \rho}{\partial t} + \frac{\partial (\rho u)}{\partial x} + \frac{\partial (\rho v)}{\partial y} + \frac{\partial (\rho w)}{\partial z} = 0 \tag{6.13}$$

**Momentum conservation laws**:

$$\frac{\partial (\rho u)}{\partial t} + \nabla.(\rho u \mathbf{V}) = \rho g_x - \frac{\partial p}{\partial x} + \frac{\partial \tau_{xx}}{\partial x} + \frac{\partial \tau_{xy}}{\partial y} + \frac{\partial \tau_{xz}}{\partial z} \tag{6.14}$$

$$\frac{\partial (\rho v)}{\partial t} + \nabla.(\rho v \mathbf{V}) = \rho g_y - \frac{\partial p}{\partial y} + \frac{\partial \tau_{xy}}{\partial x} + \frac{\partial \tau_{yy}}{\partial y} + \frac{\partial \tau_{yz}}{\partial z} \tag{6.15}$$

$$\frac{\partial (\rho w)}{\partial t} + \nabla.(\rho w \mathbf{V}) = \rho g_z - \frac{\partial p}{\partial z} + \frac{\partial \tau_{xz}}{\partial x} + \frac{\partial \tau_{yz}}{\partial y} + \frac{\partial \tau_{zz}}{\partial z} \tag{6.16}$$

where $\rho$ is density of the fluid, $\mathbf{V} = u\hat{i} + v\hat{j} + w\hat{k}$ is a velocity vector, $p$ is a pressure, and $\rho g_x$, $\rho g_y$, and $\rho g_z$ are components of the body force term along the respective axes. Normal and shear stresses are identified by the terms $\tau_{ij}$, with corresponding suffixes. Acoustic waves are associated with pressure fluctuations in a compressible fluid. In addition to audible pressure fields of moderate intensity, frequencies of ultrasonic and infrasonic waves lie beyond limits of hearing. Inviscid fluids exhibit fewer constraints to deformations than solids. The restoring forces responsible for propagating a wave are the pressure changes that occur when the fluid is compressed and expanded.

Consider a sound source kept inside a large cavity. Pressure fluctuations caused by a sound source will travel in all directions with sound speed $c$. We assume the fluid as homogeneous medium with very low coefficients of viscosity, so that the action of viscous losses can be neglected if the sound propagation is considered over a small distance. In addition, any body forces acting on the generated acoustic field are also neglected. One can divide the flow parameters into the mean and the disturbance quantities. Let us denote the mean flow parameters as $\bar{p}, \bar{\rho}, \bar{u}, \bar{v}, \bar{w}$ and the fluctuating parameters associated with sound propagation as $p', \rho', u', v', w'$. Then, the mass conservation law can be rewritten as

$$\frac{\partial (\bar{\rho} + \varepsilon \rho')}{\partial t} + \frac{\partial ((\bar{\rho} + \varepsilon \rho')(\bar{u} + \varepsilon u'))}{\partial x} + \frac{\partial ((\bar{\rho} + \varepsilon \rho')(\bar{v} + \varepsilon v'))}{\partial y} + \frac{\partial ((\bar{\rho} + \varepsilon \rho')(\bar{w} + \varepsilon w'))}{\partial z} = 0 \tag{6.17}$$

Collecting order $\varepsilon$ quantities,

$$\frac{\partial \rho'}{\partial t} + \frac{\partial(\bar{\rho}u' + \rho'\bar{u})}{\partial x} + \frac{\partial(\bar{\rho}v' + \rho'\bar{v})}{\partial y} + \frac{\partial(\bar{\rho}w' + \rho'\bar{w})}{\partial z} = 0 \qquad (6.18)$$

$$\frac{\partial \rho'}{\partial t} + \bar{\rho}\left(\frac{\partial u'}{\partial x} + \frac{\partial v'}{\partial y} + \frac{\partial w'}{\partial z}\right) + \bar{u}\frac{\partial \rho'}{\partial x} + \bar{v}\frac{\partial \rho'}{\partial y} + \bar{w}\frac{\partial \rho'}{\partial z} = 0 \qquad (6.19)$$

If we consider a case without any fluid flow (mean flow velocity is zero), Eq. (6.19) is simplified as

$$\frac{\partial \rho'}{\partial t} + \bar{\rho}\left(\frac{\partial u'}{\partial x} + \frac{\partial v'}{\partial y} + \frac{\partial w'}{\partial z}\right) = 0 \qquad (6.20)$$

If we denote the perturbation velocity field as $\mathbf{u} = u'\hat{i} + v'\hat{j} + w'\hat{k}$, Eq. (6.20) can be written in the simplified form as

$$\frac{\partial \rho'}{\partial t} + \nabla.(\bar{\rho}\mathbf{u}) = 0 \qquad (6.21)$$

Similarly one can simplify Eqs. (6.14)–(6.16) as

$$\bar{\rho}\frac{\partial u'}{\partial t} = -\frac{\partial p'}{\partial x} \qquad (6.22)$$

$$\bar{\rho}\frac{\partial v'}{\partial t} = -\frac{\partial p'}{\partial y} \qquad (6.23)$$

$$\bar{\rho}\frac{\partial w'}{\partial t} = -\frac{\partial p'}{\partial z} \qquad (6.24)$$

Equations (6.22), (6.23), and (6.24) can be written in the combined form as

$$\bar{\rho}\frac{\partial \mathbf{u}}{\partial t} + \nabla p' = 0 \qquad (6.25)$$

On differentiating Eqs. (6.22), (6.23), and (6.24) with respect to $x$, $y$, $z$ and adding, we get

$$\bar{\rho}\frac{\partial}{\partial t}\left(\frac{\partial u'}{\partial x} + \frac{\partial v'}{\partial y} + \frac{\partial w'}{\partial z}\right) = -\nabla^2 p' \qquad (6.26)$$

This is equivalent to taking the divergence of the linear momentum equation. By combining Eqs. (6.26) and (6.20), we get

$$\frac{\partial^2 \rho'}{\partial t^2} = \frac{\partial^2 p'}{\partial x^2} + \frac{\partial^2 p'}{\partial y^2} + \frac{\partial^2 p'}{\partial z^2} \qquad (6.27)$$

Acoustic wave propagation through air can be considered as isentropic process. If the thermal conductivity of the fluid is very small and the pressure fluctuations associated with acoustic waves are sufficiently small, then no appreciable thermal energy transfer occurs between adjacent fluid elements. Thermodynamic speed of sound is obtained as

$$c = \left( \sqrt{\frac{\partial p}{\partial \rho}} \right)_{\text{adiabatic}} \tag{6.28}$$

Under these conditions, entropy of the fluid remains constant and sound propagation can be considered as isentropic process.

$$p' \vartheta'^{\gamma} = \text{constant} \tag{6.29}$$

where $\gamma$ is given as the ratio of the specific heats as, $\gamma = C_p / C_v$

$$p' \rho'^{-\gamma} = \text{constant} \tag{6.30}$$

Using Eqs. (6.28) and (6.30) in Eq. (6.27), one obtains

$$\frac{\partial^2 p'}{\partial t^2} = c^2 \left( \frac{\partial^2 p'}{\partial x^2} + \frac{\partial^2 p'}{\partial y^2} + \frac{\partial^2 p'}{\partial z^2} \right) \tag{6.31}$$

Equation (6.31) is a linear, lossless wave equation for the propagation of sound in stationary fluids with phase speed $c$.

### 6.2.1  Lighthill's Equation for the Propagation of Sound Wave

Lighthill [2] proposed an inhomogeneous wave equation for sound propagation in fluids by rearranging terms in the mass and the momentum conservation of fluid flows. Left-hand side of the Lighthill's equation provides propagation of sound in a fluid while right-hand side of the equation contains the sound source term caused by the fluid flow. A brief derivation of the Lighthill's equation is discussed to understand the sound generation and propagation in fluid flow.

One can rewrite the mass conservation in a tensorial notation as

$$\frac{\partial}{\partial t} \left[ \frac{\partial \rho}{\partial t} + \frac{\partial (\rho u_i)}{\partial x_i} \right] = 0 \tag{6.32}$$

Divergence of the momentum conservation law can be represented in the tensorial notation as

$$-\frac{\partial}{\partial x_i} \left[ \frac{\partial \rho u_i}{\partial t} + \frac{\partial (\rho u_i u_j)}{\partial x_j} \right] = -\frac{\partial p}{\partial x_i} + \frac{\partial \sigma_{ij}}{\partial x_j} \tag{6.33}$$

Equations (6.32) and (6.33) can be combined together as

$$\frac{\partial^2 \rho}{\partial t^2} = \frac{\partial^2 p}{\partial x_i^2} + \frac{\partial^2 (\rho u_i u_j - \sigma_{ij})}{\partial x_i x_j} \tag{6.34}$$

Terms in Eq. (6.34) can be further rearranged as

$$\frac{\partial^2 \rho}{\partial t^2} - c^2 \frac{\partial^2 \rho}{\partial x_i^2} = \frac{\partial^2 (p - c^2 \rho)}{\partial x_i^2} + \frac{\partial^2 (\rho u_i u_j - \sigma_{ij})}{\partial x_i x_j} \tag{6.35}$$

One can rewrite Eq. (6.35) in terms of the mean and fluctuating quantities together and separate out equation for the fluctuating quantities as

$$\frac{\partial^2 \rho'}{\partial t^2} - c^2 \frac{\partial^2 \rho'}{\partial x_i^2} = \frac{\partial^2 (p' - c^2 \rho')}{\partial x_i^2} + \frac{\partial^2 (\rho u_i u_j - \sigma_{ij})}{\partial x_i x_j} \tag{6.36}$$

which further can be written in a simplified form as

$$\frac{\partial^2 \rho'}{\partial t^2} - c^2 \frac{\partial^2 \rho'}{\partial x_i^2} = \frac{\partial^2 T_{ij}}{\partial x_i x_j} \tag{6.37}$$

where $T_{ij} = \rho u_i u_j + (p' - c^2 \rho')\delta_{ij} - \sigma_{ij}$ is also known as Lighthill's stress tensor. In Eqs. (6.36) and (6.37), propagation speed of sound sources has been assumed negligible compared to speed of sound. In many engineering applications where Mach number is significantly small, it has been observed that the contribution of viscous term and pressure term in the expression for $T_{ij}$ is negligible as compared to the term $\rho u_i u_j$.

### 6.2.2 The Inhomogeneous Wave Equation

As discussed in the previous section, one needs an acoustic source or acoustic-wave-generating mechanism for the existence of an acoustic field. The wave equation, Eq. (6.31) developed in the previous section, needs two initial conditions on the disturbance component of the pressure $p'$ for obtaining the solution of an acoustic field. However, engineering problems often consist of acoustic waves generating sources either through mechanical vibration such as a speaker or through motion of the eddies in a turbulent flow. Acoustic field can also be observed through the action of time-dependent boundary conditions; however, it is convenient to write the fundamental acoustic wave propagation equations by including the acoustic source terms as shown below.

If we consider a case where mass has been injected in to the domain then one can modify the disturbance component equation for the continuity Eq. (6.20) as [1]

$$\frac{\partial \rho'}{\partial t} + \nabla.(\bar{\rho}\mathbf{u}) = G(\mathbf{r}, t) \tag{6.38}$$

where the term $G(\mathbf{r}, t)$ denotes amount of mass injected into the domain per unit volume per unit time. Similarly, in the presence of the body force, the term $\mathbf{F}(\mathbf{r}, t)$ which accounts for the body force per unit volume gets added into the linearized momentum equations as [1]

$$\bar{\rho}\frac{\partial \mathbf{u}}{\partial t} + \nabla p' = \mathbf{F}(\mathbf{r}, t) \tag{6.39}$$

Let us modify Eq. (6.38) by performing time derivative on both sides of the equation and Eq. (6.39) by taking its divergence. These modified mass and momentum conservation equations with the source terms can be combined together to form an inhomogeneous wave equation given by [1]

$$\nabla^2 p' - \frac{1}{c^2}\frac{\partial^2 p'}{\partial t^2} = -\frac{\partial G}{\partial t} + \nabla.\mathbf{F} \tag{6.40}$$

In addition to these modifications, a third modification has been proposed by Lighthill [2], to the inhomogeneous wave equation to describe the sound generation mechanism due to the fluid flow. Lighthill included the shear and the bulk viscosity effects. However, in many practical cases, contribution of viscous terms to noise production can be neglected and inhomogeneous wave equation in the tensorial notation can be derived as [1, 2]

$$\nabla^2 p' - \frac{1}{c^2}\frac{\partial^2 p'}{\partial t^2} = -\frac{\partial^2 (\rho' u_i u_j)}{\partial x_i \partial x_j} \tag{6.41}$$

The source term on the right-hand side of Eq. (6.41) accounts for the spatial rates of change of momentum flux within the fluid and corresponds to the acoustic field creation due to turbulent fluid flow. One can observe that the acoustic field can be created via injection of mass, presence of a body force term, and the turbulent fluid flow. These three different types of acoustic source mechanisms are independent of each other and a complete inhomogeneous lossless wave equation is given by [1]

$$\nabla^2 p' - \frac{1}{c^2}\frac{\partial^2 p'}{\partial t^2} = -\frac{\partial G}{\partial t} + \nabla.\mathbf{F} - \frac{\partial^2 (\rho' u_i u_j)}{\partial x_i \partial x_j} \tag{6.42}$$

The source terms present on the right-hand side of Eq. (6.42) can be related to the monopole, dipole, and the quadrupole nature of an acoustic radiation.

A monopole acoustic source radiates acoustic waves outward without any directional preference as shown in Fig. 6.5a. Alternate contraction and expansion of a solid sphere corresponds to a monopole sound source. Such source triggers acoustic waves via alternate injection and removal of fluid into the domain. At low frequencies, one can consider a boxed loudspeaker as an example of monopole sound source. The directivity pattern for a monopole acoustic source does not show any bias. In contrast, a dipole acoustic source consists of two monopole sources which are separated

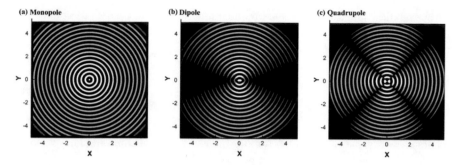

**Fig. 6.5** Instantaneous sound pressure contours for the **a** monopole source, **b** dipole source, and **c** quadrupole source have been shown. The light and the dark regions account for the compression and the rarefaction zones formed during sound propagation

by a small distance as compared to the wavelength of acoustic wave. These monopole sources have equal strength but opposite phase so that when one monopole source expands, other contracts. Such behavior introduces a directionality in the acoustic field as the fluid near the sources sloshes back and forth to produce acoustic field. One can either consider a back and forth oscillating sphere or an unboxed loudspeaker in which the front and rear portions of the diaphragm alternatively pushes and pulls the fluid. The directivity pattern for a dipole displays two regions each, where acoustic waves are very well radiated and where acoustic waves cancel each other as shown in Fig. 6.5b. A quadrupole source consists of two opposite dipoles which usually do not lie along the same line. For such acoustic source, one can consider an arrangement of four monopole acoustic sources with same amplitude and an alternating phase located at the corners of a square. The directivity pattern for a quadrupole source displays directionality. One can observe a negligible amplitude of acoustic waves along the diagonals while the horizontal and vertical wavefronts have opposite phases as shown in Fig. 6.5c.

Next, we discuss about the space and time variations of the acoustic signal and the relation between the spatial and temporal scales which is given by the physical dispersion relation. We also discuss about the signal propagation in the dispersive and non-dispersive systems.

### 6.2.3  Wave Equation and Physical Dispersion Relation

Consider a variable $u$ whose variation with respect to time $u(t)$ in the physical plane can be linked to the corresponding information in the spectral plane as

$$u(t) = \int_{-\infty}^{\infty} U(f)e^{i\omega t}\,d\omega \tag{6.43}$$

Above relation suggests that value of signal amplitude at any time $t$ is obtained by integrating contribution of all the frequencies present in the signal. If the signal has only space variation, then one can write

$$u(x) = \int_{-\infty}^{\infty} U(k)e^{ikx}dk \tag{6.44}$$

where $k = \frac{2\pi}{\lambda}$ is the wavenumber for $\lambda$ as the wavelength. If the signal has variation in both space and time, then one can write

$$u(x,t) = \int_{-\infty}^{\infty} \int_{-\infty}^{\infty} U(k,\omega)e^{i(kx\pm\omega t)}dk d\omega \tag{6.45}$$

One observes similar space and time variation in the propagation of an acoustic disturbance. Corresponding homogeneous second-order wave equation is given by

$$\frac{\partial^2 u}{\partial t^2} - c^2 \frac{\partial^2 u}{\partial x^2} = 0 \tag{6.46}$$

The general solution of Eq. (6.46) suggests presence of two functions of the arguments $(x + ct)$ and $(x - ct)$. One can split differential operators in Eq. (6.46) as

$$\left[\frac{\partial}{\partial t} - c\frac{\partial}{\partial x}\right]\left[\frac{\partial}{\partial t} + c\frac{\partial}{\partial x}\right]u = 0 \tag{6.47}$$

which simplifies to

$$\frac{\partial u}{\partial t} - c\frac{\partial u}{\partial x} = 0 \tag{6.48}$$

$$\frac{\partial u}{\partial t} + c\frac{\partial u}{\partial x} = 0 \tag{6.49}$$

Equations (6.48) and (6.49) suggest that once we provide initial disturbance, part of the initial disturbance travels toward right direction with phase velocity $c$ while the other half travels toward left with same phase speed. The amplitudes of the right going and left going disturbances do not change with time. This provides us a wave equation as a classic example of non-dispersive and non-dissipative system. We consider a first-order, one-dimensional wave equation as

$$\frac{\partial u}{\partial t} + c\frac{\partial u}{\partial x} = 0 \tag{6.50}$$

Physical nature related to disturbance propagation suggests us that the disturbance propagates toward right with phase velocity $c$, without any attenuation. Phase velocity $(V_p)$ is a rate at which phase of the wave propagated in space and is given by

$$V_p = \frac{\omega}{k} = \frac{\lambda}{T} = c \tag{6.51}$$

Variation of the signal in space and time can be given by

$$u(x, t) = \int \int U(k, \omega) e^{i(kx - \omega t)} dk d\omega \tag{6.52}$$

$$u(x, t) = \int \int U(k, \omega) e^{ik(x - ct)} dk d\omega \tag{6.53}$$

If we incorporate Eq. (6.52) in Eq. (6.50), we get

$$\int \int (-i\omega + ikc) U(k, \omega) e^{i(kx - \omega t)} dk d\omega = 0 \tag{6.54}$$

In the above equation, term in the parenthesis can only be zero, i.e., $-i\omega + ikc = 0$. We obtain the physical dispersion relation for the 1D wave equation as

$$\omega = kc \tag{6.55}$$

Dispersion relation is simply the relation between the wavenumber $k$ and circular frequency $\omega$. Equation (6.55) suggests that the ratio of circular frequency to wavenumber is always equal to the phase speed $c$. Thus, Eq. (6.50) (1D wave equation) becomes a model equation to study propagation of acoustic disturbance as given by Eq. (6.31). An acoustic signal does consist of various frequencies; however, all such components travel with speed of sound.

## 6.2.4 Dispersive and Non-dispersive System

As discussed before, an acoustic signal consists of large number of waves superimposed on each other. Let us consider a simple case where the signal is formed by superposition of two harmonic signals whose wavenumbers and circular frequencies are very close to each other. Let the two harmonic signals are given by

$$y_1 = a \cos(k_1 x - \omega_1 t); \quad c = \omega_1 / k_1 \tag{6.56}$$

$$y_2 = a \cos(k_2 x - \omega_2 t); \quad c = \omega_2 / k_2 \tag{6.57}$$

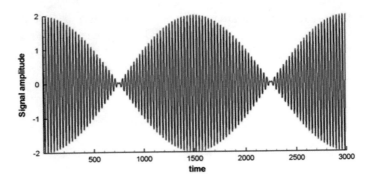

**Fig. 6.6**  Variation of superimposed signal with time

Superposition of above two signals is shown as

$$y = y_1 + y_2 = a(\cos(k_1 x - \omega_1 t) + \cos(k_2 x - \omega_2 t)) \tag{6.58}$$

$$y = 2a \cos\left[\frac{(k_1 - k_2)x}{2} - \frac{(\omega_1 - \omega_2)t}{2}\right] \times \cos\left[\frac{(k_1 + k_2)x}{2} - \frac{(\omega_1 + \omega_2)t}{2}\right] \tag{6.59}$$

If we plot variation of $y$ at a point with respect to time $t$, we observe the signal amplitude variation as shown in Fig. 6.6. Equation (6.59) shows that the superimposed signal has two components. The first component in the bracket imparts slow variation to superimposed signal causing formation of envelopes/ modulations, while the second component has phase very close to the input harmonic signals.

Depending on the governing equation, one can prescribe or evaluate a phase speed as

$$\text{Phase speed } c(k, \omega) = \frac{\omega}{k} \tag{6.60}$$

The speed with which the envelope propagates is termed as group velocity ($V_g$) and is given by

$$\text{Group velocity } V_g = \frac{\omega_1 - \omega_2}{k_1 - k_2} = \frac{d\omega}{dk} \tag{6.61}$$

For a 1D wave equation, the phase speed $c$ and the group velocity $V_g$ are both equal and are constants, i.e., independent of circular frequency and wavenumber. Such system is termed as non-dispersive and is shown in Fig. 6.7. For a non-dispersive system,

$$\frac{\omega_1}{k_1} = \frac{\omega_2}{k_2} = c = \text{ phase speed}$$

$$\frac{d\omega_1}{dk_1} = \frac{d\omega_2}{dk_2} = c = \text{ group velocity}$$

**Fig. 6.7** Representation of a non-dispersive and dispersive systems

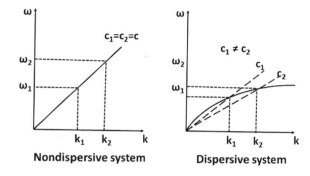

Nondispersive system          Dispersive system

However, for a dispersive system, variation of $\omega$ and $k$ is nonlinear, so that

$$c_1 = \frac{\omega_1}{k_1}; \quad c_2 = \frac{\omega_2}{k_2}; \quad c_1 \neq c_2$$

In addition, the group velocities are given as

$$V_{g_1} = \frac{d\omega_1}{dk_1} \neq c_1; \quad V_{g_2} = \frac{d\omega_2}{dk_2} \neq c_2; \quad V_{g_1} \neq V_{g_2}$$

With this introductory chapter on acoustics, we will focus our attention on the computational acoustics problems in the next chapter. We will discuss about different important properties of numerical methods and their application to solve the computational acoustics problems. Subsequently, in the last chapter, we will discuss various approaches used by different researchers to solve various aeroacoustic problems.

# References

1. L.E. Kinsler, A.R. Frey, A.B. Coppens, J.V. Sanders, *Fundamentals of Acoustics* (Wiley, New York, 2000)
2. M.J. Lighthill, On sound generated aerodynamically. I. general theory. Proc. Roy. Soc. Lond. A: Math. Phys. Eng. Sci. **211**, 564–587 (1952)
3. F.M. White, *Fluid Mechanics* (McGraw-Hill Education, 2015)

# Chapter 7
# Solutions of Computational Acoustic Problems Using *DRP* Schemes

Computational acoustics is an important and active research area [12, 14, 40, 48, 53]. Acoustic signals propagate in the form of longitudinal waves in air. Pressure fluctuations associated with acoustic signals are usually very small compared to the large background pressure field. The atmospheric pressure is around $10^5$ Pa, while the amplitude of the smallest recognizable acoustic disturbance for a human being is around $10^{-5}$ Pa. This significant gap between the hydrodynamic and acoustic scales puts severe restriction on allowable numerical error in the simulation. Effects of viscosity of the medium on the propagation of acoustic signal are negligibly small, when the signal propagates over a short distance [21]. Calculations for acoustical problems have to be performed for a long duration avoiding numerical instabilities to obtain acoustical spectra [53]. In this context, the chosen discretization schemes should neither numerically amplify nor attenuate the acoustic signal. Propagation of a computed acoustic signal strongly depends on the phase, dissipation, and the dispersion properties associated with individual wavenumber component for the used discretization schemes [12, 40, 45]. Hence, the accurate computation of the acoustic field is a challenging topic. With these challenges involved, any numerical scheme used for computing acoustic-related problems must display following features [37]:

1. The used scheme and the chosen grid must resolve all the scales present in the flow.
2. Numerical schemes must be neutrally stable. Role of added numerical diffusion in numerical stabilization must be kept as small as possible.
3. All the resolved wavenumber components must propagate with correct physical velocity and there should not be any spurious dispersion.

Computational aeroacoustic (*CAA*) problems are focused on obtaining numerical solution of formation and propagation of acoustic disturbances due to fluctuations in the fluid flow [8, 12, 54]. For an accurate numerical solution, it is necessary that the numerical scheme must resolve all the spatial and temporal scales present in the acoustic field, as well as in the flow field. Hence, the used numerical scheme must

© Springer Nature Singapore Pte Ltd. 2020
T. K. Sengupta and Y. G. Bhumkar, *Computational Aerodynamics and Aeroacoustics*,
https://doi.org/10.1007/978-981-15-4284-8_7

have a higher spectral resolution. In addition, numerical propagation speed of all the resolved scales must be identical to the corresponding physical propagation speed of the individual scale, in order to avoid dispersion error [37, 41, 56]. It is very much important to ensure that the used numerical scheme is capable of correctly estimating amplitudes and speeds of all the fluctuations in the fluid flow which act as an acoustic source. The scheme must exhibit a better dispersion relation preserving (*DRP*) nature across a large wavenumber range [37, 41] as compared to the traditional discretization schemes.

In this regard, compact schemes are widely used for solving computational aeroacoustics problems, as well as fluid flow problems [12, 14, 39, 57]. These schemes provide excellent spectral resolution as compared to the traditional explicit discretization methods, while evaluation of various derivative terms of the governing differential equations [10, 15, 24, 26, 36, 39, 40, 57]. These schemes have a relatively compact stencil, as compared to the traditional explicit schemes of same order and same spectral resolution [37]. In compact schemes, derivative of a function is obtained by either solving a tridiagonal or a pentadiagonal matrix equation [24, 39, 57]. Although compact schemes have fewer grid points in the stencil, their implicit nature accounts for the contribution from large number of nodes present in the domain, which provides a higher spectral resolution even on a coarser mesh [24, 37, 39, 40, 57, 59].

Researchers have also focused their attention on the development of new *DRP* schemes [10, 33, 37]. While designing a numerical scheme, one aims to preserve physical dispersion relation across a large wavenumber range. However, various existing numerical schemes are capable of preserving physical dispersion relation only for a limited part of the complete wavenumber spectrum. The *DRP* nature of the difference scheme is observed for the low wavenumber components, while spurious dispersion is associated with high wavenumber components [37, 46]. Spurious waves are associated with extreme form of dispersion error in which computed scales travel in a wrong direction with wrong speed are identified as the $q$-waves [30, 46]. These spurious waves are often responsible for numerical instabilities [42, 46, 57]. Numerical instabilities can be avoided by either using numerical filters [4, 5, 15, 38, 42, 57] or by using upwind scheme [10, 39, 44]. However, use of upwind schemes or multi-dimensional filters should not result into unphysical attenuation of the low wavenumber components. A new multi-dimensional adaptive filtering methodology was suggested in [4, 5] to attenuate high wavenumber components, as and when it is absolutely necessary.

Researchers working in the areas of direct numerical simulation (*DNS*) and large eddy simulation (*LES*) of complex turbulent flows have suggested alternative techniques, such as changing governing equations itself [18, 35] or the use of a staggered mesh [7, 27] to minimize spurious oscillations. In order to control the numerical instabilities associated with high wavenumber components, use of hyper-viscosity term in spectral methods [9, 17, 23] and the spectral vanishing viscosity method [19, 28, 49] has been suggested. In the particular context of the use of compact schemes in *DNS/LES* of turbulent flows, it has been suggested [22] to introduce an extra-diffusion directly with the viscous terms of the Navier–Stokes equation through a specific finite difference scheme for the computation of second derivatives. In the same work, an application for an aeroacoustic test case has also been provided.

Here, we discuss about a new coupled compact difference scheme, which has been proposed to achieve significantly improved numerical properties [32]. The full domain matrix global spectral analysis (*GSA*) technique [37, 39–41] is used to obtain the important numerical properties for the model convection equation. It is shown that the numerical properties of the newly derived coupled compact difference scheme are superior, as compared to the existing high-accuracy schemes. Applicability of the proposed scheme has been shown by solving various acoustic and model wave propagation problems, as well as by obtaining solution of the Navier–Stokes equation for the fluid flow problem. In addition to the advantages mentioned about the proposed numerical scheme, a new methodology for the adaptive numerical diffusion is also presented. The merit of the present scheme in numerically damping out unwanted reflected spurious disturbances from the domain boundaries is also shown. Thus, the present scheme can also be used to provide an artificial absorbing layer to attenuate spurious reflections, similar to the perfectly matched layer (*PML*) technique.

## 7.1  Construction and Numerical Properties of the Coupled Compact Difference Scheme

The solution of the Navier–Stokes equation needs evaluation of the first and the second derivative terms associated with flow convection and dissipation, respectively. High-accuracy compact difference schemes have been reported in [10, 11, 24, 39, 57] for the evaluation of the spatial derivative terms. Use of compact difference scheme requires special attention to suppress the numerical instabilities arising out of unphysical growth of the high wavenumber components [37]. Numerical instabilities can be avoided by using upwind schemes in which an explicit numerical diffusion [37] term is added, as expressed in the following equation:

$$\frac{\partial u}{\partial x} = \left(\frac{\partial u}{\partial x}\right)_{CD} + \alpha \left(\frac{\partial^{2n} u}{\partial x^{2n}}\right) h^{2n-1} \tag{7.1}$$

where, $\alpha$ is a diffusion coefficient, $n$ is an integer, and the subscript CD indicates derivative has been obtained using central difference scheme. The extra term $\alpha \left(\frac{\partial^{2n} u}{\partial x^{2n}}\right) h^{2n-1}$ in Eq. (7.1) adds numerical diffusion. The diffusion coefficient $\alpha$ can either be positive or negative based on the direction of propagation of information at a particular point and is also used to control the amount of added numerical diffusion. Note that the second derivative in the Navier–Stokes equations is associated with physical diffusion process and usually a fourth or higher even-order derivative terms are used for the addition of numerical diffusion [37]. In this context, a numerical scheme which evaluates the first, the second, and the fourth derivative terms in a coupled manner has been described here. A high-accuracy spectrally optimized upwind scheme has been proposed in [10], which evaluates the first and the second derivative terms together. However, this scheme does not provide an option of adaptive numerical diffusion and attenuates the solution at each time step and at all grid points. Next, we show the derivation of a new coupled compact difference scheme which has an advantage of adaptive numerical diffusion.

## 7.2  Applications of *DRP* Scheme for Model *CAA* Problems

The coupled compact difference scheme has been applied to solve the various model wave equation problems as well as 2D Navier–Stokes equation to test the efficacy of the proposed method. Various test cases discussed here display development of complex acoustic/flow field. Solutions for the problems of 1D wave equation and the spherical wave propagation are obtained to check the dispersive and dissipative nature of the present scheme. In these examples, the necessity and the corresponding methodology of adaptive numerical diffusion while solving computational acoustic/fluid flow problems have been discussed. A model computational acoustic problem related to the development of an acoustic field due to the acoustic radiations originated from an oscillating circular piston has also been solved. Use of the coupled compact difference scheme as an alternative approach to the *PML* technique has been shown by solving 2D Navier–Stokes equations for the case of a convecting vortex.

### 7.2.1  Solution of 1D Wave Equation

Here, we solve 1D wave equation as given by Eq. (3.95) and demonstrate the ability of the coupled compact difference scheme to correctly resolve the steep gradients present in the solution. We have considered a domain $0 \le x \le 200$ with 1001 equispaced grid points. Computations are performed using the coupled compact difference scheme for the spatial discretization and $RK4$ scheme for the time integration by keeping $CFL$ number as 0.01. Phase speed $c$ is kept as unity. For the present problem, the initial condition has been prescribed as

$$u(x, 0) = e^{-0.20(x-80)^2} + 0.1 \times e^{-1.60(x-50)^2} \tag{7.2}$$

The initial condition is specifically designed as a superposition of a smooth wave packet of unit amplitude located at $x = 80$ and a wave packet with sharper gradient located at $x = 50$. Traditional explicit low resolution schemes can resolve the smooth low wavenumber packet accurately. However, these schemes cannot resolve a large bandwidth of wavenumber generated at the sharp gradients or discontinuities. Insufficient resolution leads to creation of spurious waves which provide seed for numerical instability. Initial condition in Eq. (7.2) has been shown in Fig. 7.1a. Numerical solution at $t = 20$ is displayed in Fig. 7.1b. Vertical dashed lines at $x = 70$ and $x = 100$ denote the exact locations of crest for the sharp and the smooth wave packets, respectively. Horizontal dashed line at $u = 1.0$ denotes the initial amplitude of the smooth packet and is drawn for comparison. Figure 7.1b shows that the smooth as well as the sharp gradient packets are well captured by the coupled compact difference scheme. Due to the significant spectral resolution and the *DRP* nature of the coupled compact difference scheme, no significant dispersion is noted. However, only if one zooms the region close to $u = 0$ line as shown in Fig. 7.1c then the spurious $q$-waves are observed. Although the $q$-waves have small amplitude, these are capable of triggering numerical instabilities [46, 57].

One can use the information available with the evaluated fourth derivative term in coupled compact difference scheme to locate $q$-waves as shown in Fig. 7.1d. Differentiation operation magnifies the solution discontinuities [5]. Since the $q$-waves are primarily associated with the high wavenumber components, evaluation of the higher derivative term magnifies the discontinuities present in the solution. Figure 7.1d shows that the fourth derivative corresponding to the low wavenumber packet is negligible while it shoots up in the region of q-waves.

We suggest a new methodology for adaptive diffusion, based on solution gradients. Addition of numerical diffusion is decided based on the amplitude of the fourth derivative of solution variable. One can prescribe a limiting amplitude for $q$-waves above which spurious waves will be attenuated by adding numerical diffusion. Thus, in the adaptive numerical diffusion methodology, numerical diffusion is added only if the fourth derivative of the solution is above certain limiting value, which indicates presence of significant high wavenumber components. Figure 7.1d shows without addition of numerical diffusion, spurious wave amplitudes lie between $\pm 50$. In Fig. 7.2b, we have obtained the numerical result by adding numerical diffusion only for the region where the fourth derivative of the solution amplitude is greater than $\pm 1$. Addition of the numerical diffusion significantly attenuates spurious $q$-waves as observed in the zoomed view of Fig. 7.2c and the variation of the fourth derivative of solution amplitude in Fig. 7.2d. However, due to the adaptive nature associated with the numerical diffusion, low wavenumber components remain unattenuated as shown in Fig. 7.2b. This is an important advantage of the adaptive diffusion ability of the newly derived coupled compact difference scheme.

In order to compare effectiveness of proposed scheme, as compared to the available high-accuracy schemes, we have once again solved the same 1D wave equation problem by using spectrally optimized upwind *CCD* scheme of [10]. Figure 7.3a, b again shows the same initial condition given by Eq. (7.2) and the computed solution at $t = 20$, respectively. Due to the addition of the numerical diffusion, spurious waves are completely damped out as shown in the zoomed view in Fig. 7.3c and the fourth derivative of the solution amplitude in Fig. 7.3d. However, upwind nature of the scheme causes loss of neutrally stable region as shown in Fig. 3.9b and the smooth

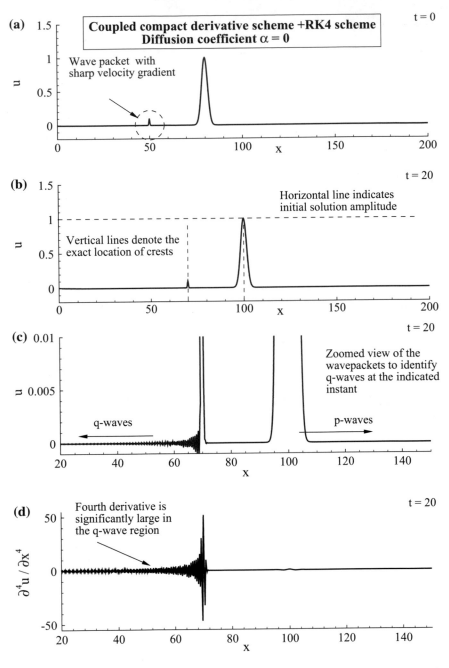

**Fig. 7.1** Initial condition for numerical solution of the 1D wave Eq. (3.95) is shown in (**a**); solution at $t = 20$ is shown in (**b**); zoomed view of the solution in **b** close to line $u = 0$ is shown in (**c**) to mark spurious waves; **d** fourth derivative of the solution amplitude is significant in the $q$−wave region as these waves are associated with high wavenumber components. Results are reproduced from [32]

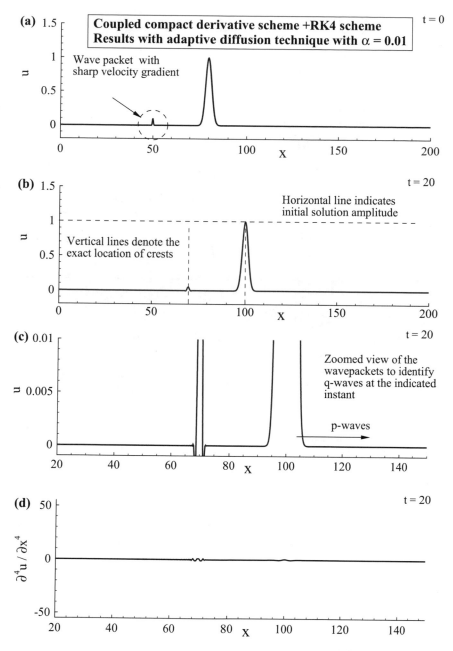

**Fig. 7.2** Initial condition for numerical solution of the 1D wave Eq. (3.95) is shown in (**a**); solution at $t = 20$ is shown in (**b**) when numerical diffusion is added in an adaptive manner in the computational domain; zoomed view of the solution in (**b**) close to line $u = 0$ is shown in (**c**) to mark attenuation of spurious waves; **d** fourth derivative of the solution amplitude is negligible as the $q$-waves are strongly attenuated. Results are reproduced from [32]

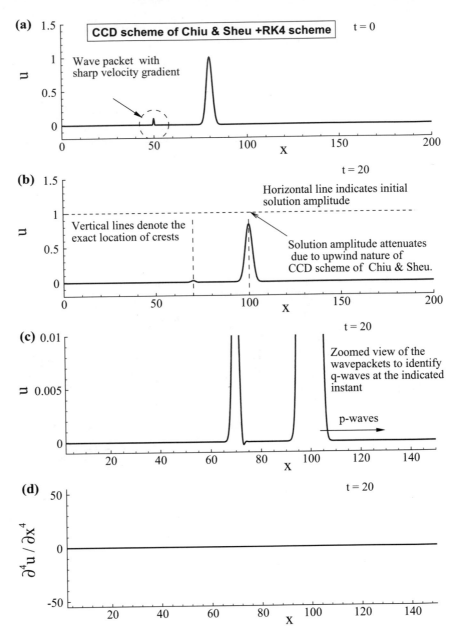

**Fig. 7.3** Initial condition for numerical solution of the 1D wave Eq. (3.95) is shown in (**a**); solution at $t = 20$ is shown in (**b**) corresponding to the upwind *CCD* scheme in [10]; zoomed view of the solution in (**b**) close to line $u = 0$ is shown in (**c**) to mark attenuation of spurious waves; **d** fourth derivative of the solution amplitude is negligible as the $q$-waves are strongly attenuated. Results are reproduced from [32]

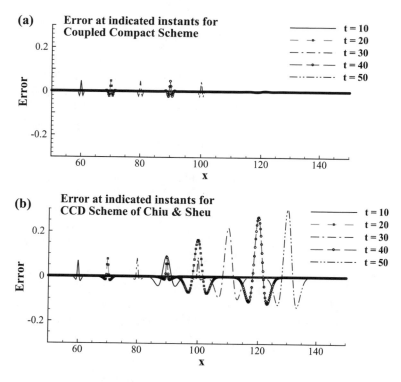

**Fig. 7.4** Comparison of the error in solving Eq. (3.95) by (**a**) coupled compact difference scheme and *RK*4 scheme and **b** *CCD* scheme in [10] and *RK*4 scheme. Note that although *CCD* scheme of [10] has higher spectral resolution and better *DRP* ability as compared to the coupled compact difference scheme. *CCD* scheme of [10] displays higher numerical error due to diffusive nature of the scheme. Results are reproduced from [32]

as well as the sharp gradient packets are severely attenuated, as shown in Fig. 7.3b. Thus, use of upwind schemes needs a careful attention.

Benefit of proposed scheme, as compared to the high-resolution scheme of [10], has been shown in Fig. 7.4, which shows comparison of the error in solving Eq. (3.95) when *RK*4 scheme has been used for time integration. The numerical error has been computed as the difference between the exact and corresponding numerical solution at a given instant. Figure 7.4 shows that error for the *CCD* scheme of [10] is one order more as compared to that of coupled compact difference scheme. Note that although *CCD* scheme of [10] has higher spectral resolution and better *DRP* ability as compared to the coupled compact difference scheme, it displays higher numerical error due to dissipative nature of the scheme.

### 7.2.2  Solution of the Spherical Wave Equation

Consider propagation of a spherical wave setup due to the prescribed monochromatic excitation in a physical domain $5 \leq r \leq 450$ at the location $r = 5$. Monochromatic excitation is given as $u = \sin \omega t$, where $\omega$ is a circular frequency. Propagation of the spherical wave is governed by the spherical wave equation given as in [48]:

$$\frac{\partial u}{\partial t} + \frac{u}{r} + \frac{\partial u}{\partial r} = 0 \tag{7.3}$$

The computational domain $5 \leq r \leq 450$ has been divided into 891 equi-spaced grid points. Numerical solutions obtained for the present problem have been compared with the exact solutions for two different monochromatic excitations. The coupled compact difference scheme has been used for the spatial discretization, while $RK4$ scheme has been used for the time integration. A time step is chosen corresponding to the *CFL* number $N_c = 0.01$.

Figure 7.5a, b shows the propagation of the spherical wave following Eq. (7.3) at the indicated instants for two different excitations of $\omega_0 = \pi/4$ and $\pi/3$, respectively. Amplitude of the wave is maximum at $r = 5$ and it decreases progressively as wave propagates outward. The exact solution for Eq. (7.3) is given in [48] as follows:

$$u(r, t) = \frac{5}{r} \sin\left[\omega_0 \{t - (r - 5)\}\right] \tag{7.4}$$

Comparison of the exact and the obtained numerical solutions in Fig. 7.5a, b shows excellent match. Thus, for the chosen excitation frequencies, use of coupled compact difference scheme produces non-dispersive and non-dissipative solution.

### 7.2.3  Reflection of Pressure Waveform

Here, we discuss propagation of a acoustic disturbance following $1D$ linearized compressible Euler equations in a dimensional form. Present problem helps to demonstrate the effectiveness of numerical scheme, while simulating acoustic wave reflection cases. The governing equations in the Cartesian coordinates are given as follows:

$$\frac{\partial u'}{\partial t} + \frac{1}{\rho_0} \frac{\partial p'}{\partial x} = 0 \tag{7.5}$$

$$\frac{\partial p'}{\partial t} + \gamma p_0 \frac{\partial u'}{\partial x} = 0 \tag{7.6}$$

We have considered a domain $-7.5 \, \text{m} \leq x \leq 7.5 \, \text{m}$ and divided it into 1501 equi-spaced grid points. Time step for the present problem has been chosen as

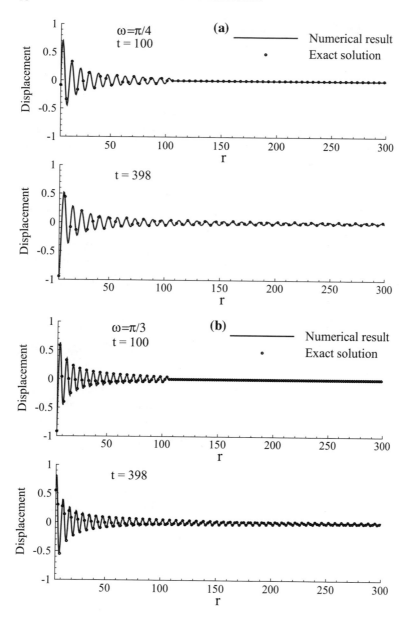

**Fig. 7.5** Wave propagation following spherical wave Eq. (7.3) is shown at the indicated instants and excitation frequencies in (**a**) and (**b**). Present results have been reproduced from [32]

$\Delta t = 0.00001$ s. Solutions for Eqs. (7.5–7.6) have been obtained using *CCS − RK4* scheme. For the present problem, the initial condition is prescribed as follows:

$$u'(t = 0) = 0; \quad p'(t = 0) = e^{-5x^2} \sin\left(\frac{40\pi x}{3}\right) \qquad (7.7)$$

The mean atmospheric pressure $p_0$ and mean atmospheric density $\rho_0$ have been prescribed as $101325\,\text{N/m}^2$ and $1.225\,\text{kg/m}^3$, respectively. The ratio of the specific heats has been indicated by $\gamma = 1.4$. Figure 7.6 shows propagation of an acoustic disturbance following Eqs. (7.5)–(7.6) at the indicated instants. In this case, solutions for Eqs. (7.5)–(7.6) have been obtained at all the grid points at every time step, with prescription of reflecting boundary condition ($\frac{\partial p'}{\partial x} = 0$) at the boundaries [53]. Top frame of Fig. 7.6 shows the initial condition. The acoustic disturbance splits into two waveforms each traveling in opposite direction. The disturbance gets reflected from the domain boundaries and recombine, as shown in the bottom-most frame. There is an excellent match between the analytical and numerical results, as shown at different instants.

### 7.2.4 Reflection of an Acoustic Pulse Over a Solid Wall

Here, we consider a case for the development of an acoustic field, in the presence of a mean flow, in a semi-infinite space. For the present problem, we have selected a two-dimensional computational domain with $-100 \leq x \leq 100$, $0 \leq y \leq 200$. The wall has been located at $y = 0$. An acoustic pulse displays reflection due to the presence of a solid wall. The Euler equations in non-dimensional form are given by [53]

$$\frac{\partial \rho'}{\partial t} + \frac{\partial (M\rho' + u')}{\partial x} + \frac{\partial v'}{\partial y} = 0 \qquad (7.8)$$

$$\frac{\partial u'}{\partial t} + \frac{\partial (Mu' + p')}{\partial x} = 0 \qquad (7.9)$$

$$\frac{\partial v'}{\partial t} + \frac{\partial (Mv')}{\partial x} + \frac{\partial p'}{\partial y} = 0 \qquad (7.10)$$

$$\frac{\partial p'}{\partial t} + \frac{\partial (Mp' + u')}{\partial x} + \frac{\partial v'}{\partial y} = 0 \qquad (7.11)$$

where $M$ denotes Mach number and for this exercise is taken as 0.5. The initial condition for the present problem is given as

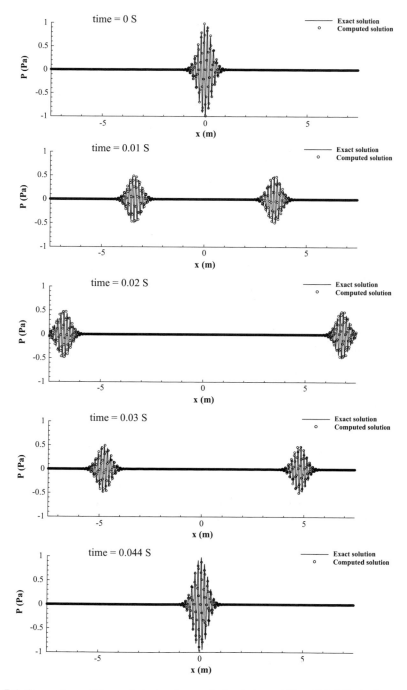

**Fig. 7.6** Comparison of the analytical and numerical solutions of Eqs. (7.5–7.7) obtained using $CCS - RK4$ scheme at the indicated instants. Comparison shows an excellent match between the analytical and numerical results and shows the applicability of the proposed scheme to compute computational acoustics problems. Results are reproduced from [31]

$$u' = v' = 0$$

$$p' = \rho' = \exp\left(-(\ln 2)\frac{x^2 + (y-25)^2}{25}\right) \tag{7.12}$$

This initial condition has been shown in the top frame of Fig. 7.7. Here, we have divided the domain with $201 \times 201$ equi-spaced grid points and computations are performed using a time step $\Delta t = 0.05$. On the bottom surface of the domain ($y = 0$), we have applied a reflecting boundary condition using a ghost point technique as described in [53]. In this technique, additional layers of grid points are added outside the physical boundary with the values having the same magnitude with opposite sign. Computations are also performed by adding numerical diffusion proportional to the fourth derivative, with diffusion coefficient 0.10 to attenuate any spurious high wavenumber components. Time evolution of the initial disturbance under the action of mean flow has been shown in the left side frames, obtained by using the present optimized schemes. Present results are compared with the results in [53] at the identified instants. Comparison shows excellent match. Thus, the present scheme successfully computes acoustic wave reflection problem which also involves mean flow.

### 7.2.5  Acoustic Scattering by a Circular Cylinder

This model problem resembles scattering of acoustic waves by the airplane fuselage [50]. Physically acoustic waves are triggered by rotation of the propeller blades. Computationally it is a challenging problem due to presence of a curved wall boundary. Figure 7.8 shows schematic of physical problem where **S** is the sound source and **A, B, C** are the three location where variation of the pressure has to be measured with respect to time. Non-dimensional equations below (in Eq. (7.13)) have been used for calculating acoustic field. Initial disturbance for the pressure is given by Eq. (7.14). Initial conditions for other dependent variables are given as $u' = v' = 0$. Objective of the problem is to find $p'(t)$ at three points $A$ ($r = 5, \theta = 90°$), $B$ ($r = 5, \theta = 135°$) and $C$ ($r = 5, \theta = 180°$) from $t = 6$ to $t = 10$.

$$\frac{\partial u'}{\partial t} + \frac{\partial p'}{\partial x} = 0 \tag{7.13a}$$

$$\frac{\partial v'}{\partial t} + \frac{\partial p'}{\partial y} = 0 \tag{7.13b}$$

$$\frac{\partial p'}{\partial t} + \frac{\partial u'}{\partial x} + \frac{\partial v'}{\partial y} = 0 \tag{7.13c}$$

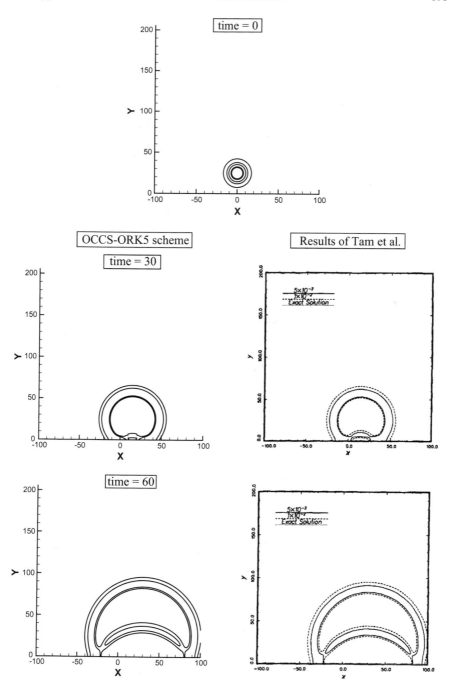

**Fig. 7.7** Reflection of an acoustic disturbance from a flat plate located at $y = 0$ following non-dimensionalized 2D Euler Eqs. (7.8–7.11) has been shown. Fluctuating pressure ($p'$) contours with levels 0.01 and 0.05 have been compared with the results in [53] at the identified instants. Comparison shows excellent match. Results are reproduced from [32]

**Fig. 7.8** Schematic diagram
of model scattering problem

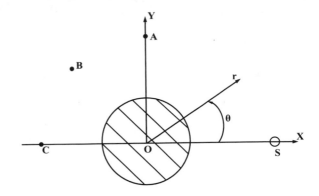

$$p' = \exp\left[-\ln 2\left(\frac{(x-4)^2 + y^2}{(0.2)^2}\right)\right] \tag{7.14}$$

Initially, the problem has been solved with non-uniform grid spacing taking dense grid near the cylinder body. Outer boundary was taken at 15 times diameter of the cylinder. Grid is taken with 361 points in the azimuthal direction and 401 points in the radial direction. Time step, $\Delta t$, is taken as 0.0001 for all the calculations. Orthogonal body-fitted grid transformation (Eq. (7.15)) has been used to convert physical domain to computational domain. Boundary conditions on the cylinder wall are prescribed as zero normal component of the velocity and zero normal pressure gradient, i.e., on the surface ($\eta = 0$) as $v'(t) = 0$, $\frac{\partial p'}{\partial \eta} = 0$. Far field boundary condition is prescribed as the radiation boundary condition [53].

$$\frac{\partial \phi}{\partial x} = \frac{1}{h_1}\frac{\partial \phi}{\partial \xi} \tag{7.15a}$$

$$\frac{\partial \phi}{\partial y} = \frac{1}{h_2}\frac{\partial \phi}{\partial \eta} \tag{7.15b}$$

where the scale factors are given by (see Chap. 9 of [37])

$$h_1 = \sqrt{x_\xi^2 + y_\xi^2}$$
$$h_2 = \sqrt{x_\eta^2 + y_\eta^2} \tag{7.15c}$$

Initial disturbance of pressure was given at four non-dimensional distance from the center of cylinder along positive $x$-axis. Gradually acoustic disturbances approach toward the cylinder surface and undergoes reflection as well as diffraction. As wavefront reaches surface of the cylinder, spurious high wavenumber components are generated which triggers numerical instability. To overcome this problem, a second-order symmetric numerical filter [57] has been applied throughout the domain in the

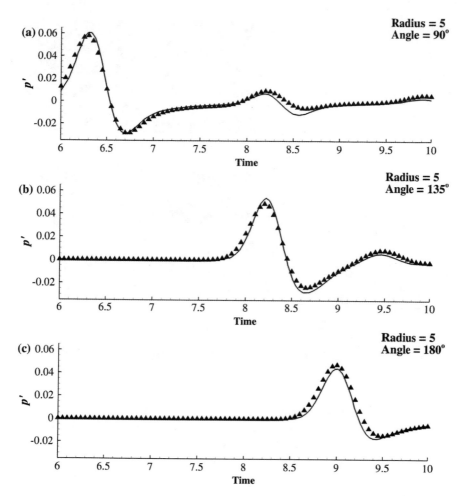

**Fig. 7.9** Results for variation of $(p')$ for locations **a** $\theta = 90°$ (A), **b** $\theta = 135°$ (B), and **c** $\theta = 180°$ (C) at radius $R = 5$ are shown. Results have been obtained by applying filters in both radial and azimuthal directions after every 0.005 non-dimensional time interval. Filled triangles represent computed results and solid lines correspond to analytical expression

radial direction as well as in the azimuthal direction. Value of filtering coefficient, $\alpha$, is chosen as 0.495 along radial direction and 0.499 along azimuthal direction. Figure 7.9 shows amplitude variation of pressure at points $A$, $B$, and $C$. Results show good agreements with analytical result [50]. Figure 7.10 shows a comparison of presently computed pressure pulse contour with that in the literature, at a non-dimensional time, $t = 7$.

**Fig. 7.10** Results for $p'$ are shown at $t = 7$. Frame (**a**) shows present numerical result and the frame (**b**) shows numerical results in [51]

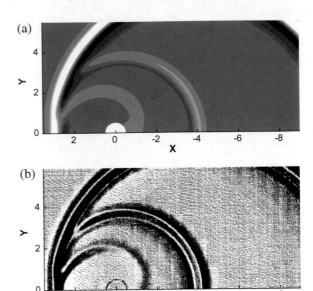

### 7.2.6   *Acoustic Radiations from an Oscillating Circular Piston*

Here, an acoustic field setup due to oscillating circular piston has been simulated. A 2D computational domain with $0 \le x \le 400, 0 \le r \le 200$ has been considered. The domain has been divided into 400 equi-spaced grid points in the $x$-direction. Grid spacings in the $x$- and $r$-directions are kept same. The wall and the piston are located at the line $x = 0$. The governing equations of this model problem can be written in the cylindrical coordinates $(r, x, \theta)$ as [14]

$$\frac{\partial u'}{\partial t} + \frac{\partial p'}{\partial x} = 0 \tag{7.16}$$

$$\frac{\partial v'}{\partial t} + \frac{\partial p'}{\partial r} = 0 \tag{7.17}$$

$$\frac{\partial p'}{\partial t} + \frac{\partial v'}{\partial r} + \frac{v'}{r} + \frac{\partial u'}{\partial x} = 0 \tag{7.18}$$

where $u'$ and $v'$ are the axial and radial components of the axi-symmetric disturbance field. The present problem has been solved with an axi-symmetric approximation as in [14] by considering a domain above the centerline of the piston $r = 0$. Equations (7.16)–(7.18) are solved in a coupled manner by using the coupled compact difference

scheme for the spatial discretization, while *RK*4 scheme has been used for time
integration. Computations are performed with a time step of $\Delta t = 0.10$. Radius of
the piston $R$ is prescribed as 10. The boundary conditions on the wall surface $x = 0$
are given as [14]

$$u' = 0, \ |r| > 10$$

$$u' = 0.0001 \sin\left(\frac{\pi t}{5}\right), \ |r| \le 10.$$

As the piston moves forward along $x$-axis, the air, immediately next to the face
of the piston, moves along with the piston. Pressure in the air, very close to piston
surface increases. This element expands in forward direction displacing the next
layer of air and causing its compression. Thus, a pressure pulse is formed, which
travels at the speed of sound. If the piston oscillates, then acoustic field composed
of compression and rarefaction zones is formed. Pressure fluctuations radiating due
to the monochromatic oscillations of the piston set up a spatio-temporal acoustic
field in the computational domain. Development of the acoustic field with time has
been shown in Fig. 7.11a, b at the indicated instants by plotting pressure contours.
Figures show that acoustic disturbances travel radially outward with time. Computed

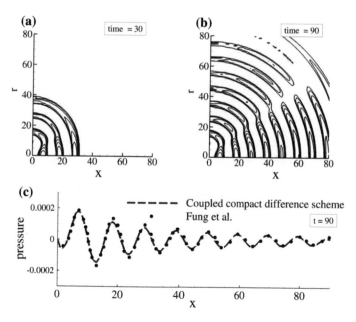

**Fig. 7.11** Instantaneous disturbance pressure (p') contours set up due to piston oscillations are
shown in (**a**) and (**b**) at the indicated times. Comparison of pressure variation on the line $r = 0$
obtained from the present results to that of results in [14] is shown in (**c**)

pressure fluctuations on the piston centerline $r = 0$ are compared with the numerical results in [14], at the indicated instant and show a good agreement between the two.

### 7.2.7 Effects of Rigid Barrier on the Development of an Acoustic Field

Next, we study the development of an acoustic field, in the presence and absence of a rigid barrier. Analysis of the acoustic field in the presence of barrier has been objective of large number of studies. Noise control, using barriers of different materials, shapes, and sizes, is an important research topic. For such a complex problem, physical phenomena such as wave reflection and diffraction occur, whose numerical prediction poses a challenge. In this case, solutions of acoustic field triggered by harmonic excitation, in the presence and absence of barrier, have been computed. Propagation of small acoustic disturbances is governed by linearized Euler equations. These equations can be written in a non-dimensional form as follows:

$$\frac{\partial \rho'}{\partial t} + \frac{\partial u'}{\partial x} + \frac{\partial v'}{\partial y} = S_1 \tag{7.19}$$

$$\frac{\partial u'}{\partial t} + \frac{\partial p'}{\partial x} = 0 \tag{7.20}$$

$$\frac{\partial v'}{\partial t} + \frac{\partial p'}{\partial y} = 0 \tag{7.21}$$

$$\frac{\partial p'}{\partial t} + \frac{\partial u'}{\partial x} + \frac{\partial v'}{\partial y} = S_1 \tag{7.22}$$

Above equations are non-dimensionalized by taking $\Delta x$ as a reference length scale, ambient sound speed $c$ as a reference velocity scale, $\Delta x/c$ as the time scale, $\rho_\infty$ as a density scale, and $\rho_\infty c^2$ as the pressure scale. The source term $S_1$ is given by

$$S_1 = \exp\left[-\ln 2\left(\frac{(x - x_0)^2 + (y - y_0)^2}{r^2}\right)\right]\cos(2\pi ft) \tag{7.23}$$

where $x_0 = y_0 = 0.50$ and $f = 5$. For the present study, we chose a two-dimensional domain $-3.5 \le x \le 5.5, 0 \le y \le 4$. Grid in the domain is equi-spaced using $901 \times 401$ points. We have chosen a time step of $\Delta t = 0.005$. We prescribe a solid wall at $y = 0$, by giving a reflecting boundary condition. Radiation boundary condition [55] ($\frac{\partial f'}{\partial t} + c\frac{\partial f'}{\partial t} = 0$) has been prescribed on the remaining three sides of the domain for the fluctuating quantities. In this study, we have considered development of an acoustic field, in the absence and presence of rigid barrier (centered at $x = 1$). For a case with an acoustic barrier, thickness of the barrier is chosen as 0.04, while the height of the barrier is 1.0. Although the choice of source resembles a monopole, which does not introduce directionality in the computed solution, prescription of reflective boundary condition at the bottom wall ($y = 0$) and the barrier surface induces directionality of the acoustic field.

Figure 7.12 shows contours of disturbance pressure at time $t = 9.8$ for the case without and with barrier in frames (a) and (b), respectively. For the case without barrier, one observes symmetric acoustic field across line $x = 0.50$. This symmetry is absent for the case with barrier. Comparison of the time histories of acoustic pressure variations at the identified points on both sides of the barrier with that of no-barrier case has been shown in frames (c) and (d). Time history at the location $x = 0$ and $y = 0.5$, situated on the left side of the barrier, shows that the case with barrier displays higher pressure amplitude variation, as compared to the no-barrier case. In contrast, the time histories for the location $x = 2.0$ and $y = 0.5$ show that the pressure amplitude variation for the case with barrier has almost three times smaller amplitude as compared to that of no-barrier case.

Figure 7.13a, b shows contours of RMS value of pressure for the cases of with and no-barrier, respectively. One can clearly observe the shadow region behind the barrier. For the case of no-barrier, contours for RMS value of pressure display symmetry about the line $x = 0.50$. Variation of RMS value of pressure at unit radial distance from the exciter location at $x = y = 0.50$ with angle $\theta$ has been shown in Fig. 7.13c. For the case of no-barrier, a main lobe at angle $\theta = 90°$ has been observed. In addition, two side lobes are also observed on either side of the main lobe for the case of no-barrier. In contrast, for the case with barrier, width of the main lobe is significantly reduced and maximum RMS value of the pressure associated with the main lobe is significantly higher as compared to the case of no-barrier.

## 7.2.8 Propagation of Acoustic Disturbance with Mean Flow

Next, propagation of acoustic waves inside a medium with mean flow is investigated. In a stationary fluid, an acoustic disturbance propagates isotropically in all directions with the velocity of sound. In a turbulent medium, sound generation and propagation are very difficult to predict. Use of Lighthill's wave equation [25] to predict acoustic field is one of the approaches used by researchers. However, this approach has limitations. Such approach is relevant in simulating an acoustic field for a homogeneous medium at rest and not suitable for finding out acoustic field for an inhomogeneous medium. Second limitation is associated with the requirement of Green function to obtain by integral formulation. Therefore, only simple geometric configurations can be studied. Another approach is to solve linearized Euler equations, which are linearized around a stationary mean flow. Researchers have also used Reynolds-averaged Navier–Stokes (*RANS*), *LES*, and *DNS* approaches to simulate the acoustic field.

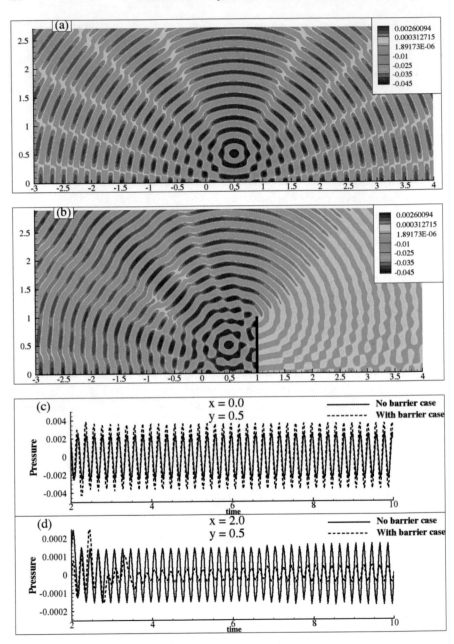

**Fig. 7.12** Solutions of the non-dimensionalized, 2D Euler Eqs. (7.19–7.22) have been shown here. Contours of disturbance pressure at time $t = 9.8$ have been compared for the case without and with barrier in (**a**) and (**b**), respectively. For the case without barrier, one observes symmetric acoustic field across line $x = 0.50$. This symmetry is absent for the case with barrier. Comparison of the time histories of acoustic pressure variations at the identified points on both sides of the barrier with that of no-barrier case has been shown in (**c**) and (**d**). Results are reproduced from [32]

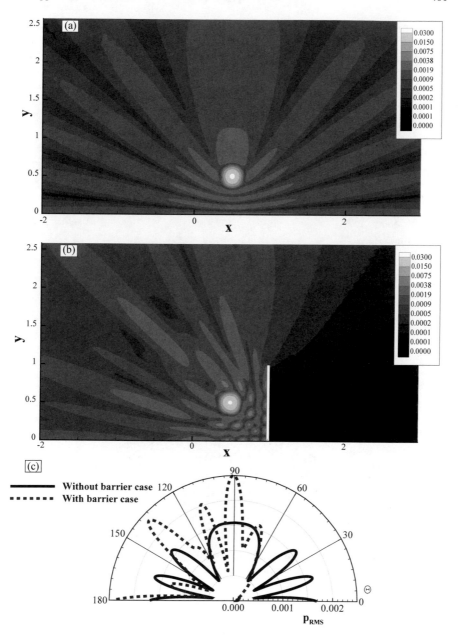

**Fig. 7.13** Solutions of the non-dimensionalized, 2*D* Euler Eqs. (7.19–7.22) have been shown here. Contours of RMS value of pressure have been shown for the case without and with barrier in (**a**) and (**b**), respectively. One can clearly observe the shadow region behind the barrier as compared to the region ahead of the barrier. Variation of RMS value of pressure at unit radial distance from the exciter location $x = y = 0.50$ for the case with and without barrier has been shown in (**c**). Results are reproduced from [32]

The linearized Euler equations around a stationary mean flow can be written as Eq. (7.24). All primed quantities ($u'$, $v'$, $p'$ and $\rho'$) are the fluctuating quantities around the mean flow. Pressure ($p_0$), density ($\rho_0$), and velocities ($u_0$ and $v_0$) are used to represent the mean flow quantities.

$$\frac{\partial \mathbf{U}}{\partial t} + \frac{\partial \mathbf{E}}{\partial x} + \frac{\partial \mathbf{F}}{\partial y} + \mathbf{H} = \mathbf{S} \tag{7.24}$$

Here, for the two-dimensional case, $\mathbf{U}$, $\mathbf{E}$, and $\mathbf{F}$ are denoted by

$$\mathbf{U} = \begin{pmatrix} \rho' \\ \rho_0 u' \\ \rho_0 v' \\ p' \end{pmatrix} \qquad \mathbf{E} = \begin{pmatrix} \rho' u_0 + \rho_0 u' \\ u_0 \rho_0 u' + p' \\ u_0 \rho_0 v' \\ u_0 p' + \gamma p_0 u' \end{pmatrix} \qquad \mathbf{F} = \begin{pmatrix} \rho' v_0 + \rho_0 v' \\ v_0 \rho_0 u' \\ v_0 \rho_0 v' + p' \\ v_0 p' + \gamma p_0 v' \end{pmatrix}$$

$$\mathbf{H} = \begin{pmatrix} 0 \\ \left(\rho_0 u' + \rho' u_0\right) \frac{\partial u_0}{\partial x} + \left(\rho_0 v' + \rho' v_0\right) \frac{\partial u_0}{\partial y} \\ \left(\rho_0 u' + \rho' u_0\right) \frac{\partial v_0}{\partial x} + \left(\rho_0 v' + \rho' v_0\right) \frac{\partial v_0}{\partial y} \\ (\gamma - 1) \left(\rho' \nabla \cdot \mathbf{u_0} - u' \frac{\partial p_0}{\partial x} - v' \frac{\partial p_0}{\partial y}\right) \end{pmatrix}$$

$\mathbf{H}$ takes care of the gradients present in the mean flow. $\mathbf{S}$ represents unsteady source present in the flow. In this case, $\mathbf{S}$ matrix is given by

$$\mathbf{S}(x, y, t) = \varepsilon \sin(\omega_0 t) e^{-\alpha(x^2 + y^2)} \begin{pmatrix} 1 \\ 0 \\ 0 \\ 1 \end{pmatrix}$$

Test cases are run for simulating propagation of acoustic disturbance in subsonic as well as supersonic mean flows. Domain size has been considered as $-200 \leq x, y \leq 200$, with $400 \times 400$ equi-spaced grid points. Prescribing correct outflow boundary condition is important for accurate simulations of all aeroacoustic problems. Inadequate outflow or radiation boundary conditions trigger spurious waves from domain boundaries, which affect the main solution. Radiation and outflow boundary conditions given in [55] are used in these simulations. Perturbation flow fields usually consist of acoustic, entropic, and vortical disturbances [55]. Schematic of the computational domain is shown in Fig. 7.14. In the present case, through the top, left, and bottom boundaries acoustic disturbances propagate only. Thus, radiation boundary condition [55] has been prescribed on these boundaries. In the case of right-hand side boundary, along with the acoustic disturbances, vorticity and entropic waves are also propagating outward. Thus, outflow boundary condition has been used on the right-side boundary.

Sound waves propagate at the speed of sound in all directions inside a steady homogeneous medium. In unsteady case with nonzero mean flow, acoustic waves

**Fig. 7.14** On the top, bottom, and left boundaries, radiation boundary condition has been applied while on the right boundary an outflow boundary condition has been prescribed

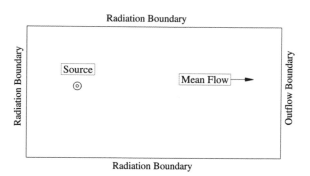

propagate in a non-uniform manner. If the velocity of mean flow is less than the sound speed, then sound propagation profile is shifted along the flow. If mean flow velocity is more than sound speed, then formation of Mach cone is observed. Here we report simulated propagation of acoustic disturbances in the subsonic, sonic, and supersonic flow environments.

For a case of acoustic wave propagation for a flow at Mach number $M = 0.5$, an acoustic source at $(0, 0)$ is prescribed. Time step is chosen as 0.0001. Theoretically, acoustic disturbance propagates faster in the downstream direction, as compared to upstream direction, due to action of mean flow. For a Mach number of 0.5, downstream velocity must be 1.5 times the speed of sound, as shown in Fig. 7.15. Present numerical result, as shown in Fig. 7.15a, has been compared with the numerical results in [2], as shown in Fig. 7.15b. One can observe the presence of spurious waves in the solution of [2] caused by inaccurate reflections from the domain bound-

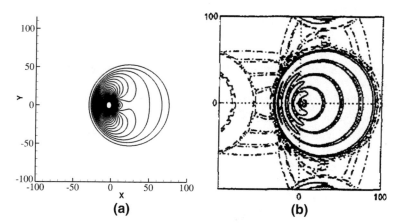

**Fig. 7.15** Contours for disturbance pressure (p') at Mach = 0.5 have been shown. (a) Present numerical result displays 61 contour levels in between $-7 \times 10^{-6}$ to $10^{-5}$. (b) Numerical result given in [2] where one observes noticeably reflections from all boundaries as indicated by dotted lines. Such spurious reflections are absent in the present computed result

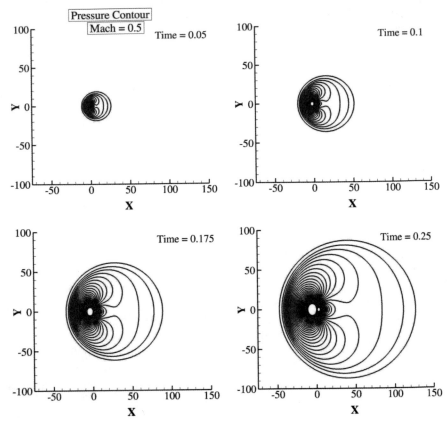

**Fig. 7.16** Different stages of disturbance pressure wave propagation at M=0.5 have been shown. Present numerical results display 61 contour levels in between $-7 \times 10^{-6}$ to $10^{-5}$

aries. Development of acoustic wavefront with time for this case has been shown in Fig. 7.16.

For higher Mach number cases with $M \geq 1$, one can check the Mach cone angle to describe the accuracy of the scheme. Theoretically, Mach angle for a case of $M = 1.5$ is given as $\mu = \sin^{-1}(\frac{1}{M}) = 41.81°$. For the present simulation results, Mach angle is around 42°, as shown in Fig. 7.17.

Two more cases are reported for $M = 1$ and $M = 2$. In these cases, time step is taken as 0.00001, so as to be in the neutrally stable zone for convection equation. At $M = 1$, it is expected that no acoustic wave should propagate in the upstream direction, and acoustic disturbances should travel twice faster than the sonic wave in the downstream direction. Obtained solutions suggest similar observations. Similarly, for $M = 2$, Mach angle has been checked and verified with the theoretical results. Mach angle is about 30°. Development of acoustic fields for $M = 1$ and $M = 2$ has been shown in Figs. 7.18 and 7.19, respectively.

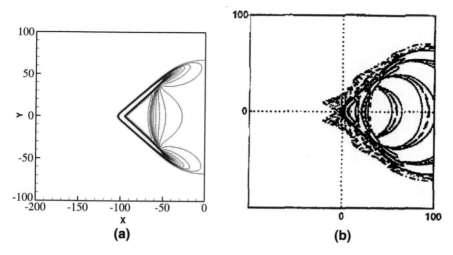

**Fig. 7.17** Disturbance pressure pulse profile for Mach = 1.5. **a** Present numerical result displays 60 contour levels in between $-10^{-5}$ to $10^{-5}$. **b** Numerical result is given in [2] which displays spurious reflections in the form of dashed lines. Here, velocity of the downstream wave is 1.5 times faster than upstream wave

## 7.2.9 Propagation of the Acoustic and the Entropic Disturbances

Next, we solve the computational acoustic wave propagation problem which consists of simultaneous propagation of acoustic and entropic disturbances.

The governing equations are already given from Eqs. (7.8) to (7.11). Here, $M = 0.50$ denoting the free-stream Mach number has been used. The initial condition ($t = 0$) for the present problem is given as

$$u' = 0.04 \, y \, \exp\left(-(\ln 2)\frac{(x-67)^2 + y^2}{25}\right)$$

$$v' = -0.04 \, (x - 67) \, \exp\left(-(\ln 2)\frac{(x-67)^2 + y^2}{25}\right)$$

$$p' = \exp\left(-(\ln 2)\frac{x^2 + y^2}{9}\right)$$

$$\rho' = \exp\left(-(\ln 2)\frac{x^2 + y^2}{9}\right) + 0.1 \exp\left(-(\ln 2)\frac{(x-67)^2 + y^2}{25}\right) \quad (7.25)$$

This case consists of an acoustic pulse generated by a Gaussian circular patch of source centered at the origin (with radius equal to 3), while the entropic disturbance is centered at $x = 67, y = 0$ with radius 5. Here the entropic disturbance refers to

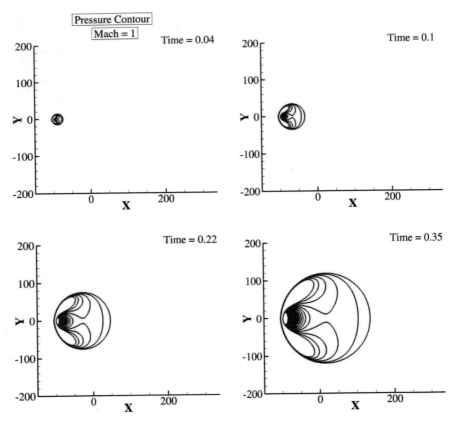

**Fig. 7.18** Different stages of disturbance pressure wave propagation at M=1 have been shown. Present numerical results display 60 contour levels in between $-10^{-5}$ to $10^{-5}$

temperature fluctuation as can be derived from the fluctuating pressure and density. Initial condition has been shown in Fig. 7.20a. The mean flow Mach number is 0.5. We have constructed the domain using (501 × 501) grid points. Downstream of the pressure pulse, at $x = 0.67$, an entropy pulse has also been superimposed. Acoustic pulse travels faster than entropic disturbances in the downstream direction, as observed in Fig. 7.20b–d, which shows development and propagation of acoustic, as well as entropic disturbances with time. We have compared the density variation on the line y=0 obtained from present simulation with that in [52], in Fig. 7.20e. Comparison shows a good match and shows accuracy and effective use of coupled compact difference scheme for obtaining high-accuracy solutions of computational acoustics problems.

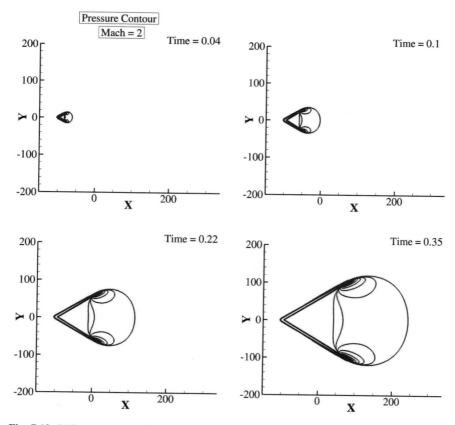

**Fig. 7.19** Different stages of disturbance pressure wave propagation at M=2 have been shown. Present numerical results display 60 contour levels in between $-10^{-7}$ to $10^{-5}$

## 7.2.10 Use of Coupled Compact Difference Scheme as a Perfectly Matched Layer (PML) Technique

While obtaining solution of 1D wave equation, merit of the coupled compact difference scheme for adaptive numerical diffusion has been demonstrated already. In the next example, we show that the coupled compact difference scheme has been further utilized as the equivalent *PML* technique [52] to remove spurious reflections from the outflow boundary. In this example, a discrete shielded vortex convecting in a uniform flow is studied. In order to solve Navier–Stokes equations, we have used the proposed coupled compact difference scheme for evaluating the first- and the second-order derivative terms, while *RK*4 scheme is used for temporal integration. A computational domain of size $-1 \le (x, y) \le 1$ is divided into 101 equi-spaced points in the respective directions. Flow enters the square domain *ABCD* from the left boundary *AD* with a uniform velocity. At the center of the computational domain, a discrete

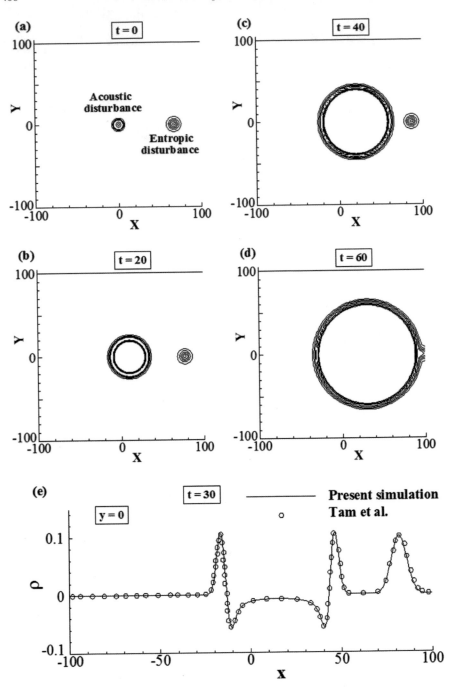

**Fig. 7.20** Initial condition and propagation of acoustic and the entropic disturbances following Eqs. (7.8)–(7.11) are shown in frames (**a**)–(**d**). Comparison of density variation on the line $y = 0$ obtained from present simulation with that of [52] is shown in (**e**)

shielded vortex [16, 47] has been prescribed as an initial condition. Schematic of the chosen problem is as shown in Fig. 7.21a. If we denote the distance between any point in the domain and the center of the vortex by symbol $r$, then the discrete shielded vortex in a non-dimensional form can be prescribed as in [16, 47]:

$$\omega = K \left(1 - 100 \, r^2\right) e^{-100 \, r^2} \tag{7.26}$$

Solution of the Navier–Stokes equations formulated in the stream function–vorticity formulation is obtained for the present problem, which is given as follows:

$$\frac{\partial^2 \psi}{\partial x^2} + \frac{\partial^2 \psi}{\partial y^2} = -\omega \tag{7.27}$$

$$\frac{\partial \omega}{\partial t} + u \frac{\partial \omega}{\partial x} + v \frac{\partial \omega}{\partial y} = \frac{1}{Re} \left[ \frac{\partial^2 \omega}{\partial x^2} + \frac{\partial^2 \omega}{\partial y^2} \right] \tag{7.28}$$

A uniform flow has been prescribed at the inflow boundary $AD$ as shown in Fig. 7.21a. Stream function and vorticity values at the remaining boundaries $BC$, $CD$, and $AB$ are updated using a convective outflow boundary condition [46]. Equations (7.27) and (7.28) are non-dimensionalized with respect to free-stream velocity $U_\infty$ as the reference velocity scale and $1/K$ as the time scale. The Reynolds number is defined as $Re = \frac{U_\infty^2}{vk}$ [4]. Simulations have been performed for $Re = 5000$, $K = 50$ with a time step of $\Delta t = 0.001$. The first and the second derivative terms in Eq. (7.28) are evaluated using the coupled compact difference scheme, while the $CD2$ scheme has been used to discretize stream function equation, Eq. (7.27) which has been iteratively solved using Bi-CGSTAB algorithm as given in [13]. Equations (7.27) and (7.28) are solved for the given initial condition and the corresponding solutions are shown in Fig. 7.21b, c, respectively. One observes that the vortex smoothly convects inside the domain; however, as it reaches the outflow boundary, it distorts and spurious components are reflected from the outflow boundary. This is essentially due to the use of inaccurate boundary condition, which results in the spurious reflections at the boundary.

Next, we have repeated the same calculation in the presence of an additional absorbing layer close to the domain boundaries marked by the dashed lines in Figs. 7.21d, e. In the region enclosed by domain boundaries and dashed lines, we have prescribed a diffusion coefficient $\alpha = 0.10$ as given in Eq. (7.1). This procedure is equivalent to the PML technique. Solution components in this region are heavily damped, which has resulted in attenuation of spurious oscillations. Figure 7.22a shows the variation of vorticity at line $y = 0$ at the indicated instant. One observes large high wavenumber fluctuations present in the solution, when no absorbing layer technique is used. Corresponding *FFT* also shows rise in amplitude of high wavenumber components for solution obtained without PML technique, as compared to the solution obtained using the equivalent PML technique. Thus, the coupled compact

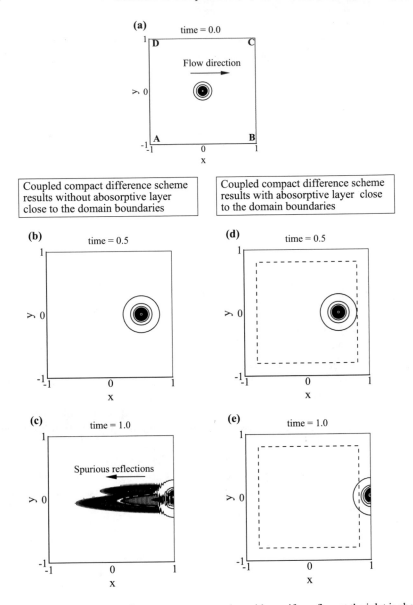

**Fig. 7.21** **a** Schematic of the discrete vortex convecting with a uniform flow at the inlet is shown. In rest of the frames, vorticity contours have been shown at the indicated instants. Spurious reflections are observed from the domain boundaries when the absorbing layer is not prescribed in (**c**) while these are absent when an absorbing layer with added numerical diffusion (equivalent PML technique) is used after the dashed box as shown in (**d**) and (**e**). Results are reproduced from [32]

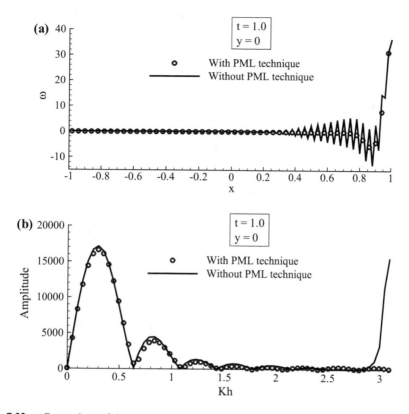

**Fig. 7.22** **a** Comparison of the variation of vorticity on the centerline $y = 0$ at time $t = 1$ in the presence and absence of the PML technique; **b** *FFT* of the vorticity data in (**a**) is compared for the results with and without equivalent PML technique. Note the presence of large amplitude of high wavenumber components for the result obtained without this equivalent *PML* technique. Results are reproduced from [32]

difference scheme has another important advantage that it can be used as a substitute to the classical PML technique used in electromagnetics [34] (p. 85 onward) and acoustics.

## 7.3   Combined Optimization of Spatial and Temporal Discretization Schemes

Here, a *DRP* scheme has been developed by combined optimization of the spatial and the multistage temporal discretization scheme to solve acoustics problems accurately. Here, the results are reproduced from [31]. The coupled compact difference scheme (*CCS*) has been spectrally optimized (*OCCS*) for accurate evaluation of the spatial derivative terms. Next, the combination of the *OCCS* scheme and the five-stage Runge–Kutta time integration (*ORK*5) scheme has been advanced to reduce numerical diffusion and dispersion error significantly. Proposed *OCCS* − *ORK*5 scheme provides accurate solutions at considerably higher *CFL* number. In addition, *ORK*5 time integration scheme consists of low storage formulation and requires less memory as compared to the traditional fourth-order Runge–Kutta schemes. Solutions of the model problems involving propagation, reflection, and diffraction of acoustic waves have been obtained to demonstrate the accuracy of the developed scheme and its applicability to solve complex problems.

In addition to the excellent spectral resolution for the spatial derivative terms, used numerical schemes must also ensure that the resolved scales in the computed solution propagate at the correct physical speed. Such schemes are classified as the *DRP* schemes [37]. The spatial and the temporal scales are linked to each other through the physical dispersion relation. Although numerical schemes solve the governing differential equations, corresponding numerical dispersion relation differs from the physical one across a complete or a band of wavenumber range due to numerical inaccuracies. The optimization of *DRP* schemes for solving computational acoustics problems was later followed in [10].

Improvements in the time integration scheme has been suggested in [3, 29] by considering the objective function in Eq. (7.29), to solve one-dimensional (1D) convection equation.

$$F = \int_0^{\alpha\pi} |G_{\text{num}} - G_{\text{exact}}|^2 \, d(kh) \tag{7.29}$$

In the above equation, $G_{\text{num}}$ and $G_{\text{exact}}$ are the numerical and the exact amplification factors, respectively. Free parameter $\alpha$ varies between 0 and 1 and $kh$ denotes non-dimensional wavenumber with $h$ as the uniform grid spacing. Value of $\alpha$ provides limiting wavenumber up to which numerical method is expected to perform as per the exact amplification factor. Function $F$ depends on the *CFL* number and coefficients of the Runge–Kutta scheme. Although the previous studies emphasized on use of neutrally stable methods, minimization of the phase and dispersion error was not considered in the optimization process as a constraint equation. These aspects are highlighted with other references in [33] and correct approach has been proposed for central explicit difference schemes.

Sengupta [37] used 1D wave equation as a model equation for the convection dominated flows and proposed the correct numerical dispersion relation [41] as $\omega_N = kc_N$, where $\omega_N$ and $c_N$ are numerical circular frequency and numerical phase speed, respectively. Using the correct numerical dispersion relation, Sengupta et al. [33] devised a new strategy to spectrally optimize the coefficients of multistage time integration schemes such that the error associated with dispersion and dissipation terms was minimized for the given spatial discretization scheme. Authors in [33] have specifically emphasized on the fact that the numerical properties of the *DRP* schemes must be evaluated by considering the spatial and the temporal discretization together [46].

So far, researchers have not optimized the spatial and the temporal discretization schemes by considering their effects together to solve 1D wave equation using correct numerical dispersion relation [41]. For non-periodic problems, improvements in numerical stability and *DRP* properties at boundary and near-boundary nodes for combined compact difference (*CCD*) scheme have been addressed in [43, 44]. In the present work, we have first optimized the spectral resolution of the spatial discretization scheme proposed in [32]. Optimized spatial discretization scheme has been referred here as *OCCS* scheme. Subsequently, we have tuned coefficients of the multistage Runge–Kutta time integration scheme so that its combination with the proposed *OCCS* scheme provides neutral stability even at higher *CFL* number along with *DRP* ability [33, 37].

## 7.3.1 Optimized OCCS Spatial Discretization Scheme

Various researchers in the past [1, 6, 10, 58] derived high-accuracy schemes by improving spectral resolution. In the present work, we use the same methodology for spectral optimization of the first derivative of the *CCS* scheme in the following way [31].

Equation (3.89) contains $a_1$, $b_1$, $c_1$, and $d_1$ as four unknown coefficients. In order to obtain these unknown coefficients, one needs four independent equations. Out of the four equations, the Taylor series expansion provides first three equations by eliminating the leading truncation error terms as

$$-2a_1 + 2d_1 = 1 \tag{7.30}$$

$$-a_1 + d_1/3 - 2b_1 = 0 \tag{7.31}$$

$$d_1/60 - b_1/3 - 2c_1 - a_1/12 = 0 \tag{7.32}$$

However, to obtain all the four unknown coefficients in Eq. (3.89), additional algebraic equation is required. Authors in [10] have chosen this additional equation so that the difference between the actual and the numerical wavenumbers has been minimized. This new equation is combined with Eqs. (7.30)–(7.32) to uniquely determine all the coefficients. Actual wavenumber for Eq. (3.89) can be given as

$$ikh \left(a_1(e^{-ikh} + e^{ikh}) + 1\right) + b_1(ikh)^2 (e^{ikh} - e^{-ikh})$$
$$+ c_1(ikh)^4 (e^{ikh} - e^{-ikh}) \simeq d_1 (e^{ikh} - e^{-ikh}) \tag{7.33}$$

All discretization schemes involve implicit filtering as the exact and numerical wavenumber differ near the Nyquist limit. Thus, different discretization schemes can be characterized by the discretization effectiveness representation in the spectral plane. Spectral error can be minimized by matching closely the exact and the numerical wavenumbers over a complete wavenumber range [10]. One can equate the numerically obtained wavenumber $k_{eq}$ to those shown on the right-hand side of Eq. (7.33).

$$\mathbf{i}k_{eq}h\left(a_1(e^{-ikh} + e^{ikh}) + 1\right) + b_1(\mathbf{i}k_{eq}h)^2 \, (e^{ikh} - e^{-ikh})$$
$$+c_1(\mathbf{i}k_{eq}h)^4 \, (e^{ikh} - e^{-ikh}) = \, d_1 \, (e^{ikh} - e^{-ikh}) \tag{7.34}$$

The quantity $k_{eq}$ is, in general, a complex quantity. Its real part is associated with the dispersion error while the imaginary part is related to the dissipation error. One can use expression for real part of $k_{eq}$ from Eq. (7.34) to minimize the dispersion error as shown in [10]. For achieving higher discretization effectiveness, $\Re[k_{eq}h]$ and $kh$ must be very close across the complete wavenumber range. Authors in [6, 10, 58] constructed an error function $E(kh)$ for optimizing resolving ability of the scheme as

$$E(kh) = \int_0^{\frac{7\pi}{8}} \left[\left(k\,h - \Re[k_{eq}\,h]\right)\right]^2 d(kh) \tag{7.35}$$

For error minimization, condition $\frac{\partial E}{\partial c_1} = 0$ has been enforced. Using this constraint equation, the four unknown coefficients for the *OCCS* scheme are determined as

$$a_1 = 0.4920325009328114, \quad b_1 = -0.0806775003109371$$
$$c_1 = 0.0012118333540624, \quad d_1 = 0.9920325009328114$$

While these are the optimized coefficients for obtaining first derivative Eq. (3.89), other coefficients for Eqs. (3.90) and (3.91) remain same as discussed before. Figure 7.23 compares variation of discretization effectiveness $\Re(k_{eq}/k)$ with respect to the non-dimensional wavenumber $kh$ for the *OCCS* scheme and the scheme in [32] for evaluating the first, the second, and the fourth derivatives. Due to error minimization constraint set by Eq. (7.35), spectral resolutions for the first, the second, and the fourth derivative terms show improvement with respect to the scheme in [32]. For the estimation of the first derivative, we have also shown the discretization effectiveness of the second-order central difference scheme (*CD2*), fourth-order central compact difference scheme (*C4*), and *CCD* scheme of [43] in Fig. 7.23. All these schemes including the newly derived *OCCS* scheme have a three point stencil. Figure 7.23 shows that *OCCS* scheme has highest resolution.

### 7.3.2 Optimized Multistage ORK5 Time Integration Scheme

Authors in [33] have derived coefficients of multistage time integration schemes by minimizing dissipation, dispersion, and phase error while solving Eq. (3.95) numer-

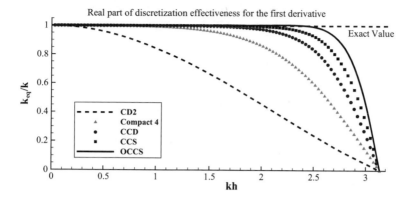

**Fig. 7.23** Comparison of the variation of the real part of discretization effectiveness with respect to the non-dimensional wavenumber $kh$ for the indicated discretization schemes has been given for the first derivatives. Results are reproduced from [31]

ically. In the present work, we have followed the same methodology for obtaining improvised time integration scheme, as suggested in [39].

Any difference scheme can be represented in the form $[A]\{u'\} = [B]\{u\}$. This representation can be further simplified as $\{u'\} = [C]\{u\}$ where $[C] = [A]^{-1}[B]$. Consider a computational domain divided into $N$ equi-spaced grid points with a grid spacing $h$, then the first derivative is given by [33, 37], $u'_j = \frac{1}{h} \sum_{l=1}^{N} C_{jl} u_l$. Same expression can be represented in the spectral plane [37] as

$$u'_j = \int \frac{1}{h} \sum C_{jl} U \, e^{ik(x_l - x_j)} \, e^{ikx_j} \, dk \qquad (7.36)$$

Sengupta et al. [41] defined the nodal numerical amplification factor as the ratio of spectral amplitudes of the computed solution at successive time steps as $G_j = U_j(k, t^{(n+1)}) / U_j(k, t^{(n)})$. For the four-stage, fourth-order Runge–Kutta time integration scheme, numerical amplification factor is given as [37, 41]

$$G_j = 1 - A_j + \frac{A_j^2}{2} - \frac{A_j^3}{6} + \frac{A_j^4}{24} \qquad (7.37)$$

where $A_j = N_c \sum_{l=1}^{N} C_{jl} \, e^{ik(x_l - x_j)}$ and $N_c$ is the *CFL* number.

Expressions for the normalized phase velocity $c_N/c$ and normalized group velocity $V_{gN}/c$ are given as [37, 41]

$$\left[\frac{c_N}{c}\right]_j = \frac{\beta_j}{\omega \Delta t} \tag{7.38}$$

$$\left[\frac{V_{gN}}{c}\right]_j = \frac{1}{N_c}\frac{\mathrm{d}\beta_j}{\mathrm{d}(kh)} \tag{7.39}$$

where $\beta_j = tan^{-1}\left[-\frac{(G_j)_{\text{imag}}}{(G_j)_{\text{real}}}\right]$. The quantities $(G_j)_{\text{real}}$ and $(G_j)_{\text{imag}}$ indicate the real and the imaginary parts of $G_j$, respectively. Thus, the choice of spatial and temporal discretization scheme decides the important numerical properties such as numerical amplification factor along with numerical phase and numerical group velocity characteristics. Optimization of the time integration scheme has been performed by considering an autonomous system of differential equations as [33]

$$\frac{\partial u}{\partial t} = f(u); \qquad\qquad t > 0 \tag{7.40}$$

For a $s$-stage Runge–Kutta method, solution advancement from time level $n$ to $n + 1$ can be given as [33, 37]

$$u^{(n+1)} = u^{(n)} + \sum_{i=1}^{s} W_i k_i$$

$$k_i = \Delta t \, f\left(u^n + \sum_{j=1}^{i-1} a_{ij} k_j\right) \tag{7.41}$$

The coefficients $a_{ij}$'s and the weights $W_i$ are chosen by first expanding terms on the left- and right-hand side of Eq. (7.41) using Taylor series approximation. Subsequently, multiple equations are obtained by matching the similar order terms on both sides of the equations so that the numerically obtained value $u^{(n+1)}$ becomes close to the exact value.

Rajpoot et al. [33] optimized classical explicit Runge–Kutta methods by enhancing important numerical properties in the spectral plane and relaxing the order of the scheme. In order to reduce the memory requirements during computations, we have optimized a low storage formulation of the time integration scheme given in [33, 37]:

$$u^{(i)} = u^{(n)} + \alpha_i \, \Delta t \, f[u^{(i-1)}]$$

$$u^{(n+1)} = u^{(n)} + \Delta t \sum_{i=1}^{s} W_i \, f[u^{(i-1)}] \tag{7.42}$$

The exact amplification factor of Eq. (3.95) is given as $G_{\text{exact}} = e^{-iN_c kh}$ [37] while the numerical amplification factor for the Runge–Kutta method with same stages and order of accuracy is [33, 37]

$$G_{\text{num}} = 1 + \sum_{j=1}^{s}(-1)^j a_j A^j \tag{7.43}$$

The time integration scheme should be designed such that the choice of the coefficients $a_j$'s minimizes the error in the wavenumber space [33, 37]. For the low storage version of RK5 method, relations between the $\alpha_i$'s, $W_i$'s, and $a_j$'s are given as [31]

$$W_1 + W_2 + W_3 + W_4 + W_5 = a_1 \tag{7.44}$$

$$W_2\alpha_1 + W_3\alpha_2 + W_4\alpha_3 + W_5\alpha_4 = a_2 \tag{7.45}$$

$$W_3\alpha_1\alpha_2 + W_4\alpha_2\alpha_3 + W_5\alpha_3\alpha_4 = a_3 \tag{7.46}$$

$$W_4\alpha_1\alpha_2\alpha_3 + W_5\alpha_2\alpha_3\alpha_4 = a_4 \tag{7.47}$$

$$W_5\alpha_1\alpha_2\alpha_3\alpha_4 = a_5 \tag{7.48}$$

In the present work, we have optimized the RK5 time integration scheme $ORK5$ such that its combination with the $OCCS$ scheme provides least error while solving the 1D wave equation (3.95). During optimization of the time integration scheme, we have fixed the coefficients $a_1 = 1$ and $a_2 = 1/2$, so that conditions for the first- and the second-order terms in the truncation error are satisfied. The coefficients $a_3$, $a_4$, and $a_5$ are fixed so that error in the spectral plane is minimized as described below.

Following the work in [33], amplification factor for the $ORK5$ scheme is given as

$$G_5 = 1 - A + \frac{A^2}{2!} - a_3A^3 + a_4A^4 - a_5A^5 \tag{7.49}$$

Objective functions for the present $OCCS - ORK5$ scheme are chosen such that dispersion and phase error are minimized in the spectral plane and at the same time neutral stability in the [wavenumber $(kh)$–CFL number $(N_c)$]-plane is ensured over a larger region. Thus, the important numerical properties are functions of unknown coefficients $a_3$, $a_4$, and $a_5$ which are optimized by setting the following constraint equations [31, 33]:

$$F_1(a_3, a_4, a_5, N_c) = \int_0^{2.0} ||G| - 1| \, d(kh) \le 0.0005 \tag{7.50}$$

$$F_2(a_3, a_4, a_5, N_c) = \int_0^{2.0} \left|\left(\frac{V_{gN}}{c}\right) - 1\right| d(kh) \le 0.1 \tag{7.51}$$

$$F_3(a_3, a_4, a_5, N_c) = \int_0^{2.0} \left|\left(\frac{c_N}{c}\right) - 1\right| d(kh) \le 0.01 \tag{7.52}$$

**Fig. 7.24** Variation of optimized coefficients $a_3$, $a_4$, and $a_5$ with respect to the *CFL* number $N_c$ has been shown in (**a**), (**b**), and (**c**), respectively. Results are reproduced from [31]

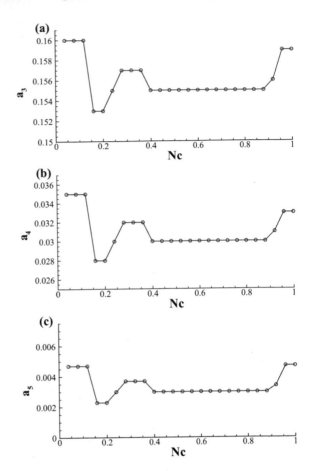

We have used the grid-search technique [33] to solve the constrained optimization problem. Variation of the optimized values of $a_3$, $a_4$, $a_5$ with respect to *CFL* number has been shown in Fig. 7.24. Although these values are function of $N_c$, we have selected these coefficients for $N_c = 0.4$, which remain more or less same for higher *CFL* numbers, as shown in Fig. 7.24. Thus, in the present exercise, we have chosen $a_3 = 0.155$, $a_4 = 0.03$ and $a_5 = 0.003$ as observed from Fig. 7.24. One can choose different combinations of weights and $\alpha$s so that Eqs. (7.44)–(7.48) are satisfied. There are total nine unknowns which include five weights and four values of $\alpha$. However, due to fewer available equations than the number of unknowns, we have fixed five values of weights as $W_1 = W_2 = W_3 = W_4 = 0$ and $W_5 = 1$. Corresponding four values of $\alpha$ coefficients are given by $\alpha_1 = a_5/a_4$, $\alpha_2 = a_4/a_3$, $\alpha_3 = a_3/a_2$, $\alpha_4 = a_2/a_1$. Thus, the algorithm for the newly optimized low storage form of the five-stage Runge–Kutta (*ORK5*) method is given as [31]

**Fig. 7.25** Variations of the numerical amplification factor contours for the solution of 1$D$ wave equation Eq. (3.95) using indicated schemes have been shown. Note that the desired neutrally stable region for the $OCCS - RK4$ scheme is very small as compared to the $OCCS - ORK5$ scheme. Results are reproduced from [31]

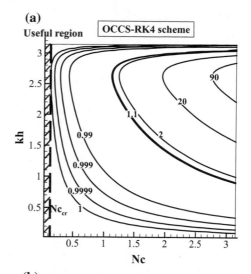

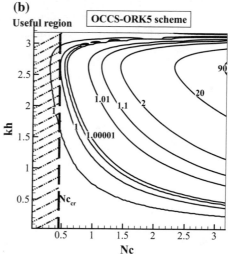

$$u^{(1)} = u^{(n)} + 0.10 \, \Delta t \, f[u^{(0)}] \tag{7.53}$$

$$u^{(2)} = u^{(n)} + \frac{0.03}{0.155} \, \Delta t \, f[u^{(1)}] \tag{7.54}$$

$$u^{(3)} = u^{(n)} + 0.31 \, \Delta t \, f[u^{(2)}] \tag{7.55}$$

$$u^{(4)} = u^{(n)} + 0.50 \, \Delta t \, f[u^{(3)}] \tag{7.56}$$

$$u^{(n+1)} = u^{(n)} + \Delta t \, f[u^{(4)}] \tag{7.57}$$

Next, we have obtained numerical properties for the $OCCS - ORK5$ scheme for the solution of Eq. (3.95). Figure 7.25 shows variations of the numerical amplification factor contours using indicated spatial and temporal discretization schemes. Note that the desired neutrally stable ($|G| = 1$) region (as identified by a hatched region) for the $OCCS - ORK5$ scheme is larges, as compared to the $OCCS - RK4$ scheme. This is an important achievement for the newly derived optimized time integration scheme so that calculations can be performed at a much larger *CFL* number, while retaining neutral stability. With a possibility of choosing higher *CFL* number for the calculations, one can use higher time step in the calculations and can perform quick computation without adversely affecting solution quality. The critical value of the *CFL* number indicating the end of the neutrally stable region for the $OCCS - RK4$ scheme is about $Nc_{cr} = 0.13$, while the same value for the $OCCS - ORK5$ scheme is about $Nc_{cr} = 0.45$. This shows that with the optimized time integration scheme, one can advance calculations more than three times faster than the traditional four-stage fourth-order Runge–Kutta time integration scheme.

Variations of the normalized numerical group velocity contours $\frac{V_{gN}}{c}$ for the solution of Eq. (3.95) using indicated spatial and temporal discretization schemes have been shown in Fig. 7.26. The region between the contour lines 0.99 and 1.01 has been identified as the *DRP* region [37]. The group velocity charts for the $ORK5$ scheme show a good physical dispersion relation preservation property, when $OCCS$ scheme has been used for the spatial discretization.

### 7.3.3   Application of OCCS − ORK5 Scheme

In this section, we have performed computation of wave propagation problem using $OCCS - ORK5$ scheme. We have demonstrated effectiveness of the optimized $OCCS - ORK5$ scheme over traditional difference schemes for the solution of 1D wave equation.

Acoustic field is composed of pressure fluctuations, which are usually very small as compared to the background pressure field. Propagation of acoustic signals through air shows minimal effects of viscosity [21]. Thus, the acoustic wave propagation in a 1D domain is non-dispersive and non-dissipative [37]. However, most of the numerical methods do not display these important properties across a complete wavenumber range, as these are dispersive as well as dissipative. Calculations for computational acoustical problems usually need to be performed over a considerable duration, as the information about the complete acoustic spectrum is required. The newly optimized $OCCS - ORK5$ discretization scheme neither numerically amplify nor attenuate the acoustic signal over a considerable *CFL* number range as observed from the properties shown in Fig. 7.25.

Here, we demonstrate the neutral stability and the *DRP* nature of the newly optimized scheme while computing at significantly higher *CFL* number. We have initially prescribed a wave packet centered at a significantly high wavenumber ($kh = 2.2$). Advantages of the newly optimized $ORK5$ time integration scheme over

**Fig. 7.26** Variations of the normalized numerical group velocity contours for the solution of $1D$ wave equation Eq. (3.95) using indicated spatial and temporal discretization schemes have been shown. Results are reproduced from [31]

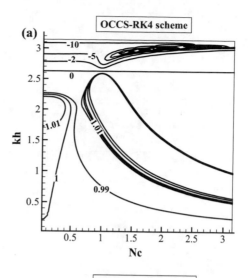

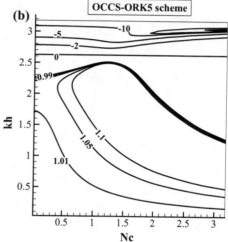

the traditional $RK4$ scheme are demonstrated by computing solutions at a high $CFL$ number ($N_c = 0.40$).

For the present problem, we have selected a domain $0 \le x \le 10$ which has been divided by 1001 equi-spaced grid points. The initial condition used for this problem is centered at $x = 5$ and is given by

$$u^0 = e^{-16(x-5)^2} \cos[kh(x-5)/h] \tag{7.58}$$

where $kh = 2.2$ is the non-dimensional wavenumber and $h$ is the constant grid spacing.

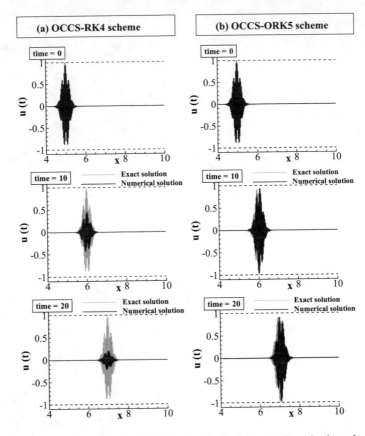

**Fig. 7.27**  Solutions of 1*D* wave Eq. (3.95) obtained by the indicated discretization schemes along with the exact solutions have been shown at different instants. Note that solutions obtained by using *OCCS − ORK*5 scheme match well with the exact solution in contrast to the solutions obtained by using *OCCS − RK*4 scheme. Results are reproduced from [31]

Solutions of Eq. (3.95) are obtained by the indicated discretization schemes in Fig. 7.27 for the initial condition given in Eq. (7.58). The exact solution at the respective instant is also shown in the corresponding frame by the dotted lines. Note that the solutions obtained using *OCCS − ORK*5 scheme match well with the exact solution. In contrast, the solutions obtained using *OCCS − RK*4 scheme show large numerical error by significant attenuation of the initial condition. Thus, the advantage of the newly derived optimized spatial–temporal discretization scheme is evident, as it has significantly less numerical error compared to the traditional discretization schemes. One can solve computational acoustics problems for long duration quickly using *OCCS − ORK*5 scheme by selecting large time step without adversely affecting accuracy of the computed solution.

Next, we demonstrate superiority of the newly derived optimized scheme over different discretization schemes which have similar stencil size. Solutions have been

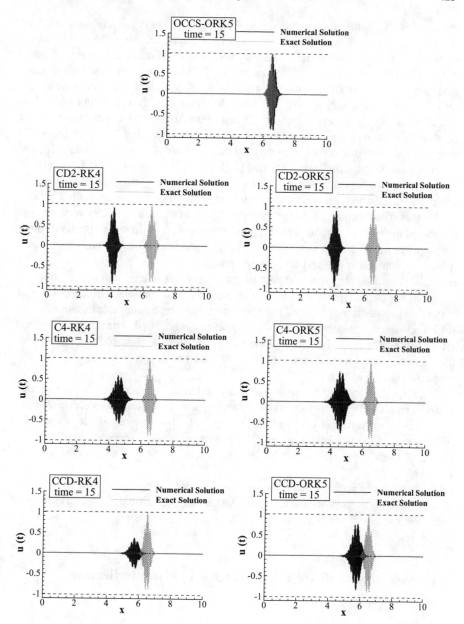

**Fig. 7.28** Solutions of 1$D$ wave Eq. (3.95) obtained by the indicated discretization schemes along with the exact solutions have been shown at *time* = 15. Note that solutions obtained by *OCCS* − *ORK*5 scheme match well with the exact solution in contrast to the solutions obtained by other spatial difference schemes with similar stencil size. Results are reproduced from [31]

obtained for the problem discussed in Fig. 7.27. Top frame in Fig. 7.28 shows the exact and the computed solutions using $OCCS - ORK5$ scheme at $t = 15$ which shows an excellent match. Bottom frames display solutions obtained at the same time using indicated difference schemes. Solutions in Fig. 7.28 have been obtained using same grid and at the same *CFL* number, as in Fig. 7.27. It should be noted that except $OCCS - ORK5$ scheme, all the difference schemes display significant numerical diffusion and dispersion error. Figure 7.28 also shows that irrespective of use of either $RK4$ or $ORK5$ time integration scheme, solutions obtained using $CD2$, $C4$, and $CCD$ compact schemes display large numerical error, where $C4$ corresponds to fourth-order central compact difference scheme [37]. Only combination of $OCCS - ORK5$ scheme displays appreciable results. This aspect clearly shows that one needs to optimize spatial and temporal discretization schemes together, as explained here.

Acoustic field is composed of pressure fluctuations which are usually very small as compared to the background pressure field. Propagation of acoustic signals through air shows minimal effects of viscosity [21]. The acoustic wave propagation in a 1D domain is non-dispersive and non-dissipative [37]. However, most of the numerical methods do not display these important properties across a complete wavenumber range, as they are dispersive as well as dissipative. Calculations for computational acoustical problems usually need to be performed over a considerable duration, as the information about the complete acoustic spectrum is required. The newly optimized $OCCS - ORK5$ discretization scheme neither numerically amplify nor attenuate the acoustic signal over a considerable *CFL* number range, as observed from the properties shown in Fig. 7.25.

Here, we demonstrate the neutral stability and the *DRP* nature of the newly optimized scheme while computing at significantly higher *CFL* number. We have initially prescribed a wave packet centered at a significantly high wavenumber ($kh = 2.2$). Advantages of the newly optimized $ORK5$ time integration scheme over the traditional $RK4$ scheme are demonstrated by computing solutions at a high *CFL* number ($N_c = 0.40$).

For the present problem, we have selected a domain $0 \leq x \leq 10$, which has been divided by 1001 equi-spaced grid points. The initial condition used for this problem is centered at $x = 5$ and is given by Eq. (7.58).

## 7.4　Assessment of Noise Attenuation by Thin Reflecting Barrier

Identification and quantification of acoustic noise sources as well as methods to reduce acoustic noise levels in the desired region have been important research topics for many decades. Acoustic-noise-related problems are regularly faced by the people living inside cities, as the community is constantly exposed to the road traffic noise as well as other sources of noise. Aural comfort has become an important design parameter while designing modern vehicles, residential buildings, transport stations, offices, and public places.

Researchers are focused on reducing acoustic noise by inserting barriers of different shapes and materials between the acoustic source and the receiver. Barriers prevent acoustic waves from directly reaching to the receiver and help to reduce noise by absorbing and deflecting away part of the incident acoustic energy. In the past, researchers have analyzed the effectiveness of the rigid as well as fully or partially noise absorbing barriers for the indoor as well as for the outdoor working environments. Here, we have used derived spectrally optimized *DRP* scheme [31, 32] to estimate modifications in acoustic field due to insertion of thin reflecting barriers.

Here, we have considered a case of diffraction of acoustic waves around a barrier for which experimental results are available in the literature. Using an analytical approach based on Keller's geometrical theory of diffraction, Kawai [20] studied the sound field produced due to incidence of a spherical sound wave on a rigid barrier. In the same work [20], analytical results were supplemented with a carefully performed experimental results. Analytical and experimental results in [20] show an excellent match. We have validated our numerical results with the experimental results available in [20].

Kawai constructed a barrier of width 0.17 m using a 24 mm-thick plywood and positioned a sound source on the left side of the barrier at a distance of 0.17 m as shown in the schematic in the top frame of Fig. 7.29. Experiments were performed for a frequency of 500 Hz using high-quality horn speakers with *YL − D5500P* driver units, for which directivity characteristics are extremely good [20]. We have performed computations for the same case as reported in [20]. Schematic of the chosen domain for simulation has been shown in the bottom frame of Fig. 7.29. Note that the reported experiments in [20] were performed in an anechoic chamber. Thus, there were no acoustic reflections from the four sides of the experimental domain. We have mimicked these experimental conditions by providing absorbing boundary condition on the sides of the computational domain and a perfectly reflecting boundary condition on the three walls of the barrier.

The domain size for the present validation study has been selected as $-7.5$ m $\leq x \leq 7.5$ m, $-3.4$ m $\leq y \leq 4.1$ m, with a grid of 1501 equi-spaced points in the $x$-direction and 751 points in the $y$-direction and a time step of $\Delta t = 0.00001$ s. Origin has been centered on the top left corner of the barrier as shown in Fig. 7.29. The governing equations in a dimensional form for the propagation of acoustic waves are given by

$$\frac{\partial u'}{\partial t} + \frac{1}{\rho_0} \frac{\partial p'}{\partial x} = 0 \qquad (7.59)$$

$$\frac{\partial v'}{\partial t} + \frac{1}{\rho_0} \frac{\partial p'}{\partial y} = 0 \qquad (7.60)$$

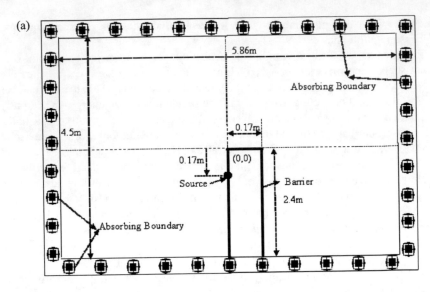

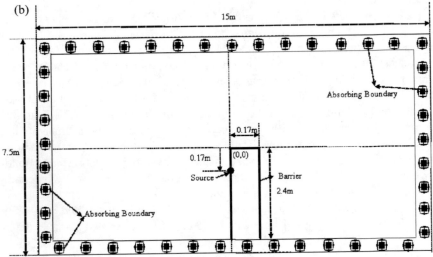

**Fig. 7.29** Figure shows the schematic of experimental [20] and present computational domains in (**a**) and (**b**), respectively

$$\frac{\partial p'}{\partial t} + \gamma P_0 \left( \frac{\partial u'}{\partial x} + \frac{\partial v'}{\partial y} \right) = 0 \qquad (7.61)$$

Computations have been carried out by prescribing monochromatic sound source on the left wall of the barrier at a distance of 0.17 m from the top edge of the barrier to mimic the experimental conditions in [20]. Equations (7.59)–(7.61) are solved using $OCCS - ORK5$ scheme to obtain acoustic field around the barrier. Figure 7.30 shows

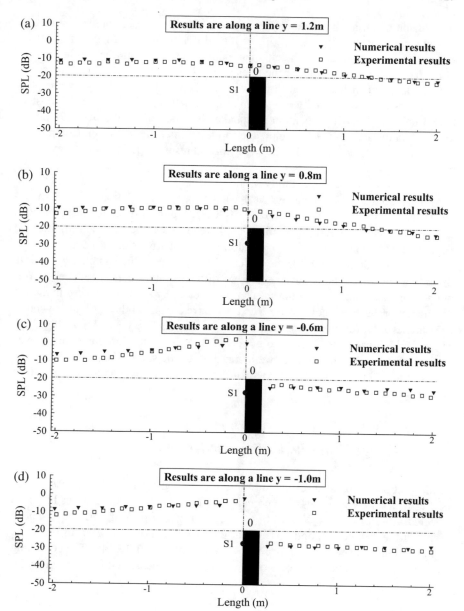

**Fig. 7.30** Comparison of variation of sound pressure levels for the present numerical results and the experimental results in [20] has been shown for four different heights as indicated. Note that respective indicated heights are measured from the top left corner of the barrier where the origin has been located. Location of speaker is indicated by point $S1$ located at a distance of 0.17 m below the top left corner of the barrier. Results are shown for a frequency of 500 Hz

comparison of variation of sound pressure levels for the present numerical results, and the experimental results in [20], for four different indicated heights. Distance associated with the indicated heights has been measured from the top left corner of the barrier. Thus, the plots for the lines $y = 1.2$ m and $y = 0.8$ m, indicate sound pressure level (SPL) distribution along a constant height above the barrier while plots for the lines $y = -0.6$ m and $y = -1.0$ m indicate *SPL* distribution along a constant height below the barrier. Location of speaker is indicated by the point $S1$ and is located at a distance of 0.17 m below the top left corner of the barrier. Comparison of present computed results with the experimental results in [20] shows an excellent match.

In order to show the applicability of the present approach to accurately compute over a wide band of frequencies, we have performed simulations for six different frequencies starting from 200 Hz to 5000 Hz. Figure 7.31 shows instantaneous acoustic pressure contours for the indicated frequencies. We have highlighted the negative pressure contour levels using dotted lines. Diffraction of acoustic waves around barrier can be noted for the indicated frequencies. One can clearly observe that the tendency of the acoustic waves to diffract around barrier reduces as frequency increases. Location of the speaker is indicated by point $S1$ marked in the top left frame for the frequency 200 Hz. Next, we have compared the variation of sound pressure level for different acoustic frequencies along a line $y = -1$ m in Fig. 7.32. Since the high-frequency waves do not diffract much, as the low-frequency waves do, noise attenuation increases as we increase, acoustic source frequency. Next, we have estimated effectiveness of barriers of different shapes over a range of one-third octave band frequencies.

### 7.4.1  *I-Shaped Thin Reflecting Barrier*

In this section, we present results for reduction in acoustic noise due to insertion of barrier with different shapes. The domain size for the present simulations has been selected as $-7.5$ m $\leq x \leq 7.5$ m, $0 \leq y \leq 7.5$ m with a grid of 1501 equi-spaced points in the $x$-direction, and 751 points in the $y$-direction as shown in Fig. 7.33. Computations have been performed with a time step $\Delta t = 0.00001$ s. The governing equations for the propagation of acoustic waves in a dimensional form are given by

$$\frac{\partial u'}{\partial t} + \frac{1}{\rho_0} \frac{\partial p'}{\partial x} = 0 \tag{7.62}$$

$$\frac{\partial v'}{\partial t} + \frac{1}{\rho_0} \frac{\partial p'}{\partial y} = 0 \tag{7.63}$$

$$\frac{\partial p'}{\partial t} + \gamma P_0 \left( \frac{\partial u'}{\partial x} + \frac{\partial v'}{\partial y} \right) = S \tag{7.64}$$

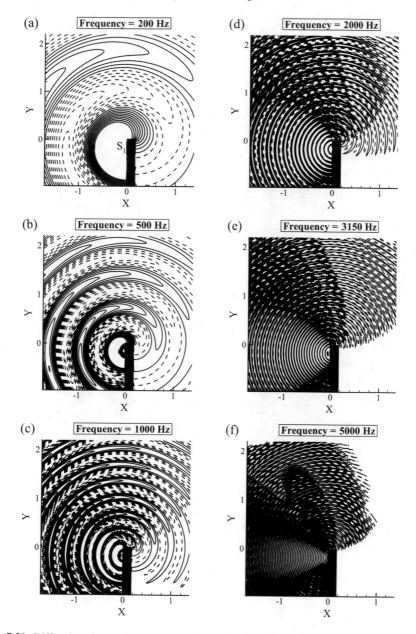

**Fig. 7.31** Diffraction of acoustic waves around barrier has been shown for the indicated frequencies. Note that as we increase frequency, tendency of acoustic wave to diffract around barrier reduces and wavelength decreases. Location of speaker is indicated by point $S1$ located at a distance of 0.17 m below the top left corner of the barrier

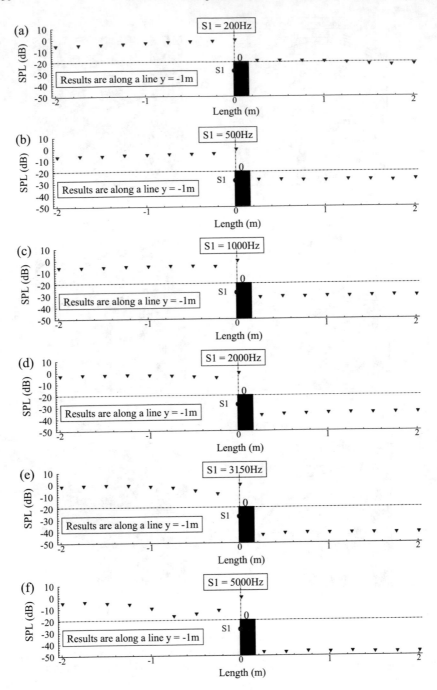

**Fig. 7.32** Variation of sound pressure level for different excitation frequencies has been compared on a line $y = -1$ m. One can clearly observe that as we increase source frequency, noise attenuation increases as lesser waves are able to diffract around the barrier

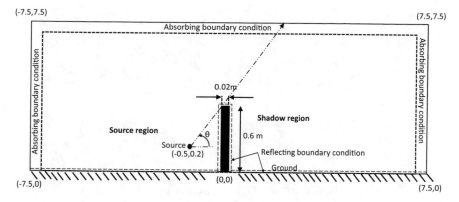

**Fig. 7.33** The schematic of the computational domain with barrier has been shown

$$S = \exp\left[-\ln(2) \times \frac{(x+0.5)^2 + (y-0.2)^2}{0.01}\right] \times \cos(\omega_0 t) \qquad (7.65)$$

where $\omega_0 = 2\pi f$ is the circular frequency. A monochromatic monopole source term $S$ has been used, while solving Eq. (7.64) to remove any directional bias introduced by the source. We have kept acoustic source centered at location $x = -0.5$ m and $y = 0.2$ m and considered nine different cases with source frequencies ($f$) in the one-third octave band ranging from 250 Hz to 1600 Hz. The term $\cos(\omega_0 t)$ in Eq. (7.23) provides monochromatic excitation at the prescribed frequency $\omega_0$ while the remaining part of the expression serves as an amplitude. As observed from Eq. (7.23), amplitude of acoustic source is maximum at the source location $x = -0.5$ m and $y = 0.2$ m and it decays exponentially away from this location. Prescription of such source does not create any discontinuity in the domain and helps computation, as high wavenumber components are not generated.

Presence of acoustic barrier causes acoustic waves to reflect and diffract, resulting in a complex superposition of acoustic waves. The barrier materials can be characterized by the reflection coefficient of the material. The rigidity of the material reflects the acoustic waves away from the barrier face. The reflections from the barrier create a diffuse sound field by interference phenomenon causing higher values of insertion loss. Diffraction is an important phenomena, as acoustic waves depending on their wavelength tend to bend around objects in their path. Apart from transmission, absorption, and reflection, acoustic waves are also diffracted from the edge of the barrier. It is observed that the low-frequency acoustic waves display higher diffraction compared to the higher frequencies. Simulation of such complex acoustic field has been performed here using $OCCS - ORK5$ scheme.

Figure 7.33 shows a schematic of the $I$-shaped barrier positioned at the middle of the domain. When a barrier is inserted between the acoustic source and the receiver, a

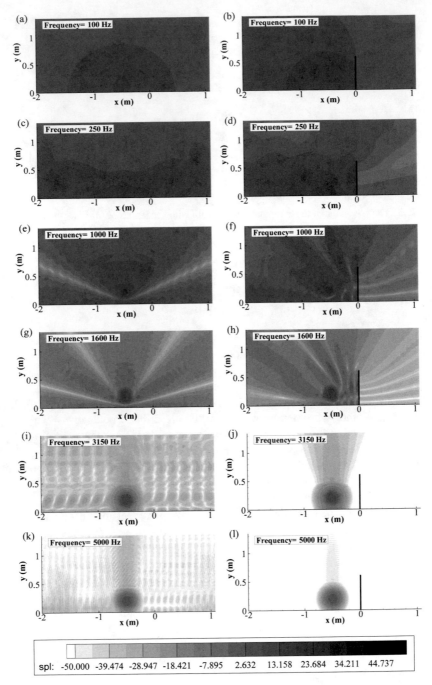

**Fig. 7.34** Distribution of sound pressure levels has been shown for the cases without barrier (left column) and with *I*-shaped barrier (right column) at the indicated frequencies. One can observe that as acoustic source frequency is increased, tendency of acoustic waves to diffract reduces causing higher noise attenuation at higher frequencies

shadow zone is formed behind the barrier. The shadow zone can be defined as the area under the straight line extended from the noise source, followed tangentially through the top of the barrier as shown in Fig. 7.33. We have prescribed completely reflecting boundary conditions at the surfaces of the barrier and at the ground ($y = 0$). On rest of the three sides of the domain we have prescribed absorbing boundary conditions. The height of the barrier has been selected as 0.6 m and width is kept as 0.02 m. Here, sound pressure levels at different locations have been numerically estimated with the assumption of reflecting ground and barrier surface. In the literature, effectiveness of barriers is usually estimated by calculating insertion loss which denotes the loss of sound pressure level at certain location due to the insertion of barrier. Insertion loss can be defined as

$$IL = 20 \log_{10} \frac{P_{\text{ref}}}{P_{\text{RMS}}} \tag{7.66}$$

where $P_{\text{RMS}}$ is the RMS value of the pressure fluctuation at a point in the domain with barrier and $P_{\text{ref}}$ is the RMS value of the pressure fluctuation at the same point without barrier case at the same source frequency. Higher value of the insertion loss indicates higher noise reduction due to acoustic barrier.

Figure 7.34 shows distribution of sound pressure for the cases without barrier and with $I$-shaped barrier at the indicated frequencies. For the case of $I$-shaped barrier, we have computed acoustic field for the source frequencies 100, 3150, and 5000 Hz, in addition to the nine different source frequencies in the one-third octave band ranging from 250 Hz to 1600 Hz. One can observe that as acoustic source frequency is increased, tendency of acoustic waves to diffract reduces causing higher noise attenuation at higher frequencies. For the source frequencies of 3150 Hz and 5000 Hz, one observes minimal contribution in the shadow region by diffraction.

In the present chapter, we have given a close look toward the numerical methods and their properties. We have also learned the way to create high-accuracy dispersion relation preserving schemes. Developed methods were used to compute model acoustic wave propagation problems. With this information, we will discuss numerical solutions obtained for the noise produced by flow past streamlined and bluff bodies in the next chapter.

# Problems

**P7.1**: Solve the 1D wave equation in a domain $-200 \le x \le 500$. The 1D wave equation and the initial condition are as follows.

$$\frac{\partial u}{\partial t} + c \frac{\partial u}{\partial x} = 0,$$

$$t = 0, \quad u = e^{-(ln2)x^2}$$

**Fig. 7.35** An ellipse of
semi-major and semi-minor
axes given by $a$ and $b$

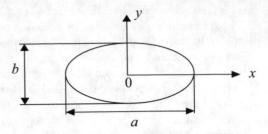

Find numerical solutions using (i) $CD2 - RK_4$ scheme, (ii) $OUCS3 - RK_4$ scheme at $t = 100, 200, 300$ and $400$. Compare these solutions by the indicated methods. Choose the grid spacing and the time step such that the *CFL* number is $0.05$. Assume phase speed $c$ as $0.1$.

**P7.2**: Solve the spherical wave propagation problem,

$$\frac{\partial u}{\partial t} + \frac{u}{r} + \frac{\partial u}{\partial r} = 0,$$

in the domain $2 \le r \le 1000$.

Initially there is no disturbance in the domain. Consider a boundary condition at $r = 2$ as, $u = \sin\omega t$. Compare the numerical solutions with exact solution for $\omega = \pi/2$ and $\omega = 3\pi/2$. Obtain solutions using $OUCS3 - RK_4$ scheme. Explain the differences between the computed and the exact solutions.

**P7.3**: Compare the scattering of sound field by ellipse of different shapes as shown in Fig. 7.35. Consider three different ellipse geometries with $a/b = 1$, $a/b = 0.75$ and $a/b = 0.5$. Consider an initial condition as, $t = 0$, $u = v = P = 0$. Treat surface of the ellipse as purely reflective.

Obtain the solution of governing linearized Euler equations which are given by,

$$\frac{\partial u}{\partial t} + \frac{\partial p}{\partial x} = 0$$

$$\frac{\partial v}{\partial t} + \frac{\partial p}{\partial y} = 0$$

$$\frac{\partial p}{\partial t} + \frac{\partial u}{\partial x} + \frac{\partial v}{\partial y} = \left[ e^{-\left( \ln \frac{(x-5)^2 + y^2}{0.1} \right)} \right] \sin \pi/2t.$$

Compare solutions at $(0,5)$, $(-5,5)$ and $(-5,0)$ at a given instant.

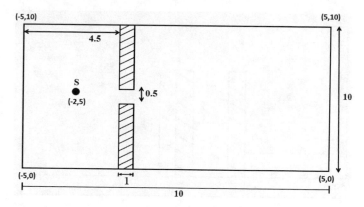

**Fig. 7.36** Schematic for problem on sound wave diffraction through a slit

**P7.4**: Consider a following domain $(-5 \leq x \leq 5, 0 \leq y \leq 10)$. Source 'S' is located at $(-2, 5)$ on the left hand side of the slit (Fig. 7.36).

(i) Assume source 'S' as $s = e^{-(\frac{(x+2)^2+(y-5)^2}{0.01})} \sin(\omega t)$ $\omega = 2\pi f, f = 1000\,\mathrm{H}_z$. The governing equations are the same as in Problem 3. Starting from rest, find solutions at $t = 0.005, 0.01, 0.015, 0.02$.

(ii) How solutions would differ if the frequency changes to $f = 5000\,\mathrm{H}_z$. Do we need to change $\Delta x$ and $\Delta t$? How can we estimate correct values of $\Delta x$ and $\Delta t$ before solving problem?

**P7.5**: Consider the following domain:
On the left boundary three pistons (A,B,C) are placed, which vibrate with time. Find the sound field for the following conditions (Fig. 7.37). Piston A,B,C vibrate as,

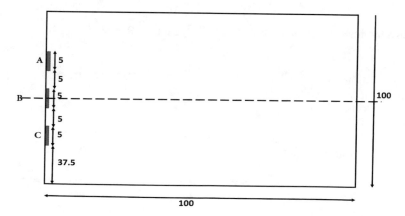

**Fig. 7.37**   Schematic of oscillating pistons A, B, C

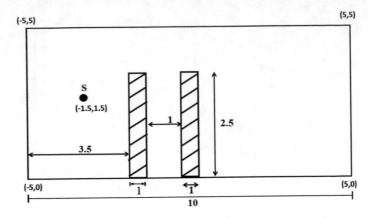

**Fig. 7.38** Here S is the source in front of two barriers as indicated. The co-ordinates of the domain are as indicated in the figure

$$U_A = \sin(\omega t)$$
$$U_B = \sin(\omega t + \phi_1)$$
$$U_C = \sin(\omega t + \phi_1 + \phi_2)$$

where, $\omega = 000\pi$, $\phi_1 = 20°$ and $\phi_2 = 40°$. Governing equations are same as in P7.3. Compute the evolution of the sound field.

**P7.6**: Compute the evolution of the sound field for the following problem, with S as the source in front of two barriers. The co-ordinates and dimensions are as marked. Assume the sound source S creates a field given as,

$$S = \left[ e^{-\left( \frac{(x-10)^2 + (y-5)^2}{0.05} \right)} \right] \sin(2000\pi t)$$

Discuss the instantaneous sound field details for this problem (Fig. 7.38).

**Fig. 7.39** A monopole is located at $S_1$ in front of the barrier fixed to the reflecting ground

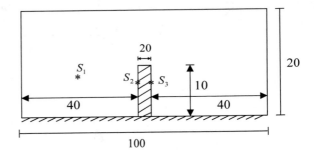

**P7.7**: Consider an open space with reflecting ground and a barrier as shown in Fig. 7.39. There is a monopole sound source at $S_1$ (20, 5) and is given by

$$S_1 = e^{-((x-20)^2+(y-5)^2)} \sin(1000\pi t)$$

One needs to reduce noise close to location $S_2$ (40, 5) and $S_3$ (60, 5) through active cancellation mechanism. Can you propose the source term for the sound sources at $S_2$ and $S_3$? Compute and plot the noise reduction after application of the active cancellation mechanism (Fig. 7.39).

# References

1. G. Ashcroft, X. Zhang, Optimized prefactored compact schemes. J. Comput. Phys. **190**(2), 459–477 (2003)
2. C. Bailly, D. Juve, Numerical solution of acoustic propagation problems using linearized euler equations. AIAA J. **38**(1), 22–29 (2000)
3. M. Bernardini, S. Pirozzoli, A general strategy for the optimization of runge-kutta schemes for wave propagation phenomena. J. Comput. Phys. **228**(11), 4182–4199 (2009)
4. Y.G. Bhumkar, *High Performance Computing of Bypass Transition*. Ph.D. Thesis, Department of Aerospace Engineering, Indian Institute of Technology, Kanpur (2012)
5. Y.G. Bhumkar, T.K. Sengupta, Adaptive multi-dimensional filters. Comput. Fluids **49** (2011)
6. Y.G. Bhumkar, T.W.H. Sheu, T.K. Sengupta, A dispersion relation preserving optimized upwind compact difference scheme for high accuracy flow simulations. J. Comput. Phys. **278**, 378–399 (2014)
7. B.J. Boersma, A staggered compact finite difference formulation for the compressible Navier-Stokes equations. J. Comput. Phys. **208**(2), 675–690 (2005)
8. C. Bogey, C. Bailly, D. Juvé, Noise investigation of a high subsonic, moderate reynolds number jet using a compressible LES. Theor. Comput. Fluid Dyn. **16**(4), 273–297 (2003)
9. V. Borue, S.A. Orszag, Local energy flux and subgrid-scale statistics in three-dimensional turbulence. J. Fluid Mech. **366**, 1–31 (1998)
10. P.H. Chiu, T.W.H. Sheu, On the development of a dispersion-relation-preserving dual-compact upwind scheme for convection-diffusion equation. J. Comput. Phys. **228**, 3640–3655 (2009)
11. P.C. Chu, C. Fan, A three-point combined compact difference scheme. J. Comput. Phys. **140**, 370–399 (1998)
12. T. Colonius, S.K. Lele, Computational aeroacoustics: progress on nonlinear problems of sound generation. Prog. Aero. Sci. **40**, 345–416 (2004)
13. H.A.V. der Vorst, Bi-CGSTAB: A fast and smoothly converging variant of Bi-CG for the solution of non-symmetric linear systems. SIAM J. Sci. Stat. Comput. **13**, 631–644 (1992)
14. K.Y. Fung, R.S.O. Man, S. Davis, A compact solution to computational acoustics, in *ICASE/LaRC Workshop on Benchmark Problems in Computational Aeroacoustics (CAA)* (1995), pp. 59–72
15. D.V. Gaitonde, J.S. Shang, J.L. Young, Practical aspects of higher-order numerical schemes for wave propagation phenomena. Int. J. Numer. Meth. Eng. **45**(12), 1849–1869 (1999)
16. S.I. Green, *Fluid Vortices: Fluid Mechanics and Its Applications* (Springer, Berlin, 1995)
17. N. Haugen, A. Brandenburg, Inertial range scaling in numerical turbulence with hyperviscosity. Phys. Rev. E **70**, 026405 (2004)
18. A.E. Honein, P. Moin, Higher entropy conservation and numerical stability of compressible turbulence simulations. J. Comput. Phys. **201**(2), 531–545 (2004)
19. G.S. Karamanos, G.E. Karniadakis, A spectral vanishing viscosity method for large-eddy simulations. J. Comput. Phys. **163**(1), 22–50 (2000)

20. T. Kawai, Sound diffraction by a many-sided barrier or pillar. J. Sound Vib. **79**(2), 229–242 (1981)
21. L.E. Kinsler, A.R. Frey, A.B. Coppens, J.V. Sanders, *Fundamentals of Acoustics* (Wiley, New York, 2000)
22. E. Lamballais, V. Fortuné, S.L. Aizet, Straightforward high-order numerical dissipation via the viscous term for direct and large eddy simulation. J. Comput. Phys. **230**(9), 3270–3275 (2011)
23. A.G. Lamorgese, D.A. Caughey, S.B. Pope, Direct numerical simulation of homogeneous turbulence with hyperviscosity. Phys. Fluids **17** (2005)
24. S.K. Lele, Compact finite difference schemes with spectral-like resolution. J. Comput. Phys. **103**, 16–42 (1992)
25. M.J. Lighthill, On sound generated aerodynamically. I. general theory. Proc. Roy. Soc. Lond. A: Math. Phys. Eng. Sci. **211**, 564–587 (1952)
26. K. Mahesh, A family of high-order finite difference schemes with good spectral resolution. J. Comput. Phys. **145**, 332–358 (1998)
27. S. Nagarajan, S.K. Lele, J.H. Ferziger, A robust high-order compact method for large eddy simulation. J. Comput. Phys. **191**(2), 392–419 (2003)
28. R. Pasquetti, Spectral vanishing viscosity method for large-eddy simulation of turbulent flows. J. Sci. Comput. **27**, 365–375 (2006)
29. S. Pirozzoli, Performance analysis and optimization of finite-difference schemes for wave propagation problems. J. Comput. Phys. **222**(2), 809–831 (2007)
30. T. Poinsot, D. Veynante, *Theoretical and Numerical Combustion*. R. T. Edwards Inc., (2005)
31. J. Pradhan, S. Jindal, B. Mahato, Y.G. Bhumkar, Joint optimization of the spatial and the temporal discretization scheme for accurate computation of acoustic problems. Commun. Comput. Phys. **24**(2), 408–434 (2018)
32. J. Pradhan, B. Mahato, S.D. Dhandole, Y.G. Bhumkar, Construction, analysis and application of coupled compact difference scheme in computational acoustics and fluid flow problems. Commun. Comput. Phys. **18**(4), 957–984 (2015)
33. M.K. Rajpoot, T.K. Sengupta, P.K. Dutt, Optimal time advancing dispersion relation preserving schemes. J. Comput. Phys. **229**, 3623–3651 (2010)
34. T. Rylander, P. Ingelström, A. Bondeson, *Computational Electromagnetics* (Springer, Berlin, 2007)
35. N.D. Sandham, Q. Li, H.C. Yee, Entropy splitting for high-order numerical simulation of compressible turbulence. J. Comput. Phys. **178**(2), 307–322 (2002)
36. T.K. Sengupta, S. Bhaumik, Y.G. Bhumkar, Direct numerical simulation of two-dimensional wall-bounded turbulent flows from receptivity stage. Phys. Rev. E **85**(2), 026308 (2012)
37. T.K. Sengupta, *High Accuracy Computing Methods: Fluid Flows and Wave Phenomena* (Cambridge University Press, USA, 2013)
38. T.K. Sengupta, Y.G. Bhumkar, New explicit two-dimensional higher order filters. Comput. Fluids **39**, 1848–1863 (2010)
39. T.K. Sengupta, G. Ganeriwal, S. De, Analysis of central and upwind compact schemes. J. Comput. Phys. **192**, 677–694 (2003)
40. T.K. Sengupta, S.K. Sircar, A. Dipankar, High accuracy schemes for DNS and acoustics. J. Sci. Comput. **26**, 151–193 (2006)
41. T.K. Sengupta, A. Dipankar, P. Sagaut, Error dynamics: Beyond von neumann analysis. J. Comput. Phys. **226**, 1211–1218 (2007)
42. T.K. Sengupta, Y.G. Bhumkar, V. Lakshmanan, Design and analysis of a new filter for LES and DES. Comput. Struct. **87**, 735–750 (2009)
43. T.K. Sengupta, V. Lakshmanan, V.V.S.N. Vijay, A new combined stable and dispersion relation preserving compact scheme for non-periodic problems. J. Comput. Phys. **228**, 3048–3071 (2009)
44. T.K. Sengupta, V.V.S.N. Vijay, S. Bhaumik, Further improvement and analysis of CCD scheme: dissipation discretization and de-aliasing properties. J. Comput. Phys. **228**(17), 6150–6168 (2009)

45. T.K. Sengupta, M.K. Rajpoot, S. Saurabh, V.V.S.N. Vijay, Analysis of anisotropy of numerical wave solutions by high accuracy finite difference methods. J. Comput. Phys. **230**, 27–60 (2011)
46. T.K. Sengupta, Y.G. Bhumkar, M. Rajpoot, V.K. Suman, S. Saurabh, Spurious waves in discrete computation of wave phenomena and flow problems. Appl. Math. Comput. **218**, 9035–9065 (2012)
47. T.K. Sengupta, Y.G. Bhumkar, S. Sengupta, Dynamics and instability of a shielded vortex in close proximity of a wall. Comput. Fluids **70**, 166–175 (2012)
48. P.L. Shah, J. Hardin, Second-order numerical solution of time-dependent, first-order hyperbolic equations, in *ICASE/LaRC Workshop on Benchmark Problems in Computational Aeroacoustics (CAA)* (1995), pp. 133–141
49. E. Tadmor, Convergence of spectral methods for nonlinear conservation laws. SIAM J. Num. Analysis **26**(1), 30–44 (1989)
50. C.K.W. Tam, J.C. Hardin, Second computational aeroacoustics (caa) workshop on benchmark problems, in *NASA Conference Publication* (1997), p. 3352
51. C.K.W. Tam, K.A. Kurbatskii, J. Fang, Numerical boundary conditions for computational aeroacoustics benchmark problems, in *NASA Conference Publication* (1997), p. 3352
52. C.K.W. Tam, H. Shen, K.A. Kurbatskii, L. Auriault, Z. Dong, J.C. Webb, Solutions of the benchmark problems by the dispersion-relation-preserving scheme, in *ICASE/LaRC Workshop on Benchmark Problems in Computational Aeroacoustics (CAA)*, vol. 1 (1995), pp. 149–172
53. C.K.W. Tam, *Computational Aeroacoustics a Wave Number Approach* (Cambridge University Press, New York, 2012)
54. C.K.W. Tam, Z. Dong, Radiation and outflow boundary conditions for direct computation of acoustic and flow disturbances in a nonuniform mean flow. J. Comput. Acoust. **4**(2), 175–201 (1996)
55. C.K.W. Tam, J.C. Webb, Dispersion-relation-preserving finite difference schemes for computational acoustics. J. Comput. Phys. **107**, 262–281 (1993)
56. R. Vichnevetsky, J.B. Bowles, *Fourier Analysis of Numerical Approximations of Hyperbolic Equations* (SIAM Stud. Appl. Math, Philadelphia, 1982)
57. M.R. Visbal, D.V. Gaitonde, On the use of higher-order finite-difference schemes on curvilinear and deforming meshes. J. Comput. Phys. **181**, 155–185 (2002)
58. C.H. Yu, Y.G. Bhumkar, T.W.H. Sheu, Dispersion relation preserving combined compact difference schemes for flow problems. J. Sci. Comput. **62**(2), 482–516 (2015)
59. Q. Zhou, Z. Yao, F. He, M.Y. Shen, A new family of high-order compact upwind difference schemes with good spectral resolution. J. Comput. Phys. **227**, 1306–1339 (2007)

# Chapter 8
# Methodologies and Solutions of Computational Aeroacoustic Problems

## 8.1 Introduction

Computational aeroacoustics (*CAA*) is an important research area which connects aerodynamics and acoustics. This research area involves simulations of flow-induced acoustic noise and can be broadly categorized into three different methodologies based on the approach used for estimating acoustic field. First two methodologies use computed fluid flow information to further calculate acoustic field details using either an acoustic analogy approach or by solving set of perturbation equations. In the last accurate but computationally expensive approach, one computes acoustic field details by solving set of compressible flow equations using the direct numerical simulation *DNS* approach. These approaches are briefly explained here, and are useful to perform *CAA* simulations for flow past bluff and streamlined bodies. Discussion on various *CAA* methodologies is followed by results for different *CAA* problems obtained using *DNS*.

### 8.1.1 *Acoustic Analogy Approach*

Prediction of sound generated due to flow past any body can be obtained using an acoustic analogy approach for certain flow conditions, as suggested by Lighthill [51], Ffowcs Williams and Hawkings [28] and Curle [19]. An acoustic analogy relates sound to surface and volume source term integrals. Thus, by evaluating source terms from computed flow field, acoustic analogies predict sound field information away from the body. In this approach, numerical simulations are carried out in two different parts, namely, the near-field and the far-field part. The near-field part deals with aerodynamics aspects and flow structures are computed using *CFD* techniques. The source terms are obtained using the near-field flow quantities. An acoustic analogy is further used to compute sound at the far-field. Ffowcs Williams and Hawkings [28] included effects of surfaces in arbitrary motions in a generalized form of the

© Springer Nature Singapore Pte Ltd. 2020
T. K. Sengupta and Y. G. Bhumkar, *Computational Aerodynamics and Aeroacoustics*,
https://doi.org/10.1007/978-981-15-4284-8_8

Lighthills acoustic analogy. Similarly, Powell [72] proposed the vortex sound theory to predict generation and propagation of the sound from compact vortical flows. If the size of the acoustic source is significantly smaller than the wavelength of the generated acoustic waves, then the source is termed as acoustically compact source.

Sound generated by uniform flow past a cylinder over a range of Reynolds number was studied by Cox et al. [18] using such hybrid approach. Based on finite volume approach, the near-field solution was obtained by solving two- and three-dimensional compressible, Favre-averaged Navier–Stokes, equations using a second-order accurate schemes in space and time. The far-field sound was obtained by Lighthill's analogy, using the Ffowcs Williams– Hawkings equations [18, 42], indicating the dipolar sound is dominant over the quadrupolar component of sound for such class of flows. Results in [18, 42] highlight that the accuracy with which the near-field flow quantities are computed by the *CFD* simulation dictates the accuracy with which sound field details can be computed. Similar hybrid approach has been used in [95, 99] to obtain sound generation due to flow past a circular cylinder and a laminar flow past a *NACA*0012 aerofoil, respectively. For a low Mach number flow, quadrupole sound was found weak compared to that of lift and drag dipoles [99].

Lighthill [51] used mass and momentum conservation equations for fluid flow to derive a single second-order partial differential equation in space and time whose left-hand side represents the wave motion while the right-hand side provides the information about acoustic source terms. Lighthill's [51] equation in terms of the fluctuating component of the density ($\rho'$) is given as [12]

$$\frac{\partial^2 \rho'}{\partial t^2} - c_0^2 \frac{\partial^2 \rho'}{\partial x_i \partial x_i} = \frac{\partial S_i}{\partial x_i} \tag{8.1}$$

Here, $c_0$ is the speed of sound and $S_i = -\partial T_{ij}/\partial x_j$ is the source term. The term $T_{ij}$ indicates Lighthill stress tensor and is given as $T_{ij} = \rho u_i u_j + (p' - c_0^2 \rho')\delta_{ij} - \tau_{ij}$, where $\tau_{ij}$ is the viscous stress tensor [12]. Thus the wave equation proposed by Lighthill is inhomogeneous equation with the first source term on the right- hand side corresponds to the Reynolds stress tensor as also explained in Eq. 6.41. Addition of the second and third terms in the source are necessary for high Mach number flows with significant viscous action.

Curle [19] extended Lighthill's acoustic analogy to include influence of solid boundaries on the radiated sound field. One can assume a compact sound source region if the dimension of the body is significantly small compared to the generated sound wavelength [89]. If one considers $y$ as a point on the body surface, $p_{ij}$ as pressure and $F_i$ as instantaneous force acting on the body in the direction of the dipole, then the Curle's solution for a fixed, rigid, compact body is given as [89]

$$c_0^2 \rho'(x, t) = \frac{1}{4\pi r^2} \frac{x_i x_j}{r c_0^2} \int_V T_{ij}(y, t - r/c_0) dV(y) - \frac{1}{4\pi r^2} \frac{x_j}{c_0^3} \frac{\partial}{\partial t} F_j(t - r/c_0) \tag{8.2}$$

One can express the Curle's exact solution for a two-dimensional flow with compact rigid body and compact acoustic source region assumption as [89]

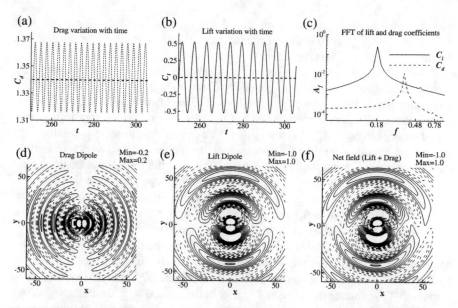

**Fig. 8.1** Time varying lift and drag coefficients for flow past a stationary cylinder at a Reynolds number of 150 has been shown in frames **a** and **b**, respectively. Fast Fourier transform (FFT) of the lift and the drag coefficient variation with time has been shown in the frame **c**. Sound radiated due to the drag and the lift dipole has been shown in frames **d** and **e**, respectively. Contribution of the lift and the drag dipole to the net sound field is shown in Fig. 8.1 (**f**)

$$p'(x_i, t) = \frac{1}{4\pi c_0} \frac{x_i}{r^2} \left[ \frac{\partial F_i}{\partial t} \right] \tag{8.3}$$

Equation (8.3) suggests that the sound generated in the far-field depends on the time variation of the fluctuating loads acting on the body. Fluctuations in the lift and drag coefficients are caused by the unsteady flow past bluff bodies. These surface force fluctuations are related to the amplitude and frequency of sound waves by various researchers in the past [42, 53, 59].

To discuss about the drag and the lift dipole information given by the Curle's acoustic analogy, here we have shown the time varying lift and drag coefficients for flow past a stationary cylinder at a Reynolds number of 150 in Fig. 8.1a, b, respectively. One can observe that the amplitude associated with the lift fluctuations is significantly larger as compared to the drag fluctuations. Fast Fourier transform (*FFT*) of the lift and drag coefficient variation with time has been shown in Fig. 8.1c. The drag coefficient varies twice as fast as compared to the lift coefficient. Information about the frequency and amplitude of the lift and drag fluctuations dictates the sound field, as given by the Curle's equation (8.3) as used in [89]. We have used Eq. (8.3) for predicting the sound radiated due to the drag and the lift dipole in Fig. 8.1d, e, respectively. Unsteady lift and drag have been used in Eq. (8.3) to compute the fluctuating acoustic pressure. One can observe that the wavelength of sound waves associated with the drag dipole are twice as small as that with the lift dipole. Contribution of the lift and the drag dipole to the net sound field is shown in Fig. 8.1f, which displays that the lift dipole dictates the radiated sound field.

Powell [72] was interested in relating the aerodynamic aspects of the flow to the corresponding acoustic field and looked for connection between the vortex motion and sound generation. He proposed that the formation of vortices in the flow is a fundamental noise-producing mechanism. Similar to acoustic analogy approach, vortex sound theory by Powell [72] assumes that the acoustic length scale is significantly larger than the flow length scale which results in higher energy levels for the fluid flow, as compared to the sound field, which decouples the sound field from the flow field. For such *CAA* problems, aeroacoustic solution depends on the aerodynamic solution, while there is no feedback from the aeroacoustic solution to the aerodynamic solution. Regions with non-zero vorticity give rise to localized acoustic source field, which is responsible for presence of acoustic waves in the far-field. Powells vortex sound theory [72] is useful for the cases, where the flow is both coherent and compact [53]. First step in this approach is to obtain the time-dependent flow field by numerically solving the incompressible flow equations. Computed velocity field is then used to evaluate the acoustic forcing. In the second step, the inhomogeneous wave equation with the acoustic forcing term on the right-hand side of the equation is solved numerically using highly accurate scheme. Powell [72], proposed approximate version of Lighthills analogy by assuming fluid as incompressible inside the source region and neglecting viscous as well as thermal losses. According to Powell [72], governing equation for the far-field acoustic pressure can be approximated for the flows with significantly lower convection velocity as

$$\frac{\partial^2 p'}{\partial t^2} - c_0^2 \nabla^2 p' = \nabla \cdot (\omega \times u) \tag{8.4}$$

where $p'$ is the normalized acoustic pressure, and $\omega$ is the vorticity vector. Authors in [53] used Powell's analogy to obtain an acoustic field for laminar flow past a rectangular cylinder. Next, we discuss about the approach involving solution of set of perturbed Euler equations.

### 8.1.2 Hydrodynamic/Acoustic Splitting Approach

In general, it is possible to split the simulation of acoustic wave propagation from the computation of acoustic sources arising due to the unsteady flow. Such separation in the analysis of the flow and the acoustic field is possible due to disparity of the turbulent and acoustic scales at low Mach numbers. This approach is also termed as hybrid approach and helps to reduce computational cost significantly. In contrast to the above-mentioned acoustic analogy-based hybrid approach, researchers also rely on second approach in which the flow quantities are represented by combination of an incompressible mean flow and a perturbation about the mean under the assumption of low Mach number. In this approach, solution of incompressible viscous equations is obtained to estimate mean flow quantities, which is further used to evaluate perturbation equations. In the near-field, the perturbation quantities account for the difference between the incompressible mean flow and the compressible flow while in the far-field, the perturbation quantities are equivalent to acoustic quantities. This approach is also called as acoustic/viscous splitting method [36] and is used to study the sound generation by a cylinder in a uniform flow [71, 87, 91]. Authors in [91] concluded that the acoustic/viscous splitting approach is an effective and convenient method of predicting acoustic fields resulting from flows with low-Mach-number, non-compact source regions.

Most of the aeroacoustic problems associated with industries involve fluid flow at low Mach numbers. *DNS* of such problems is computationally challenging due to scale disparities in the hydrodynamic and the acoustic waves. For solving such problems a hydrodynamic/acoustic splitting method is considered as a cost effective and accurate approach. In this approach, first incompressible Navier–Stokes equations are solved for obtaining flow details. Next, perturbed Euler equations (*PEE*) are solved to predict the acoustic field. PEEs are derived by subtracting the incompressible Navier–Stokes equations from the compressible ones, neglecting the viscous terms [82, 83]. In this approach, one can use different grids for flow and acoustics computation to enhance computational efficiency.

Suppose $U$, $V$, $P$, and $\rho_0$ denotes mean flow quantities indicating longitudinal and transverse velocities, fluid pressure, and density, respectively. Let us consider the quantities indicated in prime $u'$, $v'$, $p'$, $\rho'$ as acoustic disturbances. If we substitute these quantities into the compressible Navier–Stokes equations, neglecting the viscous terms and the higher order perturbations and then subtract the incompressible conservation equations, then *PEE* are obtained.

The non-linear Euler equations in Cartesian coordinates are given as

$$\frac{\partial \rho}{\partial t} + \frac{\partial (\rho u_i)}{\partial x_i} = 0 \tag{8.5a}$$

$$\frac{\partial \rho u_i}{\partial t} + \frac{\partial \left( \rho u_i u_j + p_{ij} \right)}{\partial x_j} \tag{8.5b}$$

$$\frac{\partial p}{\partial t} = c^2 \frac{\partial \rho}{\partial t} \tag{8.5c}$$

The compressible solution can be decomposed as [87]

$$u = U + u', v = V + v', p = P + p', \rho = \rho_0 + \rho' \tag{8.6}$$

If we substitute these quantities into the Eq. (8.5)

$$\frac{\partial(\rho_0 + \rho')}{\partial t} + \frac{\partial(\rho_0 + \rho')(U_i + u_i')}{\partial x_i} = 0 \tag{8.7a}$$

$$\frac{\partial(\rho_0 + \rho')(U_i + u_i')}{\partial t} + \frac{\partial\big((\rho_0 + \rho')(U_i + u_i')(U_j + u_j') + p_{ij}\big)}{\partial x_j} \tag{8.7b}$$

$$\frac{\partial(P + p')}{\partial t} = c^2 \frac{\partial(\rho_0 + \rho')}{\partial t} \tag{8.7c}$$

If we subtract incompressible mean flow from Eq. (8.7) we obtain

$$\frac{\partial \rho'}{\partial t} + \frac{\partial f_i}{\partial x_i} = 0 \tag{8.8a}$$

$$\frac{\partial f_i}{\partial t} + \frac{\partial\big[p'\delta_{ij} + f_i(U_j + u_j') + \rho_0 U_i u_j'\big]}{\partial x_j} = 0 \tag{8.8b}$$

$$\frac{\partial p}{\partial t} - c^2 \frac{\partial \rho'}{\partial t} = -\frac{\partial P}{\partial t} \tag{8.8c}$$

where

$$f_i = \rho u_i' + \rho' U_i \tag{8.8d}$$

PEEs are non-linear and involve interaction between multiple scales. There is coupling between the incompressible flow variables and the perturbed quantities [82, 83], which generates so-called perturbed vorticity [83]. One needs to be careful while handling such quantity, as it can give rise to spurious reflections from outflow boundary, as well as, can cause numerical instabilities [83]. As a result, physically inconsistent acoustic solutions may be obtained while solving the PEEs [82]. To address this issue, authors in [82] proposed perturbed compressible equations (*PCE*) to control unphysical generation and growth of perturbed vorticity. In *PCE*, perturbed viscous stresses are included in the perturbed Euler equations. The perturbed energy equation is derived from the compressible energy equation. Results in [82] show that the solution of *PCE* are as accurate as the *DNS* for solving acoustic field generated due to the flow past a circular cylinder at Re $= 200$ and Mach number of 0.3.

Cheong et al. [12] obtained solutions for the incompressible flow past a circular cylinder at high Reynolds number Re $= 1.58 \times 10^4$ and relied on hybrid *CAA* method to simulate flow-induced acoustic field. In this approach, authors [12] used the approximated form of the Lighthill's stress tensor to prescribe acoustic source term for solving linearized Euler equations using high accuracy difference scheme. If the $u_i'(X, t)$ indicates the fluctuating component of the flow field and $u_i(X, t)$ and

$\bar{u}_i(X)$ shows the instantaneous and the mean component of the velocities, respectively, then the approximated Lighthill's stress tensor ($ALST$) is given by [12]

$$S_{ALST,i} = -\frac{\partial \rho u'_i u'_j}{\partial x_j} - \frac{\partial \rho u'_i \bar{u}_j}{\partial x_j} - \frac{\partial \rho \bar{u}_i u'_j}{\partial x_j} - \frac{\partial \rho \bar{u}_i \bar{u}_j}{\partial x_j} \tag{8.9}$$

Equation (8.9) is further decomposed into three separate terms as [12]

$$S_{\text{self},i} = -\frac{\partial \rho u'_i u'_j}{\partial x_j} \tag{8.10}$$

$$S_{\text{shear},i} = -\frac{\partial \rho u'_i \bar{u}_j}{\partial x_j} - \frac{\partial \rho \bar{u}_i u'_j}{\partial x_j} \tag{8.11}$$

$$S_{0,i} = -\frac{\partial \rho \bar{u}_i \bar{u}_j}{\partial x_j} \tag{8.12}$$

Cheong et al. [12] explained that the term on right-hand side of Eqs. (8.10) and (8.11) denotes the self-noise source terms due to interaction between velocity fluctuations and shear-noise due to interaction between the mean and the fluctuating velocity components, respectively. Relative importance of the self- and the shear-noise source terms in contributing to the aeroacoustic sound generation mechanism was investigated [12]. Authors found that the contribution of the ALST in the mean flow direction and in the direction perpendicular to the mean flow are comparable and contribution due to shear-noise term is significant, over the self-noise term [12] at such high Reynolds numbers. Term $S_{0,i}$ in Eq. (8.12) does not contribute to generation of sound waves as it shows interaction between the mean quantities [12].

Acoustic source terms in Eqs. (8.10) and (8.11) are evaluated for the flow past a general elliptic cylinder at Re = 130. The eccentricity of the ellipse is prescribed by the aspect ratio (AR), which is unity for the case of a circular cylinder. Their variation in the domain has been shown in Figs. 8.2, 8.3 and 8.4 using appropriate contour levels. Figure 8.2 shows the variation of $S_{\text{self},1}$ and $S_{\text{self},2}$ terms, which accounts for the noise sources due to self-noise term as shown in Eq. (8.10) corresponding to instant, when the lift coefficient shows maximum amplitude. Figure shows that the $S_{\text{self},2}$ term is dominant over the $S_{\text{self},1}$ term. Arrangement of the noise sources in the wake region also differs for the self-noise terms. In case of $S_{\text{self},1}$ term, noise sources are arranged in a single row with alternative positive and negative structures, while for the noise sources corresponding to the $S_{\text{self},2}$ term are arranged in two parallel rows along the wake region. It is also observed that the $S_{\text{self},2}$ term displays structures stretched in the stream-wise direction as compared to $S_{\text{self},1}$ term for which structures are elongated normal to the flow direction as shown in Fig. 8.2.

Figure 8.3 shows contours for the $S_{\text{shear},1}$ and $S_{\text{shear},2}$ terms in Eq. (8.11) for flow past circular cylinder for a Re = 130. Contours have been shown corresponding to the instant at which the fluctuating lift coefficient is maximum. One observes that the

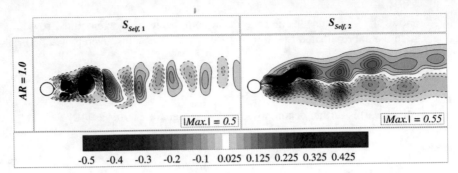

**Fig. 8.2** Contours for the $S_{\text{self},1}$ and $S_{\text{self},2}$ terms in Eq. (8.10) for flow past a circular cylinder for a Re = 130. Contours have been plotted corresponding to the instant at which the fluctuating lift coefficient is maximum

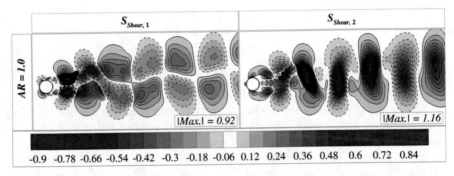

**Fig. 8.3** Contours for the $S_{\text{shear},1}$ and $S_{\text{shear},2}$ terms in Eq. (8.11) for flow past a circular cylinder for a Re = 130. Contours have been plotted corresponding to the instant at which the fluctuating lift coefficient is maximum

strength of $S_{\text{shear},1}$ and $S_{\text{shear},2}$ terms are comparable. It is also observed from Figs. 8.2 and 8.3 that the contribution of the shear-noise terms to the sound generation is almost two times larger compared to the self-noise terms. In case of shear-noise terms, $S_{\text{shear},1}$ term displays structures elongated in the stream-wise direction as compared to $S_{\text{shear},2}$ term where the contours display structures elongated perpendicular to the flow direction as shown in Fig. 8.3.

Figure 8.4 shows various frequencies and amplitudes associated with the lift and drag fluctuations for flow past a circular cylinder for a Re = 130. It is observed that the shear-noise terms predominantly fluctuate with the lift coefficient frequency while the self-noise terms predominantly fluctuate corresponding the drag coefficient frequency.

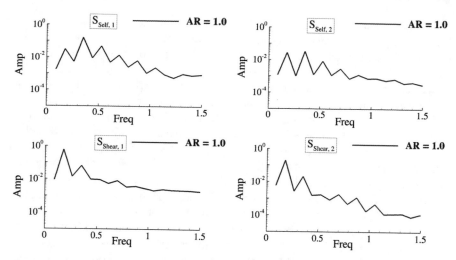

**Fig. 8.4** *FFT* of the time variation of self- and shear-noise source terms corresponding to the location $x = 1.0$ and $y = 0.5$ has been shown

### 8.1.3   DNS of CAA Problems

In a completely different approach, which does not involve any analogy, as well as splitting technique, the flow field and the acoustic field are directly computed. This methodology is termed as *DNS* approach [16, 17, 41, 42, 44–46, 49, 50]. In this approach, the Navier–Stokes equations are solved using highly space–time accurate difference schemes to accurately capture the flow field and acoustic field. This approach has ability to capture the sound generation and propagation processes in the near- and far-fields, without limiting itself to low Mach number, high Reynolds number flows, and compactness of the source region, unlike the previous two approaches. As the *DNS* involves significant computational cost, most of the *CAA* studies in the literature are based on the hybrid or acoustic/viscous splitting methods. Here, we are going to discuss flow-induced acoustic noise originating from the streamlined and bluff bodies using results based on the *DNS* approach.

## 8.2   Direct Simulation of Aeolian Tones Due to Flow Past a Bluff Body

Flow past bluff bodies, in particular flow past a circular cylinder, received significant attention after the famous work of Strouhal [92] on aeolian tones. This problem is interesting from the perspective of engineering applications. For example, engineers associated with building design, bridge design, or design of heat exchanger tubes deal with flow past cylindrical geometries. Such flows are usually unsteady and display wide band of spatial and temporal scales. In the past, number of studies have been devoted to characterize such flows using analytical, experimental, and computational approach. Work carried out by different researchers has been consolidated in the review articles [67, 102]. Along with finding out flow details, researchers have also shown keen interest to find out the sound generated by flow around a circular cylinder.

Using experimental approach, Strouhal found that the frequency $f$ of the sound radiated from flow past a cylinder of diameter $D$ is related to the free-stream flow velocity $U_\infty$ by a non-dimensional constant recognized after his name as Strouhal number St $= fD/U_\infty$. Connection between vortex shedding behind circular cylinder and aeolian tones was recognized in [33, 75]. In one of the significant contribution to the field of aeroacoustics, work of Lighthill [51] provided theoretical basis in terms of acoustic analogy to estimate sound generated aerodynamically [25–27, 51]. Using continuity and Navier–Stokes equations, Lighthill [51] obtained non-homogeneous acoustic wave equation such that the left-hand side of the equation describes wave propagation in a stationary or in a uniformly moving medium, which is triggered by a right-hand side source term. In addition to describing acoustic sources, the source term in the Lighthills equation also describes convection and refraction effects in the inhomogeneous acoustic domain [42]. As an extension of the Lighthill equation, the Ffowes Williams–Hawkings (FW–H) equation allows one to consider solid surfaces by using information of the unsteady surface pressure.

Researchers have studied the sound radiated by a flow past a cylinder using an acoustic analogy [19, 24, 70]. In another important contribution to the aeroacoustic analysis, Curle [19] extended Lighthill's acoustic analogy by including effects of solid boundaries on the generated sound field. In this work, Curle has shown the presence of monopole and dipole sources in the flow field arising out of effective mass and momentum injection into the flow via boundary terms. These sources have fundamentally greater acoustic efficiency than Lighthill's volume quadrupole [42]. Curle has shown that at the low Mach numbers, sound field is dominated by dipole sound, as compared to quadrupoles—right-hand side of the Eq. 8.1, and the sound associated with the drag force has double the frequency of that is associated with the lift force. Such dipole and quadrupole are shown in Fig. 6.5. In such sound field, the direction of the dipole is at right angles to the flow direction [19, 24, 33, 70].

### 8.2.1   Noise Radiated by Flow Past a Stationary Cylinder

Next, we discuss results for acoustic field set up due to laminar flow past a circular cylinder using *DNS* approach [42]. A schematic of flow past stationary circular cylinder fixed at origin is shown in Fig. 8.5. The flow can be initiated with free-stream velocity $U_\infty$ at Mach number $M = U_\infty/a_\infty$, with $a_\infty$ being the speed of sound at free-stream temperature $T_\infty$. The free- stream velocity is parallel to $x$-axis and the non-dimensional parameter, Reynolds number (Re) for the flow is defined as

**Fig. 8.5** Schematic of flow
past a stationary cylinder

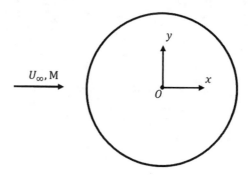

$$\mathrm{Re} = \frac{\rho_\infty U_\infty D}{\mu} \tag{8.13}$$

where $\rho_\infty$ is the free-stream density, $D$ is the diameter of cylinder and $\mu$ is dynamic viscosity of fluid under consideration.

Numerical computations are conducted at $M = 0.2$ Mach number, on O-grid topology domain having radius of $1500d$, where $d$ is the diameter of circular cylinder. The domain mainly comprises two zones, namely, acoustic zone and buffer zone, respectively. Ranges of acoustic zone and buffer zone are considered as $0.5d \le r \le 100d$ and $100d \le r \le 1500d$, respectively. The acoustic zone consists of refined grid point distribution, which is aimed at resolving flow and acoustic disturbances accurately, whereas the buffer zone contains coarsely distributed grid points in which the flow and acoustic disturbances dissipate gradually as they move outwards. It is ensured that there is no discontinuity in the grid point distribution at the interface between these two zones. As the computations are performed at low Mach number ($M = 0.2$) and for low Reynolds numbers ($100 \le Re \le 160$), the effect of temperature on transport properties is neglected [17, 42]. The wakes of flow past cylinder are found to be two dimensional for $Re \le 160$ [14, 102] and hence, two-dimensional formulation has been considered for computations. The values of Prandtl number and ratio of specific heats are considered as 0.7 and 1.4, respectively.

In the present study of flow past a stationary circular cylinder, 2D unsteady compressible fluid flow equations (i.e., Continuity, momentum, and energy equations), along with the equation of state, are solved numerically using $ORK5 - OCCD$ scheme at various Reynolds numbers ($Re = 100-160$). The governing partial differential equations are given by

$$\frac{\partial \rho}{\partial t} + \frac{\partial (\rho u)}{\partial x} + \frac{\partial (\rho v)}{\partial y} = 0 \tag{8.14}$$

$$\frac{\partial}{\partial t}(\rho u) + \frac{\partial}{\partial x}(\rho u^2 + p) + \frac{\partial}{\partial y}(\rho uv) = \frac{\partial}{\partial x}(\tau_{xx}) + \frac{\partial}{\partial y}(\tau_{xy}) \tag{8.15}$$

$$\frac{\partial}{\partial t}(\rho v) + \frac{\partial}{\partial x}(\rho u v) + \frac{\partial}{\partial y}(\rho v^2 + p) = \frac{\partial}{\partial x}(\tau_{xy}) + \frac{\partial}{\partial y}(\tau_{yy}) \qquad (8.16)$$

$$\frac{\partial}{\partial t}(\rho e_t) + \frac{\partial}{\partial x}(\rho u e_t + pu) + \frac{\partial}{\partial y}(\rho v e_t + pv) = \frac{\partial}{\partial x}(u\tau_{xx} + v\tau_{xy} - q_x)$$
$$+ \frac{\partial}{\partial y}(u\tau_{yx} + v\tau_{yy} - q_y) \qquad (8.17)$$

$$p = \rho RT \qquad (8.18)$$

The viscous stresses, $\tau_{xx}$, $\tau_{xy}$, $\tau_{yx}$, and $\tau_{yy}$ are given by

$$\tau_{xx} = 2\mu\left[\frac{\partial u}{\partial x} - \frac{1}{3}\left(\frac{\partial u}{\partial x} + \frac{\partial v}{\partial y}\right)\right] \qquad (8.19)$$

$$\tau_{xy} = \tau_{yx} = \mu\left[\frac{\partial u}{\partial x} + \frac{\partial v}{\partial y}\right] \qquad (8.20)$$

$$\tau_{yy} = 2\mu\left[\frac{\partial v}{\partial y} - \frac{1}{3}\left(\frac{\partial u}{\partial x} + \frac{\partial v}{\partial y}\right)\right] \qquad (8.21)$$

The heat fluxes, $q_x$ and $q_y$ are given by

$$q_x = -k_t\frac{\partial T}{\partial x} \qquad (8.22)$$

$$q_y = -k_t\frac{\partial T}{\partial y} \qquad (8.23)$$

where $\rho$ is the density, $u$ and $v$ are $x$- and $y$-components of fluid velocity, respectively; $p$ is the pressure, $T$ is the absolute temperature, $R$ is the universal gas constant, $\mu$ is the dynamic viscosity of fluid, $e_t$ is the total energy of the fluid per unit mass, and $k_t$ is thermal conductivity of the fluid.

As the present simulations are based on finite difference compact schemes, the governing equations are transformed to computational domain, where uniform distribution of grid points are present, and subsequently these equations are non-dimensionalized. The convective derivatives and grid derivatives are discretized using OCCD scheme, and the viscous derivatives are evaluated using second- order central difference scheme. The temporal derivatives are discretized using second-order, optimized five stage Runge–Kutta method.

During the computations, it is observed that periodic shedding is obtained before a non-dimensional time, $t = 150$ for all Reynolds numbers. The time averaging procedure is implemented from time $t = 150$ to $t = 350$, and the important flow parameters like time-averaged coefficient of pressure $(C_p)$, time-averaged drag coefficient $(C_D)$ and Strouhal frequency $(f_o)$ are evaluated. Figure 8.6a, b show the variation of average drag coefficient and Strouhal frequency with Re, respectively, and the computed results are in good agreement with the values of Henderson [38]. Thus, the present scheme is capable of resolving the flow field with good accuracy.

Figure 8.6c shows the variation of time-averaged $C_p$ on the upper surface of cylinder. The behavior of $C_p$ obtained using the present scheme is similar to the behavior of $C_p$ that was obtained in [42]. The bottom peak of $C_p$ curve decreases with increase in Reynolds number, as the fluid accelerates more on the surface with increase in Re, and thereby reducing the pressure. As the simulations are performed using higher order accurate and high resolution combined compact scheme along with RK5 time integration, the obtained flow fields also have better accuracies and resolutions.

The periodic shedding of vortices triggers the generation of acoustic disturbances from the surface of cylinder, and these disturbances propagate inside the domain. Acoustic pressure fluctuation $(p')$ at any instant inside the domain can be accurately evaluated from the obtained flow field. The instantaneous pressure fluctuation is given by

$$p'(t, x, y) = p(t, x, y) - p_m(x, y) \tag{8.24}$$

where $(p_m)$ is the time-averaged pressure and $p$ is instantaneous pressure.

In the present study, pressure fluctuations are evaluated from time, $t = 150$ to 350, at all considered Reynolds numbers. Figure 8.7 represents the contours of pressure fluctuations around the cylinder for various Reynolds numbers at time, $t = 250$. The pressure fluctuation field contains both positive and negative regions, which are distinguished using solid and dashed lines, respectively. The obtained pressure field is in accordance with that of Inoue et al. [42]. The variation of pressure fluctuation with distance $r$ ranging from the surface of cylinder $(r = 0)$ to a point inside the domain where $r = 100$ for Re = 150 at an angle $\theta = 90°$ are shown in Fig. 8.8a.

It is observed that the peak pressures decay with distance away from the surface of cylinder. The decay of pressure peaks with distance in a cycle are represented in Fig. 8.8b, and the evaluated behavior of decay of pressure peaks is in good agreement with the theoretical behavior $(p' \propto r^{-1/2})$ given by Landau and Lifshitz [48]. Figure 8.9 shows the decay of pressure peaks measured at $\theta = 90°$ for Re = 100, 120, 130, 160.

Information about the directivity associated with the sound field is equally important. One can derive the information about the directivity associated with the sound field by performing the following steps.

1. Obtain pressure variation inside the domain for multiple instants.
2. Evaluate the mean (time-averaged) and disturbance pressure field as discussed before.

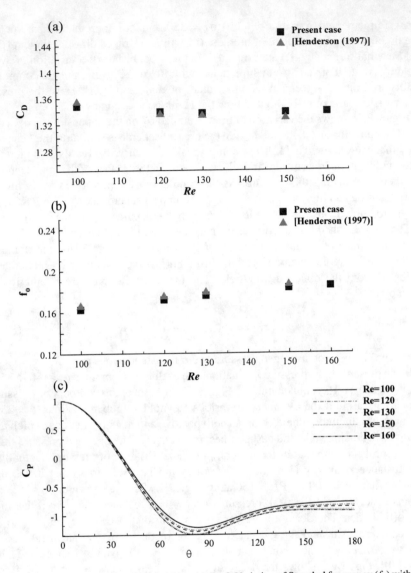

**Fig. 8.6** **a** Variation of drag coefficient ($C_D$) with Re, **b** Variation of Strouhal frequency ($f_o$) with Re, **c** Variation of coefficient of pressure ($C_p$) on the surface of cylinder. Results have been reproduced from [57]

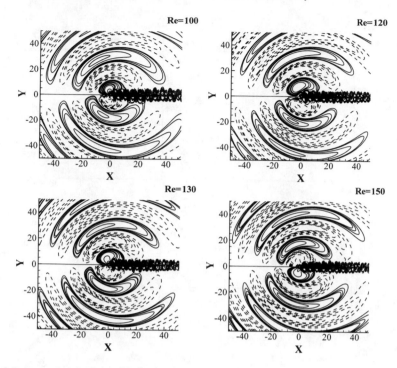

**Fig. 8.7** Pressure fluctuation field around the cylinder for various Re at time, $t = 250$. Solid line for positive pressure fluctuation and dashed line for negative pressure fluctuation. Results have been reproduced from [57]

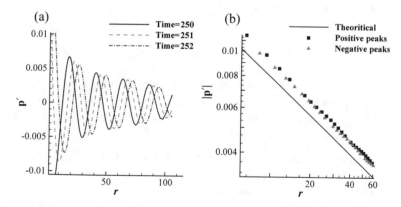

**Fig. 8.8  a** Variation of pressure fluctuations with distance $r$ at $\theta = 90$ for Re $= 150$, **b** The decay of pressure peaks with distance $r$ for Re $= 150$. Solid line for $|p'| \propto r^{-1/2}$. Results have been reproduced from [57]

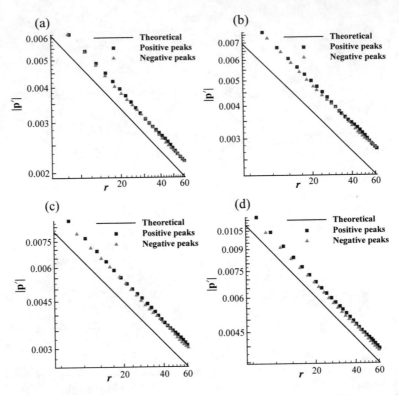

**Fig. 8.9** The decay of pressure peaks with distance $r$. **a** Re $= 100$, **b** Re $= 120$, **c** Re $= 130$, **d** Re $= 160$. Results have been reproduced from [57]

3. For the obtained fluctuating field, estimate the root mean square (*RMS*) values of pressure fluctuations inside the domain.
4. Directivity is usually shown as variation of *RMS* pressure at a constant distance from the source (in this case origin of the cylinder) in all directions. This is obtained by plotting *RMS* values of pressure fluctuations over $360°$ at a constant distance from the center of the cylinder.

The root mean square (*RMS*) values of pressure fluctuations are evaluated for a given Re, in order to obtain the directivity patterns of sound field. Figure 8.10 shows the polar plots of *RMS* values of pressure fluctuations measured at a distance of $r = 75$ from the center of cylinder for various Re, and the dipolar nature of sound field is observed. The obtained angles of directivity for Re $= 100, 120, 130, 150, 160$ are $\theta = 77.04°, 79.72°, 81.36°, 83.52°, 84.96°$, respectively. It is evident that the angle of directivity is increasing with increase in Re.

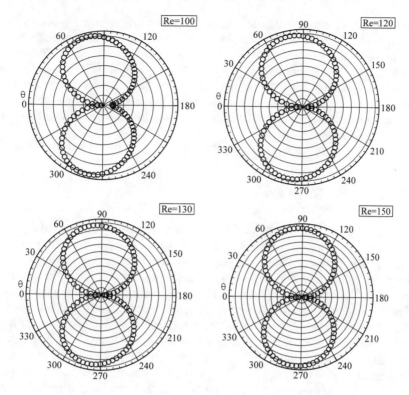

**Fig. 8.10** Polar plots of RMS values of pressure fluctuations measured at $r = 75$ for various Re. Results have been reproduced from [57]

### 8.2.1.1 Decomposition of Pressure Disturbance Field

The disturbance pressure field can be represented as [42]

$$
\begin{aligned}
p'(r, \theta, t) = A_0(r, t) \\
+ A_1(r, t) \cos(\theta) + B_1(r, t) \sin(\theta) \\
+ A_2(r, t) \cos(2\theta) + B_2(r, t) \sin(2\theta) + \cdots,
\end{aligned} \tag{8.25}
$$

where

$$
A_0 = \frac{1}{2\pi} \int_0^{2\pi} p' d\theta, \tag{8.26a}
$$

$$
A_n = \frac{1}{\pi} \int_0^{2\pi} p' \cos(n\theta) d\theta, \tag{8.26b}
$$

$$
B_n = \frac{1}{\pi} \int_0^{2\pi} p' \sin(n\theta) d\theta. \tag{8.26c}
$$

In Eq. (8.25), the monopole, dipole, and quadrapole are represented by $A_0$, $(A_1, B_1)$ and $(A_2, B_2)$, respectively. The variation of these coefficients with time measured at a distance of $r = 75$ are shown in Fig. 8.11a. The respective pressure disturbance fields represented by monopole $(A_0)$, dipole $(A_1 \cos(\theta), B_1 \sin(\theta))$, and quadrapole $(A_2 \cos 2\theta), B_2 \sin 2\theta)$ are shown in Fig. 8.11b–f respectively. It is observed that the dipole $A_1 \cos(\theta)$ resembles drag dipole and where as the dipole $B_1 \sin(\theta)$ resembles the lift dipole. However, the Doppler effect is present in these decomposed pressure fields due to the mean motion [42].

## 8.2.2  Flow Past a Wedge

Here, computations of unsteady laminar flow over a two-dimensional wedge have been reported, relating variation of flow field properties to that of the generated sound field. Results for aeolian tones generation due to laminar flow past a wedge at different angle of attacks were first published in [56] and are reproduced here with the permission of *AIP* Publishing. Figure 8.12 shows a schematic diagram for uniform flow past a two-dimensional (2D) wedge at an incidence angle $\alpha$. Length of each side has been considered as $b = 1$ m. Sides of the wedge connect to each other with a small fillet of finite radius $r_c = 0.005b$. A zoomed view of one of the corners, highlighting fillet portion has been shown in the top right position of Fig. 8.12. Computations are performed for 13 different angle of incidence cases with $\alpha$ varying from $0°$ to $60°$ in steps of $5°$. Laminar flow past a wedge has been studied for a free- stream Reynolds number of Re $= 100$ and Mach number of $M = 0.2$.

Similar to other cases involving flow past bluff bodies, periodic/multi-periodic vortex shedding has been observed for all angles of incidence. Mean as well as disturbance components of flow properties have been evaluated and analyzed. For such exercise, mean flow field variables are evaluated by conducting time averaging procedure over more than 15 cycles once a steady periodic state has been reached.

Instantaneous vorticity contours for flow past a wedge at different angles of incidence cases are shown in Fig. 8.13. Here, solid and dashed lines represent positive and negative vorticity levels, respectively. Changes in instantaneous vorticity contours for different angles of incidence $(\alpha)$ have been shown. Same contour levels have been maintained for all the cases. Strength of the shed vortices and length of the wake region increases with increase in angle of incidence. For $\alpha = 0°$ and $\alpha = 10°$ cases, vortices in the wake region quickly loose their strength and dissipate. In contrast for higher $\alpha$ ($\alpha > 40°$), vortices in the far wake region display more prominent nature.

Variations of time-averaged drag coefficient, root mean square (*RMS*) value of the lift coefficient and Strouhal number as a function of $\alpha$ have been shown in Fig. 8.14. It is observed that the *RMS* component of the lift coefficient increases with increase in $\alpha$ till $\alpha = 30°$. Subsequently the *RMS* component of the lift coefficient decreases till $\alpha = 60°$. Similar change in behavior for variation in the mean drag coefficient is also observed at $\alpha = 30°$ where the mean drag attains lowest value among all angle of incidence cases. This makes $\alpha = 30°$ a very special case as it has minimum mean

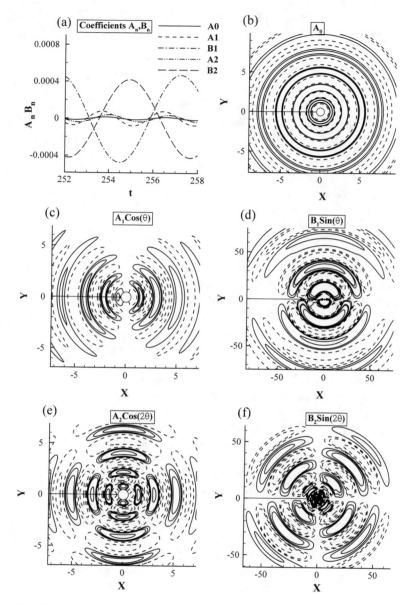

**Fig. 8.11  a** Variation of coefficients $(A_n, B_n)$ of Eq. (8.25) with time at $r = 75$, **b–f** shows the contours of pressure disturbance fields denoted by monopole, dipole, and quadrapole, respectively

drag and highest *RMS* value for the lift coefficient variations. Thus, we have analyzed cases $\alpha = 0°$, $\alpha = 30°$, and $\alpha = 60°$ in detail. Variation of the Strouhal number with $\alpha$ displays reduction in Strouhal number as $\alpha$ increases. Present computed results

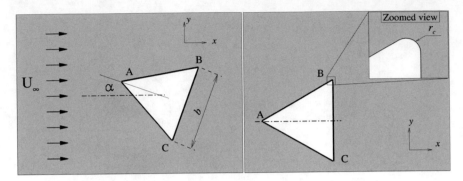

**Fig. 8.12** Schematic of free-stream flow over a two-dimensional wedge at an incidence angle $\alpha$ and zoomed view of vertex corner approximated with a fillet. Reproduced from [56], with the permission of *AIP* Publishing

match very well with the results reported by Bao et al. [4]. Results in Fig. 8.14 provide necessary validation for flow past a wedge cases.

Variations of instantaneous vorticity contours for $\alpha = 0°$ case have been shown for one complete lift cycle in the first column of Fig. 8.15. Clockwise ($CW$) and counterclockwise ($CCW$) rotating vortices are denoted using dark and light contour levels, respectively. Right and central columns represent instantaneous drag ($C_d$) and lift ($C_l$) coefficients variation with time, respectively. Figure 8.15a shows a $CW$ vortex behind a wedge which is about to detach from top corner of the wedge and get convected in the downstream direction. At this instant, drag and lift coefficients are closer to their maximum values as marked by a filled circular dot in the respective plots. Meanwhile, a $CCW$ vortex starts forming close to bottom corner of the wedge. Figures 8.15b, c show formation and subsequent growth of $CCW$ vortex accompanied by decrease in the lift and drag coefficients. Lift coefficient reaches its minimum value in frame (d), as $CCW$ vortex becomes dominant. Subsequently, in frames (e)–(g), one observes detachment of $CCW$ vortex in the wake region and growth of $CW$ vortex from the top corner of the wedge. This leads to increase in lift coefficient. One observes that during one complete cycle of lift coefficient variation, two vortices are shed; each from the top and bottom corners of the wedge causing the drag coefficient to vary twice as fast as compared to the lift coefficient. This results in the presence of the lift and the drag dipoles in the generated sound field with frequency of sound waves due to drag dipole is twice higher as compared to those due to lift dipole for flow past stationary circular cylinder [42].

The case for $\alpha = 30°$ has the least mean drag and highest $RMS$ value for the lift coefficient among all the angle of incidence cases. Thus for this interesting case, we have shown flow evolution behind the wedge over a full cycle for variation of lift coefficient in Fig. 8.16. For this case, top side of the wedge is parallel to the flow direction. We have shown pressure contours at indicated instants in the first column of Fig. 8.16, while corresponding variations of instantaneous lift ($C_l$) and drag ($C_d$) coefficients are shown in the middle and right columns, respectively. At

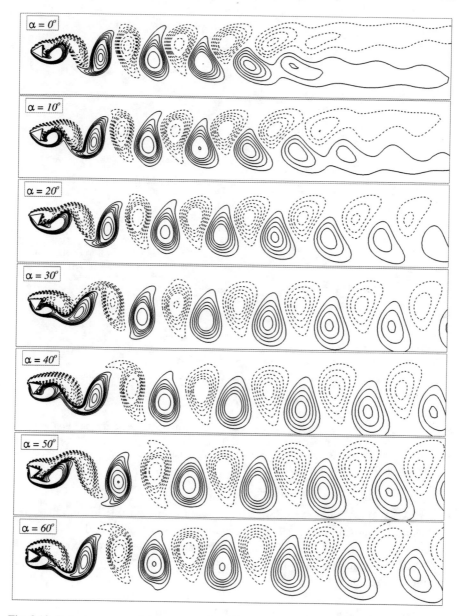

**Fig. 8.13** Instantaneous vorticity contours for different angles of incidence have been shown at time $t = 320$. Reproduced from [56], with the permission of *AIP* Publishing

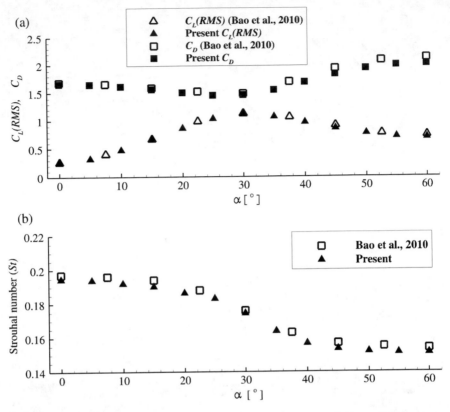

**Fig. 8.14  a** *RMS* lift ($C_{Lift}$(RMS)) and mean drag ($C_{Drag}$(mean)) coefficient variation with the angle of incidence ($\alpha$). Solid and open symbols represent present and previously computed results, respectively. **b** Variation of Strouhal number with incidence angle has been shown. Reproduced from [56], with the permission of *AIP* Publishing

time $t = 318.8$, one observes minimum drag coefficient. At this instant, detached *CW* vortex from the top surface convects away in the wake region and the formation of new *CCW* vortex takes place at the bottom corner, whose growth continues till $t = 320.6$. This results in rapid increase in the drag coefficient due to presence of large low-pressure region behind the wedge. Subsequently, the vortex gets detached from the bottom corner, convects in the downstream direction, and the drag coefficient decreases. This phenomenon gets accompanied with growth of *CW* vortex on the top right corner of the wedge. However, this *CW* vortex on the top right corner has lower strength as compared to the *CCW* vortex formed on the bottom corner of the wedge at $t = 320.6$ and does not contribute much to increase in the drag coefficient. The only change in variation of drag coefficient is noted at $t = 322.6$. As the detached vortices from top and bottom corners of the wedge continue to move in the downstream direction, drag value reduces till a new vortex starts forming at the bottom corner of the wedge at $t = 324.5$. Thus the drag and lift coefficient variations

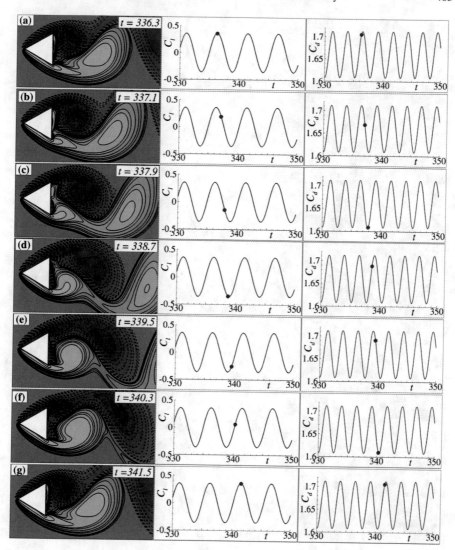

**Fig. 8.15** Variation in vorticity contours obtained for one complete lift cycle at $\alpha = 0°$. Dark and light colors in the left column show $CW$ and $CCW$ vortices, respectively. Central and right columns indicate the transient variation of lift and drag coefficients, respectively. The filled dot in the central and right columns represents the lift and drag values corresponding to the indicated instant of the flow field shown in left column. Reproduced from [56], with the permission of *AIP* Publishing

are mainly governed by vortex shedding from the bottom corner alone in contrast to the $\alpha = 0°$ and $\alpha = 60°$ cases. Due to formation of a low strength vortex at the top edge of the wedge, the mean drag coefficient value is significantly lower, as compared to other cases. It is interesting to note the presence of flow separation region on the top left corner of the wedge, which results in a low-pressure region close to it, as

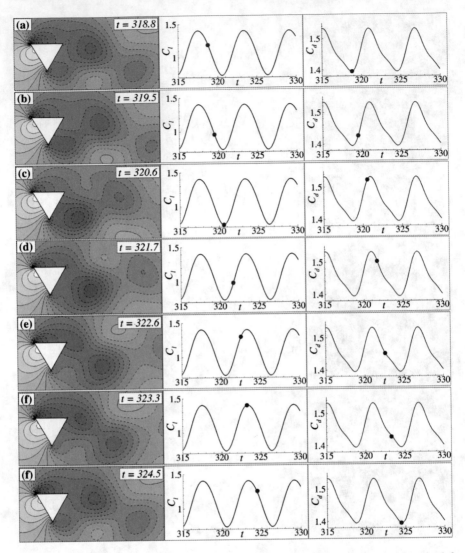

**Fig. 8.16** Pressure contours at $\alpha = 30°$ for a complete lift cycle have been shown in the left column. Dark and light colors in the left column images represent low- and high- pressure regions, respectively. Central and right columns indicate the transient variation of lift and drag coefficients, respectively. The filled dot in the central and right columns represents the lift and drag values corresponding to the indicated instant of the flow field shown in left column. Reproduced from [56], with the permission of *AIP* Publishing

observed at all instants. This leads to positive mean lift coefficient for $\alpha = 30°$ case. Thus we summarize following four important observations for $\alpha = 30°$ case which are due to formation of dominant vortex from the bottom corner of the wedge.

- The top edge of the wedge is parallel to the flow direction. Thus the flow becomes highly asymmetric and flow parameter variations are mainly governed by vortex shedding from the bottom corner.
- In contrast to $\alpha = 0°$ case, one observes same period of variation for the lift and the drag coefficients. There are multiple frequencies present in the drag coefficient variation for $\alpha = 30°$ case with the dominant frequency same as that of lift coefficient variation frequency.
- Mean drag value for $\alpha = 30°$ case is less as a dominant vortex sheds from the bottom corner of the wedge only. Vortex shedding from the top right corner of the wedge is less significant as the strength of the corresponding vortex is small.
- Mean lift coefficient for $\alpha = 30°$ case is positive as a low-pressure region is always present close to the top left corner due to flow separation.

### 8.2.2.1 Fluctuating Forces Acting on the Wedge at Different Angles of Attack

Researchers have related fluctuations in the lift and drag coefficient variations with time to the amplitude and frequency of sound waves triggered due to vortex shedding from the bluff bodies [42, 53, 59]. Thus we have evaluated fluctuations in the lift $(c_{l'})$ and the drag $(c_{d'})$ coefficients for flow past a wedge by subtracting the mean lift coefficient and drag coefficient values from the instantaneous one, respectively. Instantaneous variation of fluctuations in the lift $(c_{l'})$ and drag $(c_{d'})$ coefficients has been shown in Fig. 8.17a for flow past a wedge at $\alpha = 0°$. The amplitudes of $c_{l'}$ and $c_{d'}$ are found to be 0.361 and 0.051, respectively, and the frequency of drag coefficient variation is twice that of lift coefficient variation. As amplitude of $c_{l'}$ has been observed to be almost seven times greater than that of $c_{d'}$, and the generated

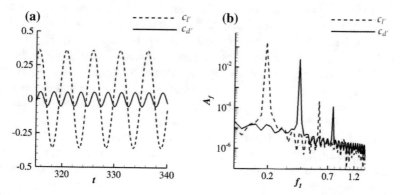

**Fig. 8.17 a** Variation of fluctuations in lift $(c_{l'})$ and drag $(c_{d'})$ coefficients with time $t$, **b** Fast Fourier transform (*FFT*) of $c_{l'}$ and $c_{d'}$ variation with time has been shown by plotting Fourier amplitude $A_f$ as a function of frequency $f_1$. Reproduced from [56], with the permission of *AIP* Publishing

pressure disturbances due to flow past a wedge are dominated by fluctuations in the lift coefficient. Later on, this fact will be made evident from the proper orthogonal decomposition (POD) of disturbance pressure field. Fast Fourier transform ($FFT$) of time varying fluctuations in lift and drag coefficients have been obtained and shown in Fig. 8.17b, by plotting variation of the Fourier amplitude ($A_f$) as a function of non-dimensional frequency ($f_1$). The first peak for coefficient of lift variation is obtained at the shedding frequency $f$ and that for drag coefficient variation at $2f$. At an incidence angle $\alpha = 0°$, the wedge is symmetric for incoming free stream which involves shedding of equal strength vortices from the top and the bottom corners of the wedge and consequently this leads to single dominant frequency peak of $f$ for the lift coefficient and $2f$ for the drag coefficient. Similar behavior in frequency peaks has been observed for the case with an incidence angle $\alpha = 60°$ will be shown subsequently in Fig. 8.19.

Time variation of $c_{l'}$ and $c_{d'}$ are evaluated for all other incidence angles and are represented in Fig. 8.18. The amplitudes of $c_{l'}$ are found to be a almost constant value, close to 0.35 for incidence angles of $\alpha \leq 30°$, whereas it varies significantly from 0.35 to 0.975 as $\alpha$ increases from $30°$ to $60°$. Amplitude of $c_{l'}$ is highest for an incidence angle $\alpha = 60°$ due to orientation of wedge such that one of its surfaces is exactly normal to incoming free stream which leads to shedding of high strength vortices from the top and bottom corners of the wedge resulting in the maximum amplitude for $c_{l'}$ fluctuations. On the other hand, amplitudes of $c_{d'}$ are observed to be much lower than that of $c_{l'}$, irrespective of incidence angle. Transient variation in $c_{d'}$ is significantly altered with incidence angle due to unequal strength of vortices shed from either side of the wedge, and consequently, presence of multiple frequencies are observed in time variation of $c_{d'}$ for all incidence angles, $\alpha$, except for $\alpha = 0°$ and $\alpha = 60°$.

Figure 8.19 represents log–log plots obtained by evaluating the $FFT$ from time variation of $c_{l'}$ and $c_{d'}$ for all incidence angles except the case $\alpha = 0°$. It is observed that first peak is obtained at the shedding frequency $f$ and other peaks are found at super harmonics of $f$ for fluctuations of lift and drag coefficients at all incidence angles except at $\alpha = 60°$. At $\alpha = 60°$, the first and secondary peaks of $c_{l'}$ are visible at $f$ and $3f$, whereas that of $c_{d'}$ are obtained at $2f$ and $4f$, respectively. For incidence angles $5° \leq \alpha \leq 55°$, the first two peaks show significant contribution for the $FFT$ of $c_{d'}$, whereas the $FFT$ of $c_{l'}$ is mostly contributed by the first peak observed at shedding frequency. As the generated sound waves depend on fluctuations in the lift and drag coefficients, it is important for us to find out the way in which $c_{l'}$ and $c_{d'}$ vary together for different angle of incidence cases.

Consider the variation of fluctuating ($c_{l'}$ and $c_{d'}$) over one vortex-shedding cycle, for all incidence angles, as shown in Fig. 8.20. Each trajectory represents fluctuation of lift and drag forces as a load vector acting on the surface of a wedge over one vortex-shedding cycle. As explained before, fluctuating aerodynamic forces have been significantly dominated by the magnitudes of fluctuations in the $c_{l'}$ rather than fluctuations in the $c_{d'}$. This aspect is useful in determining contribution of lift and drag dipole to the resultant amplitude of sound field variation. Frequency of generated sound waves depends on the frequency with which fluctuations in the lift and drag

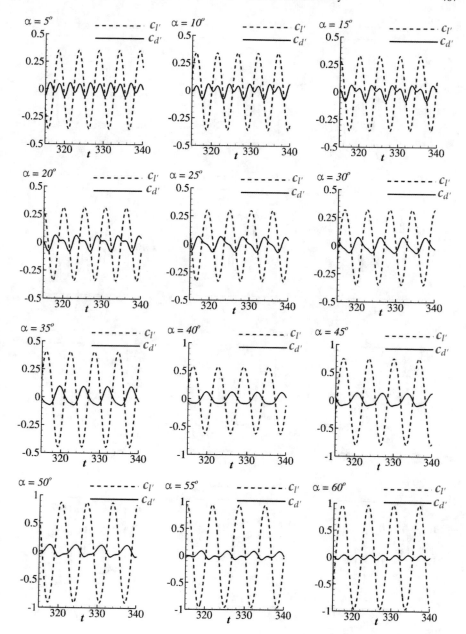

**Fig. 8.18** Variation of fluctuations in lift ($c_{l'}$) and drag ($c_{d'}$) coefficients with time $t$ for all incidence angles except for the case $\alpha = 0°$. Reproduced from [56], with the permission of *AIP* Publishing

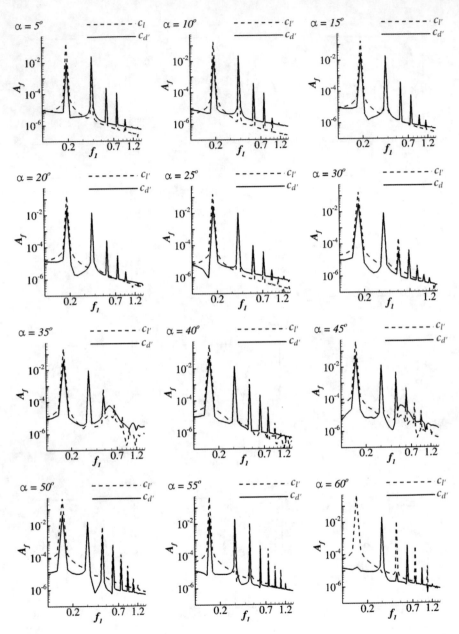

**Fig. 8.19** Fourier amplitude $A_f$ as a function of frequency $f_1$ is obtained by evaluating Fast Fourier transform (*FFT*) for time variation of $c_{l'}$ and $c_{d'}$ for all incidence angles except for the case $\alpha = 0°$. Reproduced from [56], with the permission of *AIP* Publishing

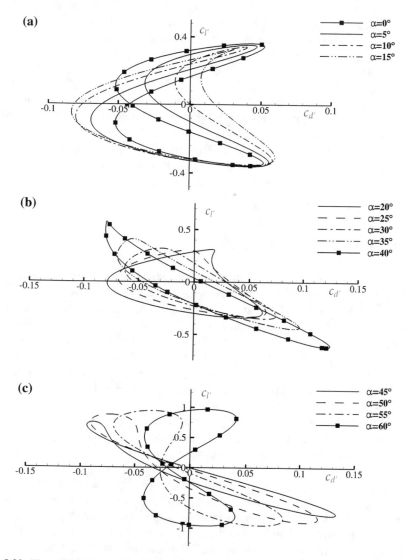

**Fig. 8.20** Fluctuations of aerodynamic loads over one vortex-shedding cycle have been shown for various incidence angles. Reproduced from [56], with the permission of *AIP* Publishing

coefficients vary. For this purpose, we have categorized the sound field in to three broad categories. In the first category, the drag coefficient varies twice as fast as compared to the lift coefficient at low angle of incidence cases $0 \leq \alpha \leq 15°$, and one observes the cross over of these two fluctuating coefficients in the top frame of the Fig. 8.20. In contrast, in the second category, with angle of incidence ranging from $20 \leq \alpha \leq 40°$, each trajectory has not crossed itself due to the fact that frequency of both lift and drag fluctuations are almost identical and dominated by the first frequency peak corresponding to the shedding frequency. Thus for this category, resultant sound field will be dominated by shedding frequency alone. In the third category, with angle of incidence varying from $45 \le \alpha \le 60°$, one again observes cross over of the fluctuating lift and drag coefficients indicating importance of both lift and drag dipoles in defining sound field. However for these high angle of incidence cases, wedge displays higher bluff body nature compared to that of low angle of incidence cases. At $\alpha = 60°$, maximum amplitude of lift fluctuation is almost double as compared to that at $\alpha = 0°$, and this has been reflected well in the directivity patterns of sound field explained later. Though there has been variations in $c_{d'}$, the amplitudes of $c_{l'}$ dominated the fluctuating load vector and consequently the directivity plots and acoustic fields are supposed to be significantly affected by lift dipoles for all incidence angles, as explained next.

### 8.2.2.2  Disturbance Pressure Fields

Similar to the case of circular cylinder, the fluctuations in aerodynamic forces due to vortex shedding resulted in triggering disturbance pressure pulses from top and bottom corners of wedge. These disturbances propagate in all directions, as time progresses. Figure 8.21 represents transient evolution and propagation of disturbance pressure pulses from the top and bottom surfaces of the wedge, at an incidence angle $\alpha = 0°$, over one vortex shedding cycle. In this figure, positive and negative values of pressure disturbances are denoted by solid and dashed lines, respectively. Although, the fluctuations exist in both lift and drag forces, it is observed that pressure pulses seem to be more inclined toward vertical direction. This is due to dominating amplitudes of lift fluctuations over drag fluctuations as explained before.

Figure 8.22 represents the variation in acoustic field behavior obtained at time $t = 322$ for various incidence angles ranging from $\alpha = 10°$ to $\alpha = 60°$. For all incidence angles shown in Fig. 8.22, the contour levels range from $-0.1$ to $0.1$, with 10 equi-spaced sub-levels. It has been observed that the difference between maximum and minimum fluctuating pressure levels ($\Delta p'_{max}$) remains similar till $\alpha \le 30°$, and thereafter increases significantly with increase in incidence angle till $\alpha = 60°$. An exactly similar kind of variation in lift amplitudes is observed, shown in Fig. 8.18, and also in Fig. 8.23a. Although it can be concluded that acoustic fields and their variation with incidence angles are significantly dominated by fluctuations in lift amplitudes, a better picture of the same has been explained quantitatively using directivity patterns based on $RMS$ values of pressure fluctuations as shown in Fig. 8.23.

As it has been evident that amplitudes of lift coefficient fluctuations are greater than that of drag coefficient, difference between the peak to peak amplitude of lift coefficient ($C_{L_A}$) is considered for explaining directivity patterns. Parameter $C_{L_A}$ is plotted as a function of incidence angle ($\alpha$), as shown in Fig. 8.23a. With increase in incidence angle from $0°$ to $30°$, slight reduction in magnitude of $C_{L_A}$ is observed. Subsequently at higher $\alpha$, $C_{L_A}$ value increases significantly with a maximum at $\alpha = 60°$.

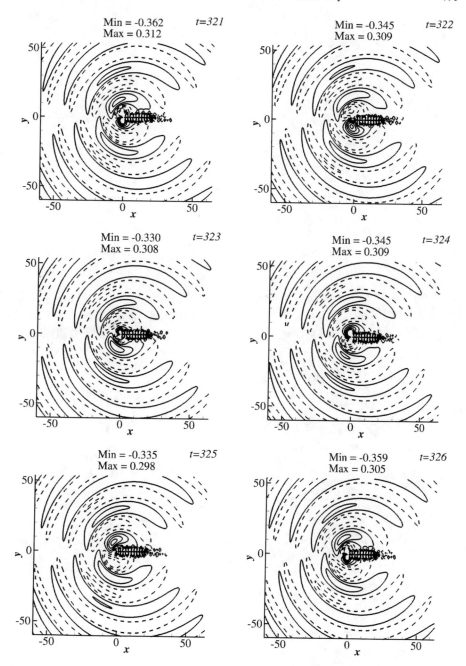

**Fig. 8.21** Evolution and propagation of disturbance pressure pulses from surface of wedge into the domain at $\alpha = 0°$ over one shedding cycle. Solid and dashed lines denote positive and negative pressure fluctuations, respectively. Contour levels of disturbance pressure ranges from $-0.1$ to $0.1$ with 10 equi-spaced sub-levels. Reproduced from [56], with the permission of *AIP* Publishing

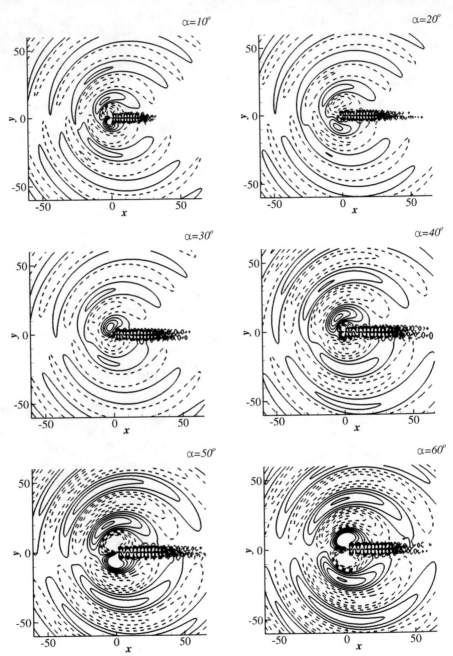

**Fig. 8.22** Variation of disturbance field in terms of pressure disturbance ($p'$) with incidence angle $\alpha$ at time $t = 322$. Solid and dashed lines represent positive and negative pressure fluctuations respectively. Contour levels of disturbance pressure ranges from $-0.1$ to $0.1$ with 10 equi-spaced sub-levels. Reproduced from [56], with the permission of *AIP* Publishing

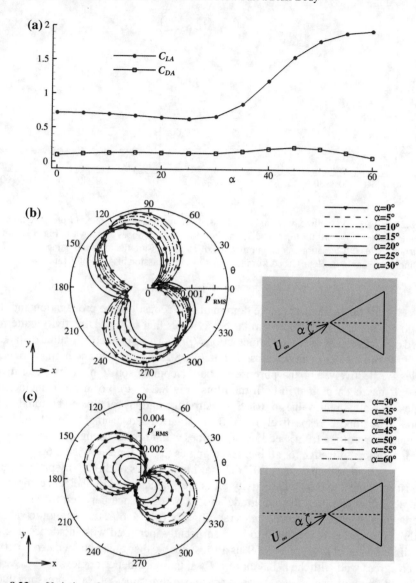

**Fig. 8.23** **a** Variation of peak to peak amplitude of lift coefficient ($L_A$) and drag coefficient ($D_A$) as a function of incidence angle ($\alpha$) has been shown; **b** and **c** Behavior of directivity patterns based on *RMS* values of pressure fluctuations at a radial distance $r = 75$ for various incidence angles. Reproduced from [56], with the permission of *AIP* Publishing

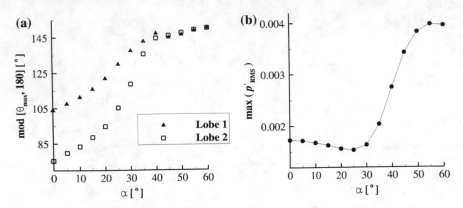

**Fig. 8.24  a** Variation in directivity angle $\theta_{max}$ based on maximum value of *RMS* pressure fluctuations evaluated at $r = 70$ for top (Lobe 1) and bottom (Lobe 2) lobes as shown in Fig. 8.23 with incidence angle $\alpha$. **b** Variation in maximum value of *RMS* pressure fluctuation obtained at $r = 70$ with incidence angle $\alpha$. Reproduced from [56], with the permission of *AIP* Publishing

As the lift fluctuations are more dominant, the acoustic noise propagation and its directivity patterns must show similar behavior as that of $C_{L_A}$. Directivity patterns based on RMS values of pressure fluctuations ($p'_{RMS}$) are obtained at a radial distance $r = 75$ for all incidence angles, as shown in Fig. 8.23b, c, for low and high incidence angles, respectively. For the purpose of better comparison, directivity pattern of $\alpha = 30°$ has been plotted in both the plots (Fig. 8.23b, c). For a given incidence angle $\alpha$, the maximum values of RMS pressure fluctuations for top and bottom lobes (Lobe-1 and Lobe-2, respectively) are different due to unequal amplitudes of pressure fluctuations resulting from shedding of vortices with unequal strengths from top and bottom corners of wedge for all incidence angles except for $\alpha = 0°$ and $60°$.

At low incidence angle $\alpha \leq 30°$, the maximum values of directivity patterns are almost equivalent to 0.002. Directivity plots are characterized by incidence angle $\alpha$, Doppler effect and both the amplitude and phase difference between lift and drag fluctuations. At high incidence angles ranging from $30°$ to $60°$, the maximum values of directivity patterns are found to be significantly increased with incidence angle. Maximum value of $p'_{RMS}$ at $\alpha = 60°$ is almost double the value obtained at $\alpha = 30°$, which agrees well with the behavior of $CL_A$ at these incidence angles.

The directivity angles ($\theta_{max}$) based on maximum values of *RMS* pressure fluctuations at $r = 75$ are obtained for both lobes (Lobe-1 and Lobe-2). The parameter  mod [$\theta_{max}$, 180] is represented as a function of incidence angle $\alpha$, shown in Fig. 8.24a. The difference between angles obtained for Lobe-1 and Lobe-2 is highest at $\alpha = 0°$ and this difference reduces with increase in $\alpha$ with zero difference at $\alpha = 60°$. Also, maximum value of *RMS* pressure fluctuation obtained at $r = 75$ is plotted as a function of incidence angle $\alpha$, as shown in Fig. 8.24b, and its behavior is exactly similar to that of variation of lift amplitude $C_{L_A}$ with incidence angle $\alpha$, as shown in Fig. 8.23.

It is clear that generated sound fields around wedge are dominated by fluctuating lift forces. Present observation is in accordance with the similar observations reported for flow past bluff bodies [42, 53, 59].

### 8.2.2.3 Proper Orthogonal Decomposition (POD) of Disturbance Pressure Field

While conducting research related to complex flow problems using either experimental or numerical approach, identification of coherent structures in the flow is important. This task involves construction of an empirical mode base which captures dominant flow structures. We have obtained information about various time varying spatial modes in disturbance pressure fields using *POD* technique, as explained in previous studies involving POD implementation [40, 81, 85].

Reported computations are based on the space–time accurate solutions of unsteady compressible Navier–Stokes equations and no flow modeling is adopted. The *POD* analysis has been carried out using the method of snapshots [88]. In the present work, the information about *POD* modes has been derived in a similar way as mentioned in the literature [77, 81, 85]. Here, the disturbance pressure field is given as

$$p'(x, y, t) = \sum_{k=1}^{N} a_k(t)\, \phi_k(x, y) \qquad (8.27)$$

where $N$ is number of solution files (snapshots) as considered in the method of snapshots. In POD, one tries to represent the stochastic dynamics by low dimensional deterministic projections [47] by posing this as a variational calculus problem with the help of correlation matrix of the variable. This leads to an eigenvalue problem. When this is done by minimizing the number of modes with respect to kinetic energy representation [66] the minimal representation produces eigenvalues whose cumulative sum indicates the amount of translational energy content. This approach has problems with respect to pressure velocity coupling issues and most often this is neglected. In contrast when one uses vorticity transport equation as a governing equation and poses the problem in terms of enstrophy, no such problems arise [77, 81, 85]. In problems of acoustics, one is confronted with representing fluctuating pressure. Thus, in the presented approach we advocate the use of mean square variation of pressure which is optimally represented by POD modes. Time-dependent *POD* amplitudes are denoted by $a_k(t)$ and $\phi_k(x, y)$ represent *POD* eigenfunctions of a covariance matrix $R_{i,j}$ which is defined as

$$R_{i,j} = \frac{1}{N} \int \int_S p'(x, y, t_i)\, p'(x, y, t_j)\, dS \qquad (8.28)$$

with $i, j = 1, 2, \ldots N$ are defined over all the snapshots for all the points in the domain $S$. Linear combination of the $a_k(t)$ and $\phi_k(x, y)$ provides instantaneous disturbance pressure field triggered by flow past a wedge. The *POD* analysis helps to decompose disturbance pressure field based on significance or amount of "energy" carried by different modes. This is an important advantage as one is able to reduce the order of the system by considering few important dominant modes contributing to most of the "energy" [81, 85]. For laminar flow past a bluff body, one needs to analyze only few modes [81, 85], as compared to hundreds of modes for transitional and turbulent flow fields. The *POD* analysis helps to identify coherent structures in the field which in the present analysis represent sound waves. The *POD* analysis also quantifies importance of various modes based on respective "energy/enstrophy/mean-square-fluctuating-pressure" content as given by eigenvalues of matrix $R_{i,j}$ as stated in variational calculus given by

$$\int_{\text{region}} R_{i,j}(r, r')\phi_j(r')dr' = \lambda_i \phi_i(r)$$

where $\phi_i$ are the eigenvector for the eigenvalue $\lambda_i$. The time varying *POD* amplitudes are obtained from eigenfunctions $\phi_k$ and disturbance pressure $p'(x, y, t)$ as

$$a_k(t) = \left( \int \int_S p'(x, y, t)\phi_k(x, y) \, dS \right) / \left( \int \int_S \phi_i^2 \, dS \right) \qquad (8.29)$$

For flow past a cylinder, researchers had previously noted presence of a pair of modes $\phi_1$ and $\phi_2$ which are phase shifted by 90° and were considered to compose the real and imaginary parts of the dominant instability mode. Similar pair of *POD* modes has also been reported for flow past a cylinder and flow inside a lid-driven cavity which forms a conjugate pair with a phase shift of 90° [85]. Such pair of modes was held responsible for triggering flow instabilities and has been termed as regular or *R* modes [81, 85]. *POD* modes have been classified into regular and anomalous modes based on the time variations of the *POD* amplitude functions $a_k(t)$ [81, 85].

The *POD* analysis is performed here for more than 15 periodic shedding cycles in order to obtain different dominant coherent structures/modes which show significant contribution to the disturbance pressure field. The *POD* analysis has been performed at three incidence angles $\alpha = 0°, 60°$, and 30°, with the significant *POD* modes along with their time varying *POD* amplitudes, are shown in Figs. 8.25, 8.26, and 8.27, respectively. At $\alpha = 0°$, with the shedding process being along the positive $x$-direction, 99.8% of the total acoustic disturbance pressure field has been contributed from first six modes and almost 90% contributions comes from first two modes (based on eigenvalues $\lambda_k$), as shown in Fig. 8.25a. Propagation of pressure pulses are almost along vertical direction in first two modes (Mode 1 and Mode 2) and along horizontal direction in third and fourth modes (Mode 3 and Mode 4). The frequency of pressure pulses found to be almost double in third and fourth modes as compared to that of first and second modes. The frequencies of POD amplitudes $a_k$ obtained for third and fourth modes are found to be equal and two times higher as compared to that of first two modes, as shown in Fig. 8.25b. It has also been identified

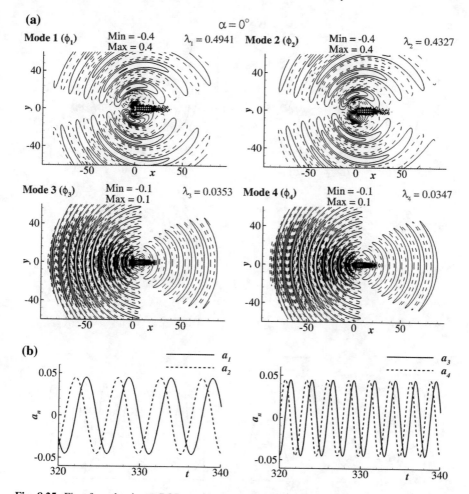

**Fig. 8.25** First four dominant *POD* modes and time varying *POD* amplitudes are obtained for disturbance pressure fields at $\alpha = 0°$. In Fig. 8.25a, Modes 1 and 2 forming a pair are contributed from the fluctuations in lift, whereas Modes 3 and 4 forming a pair are contributed from fluctuations in drag. In Fig. 8.25b, frequencies of $a_1$ and $a_2$ and $a_3$ and $a_4$ are equal to $f$ and $2f$, respectively. Reproduced from [56], with the permission of *AIP* Publishing

that the evaluated frequencies of POD amplitudes for the first two modes are exactly equal to the frequencies of vortex shedding cycle, whereas for the third and fourth modes, the frequencies are found to be double of the vortex shedding cycle, as shown in Fig. 8.25b. Based on these observations, it is inferred that the contribution of first two modes are from the fluctuations of the lift dipole and that of the next two modes are contributions from the drag dipole.

It is clearly evident that the contributions to the disturbance pressure field from lift fluctuations (Mode 1 and Mode 2) is more dominant than the drag fluctuations,

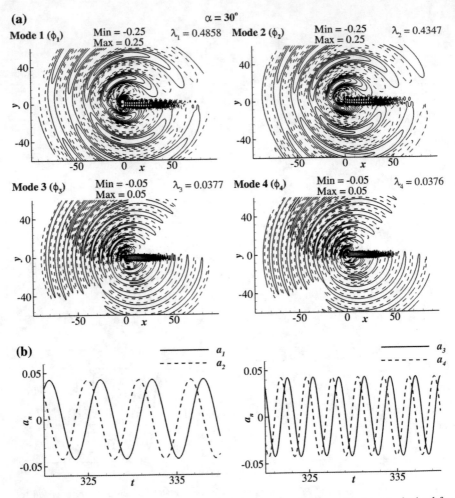

**Fig. 8.26** First four dominant *POD* modes and time varying *POD* amplitudes are obtained for disturbance pressure fields at $\alpha = 60°$. In Fig. 8.26a, Modes 1 and 2 forming a pair are contributed from the fluctuations in lift, whereas Modes 3 and 4 forming a pair are contributed from fluctuations in drag. In Fig. 8.26b, frequencies of $a_1$ and $a_2$ and $a_3$ and $a_4$ are equal to $f$ and $2f$, respectively. Reproduced from [56], with the permission of *AIP* Publishing

(Mode 3 and Mode 4). The major contribution to the disturbance pressure field has been found to be from lift fluctuations for an incidence angle $\alpha = 0°$, which agrees well with the directivity patterns shown in Fig. 8.23. Similarly, for $\alpha = 60°$, case *POD* modes are shown in Fig. 8.26a. The contribution from first two modes is more than 92%. The transient variation of *POD* amplitudes and their corresponding frequencies for the case of $\alpha = 60°$ are shown in Fig. 8.26b. The last two *POD* modes (Mode 3 and Mode 4) shown in Figs. 8.25a and 8.26a represent mostly the disturbances present in the wake. Also, pairing of POD modes with a phase difference of 90° are

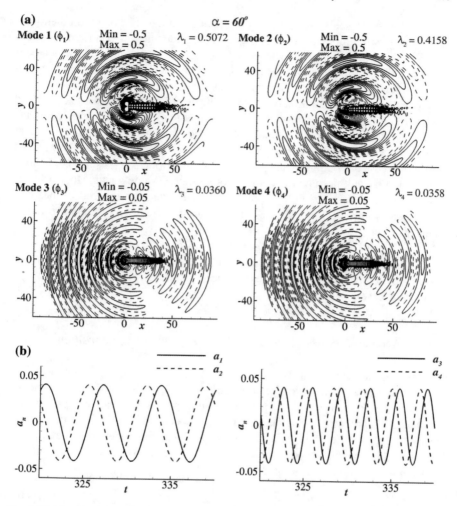

**Fig. 8.27** First four dominant *POD* modes and time varying *POD* amplitudes are obtained for disturbance pressure fields at $\alpha = 30°$. In Fig. 8.27a, both the mode pairs are contributed from both lift and drag fluctuations. In Fig. 8.27b, frequencies of $a_1$ and $a_2$ and $a_3$ and $a_4$ are equal to $f$ and $2f$, respectively. Reproduced from [56], with the permission of *AIP* Publishing

observed as shown in Figs. 8.25 and 8.26. Those modes are shown as Mode 1 and Mode 2 and Mode 3 and Mode 4. However, an anomalous mode (Mode 5) which doesn't form a pair has also been observed in both figures Figs. 8.25a and Fig. 8.26a. *POD* analysis performed by Sengupta et al. [81, 85] relates modes occurring in pairs with the flow instabilities while the anomalous mode has been made responsible for providing necessary mean value to the field under investigation. For the first time, we have shown presence of anomalous mode associated with the sound field.

Figure 8.27 represents the first four dominant *POD* modes along with their time varying *POD* amplitudes for $\alpha = 30°$ case. These four modes contribute to more than 99% of the total disturbance pressure field. This is a special case in which both the lift and the drag fluctuations are associated with the shedding frequency and its super-harmonics, as well. For the pair formed by Mode 1 and Mode 2, although the frequencies of *POD* amplitudes shown in Fig. 8.27b are equal to vortex shedding cycle, the first two *POD* modes in Fig. 8.27a are not completely contributed by lift fluctuations alone, as drag fluctuations also contribute to the same frequency, as that of shedding cycle (as shown in Figs. 8.18 and 8.19). Hence, the first two POD modes are contributed from both drag and lift fluctuations. Mode 3 represents anomalous mode, which has not been paired with any other mode. Mode 4 and Mode 5 forms a pair with frequencies of *POD* amplitudes are double than the vortex shedding frequency as shown in Fig. 8.27b. Modes 4 and 5 are also contributed from the fluctuations in lift and drag forces.

Next, we use Doak's decomposition [22] to differentiate between the hydrodynamic, acoustic, and thermal modes, as explained in the next subsection. This analysis allows us to compare the hydrodynamic and acoustic modes present for various angle of incidence cases.

### 8.2.2.4   Acoustic, Hydrodynamic, and Thermal Mode Decomposition

Here, momentum density has been decomposed as a linear superposition of hydrodynamic, acoustics, and thermal components using Doak's decomposition theory [22]. Doak's used the Helmholtz decomposition technique to separate out the mean, fluctuating solenoidal and fluctuating irrotational components for the momentum density, as given in Eq. (8.30). Fluctuating irrotational component has been further decomposed into acoustic and thermal modes. According to the Doak's decomposition theory, momentum density is written as [22, 98]

$$\rho \mathbf{u} = \overline{\mathbf{B}} + \mathbf{B}' - \nabla \psi' \tag{8.30}$$

where $\overline{\mathbf{B}}$ and $\mathbf{B}'$ are the mean and fluctuating solenoidal components, and $\psi'$ is fluctuating scalar potential representing an irrotational field. Here, $\nabla \cdot \overline{\mathbf{B}} = 0$, $\nabla \cdot \mathbf{B}' = 0$ and, for time stationary flows, mean scalar potential $(\overline{\psi})$ has been assumed to be zero. Fluctuating solenoidal $\mathbf{B}'$ field represents hydrodynamic component associated with the momentum density field.

The above-mentioned fluctuating scalar potential field has been evaluated by using the Poisson equation [98], which is obtained by taking a divergence of Eq. (8.30) as

$$\nabla^2 \psi' = \frac{\partial \rho'}{\partial t} \tag{8.31}$$

Now, the irrotational field is split into linear combination of thermal and acoustics modes (Eq. 8.32a) [98]. These modes are further estimated by solving corresponding Poisson equations (Eqs. (8.32b) and (8.32c)), respectively.

$$\psi' = \psi'_T + \psi'_A \tag{8.32a}$$

$$\nabla^2 \psi'_A = \frac{1}{c^2} \frac{\partial p'}{\partial t} \tag{8.32b}$$

$$\nabla^2 \psi'_T = \frac{\partial \rho'}{\partial S} \frac{\partial S}{\partial t} \tag{8.32c}$$

The irrotational component $\psi'$ in Eq. (8.30) is obtained by solving corresponding Poisson equation from the obtained flow field. As the radius of outer boundary is $r = 1500$, flow has been assumed as purely irrotational at the outer boundary. This assumption leads to negligible value of fluctuating solenoidal component near the outer boundary, hence the fluctuation component of Eq. (8.30) has been reduced to $(\rho \mathbf{u})' = -\nabla \psi'$. This reduced equation has been integrated along the outer boundary to evaluate $\psi'$, and subsequently provided as an outer boundary condition to solve the Poisson equation Eq. (8.31). Zero-gradient Neumann condition of $\psi'$ has been applied on the wedge surface. Similarly, the acoustic component $\psi'_A$ has been evaluated by solving corresponding Poisson equation given by Eq. (8.32b). At the outer boundary thermal fluctuations are assumed to be zero, hence, Eq. (8.32a) has been simplified as $\psi' = \psi'_A$. Using this simplified equation, a Dirichlet boundary condition at the outer boundary has been prescribed for the Eq. (8.32b). On the other hand, zero-gradient Neumann condition has been prescribed on the surface of wedge. As the mean solenoidal component $\overline{\mathbf{B}}$ is known from the flow field, fluctuating solenoidal component, $\mathbf{B}'$ has been obtained by using Eq. (8.30).

This method of decomposition has been implemented for solution data obtained at incidence angles $\alpha = 0°$, $30°$ and $60°$. As our main focus is to understand the sound generated around the wedge, we restrict our discussion to fluctuating irrotational $\|\nabla \psi'\|$, acoustic $\|\nabla \psi'_A\|$ and solenoidal $\|\mathbf{B}'\|$ components of momentum density function.

The solenoidal component $\|\mathbf{B}'\|$ has been evaluated for $\alpha = 0°$, $30°$ and $60°$, as shown in Fig. 8.28. The solenoidal component is observed only in the wake region where the hydrodynamic modes are found to be dominant. Similar behavior is also observed at the center line of supersonic cold jet as analyzed by Unnikrishan and Gaitonde [98]. These hydrodynamic modes in the wake exist for longer distances for $\alpha = 30°$ and $\alpha = 60°$, whereas, these disappear in the wake after a short distance for $\alpha = 0°$.

Figure 8.29 represents the contours obtained for fluctuations in total irrotational component $\|\nabla \psi'\|$ of momentum density for above-mentioned incidence angles $\alpha$. The intensity of $\|\nabla \psi'\|$ is almost same for $\alpha = 0°$ and $\alpha = 30°$ cases, whereas highest intensity level is observed at $\alpha = 60°$. The observed contours are dominantly directed toward vertical directions which happened due to the fact that amplitudes of fluctuations in lift are significantly greater than those of drag for all incidence angles $\alpha$.

Figure 8.30 shows the contours of sound fluctuations $\|\nabla \psi'_A\|$ obtained for different incidence angles. These modes completely represent the sound generated during the

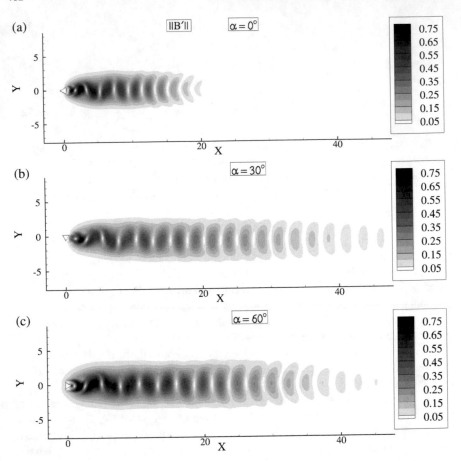

**Fig. 8.28** Contours of solenoidal fluctuations $||B'||$ obtained for incidence angles $\alpha =$ $0°$, $30°$, and $60°$ are shown at time $t = 330.9$. Reproduced from [56], with the permission of *AIP* Publishing

flow past a wedge. The intensity of these acoustic modes are almost same for $\alpha = 0°$ and $\alpha = 30°$ cases, whereas as highest intensity has been observed for $\alpha = 60°$ case. These noise patterns observed are significantly dominated by lift fluctuations. Also, the variation of sound intensity with incidence angle is in accordance with results shown by fluctuations in lift coefficient, *RMS* pressure fluctuations (shown in Figs. 8.23 and 8.24b) and disturbance pressure fields (shown in Figs. 8.21b and 8.22) as mentioned in previous subsections.

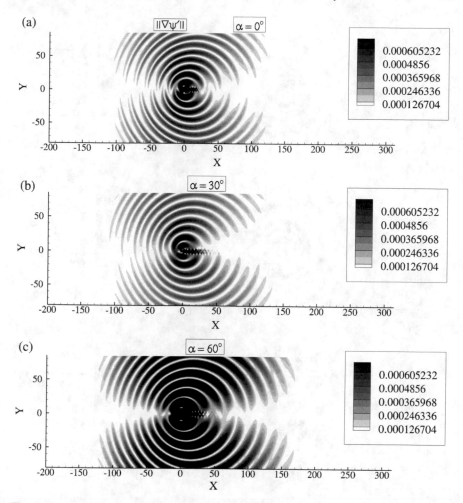

**Fig. 8.29** Contours of irrotational fluctuations $||\nabla \psi'||$ obtained for incidence angles $\alpha = 0°$, $30°$, and $60°$ are represented at time $t = 330.9$. Reproduced from [56], with the permission of *AIP* Publishing

### 8.2.3 Flow Past a Cylinder Performing Rotary Oscillations

Flow past bluff bodies can be controlled by altering wake geometry. Wake of a circular cylinder gets altered due to imposed in-line, transverse or rotary oscillations. Researchers have studied flow modifications due to in-line, or transverse rectilinear oscillations [5, 6, 34, 93, 102]. Motivation behind most of the previous studies is to achieve drag reduction by wake alteration through imposed oscillations. Researchers have also studied modification in flow field due to steady rotation of a circular cylinder [2, 9, 10, 79]. Complete suppression of vortex shedding has been reported at high

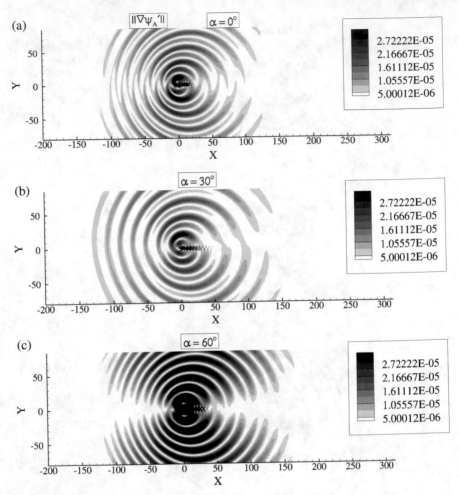

**Fig. 8.30** Contours of acoustic fluctuations $||\nabla \psi'_A||$ obtained for incidence angles $\alpha = 0°$, $30°$, and $60°$ are shown at time $t = 330.9$. Reproduced from [56], with the permission of *AIP* Publishing

rotation rates and moderate Reynolds numbers for flow past a cylinder undergoing steady rotation [21].

Researchers have used experimental approach to study modifications in the wake region for flow past a cylinder performing rotary oscillations [29, 68, 96, 97]. Okajima et al. [68] measured forces acting on a cylinder performing rotary oscillations over a wide range of Reynolds numbers. Authors discussed the influence of Reynolds number on the aerodynamic parameters and the synchronization phenomenon in which vortex shedding occurs at the forcing frequency [68]. Taneda [96] discussed the qualitative modifications in vortex shedding behind circular cylinder performing rotary oscillation. Alterations in vortex shedding were also reported by Filler et al. [29] over a range of Reynolds number for different cylinder peripheral speed. Drag reduction for such flows was also observed by Thiria et al. [97] and the phase lag between the vortex shedding and the imposed rotary motion was shown responsible for reduction in wake fluctuations [97]. In most of these studies, the forcing parameters are considered in the synchronization region, where shedding patterns are governed by forcing frequency alone.

Researchers have also simulated flow past a cylinder performing rotary oscillation over a range of Reynolds number, amplitude of oscillation and forcing frequency [3, 7, 11, 13, 23, 37, 55, 74, 78, 80]. Laminar flow over the circular cylinder experiences aerodynamic tripping due to imposed rotary oscillations [7, 78, 80]. While simulating flow field past cylinder performing forced oscillations at Re = 150, it has been concluded that the forcing results in the mean flow alteration by the action of the divergence of Reynolds stress tensor [73].

Here we have simulated sound generation due to two-dimensional, unsteady, laminar flow past a circular cylinder performing rotary oscillations. Although large number of researchers have paid a close attention toward flow control aspects by imparting rotary oscillations to a circular cylinder, corresponding modifications in the radiated sound field have not been studied in detail. It is important to understand different forcing conditions which cause changes in frequency, amplitude, and the directivity of radiated sound field. Analysis has been carried out to address whether one can reduce overall acoustic noise and change frequency content as well as directivity of the radiated sound field. Results for aeolian tones generation due to laminar flow past a circular cylinder performing rotary oscillation were first published in [65] and are reproduced here with the permission of *AIP* Publishing.

### 8.2.3.1 Problem Description

The schematic of a two-dimensional, uniform flow over a circular cylinder performing rotary oscillations has been shown in Fig. 8.31. The center of circular cylinder of diameter $D$ is located at origin $O$ such that the free-stream and transverse directions are aligned along positive $x$- and $y$-axes, respectively. Let $\Theta$ denotes angle associated with rotary oscillation of a circular cylinder.

The free-stream values of density $(\rho_\infty)$ and temperature $(T_\infty)$ are taken as $1.12 \, \text{kg/m}^3$ and 287 K, respectively. The free-stream Prandtl number $(Pr_\infty)$ and the ratio of specific heats $(\gamma)$ are prescribed as 0.7 and 1.4, respectively. The values of free-stream Mach number $(M)$ and the diameter of cylinder are considered as 0.2 and 1 m, respectively. Variation of the properties like thermal conductivity and molecular viscosity due to change in temperature has been assumed to be insignificant at this low Mach number [17, 42].

Strouhal number for the flow past a stationary cylinder is given by $S_0 = f_0 D/U_\infty$, where $f_0$ is the shedding frequency of a stationary cylinder and $U_\infty$ is the free-stream velocity. In addition, Strouhal number for the rotary oscillation case is given as $S_f = f_f D/U_\infty$, where $f_f$ denotes the forcing frequency of rotary oscillation. Maximum

**Fig. 8.31** Schematic of
uniform flow past a circular
cylinder subjected to rotary
oscillations. Reproduced
from [65], with the
permission of *AIP*
Publishing

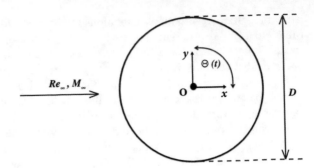

rotation rate $\Omega_1$ and maximum angular displacement $\Theta_0$ are related as $\Omega_1 = 2\pi f_f \Theta_0$.
One can prescribe the instantaneous rotation rate $\Omega(t)$ as

$$\Omega(t) = \Omega_1 \sin(2\pi S_f t) \tag{8.33}$$

Rotary oscillating motion for a circular cylinder has been prescribed using a
non-dimensional forcing frequency-ratio $f_r = f_f / f_0$ along with the non-dimensional
surface speed $A$ which is given as $A = \Omega_1 D / 2U_\infty$. In this work, we have reported
results with the non-dimensional forcing frequency-ratio $f_r$ and amplitude $A$ so as to
validate our results with results of experiments performed at Reynolds number Re =
150 [97]. The present range of forcing parameters is found to be in synchronization
region [97].

### 8.2.3.2  Effects of Rotary Oscillations on Flow Field

It has been confirmed by researchers that the fluctuating forces acting on the body are
caused by the fluctuations in the flow variables which are also responsible for sound
generation and propagation [42, 56]. Thus in the subsequent analysis, fluctuations
in the lift and the drag coefficients with time have been discussed in detail to study
corresponding effects on generated sound field. Computations for the unsteady, two-
dimensional flow past a circular cylinder performing rotary oscillations have been
carried out for Re = 150 and $M = 0.2$ for the following forcing conditions.

1. Effects of variation in forcing frequency on generated sound have been simulated
   for the forcing amplitude $A = 2$ by obtaining results for 10 different values of
   forcing frequency-ratio $f_r$ varying from 0.0 to 3.0.
2. Effects of variation in forcing amplitude on generated sound have been simulated
   for the forcing frequency-ratio $f_r = 1$ and six different forcing amplitudes $A =$
   2, 3, 4, 5, 7.5 and 10.

In the present study, the forcing parameters belong to synchronization region.
Here, attention has been focused on understanding the behavior of fluctuating properties of flow and sound fields with change in forcing amplitude, $A$, and frequency-

ratio, $f_r$. Effects of fluctuating flow properties on sound generation and its propagation have been studied in detail. The case with $f_r = 0.0$ represents flow past a stationary circular cylinder.

### 8.2.3.3   Effects of Variation in Forcing Frequency

Here, we investigate effects of forcing frequency-ratio $f_r$ on the flow field parameters for Re $= 150$, $M = 0.2$ with forcing amplitude $A = 2$.

Figure 8.32 represents the comparison of wake features obtained from present numerical study with experimental results at $f_r = 1.5$ and $A = 2.0$ [97]. Left column represents the instantaneous vorticity contours at three different instants based on present computations, whereas the right column shows wake patterns obtained from experiments for the same forcing conditions [97]. Present results show an excellent qualitative agreement with experimental results [97].

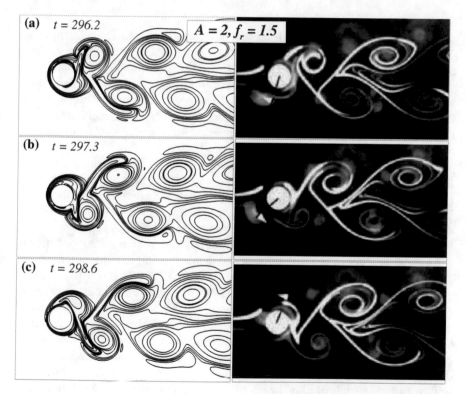

**Fig. 8.32** Instantaneous vorticity contours based on present computations are shown in left column for the case with $A = 2.0$ and $f_r = 1.5$ at three different instants in a given vortex shedding cycle. Right column displays wake patterns extracted from experimental results [97] for same forcing parameters. Present results are in good agreement with the experimental results [97]. Reproduced from [65], with the permission of *AIP* Publishing

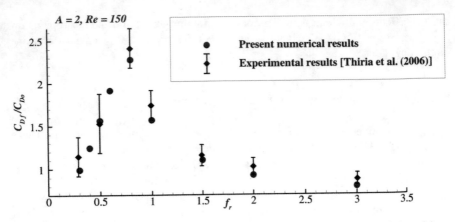

**Fig. 8.33** Validation of present numerical results has been shown by plotting the variation of time-averaged normalized drag coefficient $C_{Df}/C_{Do}$ with forcing frequency. Reproduced from [65], with the permission of *AIP* Publishing

Time-averaged drag coefficients for flow past a stationary and an oscillating cylinders are denoted as $C_{Do}$ and $C_{Df}$, respectively. Figure 8.33 represents the variation of time-averaged normalized drag coefficient ($C_{Df}/C_{Do}$) with frequency-ratio $f_r$ for a forcing amplitude $A = 2.0$. Present numerical results display an excellent match with the available experimental results [97]. These results also provide necessary validation for the present rotary oscillation cases. Time-averaged lift coefficient ($C_L$) is found zero for all the values of $f_r$, as the cylinder performs periodic rotary oscillations resulting in zero mean lift.

Fluctuations in the lift ($c_{l'}$) and drag coefficients ($c_{d'}$) are evaluated as the difference between the instantaneous and mean values of the lift and drag coefficients, respectively. Several studies on flow past bluff bodies have proposed that time varying fluctuating loads (fluctuations in $c_{l'}$ and $c_{d'}$) acting on the surface of bluff body are predominantly responsible for generation and propagation of sound waves [42, 53, 56, 59]. Alternately shed vortices from the top and bottom surfaces of stationary circular cylinder have equal strength due to symmetric geometry of the circular cylinder for the incoming uniform flow.

Time variation for the lift $c_{l'}$ and drag $c_{d'}$ fluctuations for flow past a stationary circular cylinder case $f_r = 0.0$ has been shown in Fig. 8.34a. The frequency associated with $c_{d'}$ variation is double as that of $c_{l'}$ variation. It is observed that the maximum amplitude of $c_{l'}$ is almost 20 times larger as compared to that of $c_{d'}$. Hence, for flow past a stationary cylinder case, the generated sound field is dominated by the lift dipole as identified in the previous *DNS* study [42]. Fast Fourier Transform (*FFT*) of time varying signals of $c_{l'}$ and $c_{d'}$ for flow past a stationary circular cylinder has been obtained and the Fourier amplitude ($A_f$) as a function of Fourier frequency ($f_1$) has been represented in Fig. 8.34b. It is observed that the dominant peak of $c_{l'}$ is obtained at a shedding frequency $f_1 = f_o$. Although many experimental studies have found that the shedding phenomena is only governed by a single frequency

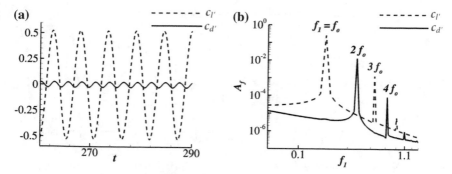

**Fig. 8.34 a** Variation of fluctuations in lift ($c_{l'}$) and drag ($c_{d'}$) coefficients with time, **b** Fast Fourier transform *FFT* obtained for time varying fluctuations in lift ($c_{l'}$) and drag ($c_{d'}$) coefficients. Reproduced from [65], with the permission of *AIP* Publishing

corresponding to the vortex shedding frequency $f_o$, secondary and tertiary peaks for $c_{l'}$ are also observed in the present results at $f_1 = 3f_o$ and $f_1 = 5f_o$, respectively. The magnitudes of secondary and tertiary peaks are negligibly small as compared to that of primary peak. Furthermore, the dominant peak of $c_{d'}$ is observed at $f_1 = 2f_o$ as the frequency of $c_{d'}$ is twice that of $c_{l'}$. The secondary and tertiary peaks of $c_{d'}$ have very low magnitude and are identified at $f_1 = 4f_o$ and $f_1 = 6f_o$, respectively.

Figure 8.35 shows the time variation of fluctuations in $c_{l'}$ and $c_{d'}$ obtained for flow past a cylinder performing rotary oscillations. Results are shown for 9 different values of forcing frequency-ratios ($0.3 \leq f_r \leq 3.0$) with forcing amplitude $A = 2.0$. As the value of $f_r$ increases from 0.3 to 0.6, the amplitude of $c_{l'}$ does not show much variation, and then it decreases continuously, as $f_r$ increases further achieving minimum value at $f_r = 3.0$. Beyond $f_r = 0.8$, the drop in maximum amplitude of $c_{l'}$ is significant. The amplitude of $c_{l'}$ is observed to be greater than that of a stationary cylinder case ($f_r = 0.0$) in the range $0.3 \leq f_r \leq 1.0$. The amplitude of $c_{l'}$ at $f_r = 1.5$ is approximately equal to that of a stationary cylinder case ($f_r = 0.0$) and reduces further in the forcing frequency range $1.5 < f_r \leq 3.0$. The frequency associated with $c_{l'}$ variation increases with $f_r$ in the range $f_r = 0.3$ to $f_r = 3.0$. Further the amplitude of fluctuating $c_{d'}$ is very much lower than that of $c_{l'}$ in the range $0.3 \leq f_r \leq 0.8$ and $1.5 < f_r \leq 3.0$. For $f_r = 1.0$ and 1.5 cases, the amplitude of $c_{d'}$ is still smaller, but comparable in magnitude with that of $c_{l'}$. Also, the frequency of $c_{d'}$ is observed to be twice as that of $c_{l'}$ irrespective of the case considered. For the case of $f_r = 3.0$, the positive and negative peaks of $c_{l'}$ are almost identified at the same time where $c_{d'}$ has its positive peaks and zero value of $c_{l'}$ has been obtained at the same instant, where $c_{d'}$ has its negative peak.

The *FFT* of time varying fluctuations in $c_{l'}$ and $c_{d'}$ have been obtained for all the values of $f_r$. Figure 8.36 shows plots of the Fourier amplitude $A_f$ as a function of Fourier frequency-ratio $f_{r1} = f_1/f_o$, at different values of $f_r$. The present computations are performed at different values of frequency-ratio $f_r$, which belong to synchronization region [97]. So, irrespective of the case with frequency-ratio $f_r$, the

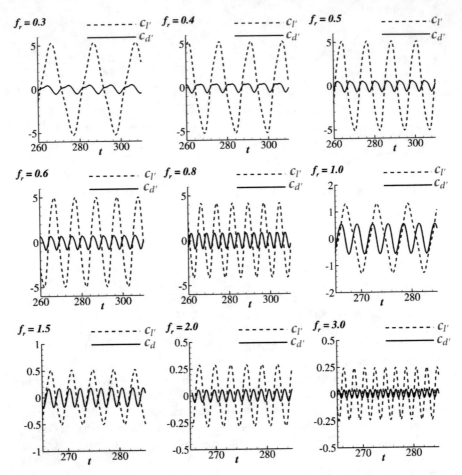

**Fig. 8.35** Variation of fluctuations in lift and drag coefficients with time for the frequency range $0.3 \le f_r \le 3.0$ with forcing amplitude $A = 2$. Reproduced from [65], with the permission of *AIP Publishing*

most dominant peak of $c_{l'}$ has been identified at the corresponding forcing frequency-ratio, $f_{r1} = f_r$. On the other hand, most dominant peak of $c_{d'}$ is found at $f_{r1} = 2 f_r$ for a given $f_r$. Modulation in fluctuations of $c_{d'}$ are more significant for the low values of $f_r$ ($0.3 \le fr < 1.0$) cases and beyond this range, the number and magnitude of higher frequency peaks show decreasing trend in Fig. 8.35. Let us consider a low frequency case $f_r = 0.3$. Although, the first peak of $c_{l'}$ and $c_{d'}$ are identified at the same frequency-ratio $f_{r1} = f_r = 0.3$, but the peak magnitude of $c_{d'}$ is several order lesser than that of $c_{l'}$. An exactly similar kind of first peak in $c_{d'}$ containing very low magnitude has also been observed in other frequency-ratio cases except for $f_r = 0.4$ and $1.0$. Further the secondary peaks of $c_{l'}$ and the most dominant peaks of $c_{d'}$ are found to be of comparable magnitudes only in the range $0.3 \le f_r \le 0.5$.

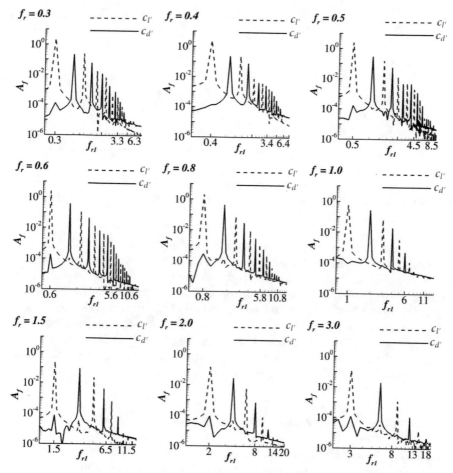

**Fig. 8.36** Fast Fourier transform *FFT* of time varying fluctuations in lift and drag coefficients have been obtained for the frequency range $0.3 \leq f_r \leq 3.0$ with forcing amplitude $A = 2$. Reproduced from [65], with the permission of *AIP* Publishing

### 8.2.3.4 Effects of Variation in Forcing Amplitude

Next, we discuss effects of variation of the forcing amplitude ($A$) on generated sound field. For this analysis, we have fixed the forcing frequency-ratio as $f_r = 1.0$ and the simulations are performed for six different forcing amplitudes $A = 2, 3, 4, 5, 7.5$ and 10 for the same Reynolds number Re $= 150$ and Mach number $M = 0.2$. Figure 8.37 represents the variation of normalized drag coefficient $C_{Df}/C_{Do}$ with forcing amplitude $A$. It is clearly visible that $C_{Df}/C_{Do}$ increases gradually with increase in $A$ reaching maximum at $A = 5$ and then drops for higher forcing amplitudes.

Figure 8.38a represents the time variation of fluctuations in lift and drag coefficients for the indicated values of forcing amplitudes. It is observed that amplitudes of

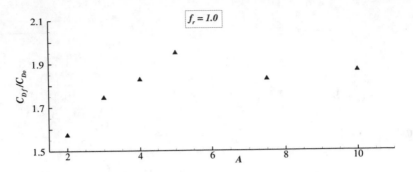

**Fig. 8.37** Variation of normalized drag coefficient $C_{df}/C_{do}$ with forcing amplitude $A$ at $f_r = 1.0$. Reproduced from [65], with the permission of *AIP* Publishing

$c_{l'}$ and $c_{d'}$ are of almost comparable magnitudes, and hence the generated sound fields are equally influenced by fluctuations in both lift and drag coefficients, as proven later. As the magnitudes of $c_{l'}$ and $c_{d'}$ are found minimum for the case $A = 5$, corresponding acoustic field intensity is minimum as discussed subsequently. Figure 8.38b represents the *FFT* of $c_{l'}$ and $c_{d'}$ for various values of $A$. The first peak of $c_{l'}$ is identified at the forcing frequency-ratio $f_{r1} = f_r = 1.0$ and the dominant peak of $c_{d'}$ is found at $f_{r1} = 2f_r$. The dominant peak of $c_{d'}$ has Fourier amplitude greater than that of $c_{l'}$ for all the cases except for the case $A = 2$. Thus the generated sound fields must display similar contribution due to fluctuations in the drag and lift coefficients.

### 8.2.3.5  Disturbance Pressure Fields

Fluctuations in the lift and drag forces are caused by vortex shedding behind stationary circular cylinder, which has been found responsible for the generation of acoustic pressure pulses [42]. Here, we have evaluated the disturbance pressure field from the mean and instantaneous pressure fields.

Consider the case with $f_r = 2.0$ and $A = 2.0$ in order to examine the transient evolution of disturbance pressure pulses. Time variation of lift and drag coefficients are shown in Fig. 8.39a1, a2, respectively. Here, several instants in a given vortex shedding cycle are denoted by circular dots. Figure 8.39b1, b2 represent the vorticity contours in the cylinder wake denoted by lines at two different instants corresponding to minima and maxima of lift coefficient. Solid and dashed lines indicate positive and negative values of vorticity. The background gray scale flood represents instantaneous pressure field at those instants. Dark and bright portions denote high- and low-pressure regions, respectively. All the vortex centers are observed to be of low-pressure zones. Instantaneous disturbance pressure fields are evaluated and represented in Fig. 8.39c1–c4 in a given vortex shedding cycle. These disturbance fields are obtained at the instants denoted by dots, as shown in Fig. 8.39a1, a2. There is generation of positive and negative disturbance pressure pulses from the

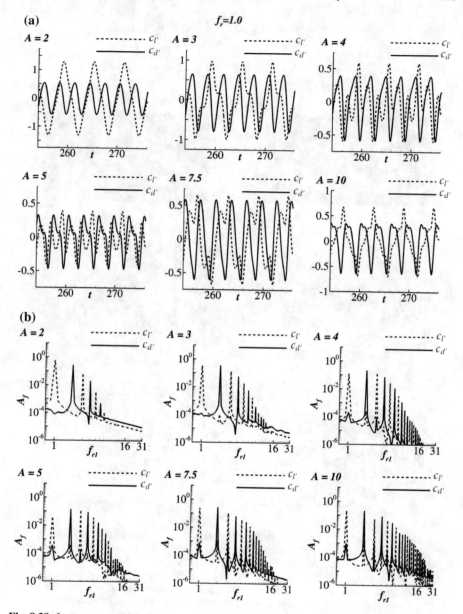

**Fig. 8.38** Instantaneous lift and drag coefficients are obtained for various forcing amplitudes $A =$ 2, 3, 4, 5, 7.5 and 10 with $f_r = 1.0$. **a** Variation of fluctuations in lift ($c_{l'}$) and drag ($c_{d'}$) coefficients with time $t$. **b** Fast Fourier transform (*FFT*) of time varying fluctuations in lift and drag coefficients is shown. Reproduced from [65], with the permission of *AIP* Publishing

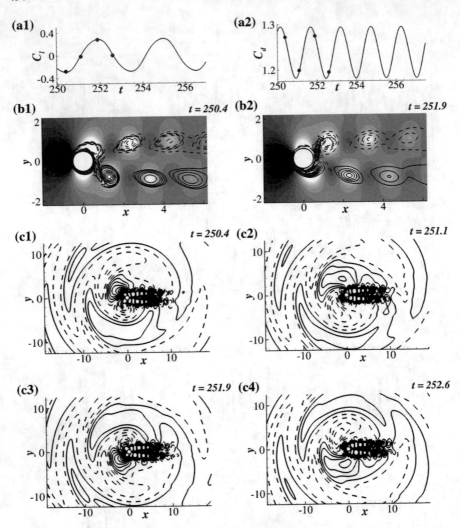

**Fig. 8.39** For $f_r = 2.0$ and $A = 2$ case, frames **a1** and **a2** show time varying lift and drag coefficients with filled red dots denoting several instants in a shedding cycle. Frames **b1** and **b2** represent vorticity contours using lines at indicated instants. The flooded Grey scale in the background represents instantaneous pressure at those instants with low- and high-pressure regions denoted as bright and dark zones. Frames (**c1**–**c4**) demonstrate the transient evolution and propagation of disturbance pressure pulses from the cylinder surface in one shedding cycle. Reproduced from [65], with the permission of *AIP* Publishing

surface of cylinder in response to the vortex shedding phenomenon. These pressure pulses propagate away from the cylinder with increase in time. Hence, this results in the generation of disturbance pressure field around a cylinder performing rotary oscillations.

### 8.2.3.6  Effects of Forcing Frequency on Disturbance Pressure Fields

Figure 8.40 illustrates the instantaneous disturbance pressure fields obtained for various values of $f_r$ with $A = 2$ at time $t = 260$. The maximum and minimum values of represented contour levels are also shown for each $f_r$. The case with $f_r = 0.0$ represents flow past a stationary cylinder case. As the frequency-ratio increases from 0.3 to 3.0, the number of pressure pulses present in the given domain increases. As the computations are performed in the synchronization region, the obtained disturbance pressure fields are governed by corresponding forcing frequency-ratio $f_r$. One of the important observations of the present analysis is that the forcing frequency of rotary oscillation dictates the frequency content of the generated sound field. Low-frequency rotary excitation triggers sound waves with low frequencies and large wavelengths. As rotary forcing frequency increases, corresponding sound field displays shorter wavelengths.

Figure 8.41a shows time varying disturbance pressure $p'$ extracted at three different radial locations along $\theta = 90°$ for $f_r = 2.0$ and $A = 2$ case. One can observe that the amplitude of $p'(t)$ is reduced with increase in radial distance. In addition, the wavelength of these signals are observed to be equal. Further, FFTs of these time varying disturbance pressure profiles have been evaluated and the Fourier amplitude $A_f$, as a function of frequency-ratio $f_{r1} = f_1/f_o$ has been shown in Fig. 8.41b. The dominant primary peak has been observed at forcing frequency-ratio $f_{r1} = f_r = 2.0$ for every radial location considered here. The secondary, tertiary, and quaternary peaks are identified at $2f_r$, $3f_r$, and $4f_r$, respectively. These frequency locations are identical to those observed for time varying lift and drag fluctuations of $f_r = 2.0$ and $A = 2$ case (shown in Fig. 8.36).

It has been found that in case of an unconfined flow past a stationary circular cylinder, the decay of acoustic pressure peaks agrees well with the theoretical decay rate ($p' \propto r^{-0.5}$) as shown in Fig. 8.42e [42]. To demonstrate the accuracy and resolution of the obtained disturbance pressure fields $p'(x, y, t)$ for $A = 2$ and all the cases of $f_r$, a special exercise has been conducted. The disturbance pressure fields are evaluated for more than 10 cycles with sufficient sampling. At an azimuthal angle $\theta = 90°$, the time varying disturbance pressures are extracted at all the grid point locations present inside the sound zone ($0.5 \leq r \leq 100$). FFT of these time varying disturbance pressure profiles are obtained at all these grid point locations. Multiple frequency peaks (primary, secondary, tertiary, and quaternary peaks) are observed at every grid point location and the magnitude of each frequency peak starts decaying as the radial distance increases (as shown in Fig. 8.42).

The Fourier amplitudes of primary peaks $A(p')$ for disturbance pressure are extracted along the radial distance $r$, for all the values of frequency-ratio $f_r$ at $\theta = 90°$. Figure 8.43a, b show the decay of Fourier amplitudes obtained for primary peaks along a radial distance $r$ on a log–log scale. Irrespective of the frequency-ratio $f_r$, the amplitude of disturbance pressure $A(p')$ decays and it follows the theoretical decay rate of acoustic pressure fluctuation $p' \propto r^{-0.5}$, as shown for flow past a stationary cylinder case [42]. It is observed that the magnitude levels of $A(p')$ are higher in the range $0.3 \leq f_r \leq 0.8$ (as shown in Fig. 8.43a) which has been justified

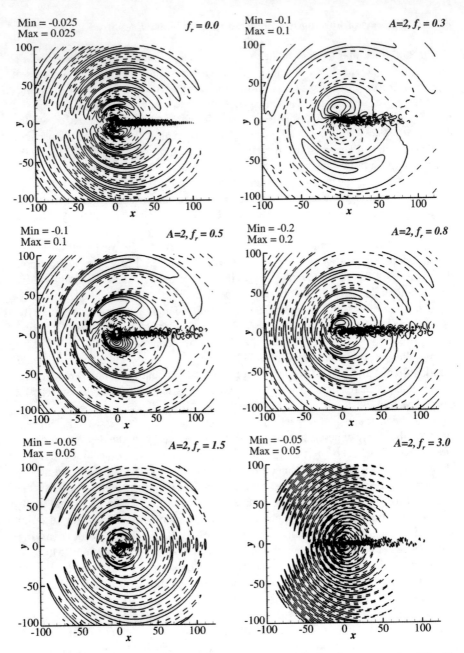

**Fig. 8.40** Disturbance pressure fields obtained for indicated forcing frequency ratios ($f_r$) with forcing amplitude $A = 2.0$. The contour levels vary from the represented values of minimum (Min) to maximum (Max) with 30 equi-spaced sub-levels. As we increase $f_r$ from 0.3 to 3.0, the frequency of disturbance pressure pulses increases and corresponding wavelength reduces. Reproduced from [65], with the permission of *AIP* Publishing

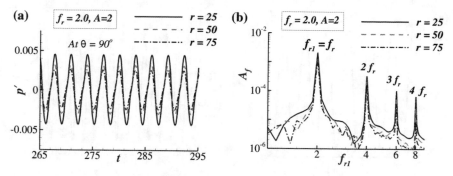

**Fig. 8.41 a** Transient variation of disturbance pressure at indicated radial locations considered along $\theta = 90°$ for $f_r = 2.0$ and $A = 2$ case. **b** FFTs of these time varying disturbance pressure profiles are shown. Reproduced from [65], with the permission of *AIP* Publishing

as the magnitudes of lift coefficient fluctuations ($c_{l'}$) are very high in this range. On the other hand, the magnitude levels of $A(p')$ are comparatively less in the range $1.0 \leq f_r \leq 3.0$ (as shown in Fig. 8.43b) due to lower magnitudes of fluctuations in ($c_{l'}$) and ($c_{d'}$). Important observation from Fig. 8.43 is the Fourier amplitude of disturbance pressure which is found to be maximum and minimum for $f_r = 0.6$ and $f_r = 2.0$, respectively. Hence, the generated sound fields should show similar trends in sound intensities at the respective forcing frequencies. This is also observed in the directivity patterns based on *RMS* values of pressure fluctuations explained later for Fig. 8.44. Figure 8.43c shows the decay of Fourier amplitude $A(p')$ obtained for secondary peaks along a radial distance $r$ at the indicated values of $f_r$ (log−log scale). It is clear that the secondary peaks also follow the theoretical decay rate $p' \propto r^{-0.5}$. The decay of tertiary peaks has found to follow the theoretical decay rate, as well. This signifies the fact that although the generated sound field constitutes multiple frequencies, each individual frequency component of the disturbance pressure signal obeys the theoretical decay rate. Hence irrespective of the forcing frequency $f_r$, variation of disturbance pressure amplitude displays the theoretical decay rate.

Next, the variation of sound intensities with $f_r$ has been estimated using the directivity patterns based on *RMS* value of disturbance pressure fields. Root mean square values of pressure fluctuations ($p'_{RMS}$) have been evaluated using time varying disturbance pressure fields for all values of $f_r$. The $p'_{RMS}$ has been extracted at a radial distance $r = 75$ and it has been represented as a function of azimuthal angle $\theta$ at this location, as shown in Fig. 8.44a, b. It is observed that the magnitude of $p'_{RMS}$ is higher in the range $0.3 \leq f_r \leq 0.8$ (as shown in Fig. 8.44a), which is due to high magnitude of fluctuations in lift coefficient ($c_{l'}$) in this range. On the other hand, magnitude of $p'_{RMS}$ is comparatively lower in the range $1.0 \leq f_r \leq 3.0$ (as shown in Fig. 8.44b), which results from smaller magnitude of fluctuations in the lift and drag coefficients.

In the low frequency range $0.3 \leq f_r \leq 0.8$, the directivity patterns are dominated by lift dipole as the magnitude of lift coefficient fluctuations are found to be significantly more than that of drag fluctuations (as shown in Figs. 8.35, 8.36). In the rest

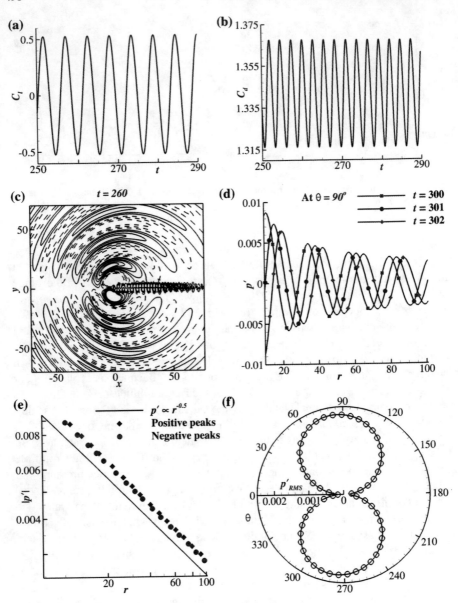

**Fig. 8.42** For flow past a stationary circular cylinder $f_r = 0.0$, **a** and **b** represent time variation of lift and drag coefficients, respectively. **c** Disturbance pressure field ($p'$) at time $t = 260$. **d** Variation of $p'$ along radial distance $r$ considered at $\theta = 90°$. **e** Decay of disturbance pressure peaks with radial distance $r$ considered at $\theta = 90°$. Solid line indicating $p' \propto r^{-0.5}$ represents theoretical decay rate of acoustic pressure fluctuations [42]. **f** Directivity pattern based on root mean square (*RMS*) values of disturbance pressure is evaluated at $r = 75$. Reproduced from [65], with the permission of *AIP* Publishing

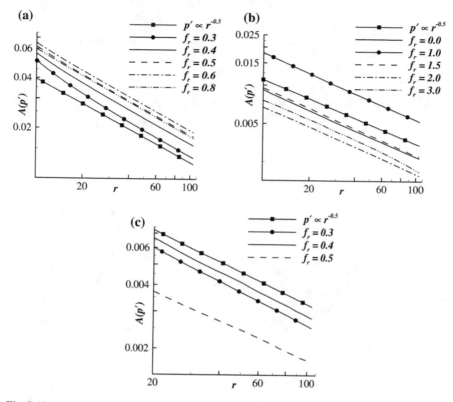

**Fig. 8.43** Fourier amplitudes $A(p')$ associated with multiple frequency peaks (Primary, secondary and tertiary peaks) have been evaluated for time varying disturbance pressures at all radial locations having an azimuthal angle $\theta = 90°$. **a** and (**b** represent the decay of $A(p')$ identified at primary frequency peak with $r$ for all the values of $f_r$. **c** represents the decay of $A(p')$ identified at secondary frequency peak with $r$ for indicated values of $f_r$. Theoretical decay rate of acoustic pressure fluctuations are represented by $p' \propto r^{-0.5}$ [42]. Reproduced from [65], with the permission of *AIP Publishing*

of the frequency range $1.0 \leq f_r \leq 3.0$, the directivity patterns are not significantly dominated by the lift dipole which is due to the fact that magnitude of $c_{l'}$ has been found to be not significantly higher than that of $c_{d'}$ (as shown in Figs. 8.35, 8.36). One interesting feature has been observed at $f_r = 1.0$, where the directivity pattern shows that the radiated sound is more evenly distributed in all directions and this has been justified because the amplitude of fluctuating drag coefficient is just half the magnitude, as compared to that of fluctuating lift coefficient, as shown in Fig. 8.35. At $f_r = 3.0$, the trajectory is oriented at an angle of $60°$ approximately, with the negative $x$-axis.

In order to find the net acoustic power output, acoustic power has been evaluated as [59],

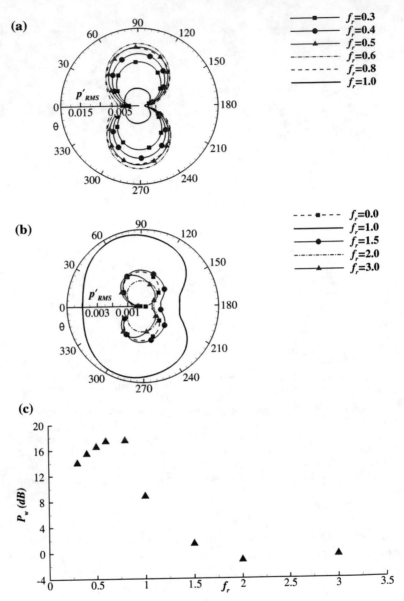

**Fig. 8.44** Directivity patterns based on *RMS* values of disturbance pressure ($p'_{RMS}$) are obtained for $A = 2$ and indicated values of $f_r$. **a** and **b** represent variation of $p'_{RMS}$ with azimuthal angle $\theta$ at radial distance $r = 75$ for indicated values of $f_r$, respectively. **c** Variation of acoustic power ($P_w$) obtained at a radial distance $r = 75$ with $f_r$ has been shown here. Reproduced from [65], with the permission of *AIP* Publishing

$$W = \int_{r=75} I_a(r, \theta) r d\theta \qquad (8.34)$$

where $I_a = \frac{p'_{RMS}}{\rho_\infty a_\infty}$. The acoustic power in decibels (dB) is evaluated as

$$P_w = 20 \log_{10} \frac{W}{W_o} \qquad (8.35)$$

where $W_o$ corresponds to acoustic intensity flux obtained for flow past a stationary circular cylinder for the same free-stream conditions. Figure 8.44c shows the variation of acoustic power $P_w$ (in dB) obtained at $r = 75$, with forcing frequency-ratio $f_r$, for a given forcing amplitude $A = 2$. The power $P_w$ increases in the range $0.3 \leq f_r \leq 0.8$ with its maximum around $f_r = 0.6$ and $0.8$, and then it drops gradually as $f_r$ increases. For the cases with $f_r = 2$ and $f_r = 3$, the acoustic power $P_w$ is negative, and this signifies that the acoustic power obtained for these cases is lesser than that obtained for a stationary cylinder case under the same free-stream conditions. This shows the acoustic noise has been reduced due to the imposed rotary oscillations.

From the directivity patterns based on $p'_{RMS}$ and behavior of Fourier amplitudes of disturbance pressure $A(p')$, it is clearly evident that the generated sound fields are governed by the relative amplitudes and phase difference of fluctuating lift and drag forces. In the low frequency region $0.3 \leq f_r < 1.0$, the disturbance pressure fields are significantly dominated by lift fluctuations alone, whereas in the rest of frequency range $1.0 \leq f_r \leq 3.0$, the directivity of disturbance pressure fields is affected by both lift and drag fluctuations.

### 8.2.3.7 Effects of Forcing Amplitude on Disturbance Pressure Fields

Disturbance pressure fields are obtained for various values of forcing amplitude $(A)$. Figure 8.45 shows the instantaneous disturbance pressure fields obtained for the indicated values of $A$ at frequency-ratio $f_r = 1.0$. Represented contour levels vary from $-0.1$ to $0.1$ with 30 equi-spaced sub-levels. It is observed that the disturbance pressure intensity is found decreasing from $A = 2.0$ to $5.0$, and then again increases up to $A = 10$. The minimum and maximum intensities of the obtained disturbance pressure fields are identified at $A = 5$ and $A = 10$, respectively. For the cases with $A = 7.5$ and $A = 10$, it has been identified that the disturbance pressure pulses are found to be of more intense toward negative $x$-direction. This has not been observed during the study on the effects of variation in frequency-ratio $f_r$ on radiated sound field, where the majority of the cases are dominated by the lift dipole.

Figure 8.46a shows the directivity patterns obtained for different values of $A$ based on the RMS values of pressure fluctuations $p'$ at a radial distance $r = 75$. The value of $p'_{RMS}$ is found to be maximum for the case $A = 10$, which can also be observed from Fig. 8.45. It is clearly visible that directivity patterns are governed by both lift and drag fluctuations whose magnitudes are comparable in nature (as shown in Fig. 8.38a),

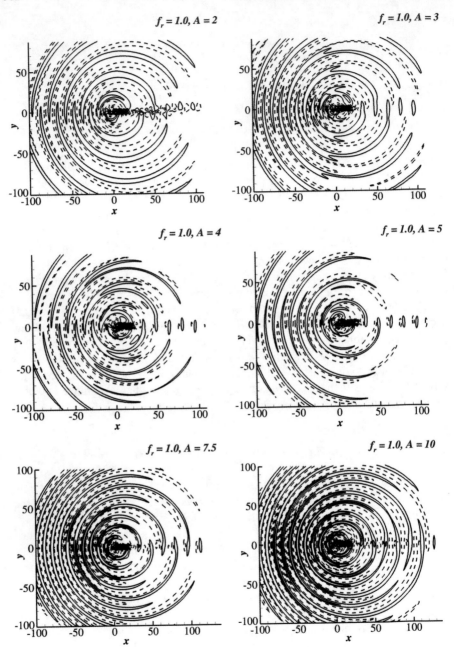

**Fig. 8.45** Disturbance pressure fields obtained for $A = 2, 3, 4, 5, 7.5$ and $10$ with $f_r = 1.0$. The contour levels vary from $-0.1$ to $0.1$ with 30 equi-spaced sub-levels. Reproduced from [65], with the permission of *AIP* Publishing

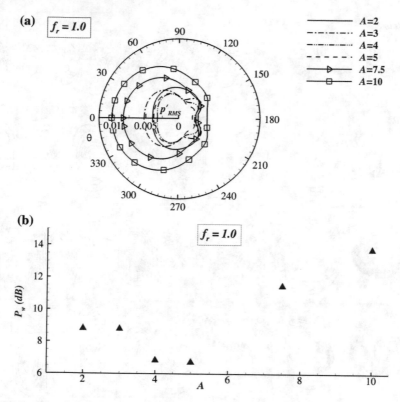

**Fig. 8.46  a** Directivity patterns based on *RMS* values of pressure fluctuations $p'$ obtained at a radial distance of $r = 75$ for indicated values of $A$ at $f_r = 1.0$. **b** Variation of acoustic power ($P_w$) of the sound generated with forcing amplitude $A$ for $f_r = 1.0$ at $r = 75$. Reproduced from [65], with the permission of *AIP* Publishing

unlike the cases with $f_r \leq 0.8$ and $A = 2$, where directivity patterns are dominated by lift dipole (as shown in Fig. 8.44). The directivity patterns of $A = 2$ and 3 and $A = 4$ and 5 have similar variation. Also, the directivity patterns of $A = 7.5$ and $A = 10$ cases are dominated by drag fluctuations as maximum values of $p'_{RMS}$ are found along negative $x$-direction and this has also been shown with the analysis of *POD* modes based on disturbance pressure fields, as explained later. Figure 8.46b displays the variation of acoustic power $P_w$ with $A$ measured at $r = 75$. The power $P_w$ is found minimum at $A = 5$ and maximum at $A = 10$ and the similar behavior has also been observed from the Figs. 8.46a and 8.45.

Next, observations regarding radiated sound field are supplemented using Doak's decomposition theory which decomposes instantaneous flow field information into acoustic, entropic and hydrodynamic modes.

Figure 8.47 shows the instantaneous vorticity contours for the indicated values of $f_r$ at time $t = 260$. Contours have been shown for the vorticity values ranging from $-1$ to 1 with 30 equi-spaced sub-levels. The case $f_r = 0.0$ represents flow

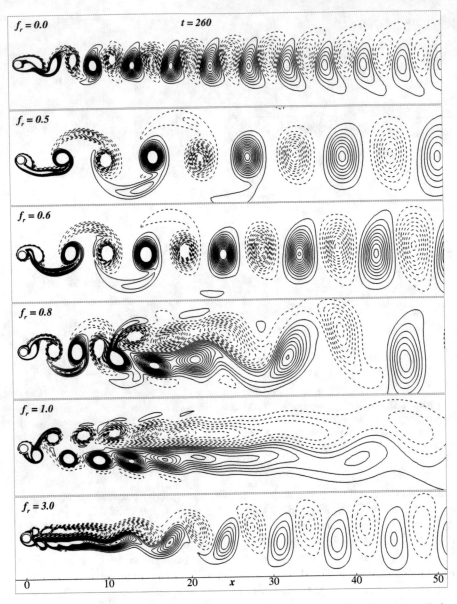

**Fig. 8.47** Instantaneous vorticity contours in the wake of rotationally oscillating circular cylinder are shown for different values of $f_r$ at indicated time $t$ with $A = 2.0$. Reproduced from [65], with the permission of *AIP* Publishing

past a stationary cylinder. The vorticity contours are shown over a distance $r = 50$. Positive and negative values of vorticity contours are represented by solid and dashed lines, respectively. These wake patterns are strongly governed by applied forcing frequency as it belongs to synchronization region [97]. Spacings between the shed vortices in the near wake region are also strongly influenced by $f_r$. For low values of $f_r$, spacing between the shed vortices is relatively large compared to high values of $f_r$. The forcing causes change in wake patterns over a certain distance in the wake, beyond which the behavior of wakes behind cylinder performing rotary oscillations is similar to the wake patterns of flow past a stationary cylinder. Obtained wake patterns are qualitatively similar to those observed in the literature under the same forcing conditions [97]. With the help of accurate numerical schemes and a refined mesh, wake patterns are accurately resolved up to a radial distance of $r = 100$.

### 8.2.3.8  Decomposition of the Sound Field

The obtained *DNS* results are further analyzed using Doak's decomposition theory [22, 98]. This decomposition analysis has been implemented for the six cases with $f_r = 0.3,\ 0.6,\ 0.8,\ 1.0,\ 2.0$ and $3.0$ for forcing amplitude, $A = 2.0$. In the present discussion, contours of acoustic $||\nabla \psi'_A||$ and solenoidal $||\mathbf{B}'||$ components are compared for these six cases. Figure 8.48 shows the contours of solenoidal fluctuations (hydrodynamic fluctuations) represented by $||\mathbf{B}'||$ at time $t = 280$. These hydrodynamic modes are found to be distinctive based on the value of $f_r$. This is due to occurrence of different wake patterns for various values of $f_r$ as shown in Fig. 8.47. For the $f_r = 0.6$ and $0.8$ cases, one observes prominent $||\mathbf{B}'||$ region, suggesting maximum hydrodynamic fluctuations in the wake, while for the $f_r = 2.0$ and $3.0$ cases, one observes significant attenuation in hydrodynamic $(||\mathbf{B}'||)$ fluctuations.

Figure 8.49 shows the contours of acoustic component represented by $||\nabla \psi'_A||$ at time $t = 280$. It is observed that the intensity of $||\nabla \psi'_A||$ is maximum for the $f_r = 0.6$ and $0.8$ cases and minimum at $f_r = 2.0$. For the higher forcing frequency-ratio $f_r = 2.0$ case, intensity decreases considerably. These acoustic intensity patterns agree well with the trends shown for *RMS* values of pressure fluctuations in Fig. 8.44. Wavelength of $||\nabla \psi'||$ is observed to be governed by forcing frequency-ratio $f_r$. This validates variation in intensity of acoustic field with frequency-ratio.

Next the disturbance pressure fields are further analyzed using *POD* analysis. In particular, we have focused our attention on two cases as, $f_r = 1.0$ and $A = 2.0; f_r = 1.0$ and $A = 7.5$. The evaluated dominant coherent mode structures of disturbance pressure fields are related to fluctuations in the lift and drag coefficients.

Figure 8.50 represents first six dominant *POD* modes, and their respective time varying *POD* amplitudes $(a_k)$ for a frequency-ratio of $f_r = 1.0$ and $A = 2$. Based on the values of $\lambda_k$, the first six modes contribute more than 96% to disturbance pressure field. The pressure pulses are found to propagate in the vertical direction for the first mode pair (Mode 1 and Mode 2). On the other hand they propagate in the horizontal direction for the second mode pair (Mode 3 and Mode 4). The wavenumber of second mode pair is almost twice higher than that of first pair. The time varying

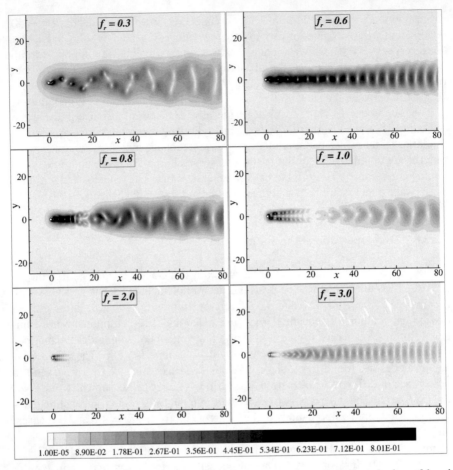

**Fig. 8.48** Contours of solenoidal fluctuations $||\mathbf{B}'||$ have been shown for indicated values of $f_r$ and $A = 2$ at time $t = 280$. Reproduced from [65], with the permission of *AIP* Publishing

*POD* amplitudes $a_k$ for $f_r = 1.0$ are shown in Fig. 8.50b. The frequencies of $a_1$ and $a_2$ are identified at the same forcing frequency-ratio $f_{r1} = f_r$, where the most dominant peaks of $c_{l'}$ have been identified. This shows that the first mode pair of $f_r = 1.0$ case is determined by lift fluctuations. Similarly, frequencies of $a_3$ and $a_4$ and $a_5$ and $a_6$ are found to be $2f_r$ and $3f_r$, respectively. Hence, the second mode pair (Mode 3 and Mode 4) is determined by drag fluctuations, whereas third mode pair (Mode 5 and Mode 6) has been contributed from lift fluctuations.

Next, effects of relatively larger forcing amplitude on the generated sound field have been studied by performing *POD* analysis for $A = 7.5$ and the frequency-ratio $f_r = 1.0$. The *POD* results for the case $f_r = 1.0$ for $A = 7.5$ have been shown in Fig. 8.51. Based on the magnitudes of eigenvalues $\lambda_k$, around 98% of the disturbance pressure field has been contributed by the first six dominant *POD* modes as shown in

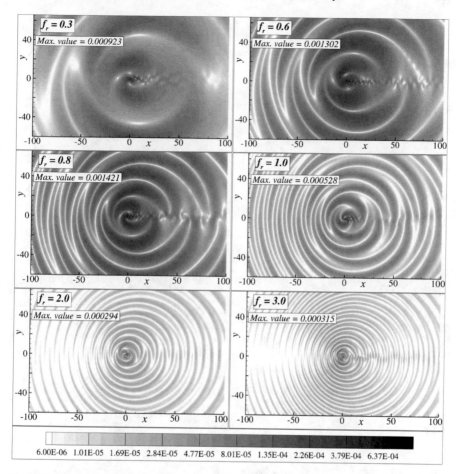

**Fig. 8.49** Contours of acoustic fluctuations $||\nabla \psi'_A||$ have been shown for indicated values of $f_r$ and $A = 2$ at time $t = 280$. One observes that at higher forcing frequencies, wavelength of acoustic waves decreases while at lower forcing frequencies, acoustic waves have relatively higher wavelengths. Reproduced from [65], with the permission of *AIP* Publishing

Fig. 8.51a. The first four modes have almost similar contribution to the disturbance pressure field based on values of $\lambda_k$. The frequencies of $(a_1, a_2)$, $(a_3, a_4)$, and $(a_5, a_6)$ are found at $2f_r, f_r$, and $3f_r$, respectively. Hence, first mode pair (Mode 1 and Mode 2) is associated with a drag fluctuations acting on the cylinder. The second pair (Mode 3 and Mode 4) and third mode pair (Mode 5 and Mode 6) are associated with the lift fluctuations acting on the cylinder. One can observe an interesting behavior in the generated sound field due to the second mode pair (Mode 3 and 4). Although this pair corresponds to the lift dipole, corresponding contour plots display a kink as shown in Fig. 8.51a, which resemble to quadrupole kind of directivity pattern. This is an unique behavior that has been observed for the case of higher forcing amplitude. Thus, by

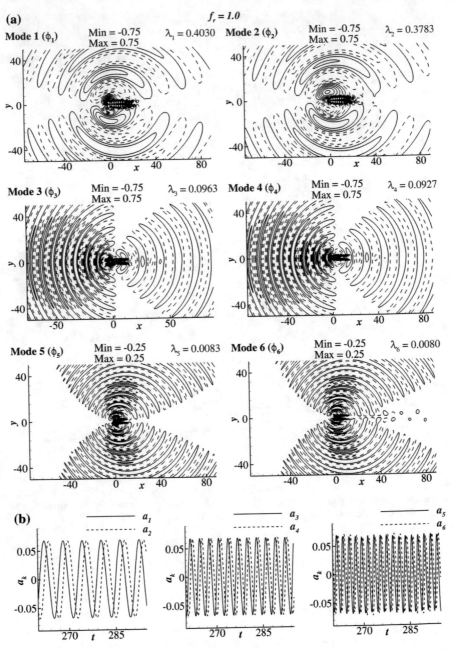

**Fig. 8.50** First six dominant *POD* modes and time varying *POD* amplitudes are obtained for disturbance pressure fields at $A = 2$ and $f_r = 1.0$. In Fig. 8.50a, First mode pair (Modes 1 and 2) and third mode pair (Modes 3 and 4) are contributed from the fluctuations in lift forces, whereas Modes 3 and 4 forming a pair are contributed from fluctuations in drag forces. In Fig. 8.50b, frequencies of time varying *POD* amplitudes $a_1$ and $a_2$, $a_3$ and $a_4$, and $a_5$ and $a_6$ are observed to be $f_r$, $2f_r$, and $3f_r$, respectively. Reproduced from [65], with the permission of *AIP* Publishing

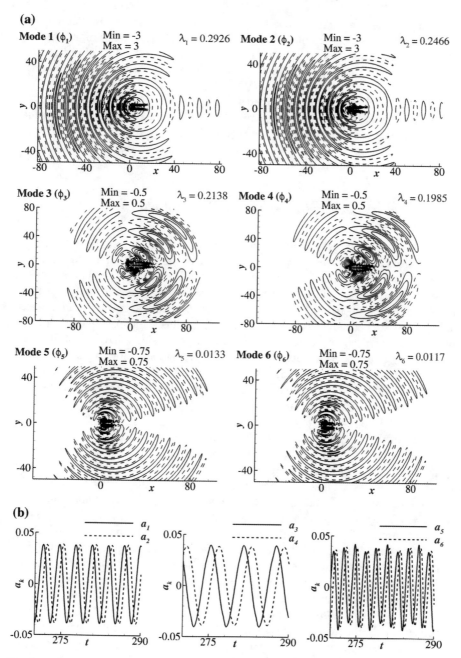

**Fig. 8.51** First six dominant *POD* modes and time varying *POD* amplitudes are obtained for disturbance pressure fields for $f_r = 1.0$ and $A = 7.5$. In Fig. 8.51a, Modes 1 and 2 forming a pair are contributed from fluctuations in drag forces, whereas second mode pair (Modes 1 and 2) and third mode pair (Modes 3 and 4) are contributed from the fluctuations in lift forces. In Fig. 8.51b, frequencies of time varying *POD* amplitudes $a_1$ and $a_2$, $a_3$ and $a_4$ and $a_5$ and $a_6$ are observed to be $2f_r, f_r$ and $3f_r$, respectively. Reproduced from [65], with the permission of *AIP* Publishing

imparting higher forcing amplitude to cylinder performing rotary oscillations, one can distribute sound energy more evenly in the whole domain. For higher forcing amplitude cases, lift dipole does not remain dominant over the drag dipole and the amplitudes, as well as, frequencies associated with drag dipole also dictate generated sound field.

Thus, the *POD* analysis has confirmed that the generated sound fields are governed by fluctuations in lift and drag forces resulting from vortex shedding phenomena. However, the vortex shedding patterns are in turn dictated by frequency and amplitude of rotary oscillations.


## 8.3 Tonal Noise Due to Flow Past Streamlined Body

It is observed that for certain conditions, flow past an aerofoil gives rise to a piercing whistle, usually $20-30$ dB above background acoustic field with a dominant discrete frequency. This phenomenon is termed as 'tonal noise' [61] and is commonly observed for flow past gliders and small aircrafts. For such flows, usually Reynolds number ranges from $10^5$ to $10^6$. Tonal noise has also been observed for underwater applications such as flow past hydrofoils and propellers. Flow past helicopter rotor blades and fans is complex, as compared to flow past a single fixed aerofoil, since the wake of one blade pass over the other, triggering early flow transition which changes acoustic field. Study of acoustic field around such engineering applications is important from noise reduction point of view [54, 76]. Thus, many researchers in the past studied a simple case of tonal noise generation from a single aerofoil [1, 8, 20, 60, 61, 64, 69, 94].

In one of the early research work, Paterson et al. [69] carried out experimental study in an anechoic open jet wind tunnel for studying tonal noise emission from symmetric *NACA*0012 and *NACA*0018 aerofoils for Reynolds number in the range of $10^5$ to $10^6$ corresponding to the flow conditions as encountered by a full scale helicopter rotor, for several angles of attack. In this study, authors kept a hot-wire probe in the aerofoil wake region and observed that the wake fluctuations have the same frequency as that of acoustic tone. Authors proposed that the vortex shedding behind the trailing edge of the aerofoil is responsible for the tonal noise generation and is a dominant source of noise generation as compared to other noise sources related to turbulent boundary layer. Paterson et al. [69] observed that the variation of frequency $f$ of the tones against the free-stream velocity $U_\infty$ has a "ladder-like" structure with "rungs" of curves proportional to $U_\infty^{0.8}$ and suggested that the tonal frequency $f$ is dependent on the the free-stream velocity $U_\infty$ and boundary layer thickness, $\delta$. Average behavior relating the dependence of the tonal noise on the free-stream velocity is given by the scaling law $f \sim U_\infty^{3/2}$ [69]. Paterson et al. [69] also found that velocity range over which tonal noise was present increased with angle of incidence until the aerofoil was stalled.

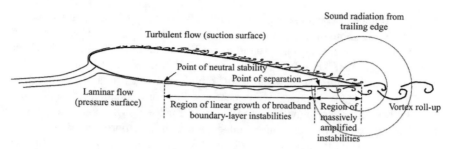

**Fig. 8.52** Schematic showing flow past an aerofoil [64]

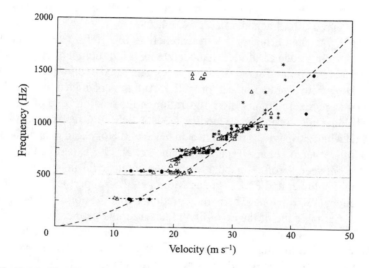

**Fig. 8.53** Tonal noise generation for flow past NACA0012 aerofoil with freestream velocity has been shown from Nash et al. [64]

Theoretical prediction about effects of laminar flow separation over the pressure surface of *NACA0012* aerofoil on tonal noise generation over a range of Reynolds number was confirmed by extensive experimental surface flow visualization by Nash et al. [64]. Tonal noise generation mechanism was further developed by Nash et al. [64], which consists of three Parts, as shown in Fig. 8.6. In the first stage, T-S instability is developed inside a laminar boundary layer on the pressure surface of the aerofoil. In second part, these instabilities are largely amplified due to the inflectional profiles of the separating shear layer close to the trailing edge. In third stage, amplified instabilities pass over the trailing edge and the sound is radiated often displaying a "ladder-like" structure (Figs. 8.52 and 8.53).

The relationship between peak tonal frequencies and wind tunnel velocity is shown in Fig. 8.7 [64] and the results are compared with empirical relation proposed by Paterson et al. [69].

The observed tonal frequency dependence on free-stream velocity for flow past a streamlined body contrasted the earlier observation for flow past bluff bodies, where the frequency was observed approximately constant. With the advancement in understanding about flow transition over streamlined bodies, sound generation due to Karman vortex street has been found less effective and flow transition has been noted as an important issue. Indeed, Karman vortex street is a dominant mechanism for the sound generation in flow past bluff bodies and is a primary source behind aerodynamic noise. In contrast, for the case of flow past streamlined bodies like aerofoils, Karman street plays an important role for Reynolds number lower than $10^5$ or for aerofoils with blunt trailing edges, as has been reported by Brooks et al. [8]. The flow instability has greater role in the generation of trailing edge noise. At sufficiently high Reynolds numbers, the instability waves generated inside the boundary layer cause flow transition from laminar to turbulent flow for flow past an airfoil as observed experimentally in [1, 54, 64, 69].

Clark [15] was one of the first few researchers to report discrete frequency tone from a sharp trailing edge aerofoil for a moderate Reynolds numbers. Hersh and Hayden [39] observed similar discrete acoustic tones generated by flow past NACA0012 aerofoil and by flow past two-bladed propeller. Authors concluded that the discrete frequencies were due to laminar flow separation at the leading edge of the aerofoil and used leading edge serrations to modify acoustic field. It is reported that all noise disappeared when a trip wire is placed at a distance of 0.8 of chord in laminar boundary layer [39]. Paterson et al. [69] and Sunyanch et al. [90] experimentally studied relationship between aerofoil boundary layer and acoustic far-field measured using microphone. Schlinker and Fink [84] and Munin et al. [62] performed experiments using directional microphone. Their study predicted that the noise is generated very close to the trailing edge of the aerofoil and has a dipole nature. Munin et al. [62] studied effects of free-stream velocity and turbulence on the acoustic field generation and concluded that tone's strength diminish and eventually disappear as one increases free-stream velocity and free-stream turbulence.

Tam [94] suggested that the aerofoil is a streamlined body and corresponding acoustic field should have different nature compared to that triggered by flow past bluff bodies. He proposed the idea that there exists a self-excited feedback loop between a point on the trailing edge of the aerofoil and point in the wake region which is responsible for tonal noise generation [94]. He suggested that the boundary-layer instabilities in the form of convecting vortices grow and results in lateral vibrations of the wake which are responsible for emission of acoustic waves close to the trailing edge of the aerofoil. Thus, Tam's aeroacoustic feedback loop provided correlation between hydrodynamic instability in a boundary layer and resultant tonal-noise emission. According to this feedback model, initiation of the disturbances at the trailing edge of the aerofoil is continuous.

This feedback loop model was later modified in [1, 30, 54, 103]. Sound generation was thought to be associated with the interaction between sharp trailing edge of the aerofoil and fluctuating pressure distribution on the aerofoil surface caused by aerodynamic disturbances. It has been also observed that the tonal noise depends on shape of the aerofoil trailing edge whether it is sharp or it has a blunt edge. Blunt edge of the aerofoil results in formation of quasi-periodic vortex street. Usually amplitude of the tonal noise for a blunt edged aerofoil is higher compared to sharp-edged aerofoil [61]. Most of the researchers have associated tonal noise with the laminar boundary layer extending up to the trailing edge on the pressure side of the aerofoil. Results in [61] suggested that tonal noise is dependent on the existence of laminar separation bubble on the pressure surface of the aerofoil where large fluctuations occur around the separation bubble at the frequency of tonal noise.

Wright [103] investigated both discrete, as well as broadband noise originating from a wide range of rotors including helicopter rotors, propellers, and fans. Tonal noise was associated with the T-S instabilities for blades at low angles of incidence with small chord length, large thickness to chord ratios, and low operating speeds. Wright proposed an aeroacoustic feedback loop between the radiation point that is the trailing edge of the aerofoil, and the source, which is the region upstream on the aerofoil surface. He assumed that flow does not undergo transition from laminar to turbulent region up to the trailing edge of the aerofoil. He proposed that the convection of disturbances in boundary layer to the trailing edge along the aerofoil surface is responsible for tonal noise generation. Pressure disturbance leaving the aerofoil surface at the trailing edge are induced by these boundary layer waves. The resulting radiation propagates upstream to reinforce their original disturbance and completes the feedback loop.

Results in [61] further concluded that the distinctive "piercing whistle" radiated by the aerofoil are due to coupling between the boundary layer and wake instabilities. Boundary layer instability waves on the pressure side of the aerofoil determines vortex shedding in the wake region.

Fink et al. [31] did experimentation in an acoustic open jet wind tunnel on NACA0012 aerofoil with test models having constant, as well as tapered chord. The constant chord model was observed to generate discrete multiple tones, which were several hundred Hertz apart. This was associated with the span-wise irregularities in the wing. The tapered chord model radiated discrete multiple tones, as well. However, more tones were recorded having a wider band of frequencies, as compared to the model with constant chord. There was a strong agreement of results in [31] with the results of Paterson et. al. [69]. Fink et. al. [31] also proposed a model for $T$-$S$ instability process theoretically, in order to provide a description of the proposed feedback process, but it failed to agree with the experimental data.

To display tonal noise behavior for the flow past a streamlined body, here we have computed flow past an elliptic cylinder with ratio of major to minor axis as 10. Reynolds number for the present calculation is $10^4$ and the angle of attack is prescribed as $4°$. Mach number based on free-stream velocity is 0.2. An elliptic, body fitted grid with 501 points in the azimuthal direction and 900 points in the wall normal direction has been generated. We have performed direct simulation of an acoustic field by solving $2D$ compressible Navier–Stokes equations as discussed for the case of flow past a cylinder.

Figure 8.54 shows variation of lift and drag coefficient with time in the top and middle frame, respectively, while the bottom frame shows vorticity contours. One can observe vortex shedding in the wake region, which is responsible for the periodic variation of the lift and the drag coefficient. Such periodic shedding also triggers acoustic field, as shown in Fig. 8.55 at various instants. One can observe propagation

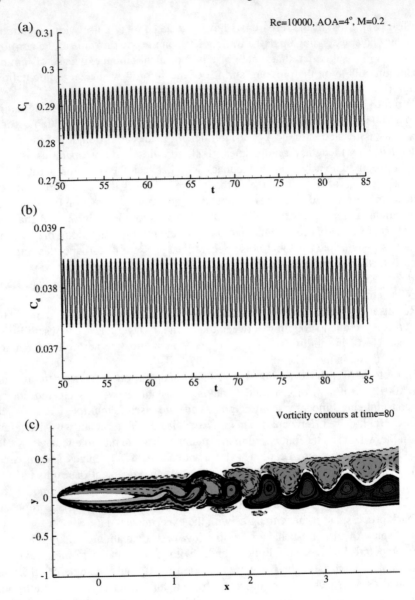

**Fig. 8.54** Variation of the lift and drag coefficient with time along with instantaneous vorticity contours for flow past elliptic cylinder at $4°$ *AOA* and Re $= 10^4$

of acoustic waves in a radially outward direction. Thus, the present calculations support the observations made by numerous experimental and numerical studies about presence of tonal noise (Fig. 8.55).

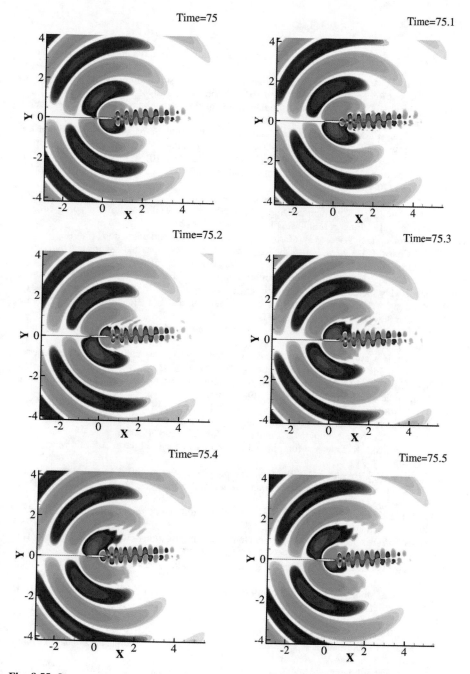

**Fig. 8.55** Instantaneous sound pressure contours for flow past elliptic cylinder at 4° *AOA* and Re = 10⁴

# References

1. H. Arbey, J. Bataille, Noise generated by airfoil profiles placed in a uniform laminar flow. J. Fluid Mech. **134**, 33–47 (1983)
2. H.M. Badr, M. Coutanceau, S.C.R. Dennis, C. Menard, Unsteady flow past a rotating circular cylinder at Reynolds numbers $10^3$ and $10^4$. J. Fluid Mech. **220**, 459–484 (1990)
3. S.J. Baek, H.J. Sung, Numerical simulations of the flow behind a rotary oscillating circular cylinder. Phys. Fluids **10**, 869–876 (1998)
4. Y. Bao, D. Zhou, Y.J. Zhao, A two-step Taylor-characteristic-based Galerkin method for incompressible flows and its application to flow over triangular cylinder with different incidence angles. Int. J. Numer. Meth. Fluids **62**, 1181–1208 (2010)
5. P.W. Bearman, Vortex shedding from oscillating bluff bodies. Annu. Rev. Fluid Mech. **16**, 195–222 (1984)
6. E. Berger, R. Willie, Periodic flow phenomenon. Annu. Rev. Fluid Mech. **4**, 313–340 (1972)
7. Y.G. Bhumkar, T.K. Sengupta, Drag reduction by rotary oscillation for flow past a circular cylinder. Int. J. Emerg. Multidisc. Fluid Sci. **1**(4), 269–298 (2010)
8. T.F. Brooks, D.S. Pope, M.A. Marcolini, Airfoil self-noise and prediction (1989)
9. C.C. Chang, R.L. Chern, Vortex shedding from an impulsively started rotating and translating circular cylinder. J. Fluid Mech. **233**, 265–298 (1991)
10. Y.M. Chen, Y.R. Ou, A.J. Pearlstein, Development of the wake behind circular cylinder impulsively started into rotary and rectilinear motion. J. Fluid Mech. **253**, 449–484 (1993)
11. M. Cheng, Y.T. Chew, S.C. Luo, Numerical investigation of a rotationally oscillating cylinder in mean flow. J. Fluids Struct. **15**, 981–1007 (2001)
12. C. Cheong, P. Joseph, Y. Park, S. Lee, Computation of aeolian tone from a circular cylinder using source models. Appl. Acoust. **69**(2), 110–126 (2008)
13. S. Choi, H. Choi, S. Kang, Characteristics of flow over a rotationally oscillating cylinder at low Reynolds number. Phys. Fluids **14**(8), 2767–2777 (2002)
14. J.M. Cimbala, H.M. Nagib, A. Roshko, Large structure in the far wakes of two-dimensional bluff bodies. J. Fluid Mech. **190**, 265–298 (1988)
15. L.T. Clark, The radiation of sound from an airfoil immersed in a laminar flow (ASME, 1971)
16. T. Colonius, S.K. Lele, P. Moin, The scattering of sound waves by a vortex: numerical simulation and analytical solutions. J. Fluid Mech. **260**, 271–298 (1994)
17. T. Colonius, S.K. Lele, P. Moin, Sound generation in a mixing layer. J. Fluid Mech. **330**, 375–409 (1997)
18. J.S. Cox, K.S. Brentner, C.L. Rumsey, Computation of vortex shedding and radiated sound for a circular cylinder: subcritical to transonic Reynolds numbers. Theor. Comput. Fluid Dyn. **12**, 233–253 (1998)
19. N. Curle, The influence of solid boundaries upon aerodynamic sound. Proc. R. Soc. Lond. A **231**, 505–514 (1955)
20. G. Desquesnes, M. Terracol, P. Sagaut, Numerical investigation of the tone noise mechanism over laminar airfoils. J. Fluid Mech. **591**, 155–182 (2007)
21. F. Diaz, J. Gavalda, J.G. Kawall, J.F. Keffer, F. Giralt, Vortex shedding from a spinning cylinder. Phys. Fluids **26**(12), 3454–3460 (1983)
22. P.E. Doak, Momentum potential theory of energy flux carried by momentum fluctuations. J. Sound Vib. **131**(1), 67–90 (1989)
23. S.C.R. Dennis, P. Nguyen, S. Kocabiyik, The flow induced by a rotationally oscillating and translating circular cylinder. J. Fluid Mech. **407**, 123–144 (2000)
24. B. Etkin, G.K. Korbacher, R.T. Keefe, Acoustic radiation from a stationary cylinder in a finite stream (aeolian tones). J. Acoust. Soc. Am. **29**, 30–36 (1957)
25. J.E. Ffowcs Williams, Hydrodynamic noise. Annu. Rev. Fluid Mech. **1**, 197–222 (1969)
26. J.E. Ffowcs Williams, Aeroacoustics. Annu. Rev. Fluid Mech. **9**, 447–468 (1977)
27. J.E. Ffowcs Williams, Aeroacoustics. J. Sound Vib. **190**, 387–398 (1996)
28. J.E. Ffowcs Williams, D.L. Hawkings, Sound generated by turbulence and surfaces in arbitrary motion. Philos. Trans. the R. Soc. Lond. Ser. A, **264**, 321–342 (1969)

29. J.R. Filler, P.L. Marston, W.C. Mih, Response of the shear layers separating from a circular cylinder to small-amplitude rotational oscillations. Fluid Mech. **231**, 481–499 (1991)
30. M.R. Fink, Fine structure of airfoil tone frequency. J. Acoust. Soc. Am. **63**(S1), S22–S22 (1978)
31. M.R. Fink, R.H. Schlinker, R.K. Amiet, Prediction of rotating-blade vortex noise from noise of nonrotating blades (1976)
32. W.K. George, Some thoughts on similarity, the POD, and finite boundaries, in *Fundamental Problematic Issues in Turbulence*, ed. by A. Gyr, W. Kinzelbach, A. Tsinober (Birkhauser Verlag, Basel, 1999)
33. J.H. Gerrard, Measurements of the sound from circular cylinders in an air stream. Proc. Phys. Soc. Lond. B **68**, 453–461 (1955)
34. O.M. Griffin, M.S. Hall, Vortex shedding lock on and flow control in bluff body wakes. J. Fluids Eng. **113**, 526–537 (1991)
35. J.C. Hardin, S.L. Lamkin, Aeroacoustic computation of cylinder wake flow. AIAA J. **22**, 51–57 (1984)
36. J.C. Hardin, D.S. Pope, An acoustic/viscous splitting technique for computational aeroacoustics. Theor. Comp. Fluid Dyn. **6**, 323–340 (1994)
37. J.W. He, R. Glowinsky, R. Metcalfe, A. Nordlander, J. Periaux, Active control and drag optimization for flow past a circular cylinder. J. Comput. Phys. **163**, 83–117 (2000)
38. R.D. Henderson, Nonlinear dynamics and pattern formation in turbulent wake transition. J. Fluid Mech. **352**, 65–112 (1997)
39. A.S. Hersh, R.E. Hayden, Aerodynamic sound radiation from lifting surfaces with and without leading-edge serrations. Public Domain (1971)
40. P. Holmes, J.L. Lumley, G. Berkooz, C.W. Rowley, *Turbulence, Coherent Structures, Dynamical Systems and Symmetry*, 2nd edn. (Cambridge University Press, Cambridge, 2012)
41. O. Inoue, Propagation of sound generated by weak shock-vortex interaction. Phys. Fluids **12**, 1258–1261 (2000)
42. O. Inoue, N. Hatakeyama, Sound generation by a two-dimensional circular cylinder in a uniform flow. J. Fluid Mech. **471**, 285–314 (2002)
43. O. Inoue, Y. Hattori, Sound generation by shock-vortex interactions. J. Fluid Mech. **380**, 81–116 (1999)
44. O. Inoue, Y. Hattori, T. Sasaki, Sound generation by coaxial collision of two vortex rings. J. Fluid Mech. **424**, 327–365 (2000)
45. O. Inoue, Y. Takahashi, Successive generation of sounds by shock-strong vortex interaction. Phys. Fluids **12**, 3229–3234 (2000)
46. O. Inoue, T. Yamazaki, Secondary vortex streets in two-dimensional cylinder wakes. Fluid Dyn. Res. **25**, 1–18 (1999)
47. D.D. Ksambi, Statistics in function space. J. Indian Math. Soc. **7**, 76–88 (1943)
48. L.D. Landau, E.M. Lifshitz, Fluid mechanics. Course Theor. Phys. **6** (1987)
49. B.E. Lele, S.K. Lele, P. Moin, Direct computation of the sound from a compressible co-rotating vortex pair. J. Fluid Mech. **285**, 181–202 (1995)
50. B.E. Lele, S.K. Lele, P. Moin, Direct computation of the sound generated by vortex pairing in an axisymmetric jet. J. Fluid Mech. **383**, 113–142 (1999)
51. M.J. Lighthill, On sound generated aerodynamically: I. General theory. Proc. R. Soc. Lond. A **221**, 564–587 (1952)
52. C.C. Lin, *The Theory of Hydrodynamic Stability*, vol. 155 (Cambridge University Press, Cambridge, 1955), p. 22s 6d
53. Y.S.K. Liow, B.T. Tan, M.C. Thompson, K. Hourigan, Sound generated in laminar flow past a two-dimensional rectangular cylinder. J. Sound Vib. **295**, 407–427 (2006)
54. R.E. Longhouse, Vortex shedding noise of low tip speed, axial flow fans. J. Sound Vib. **53**(1), 25–46 (1977)
55. X.Y. Lu, J. Sato, A numerical study of flow past a rotationally oscillating circular cylinder. J. Fluids Struct. **10**(8), 829–849 (1996)

56. B. Mahato, N. Ganta, Y.G. Bhumkar, Direct simulation of sound generation by a two-dimensional flow past a wedge. Phys. Fluids **30**(9), 096101 (2018)
57. B. Mahato, N. Ganta, Y.G. Bhumkar, Computation of aeroacoustics and fluid flow problems using a novel dispersion relation preserving scheme. J. Theor. Comput. Acoust. **26**(4), 1850063(1–30) (2018)
58. R.R. Mankbadi, M.H. Hayder, L.A. Povinelli, Structure of supersonic jet flow and its radiated sound. AIAA J. **32**, 897–906 (1994)
59. F. Margnat, Hybrid prediction of the aerodynamic noise radiated by a rectangular cylinder at incidence. Comput. Fluids **109**, 13–26 (2015)
60. O. Marsden, C. Bogey, C. Bailly, Direct noise computation of the turbulent flow around a zero-incidence airfoil. AIAA J. **46**(4), 874 (2008)
61. A. McAlpine, E.C. Nash, M.V. Lowson, On the generation of discrete frequency tones by the flow around an aerofoil. J. Sound Vib. **222**(5), 753–779 (1999)
62. A.G. Munin, A.G. Prozorov, A.V. Toporov, J.S. Wood, Experimental study of noise generated by an airfoil in a low-velocity flow. Sov. Phys. Acoust. **38**(1), 55–57 (1992)
63. E.C. Nash, AIAA 94-0358 laminar boundary layer aeroacoustic instabilities (1994)
64. E.C. Nash, M.V. Lowson, A. McAlpine, Boundary-layer instability noise on aerofoils. J. Fluid Mech. **382**, 27–61 (1999)
65. N. Ganta, B. Mahato, Y.G. Bhumkar, Analysis of sound generation by flow past a circular cylinder performing rotary oscillations using direct simulation approach. Phys. Fluids **31**, 026104 (2019)
66. B.R. Noack, K. Afanasiev, M. Morzynski, G. Tadmor, F. Thiele, A hierarchy of low-dimensional models for the transient and post-transient cylinder wake. J. Fluid Mech. **497**, 335–363 (2003)
67. H. Oertel Jr., Wakes behind blunt bodies. Annu. Rev. Fluid Mech. **22**, 539–564 (1990)
68. A. Okajima, H. Takata, T. Asanuma, Viscous flow around a rotationally oscillating cylinder. Inst. Space Aeronaut. Sci. **532**, 311–318 (1975)
69. R.W. Paterson, P.G. Vogt, M.R. Fink, C.L. Munch, Vortex noise of isolated airfoils. J. Aircr. **10**(5), 296–302 (1973)
70. O.M. Phillips, The intensity of aeolian tones. J. Fluid Mech. **1**, 607–624 (1956)
71. D.S. Pope, A viscous/acoustic splitting technique for aeolian tone prediction, in *Proceedings of Second Computational Aeroacoustics (CAA) Workshop on Benchmark Problems, NASA CP-3352* (1997), pp. 305–318
72. A. Powell, Theory of vortex sound. J. Acoust. Soc. Am. **36**, 177–195 (1964)
73. B. Protas, J.E. Wesfreid, Drag force in the open-loop control of the cylinder wake in the laminar regime. Phys. Fluids **14**(2), 810–826 (2002)
74. B. Protas, A. Styczek, Optimal rotary control of the cylinder wake in the laminar regime. Phys. Fluids **14**(7), 2073–2087 (2002)
75. L. Rayleigh, *The Theory of Sound*, vol. I and II (Macmillan, London, 1896)
76. M. Roger, S. Moreau, Broadband self-noise from loaded fan blades. AIAA J. **42**(3), 536–544 (2004)
77. T.K. Sengupta, S. Bhaumik, Y.G. Bhumkar, Nonlinear receptivity and instability studies by POD, in *AIAA Conference on Theoretical Fluid Mechanics* (AIAA, 2011), pp. 2011–3293
78. T.K. Sengupta, K. Deb, S.B. Talla, Control of flow using genetic algorithm for a circular cylinder executing rotary oscillation. Comput. Fluids **36**, 578–600 (2007)
79. T.K. Sengupta, A. Kasliwal, S. De, M. Nair, Temporal flow instability for Magnus-Robins effect at high rotation rate. J. Fluids Struct. **17**, 941–953 (2003)
80. T.K. Sengupta, G. Kumar, Bluff-body flow control by aerodynamic tripping, in *ASME PVT 2006/ICPVT-11 Conference on Flow-Induced Vibration*, Vancouver, Canada (2006)
81. T.K. Sengupta, N. Singh, V.K. Suman, Dynamical system approach to instability of flow past a circular cylinder. J. Fluid Mech. **656**, 82–115 (2010)
82. J.H. Seo, Y.J. Moon, Perurbed compressible equations for aeroacoustic noise prediction at low mach numbers. AIAA J. **43**(8), 1716–1724 (2005)

83. J.H. Seo, Y.J. Moon, Linearized perturbed compressible equations for low mach number aeroacoustics. J. Comput. Phys. **218**, 702–719 (2006)

84. R.H. Schlinker, R. Fink, Vortex noise fron nonrotating cylinders and airfoils, in *14th Aerospace Sciences Meeting*, Washington, DC, USA (1976)

85. T.K. Sengupta, V.V.S.N. Vijay, N. Singh, Universal instability modes in internal and external flows. Comput. Fluids **40**(1), 221–235 (2011)

86. S. Shen, Calculated amplified oscillations in plane poiseuille and blasius flows. J. Aeronaut. Sci. **21**, 222–224 (1954)

87. W.Z. Shen, J.N. Sørensen, Comment on the aeroacoustic formulation of hardin and pope. AIAA J. **37**, 141–143 (1999)

88. L. Sirovich, Turbulence and the dynamics of coherent structures, part I-III. Q. Appl. Math. **45** (1987)

89. S.R.L. Samion, M.S.M. Ali, A. Abu, C.J. Doolan, R.Z.-Y. Porteous, Aerodynamic sound from a square cylinder with a downstream wedge. Aerosp. Sci. Technol. **53**, 85–94 (2016)

90. M. Sunyach, H. Arbey, D. Robert, J. Bataille, G. Comte-Bellot, Correlations between far field acoustic pressure and flow characteristics for a single airfoil **1974**, 12 (1974)

91. S.A. Slimon, M.C. Soteriou, D.W. Davis, Computational aeroacoustics simulations using the expansion about incompressible flow approach. AIAA J. **37**, 409–416 (1999)

92. V. Strouhal, Ueber eine besondere art der tonerregung. Annu. Phys. Chem. (Wied. Annu. Phys.) **5**, 216–251 (1878)

93. B.M. Sumer, J. Fredsøe, *Hydrodynamics Around Cylindrical Structures* (World Scientific, Singapore, 1997)

94. C.K.W. Tam, Discrete tones of isolated airfoils. J. Acoust. Soc. Am. **55**(6), 1173–1177 (1974)

95. C.K.W. Tam, J.C. Hardin, in *Proceedings of Second Computational Aeroacoustics (CAA) Workshop on Benchmark Problems. NASA CP-3352* (1997)

96. S. Taneda, Visual observations of the flow past a circular cylinder performing a rotary oscillation. J. Phys. Soc. Jpn. **45**, 1038–1043 (1978)

97. B. Thiria, S. Goujon-Durand, J.E. Wesfreid, The wake of a cylinder performing rotary oscillations. J. Fluid Mech. **560**, 123–147 (2006)

98. S. Unnikrishnan, D.V. Gaitonde, Acoustic, hydrodynamic and thermal modes in a supersonic cold jet. J. Fluid Mech. **800**, 387–432 (2016)

99. M. Wang, S.K. Lele, P. Moin, Computation of quadrupole noise using acoustic analogy. AIAA J. **34**, 2247–2254 (1996)

100. M. Wang, S.K. Lele, P. Moin, Sound radiation during local laminar breakdown in a low mach number boundary layer. J. Fluid Mech. **319**, 197–218 (1996)

101. C.H.K. Williamson, R. Govardhan, Vortex-induced vibrations. Annu. Rev. Fluid Mech. **36**, 413–455 (2004)

102. G.A. Williamson, B.D. McGranahan, B.A. Broughton, R.W. Deters, J.B. Brandt, M.S. Selig, Department of Aerospace Engineering, University of Illinois at Urbana-Champaign. Summ. Low-Speed Airfoil Data **5** (2012)

103. S.E. Wright, The acoustic spectrum of axial flow machines. J. Sound Vib. **45**(2), 165–223 (1976)

# Index